AF893497
ausgesondert
Fachinformation

G. Müller (Editor)

Ceramics – Processing, Reliability, Tribology and Wear

EUROMAT 99 – Volume 12

DGM

WILEY-VCH

EUROMAT 99 was the biannual meeting of the Federation of European Materials Societies (FEMS) with its 21 member societies across Europe.

The program of the EUROMAT 99 congress was divided into 12 topics. The scientific coordination was managed by topic coordinators. The responsible experts for the individual topics were:

Topic A – Materials for Information Technology
K. Grassie, Philips GmbH, Aachen (D)
Topic B – Materials for Transportation Technology
P. J. Winkler, DaimlerChrysler AG, München (D)
Topic C – Materials for Electrical Power Generation and Energy Conversion
E. Tenckhoff, Siemens AG, Erlangen (D)
Topic D – Materials for Medical Engineering
H. Stallforth, Aesculap AG, Tuttlingen (D)
P. Revell, University College London (UK)
Topic E – Materials for Buildings and Structures
F. H. Wittmann, ETH Zürich (CH)
Topic F1 – Characterization Methods and Procedures
Y. Bréchet, INPG, Grenoble (F)
Topic F2 – Metals
R. Wagner, FZ. Jülich (D)
Topic F3 – Ceramics
G. Müller, Fraunhofer-Inst. f. Silicatforschung ISC, Würzburg (D)
Topic F4 – Polymers
G. Wegner, MPI für Polymerforschung, Mainz (D)
P. J. Lemstra, Univ. of Eindhoven (NL)
Topic F5 – Interface Controlled Materials
M. Rühle, MPI für Metallforschung, Stuttgart (D)
Topic G – Surface Technology
H. Dimigen, Fraunhofer-Inst. f. Schicht- und Oberflächentechnik IST, Braunschweig (D)
Topic H – Nano- and Microtechnology
J. Haußelt, Forschungszentrum Karlsruhe (D)

G. Müller (Editor)

Ceramics – Processing, Reliability, Tribology and Wear

EUROMAT 99 – Volume 12

Deutsche Gesellschaft
für Materialkunde e.V.

WILEY-VCH

Weinheim · New York · Chichester
Brisbane · Singapore · Toronto

Editor:
Prof. Dr. G. Müller
Fraunhofer-Institut für Silicatforschung
Neunerplatz 2
97082 Würzburg
Germany

This book was carefully procuced. Nevertheless, authors, editor and publisher do not warrant the information contained therein to be free of errors. Readers are advised to keep in mind that statements, data, illustrations, procedural details or other items may inadvertently be inaccurate.

18.05.10
DORNIER GMBH
- FachinformationsCenter -
66899,00

Library of Congress Card No. applied for.

A catalogue record for this book is available from the British Library.

Deutsche Bibliothek Cataloguing-in-Publication Data:
A catalogue record for this publication is available from Die Deutsche Bibliothek
ISBN 3-527-30194-1

© WILEY-VCH Verlag Gmbh, D-69469 Weinheim (Federal Republic of Germany), 2000

Printed on acid-free and chlorine-free paper.

All rights reserved (including those of translation in other languages). No part of this book may be reproduced in any form – by photoprinting, microfilm, or any other means – nor transmitted or translated into machine language without written permission from the publishers. Registered names, trademarks, etc. used in this book, even when not specifically marked as such, are not to be considered unprotected by law.

Composition: WGV Verlagsdienstleistungen GmbH, Weinheim
Printing: betz-druck, Darmstadt
Bookbinding: Buchbinderei Osswald, Neustadt/Wstr.
Printed in the Federal Republic of Germany

Preface

Engineering progress essentially depends on the availability and the intelligent use of materials. For many key industry areas, Europe constitutes a premier place for the development of new materials and their applications. EUROMAT 99, the biannual meeting of the Federation of European Materials Societies with its 21 member societies across Europe set out to become the most comprehensive European event to demonstrate the wide range of the interdisciplinary performance of materials.

EUROMAT was essentially focused on applications of materials with high innovation potential. At the same time, fundamental approaches and processing related aspects for unconventional materials were addressed. In the frame of the 12 conference topics, 1650 papers were originally submitted to the 52 symposia. A total of 655 have been selected as oral presentation by the international group of chairpersons and were presented in 161 sessions. Further, the chairpersons have selected 65 renowned experts for keynote lectures in the frame of their symposium. Roughly 700 papers were displayed as posters.

The scope of EUROMAT was truly international. Papers originated from 57 countries. Among them the major industrial countries of the world have contributed considerably to the wealth of the programme. An overwhelming Eastern European contingent shows that there is a strong interest of these countries in international cooperation.

EUROMAT 99 represents a showcase of the competence of the European materials societies. Various European sister societies and federations act as cosponsors of the event. Joining with FEMS, they are about to establish the network MatNet in order to promote and facilitate their communication and cooperation. They have started a dialogue with the European Commission in order to discuss programme goals and priorities for maintaining Europe´s global competitiveness. In view of this promising international perspective, the European Community has agreed to sponsor EUROMAT 99 generously for which we are very grateful. EUROMAT 99 was focused to a large extent on the aims of the closing 4th Framework Programme many projects of which were presented.

EUROMAT 99 was hosted by WERKSTOFFWOCHE, a multisociety joint conference project established in Germany in 1996. Among its initiators is the Deutsche Gesellschaft für Materialkunde, one of the founding member societies of FEMS and technical organiser of this year´s EUROMAT.

EUROMAT 99 represented an outstanding success. As the President of FEMS, I would hope that it will serve as a model for future meetings, both in terms of organisation and international cooperation. I would like to extend my gratitude to the scientists, chairpersons and coordinators as well as to the various organisations and particularly to the Messe München who have made this success possible.

Dr. Paul Costa
President of the Federation of European Materials Societies

Contents

I. Ceramics Processing

I Ceramics Processing

Supercritical Extraction of Binders from Green Ceramic Compacts

T. Chartier[1], E. Delhomme[1], J.F. Baumard[1], Ph. Marteau[2], R.Tufeu[2]
[1] SPCTS-UMR 6638 CNRS, ENSCI, Limoges, France
[2] LIMHP-UPR CNRS, Institut Galilée, Villetaneuse, France

1 Abstract

The removal of organic additives from ceramic green parts remains one of the most critical stage in ceramic processing. An original method of extraction of hydrocarbon-type binders, based on the unique dissolving characteristics and transport properties of supercritical fluids, is presented. The binder extraction by supercritical fluid is controlled by two mechanisms, namely the solubilisation of the molecules and the diffusion of dissolved species within the porosity of the green sample. Solubility of a paraffin binder is experimentally determined and a model is developed that takes account of its composition. A second model allows to estimate the kinetics of paraffin binder extraction by supercritical CO_2. This original method offers the great advantage of removing binders from ceramic green pieces without melting and without any thermal degradation.

2 Introduction

Most high technology ceramic processing like dry-pressing, tape-casting, injection or extrusion-moulding require the use of organic compounds as dispersants, binders, plasticizers and so on, to confer such properties as cohesion, flexibility and workability in the green state. Amounts as large as 50 vol.% of organics are sometimes added to the ceramic powder during the forming step, and have to be removed prior to sintering. An effort to understand the many physico-chemical processes that occur during thermal debinding, which is the most widely used method for binder removal, was done the last few years [1-7]. Binder removal by thermal treatment involves i) chemical mechanisms with the thermal degradation of organic compounds into volatile species and, ii) physical mechanisms like the diffusion of these species to the surface as well as the changes in the binder distribution within the green body. Modifications of the binder distribution, governed by diffusion and capillary migration, and the magnitude of the mass- and heat-transfer limitations are critical parameters for the design of an efficient heating cycle to prevent stresses and the formation of defects in ceramic parts [5]. Thus, thermal debinding remains one of the most critical steps of ceramic processing, especially in the case of large size parts. Defects are generated which affect properties of the sintered pieces. It is very time-consuming and residues of pyrolytic degradation can be detrimental to the subsequent sintering stage.

This is why alternative techniques are needed to produce defect-free green bodies and to reduce debinding time. We have used an original method of extraction of organic compounds based on the unique dissolving characteristics and transport properties of supercritical fluids [8-10]. The binder extraction by supercritical fluid is controlled by the solubilisation of the molecules and the diffusion of dissolved species within the porosity of the sample. In a first part, the solubility of paraffins in supercritical CO_2, using an infrared spectroscopy method, is presented. In a second part, results concerning the kinetic aspects of binder removal are given and the influence of the pore size of the ceramic pieces on the debinding rate is investigated.

3 Experimental

3.1 Samples

We have first chosen an organic formulation, based on paraffin binders, both suitable for the ceramic forming technique and for CO_2 extraction. The samples used to determine the extraction rates are injection-moulded bars (10×25×40 mm^3) made from a silica powder and a mixture of two paraffins, one melting at 52°C (P52) and one at 62°C (P62). The samples used to study the influence of the microstructure are porous pre-sintered alumina samples prepared by dry-pressing. Four alumina powders with different mean particle sizes were used in order to obtain samples with different mean pore diameters as well as a narrow pore size distribution (table 1). These pre-sintered preforms are filled with a paraffin melting at 42 °C by infiltration under vacuum.

Table 1. Mean pore diameter and porosity of the pre-sintered ceramic porous bodies (mercury porosimetry).

Alumina grade (Péchiney, F)	P172SB	P152SB	P152	P122B
Porosity (vol. %)	39.6	39.0	46.4	43.3
Mean pore diameter (μm)	0.12	0.27	0.70	0.80

3.2 Removal of paraffin

The extraction of paraffin from the injection-moulded bars, using CO_2 in supercritical state (critical point : Tc=31°C, Pc=7.37 MPa), was performed from 12 to 30 MPa, at temperatures ranging from 40 to 120°C. The extraction of paraffin from the alumina porous pre-sintered preforms was performed under 20 and 28 MPa, at 40 and 70 °C, for 30 minutes.

The schematics of the equipment used during the present work are shown in figure 1. Green samples to be debinded are introduced in a cylindrical extractor. Carbon dioxide, taken from a storage tank, is cooled down to -10°C. The resultant liquid carbon dioxide is pressurized up to the working pressure. Then, the pressurized liquid is heated up to the working temperature (40 to 120°C), leading to supercritical carbon dioxide. The carbon dioxide, in the supercritical state, is flowing through the extractor, dissolving binders from the green sample. The dissolved binders are recovered by means of three separators and the carbon dioxide is allowed to flow again through the extraction vessel. The CO_2 flow rate used during the experiments was set to 2.5 $l.h^{-1}$, which avoids any problem of confinement. The experimental extraction rate was estimated from the weight change of the sample during the supercritical treatment.

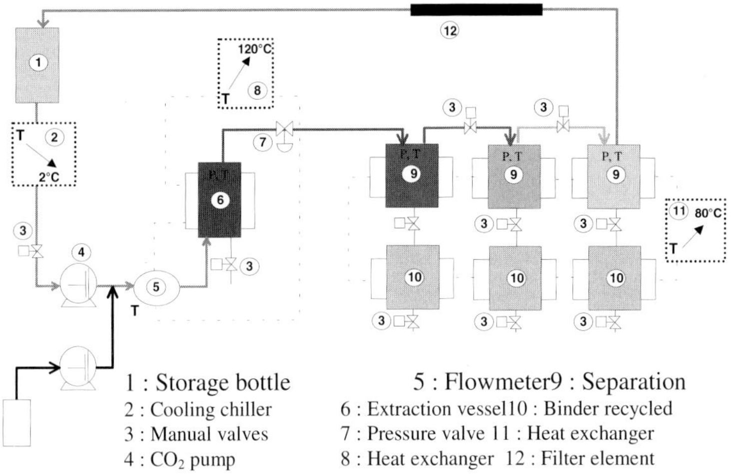

Figure 1. Schematic diagram of the extraction device.

3.3 Solubility measurements

The compositions of paraffins P52 and P62 were determined by gas phase chromatography (GC 6000, Carlo Erba). Paraffins are mixtures of alcanes and of a low concentration of isomers. We assume that the paraffin waxes are only composed of n-alcanes by combining n-alcanes and isomers of the same molar mass. According to the compositions of paraffins, the solubility of seven pure alcanes (C_{17}, C_{19}, C_{21}, C_{24}, C_{26}, C_{28} and C_{32}) was measured. CO_2-paraffin mixtures and CO_2-alcane mixtures are compressed and heated in a high pressure cell fitted with two sapphire windows and a piston [11]. Pressure is adjusted by varying the internal volume of the cell with the piston. The pressure is measured with an accuracy of ± 0.05 MPa. The cell is inserted in a furnace and the temperature is stabilised within ± 0.2 °C. The infrared light beam goes through the sapphire windows. The cell and the furnace can be rotated around the optical axis, in order to observe either the gas phase or the liquid phase, and also to mix the solute-solvent system to reach thermodynamic equilibrium. The infrared spectra are recorded with a spectrometer (Bomem MB155) fitted with an InAs detector.

The concentration, c, of each specie in the CO_2 rich phase is deduced from these spectra in a simple way as the absorption bands of the alcanes are well separated from those of CO_2. Under these conditions, the concentration is derived from the absorption intensities, without any further mathematical treatment, using the Beer-Lambert law :

$$c = \frac{1}{\alpha(\nu) l_a} \times \ln \frac{I_0(\nu)}{I(\nu)} \qquad (1)$$

where l_a is the absorption pathlength, $\alpha(\nu)$ the absorption coefficient at the frequency ν, $I_0(\nu)$ and $I(\nu)$ the transmitted intensities through the empty cell and through the sample, respectively. Absorption intensity measurements are performed on the absorption bands located at 4860.5 and 5109 cm^{-1} for the carbon dioxide and at 4180 cm^{-1} for the organic species. It was before established from the observation of the spectra of pure CO_2, pure alcanes and paraffin waxes, that the absorption frequency ranges of the solvent and of the solutes were clearly separated. We also noted that all the n-alcanes and paraffin waxes have rigorously the same absorption spectra.

4 Results and Discussion

4.1 Solubility of paraffin in supercritical CO_2

In order to model the extraction of the paraffin, the knowledge of the total solubility of the paraffin in supercritical CO_2 is not sufficient. Indeed, it is necessary to know the evolution of the paraffin composition both in the CO_2 rich phase and in the paraffin rich phase which will remain in the green part during extraction. Then, one must determine a partition coefficient K_n for each n-alcane between the CO_2 rich phase and the paraffin rich phase. K_n is defined as the ratio of the mass fraction (W_n^s) of the alcane n in the CO_2 rich phase to the mass fraction (W_n^p) of the same alcane n in the paraffin rich phase considering, in a first approximation, the paraffin rich phase without solvent :

$$K_n = \frac{W_n^s}{W_n^p} \tag{2}$$

Then, in an equilibrium taking only into account the binary alcane-CO_2 system, $W_n^p = 1$ and K_n is equal to the mass fraction of the alcane n in the CO_2 rich phase.

The K_n values, experimentally determined, may be expressed by the following expression :

$$\ln(K_n) = -\alpha n + \beta \tag{3}$$

where α and β are functions of the supercritical fluid density then, of the pressure and of the temperature (figure 2).

Paraffin being mixtures of linear hydrocarbons, a model was developed by analogy with the model of the solubility of a polydisperse polymer in a solvent. This model relates, at given pressure and temperature, the solubility of the n-alcanes to the solubility of the paraffin (W) in the supercritical CO_2 :

$$W = \sum_n K_n X_n \tag{4}$$

where X_n is the concentration of the alcane n in the paraffin determined by the gas phase chromatography. This model of solubility is in good agreement with experimental values (figure 3).

This model allows the determination of the evolution of the n-alcane distributions in the CO_2 rich phase and in the paraffin rich phase remaining in the green sample during extraction.

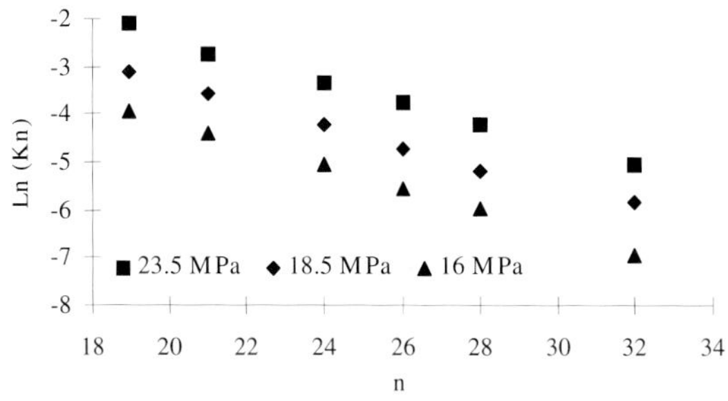

Figure 2. Solubility of n-alcanes in supercritical CO_2 for three pressures (T=70°C).

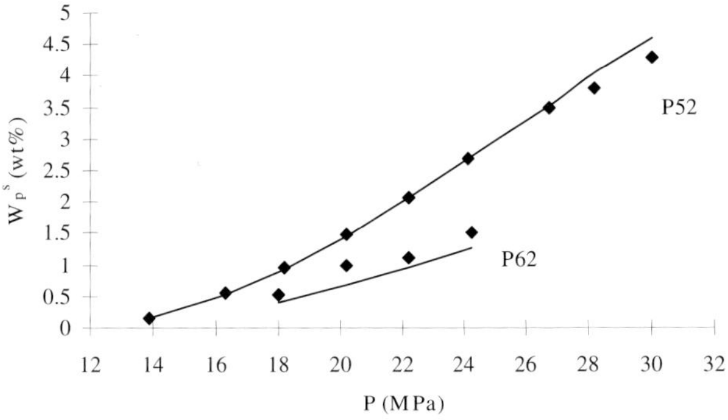

Figure 3. Experimental and theoretical solubilities of the two paraffins used (P52 and P62) in supercritical CO_2 (70°C) under various pressures.

In order to control the validity of our model, a sample of the CO2 rich phase, obtained from a mixture of paraffin 52 and CO_2 (P = 25.2 MPa, T = 70°C), was collected from the optical cell and analysed by gas phase chromatography to determine its n-alcane distribution. This distribution was compared to the calculated distribution (fig. 4). Taking into account the imprecision on the chromatographic analysis, experimental and calculated distributions are in good agreement. Furthermore, the evolution of the mean molar weight of the paraffin in the CO2 rich phase (tab. 2) and of the paraffin remaining in the green sample during extraction is also correctly predicted by the model.

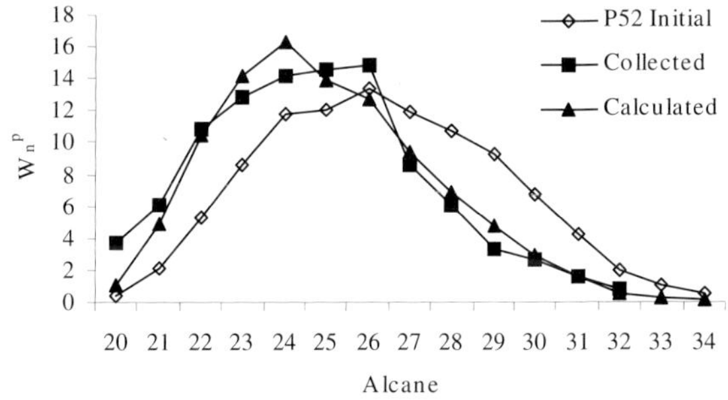

Figure 4. Experimental and calculated n-alcane distributions in the CO_2 rich phase (CO_2-paraffin 52, 70 °C, 25.2 MPa).

Table 2. Average molar weights of the initial paraffin wax P52 and of the paraffin wax in the CO_2 rich phase (experimental and calculated)

Average molar weight in CO_2 rich phase (g.mol^{-1})	P52
Initial	371.1
Experimental	352.5
Calculated	352.2

4.2 Polymer diffusion through a porous medium containing pressurised CO_2

Once solubilised, dissolved molecules diffuse through the porous medium. As already shown [12,13], diffusion is the limiting step of soluble paraffin extraction and a non-planar diffusing front is created in the sample during extraction. In the case of a parallelepipedic sample (L,l,h) with an initial organic concentration ϕ_0, and under the assumption that the diffusivity D of the dissolved species remains constant, the problem is similar to that of heat conduction [14]. This diffusivity depends on the diffusion of dissolved species and on the microstructure of the green piece (porosity, tortuosity). The mean concentration $\phi(t)$ in organic at a given time t is expressed by :

$$\phi(t) = \frac{512}{\pi^6}\phi_0 \sum_{i=0}^{\infty}\sum_{j=0}^{\infty}\sum_{k=0}^{\infty} \frac{\exp(-\alpha_{i,j,k}t)}{((2i+1)(2j+1)(2k+1))^2} \tag{5}$$

with $\alpha_{i,j,k} = D\pi^2\left(\frac{(2i+1)^2}{L^2} + \frac{(2j+1)^2}{l^2} + \frac{(2k+1)^2}{h^2}\right)$

Samples prepared with paraffin 52 were treated under various experimental conditions. Theoretical curves (figure 5), plotted using calculated values of diffusivity (Eqn. 5) are in good agreement with experimental data, suggesting that paraffin extraction is controlled by diffusion of dissolved species. The model was also used to plot the profiles of concentration of paraffin $\phi(x,y,z,t)$, from the centre to the surface of the sample, as a function of treatment time (figure 6). The porosity becomes entirely opened and interconnected after an extraction of 40% of paraffin. Remaining paraffin can be theoretically removed, during sintering at a higher heating rate, without causing stresses.

Though the mean calculated values of diffusivity are sufficient to estimate the time necessary for the removal of a paraffin binder from a ceramic green part, we have to take in mind, in the present model, that the diffusivity coefficient was supposed to remain constant during extraction, that does not represent the real extraction process. First, paraffin waxes are mixtures of n-alcanes that will not diffuse with the same velocity through the sample. Second, the composition of the paraffin varies during extraction, the lighter n-alcanes being preferentially removed at the beginning of the debinding step. Third, the pores of the sample, initially filled with the binder phase, become filled with the supercritical phase in which solubilised molecules diffuse more rapidly during the experiment. Finally, the ratio between the pore size and the size of the solubilised molecules are particularly concerned with the kinetics. For all these reasons, the diffusivity coefficient varies during the experiment and we began a first series of experiments in order to estimate the influence of the pore size on the kinetics of extraction by using porous preforms impregnated with paraffin 42.

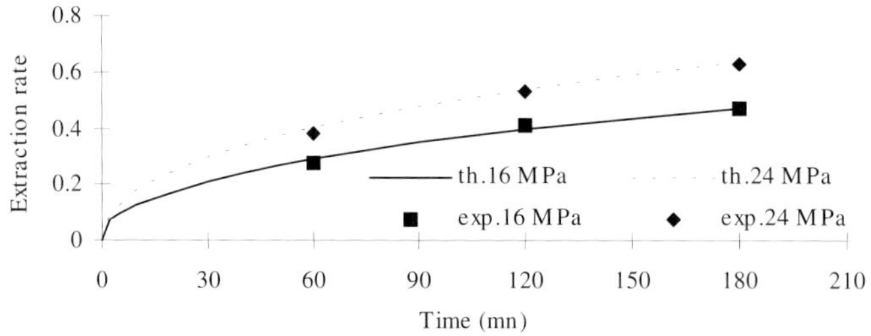

Figure 5. Kinetics of extraction of paraffin 52 (50°C).

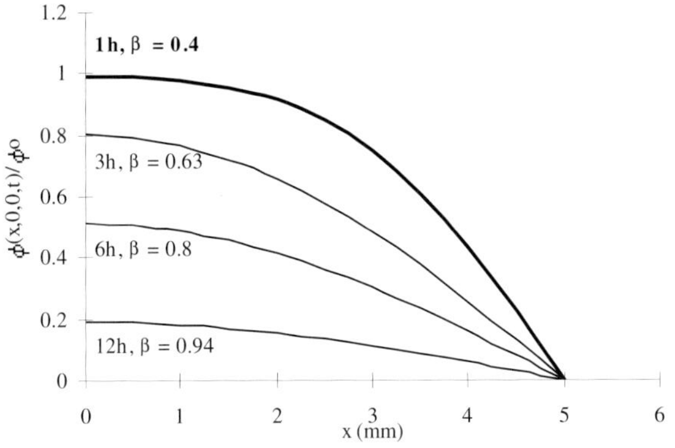

Figure 6. Profiles of paraffin 52 concentration during debinding (50°C-24 MPa).

4.3 Influence of the pore size of the ceramic samples

The first results concerning the influence of the mean pore size on the experimental extraction rate are presented in figure 7. The extraction rate becomes roughly constant for a pore size larger than 0,7 μm. We can then suppose that diffusion is no more reduced by a microstructural effect above this critical value. In order to precise this effect, it would be therefore interesting to prepare porous samples with i) a mean pore sizes below 0.1 μm to determine the diffusion rate-reducing effect of the microstructure and, ii) a mean pore size above 10 μm to determine the effective diffusion coefficient of the solubilised molecules.

Furthermore, as shown previously, the solubility of the binder depends on its n-alcane distribution, which varies with time of extraction. Thus, a fully developed diffusion model should also depend on the alcane distribution. We can suppose that each type of alcane diffuses at a specific velocity. Porous preforms infiltrated with some pure alcanes will allow to experimentally determine the diffusion of solubilised alcanes in supercritical CO_2 and then, to develop such a diffusion model in which the diffusivity of a paraffin wax is related to the diffusivity of the alcanes it contains.

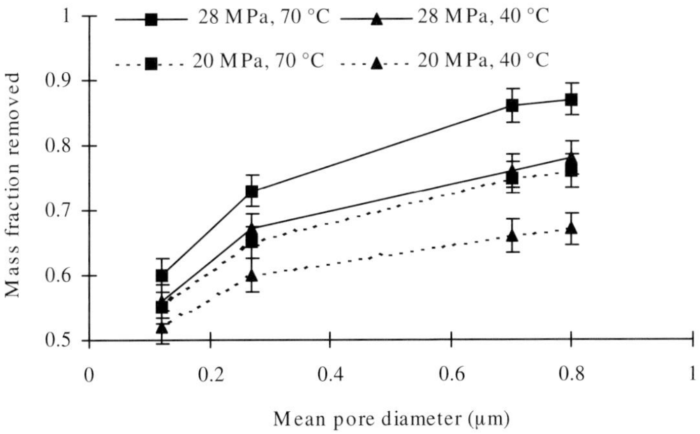

Figure 7. Influence of the mean pore diameter on the extraction rate of paraffin 42 at 40 °C and 70 °C (30 min)

5 Conclusion

Organics removal from ceramic green parts is a critical stage in ceramic processing. An original method based on unique properties of supercritical fluids was developed. This low temperature technique (no melting and no degradation of organic binders) offers great advantages in comparison to the classical technique of pyrolysis : absence of deformation and of stresses in the green part, absence of residues of pyrolytic degradation, high debinding rate.

Two mechanisms control the extraction, namely the solubilisation of the binder and the diffusion of dissolved species. A first model allows to predict the solubility, in supercritical CO_2, of a paraffin used as ceramic forming additive. A second model allows to estimate the diffusivity of dissolved species through the porous ceramic. These two models are in good agreement with experimental results. Finally, a more complex kinetic model that take into account the evolution of the effective diffusivity of solubilised molecules in the porosity of the green ceramic part and the composition of the remaining binder during extraction has to be developed.

6 References

1. German, R.M., Theory of Thermal Debinding, Int. J. Powder Metall., 23 (4) (1987) 237-45
2. Calvert, P.D. and Cima, M.J., Theoretical Models for Binder Burnout, J. Am. Ceram. Soc., 73 (3) (1990) 575-79
3. Evans, J.R.G., Edirisinghe, M.J., Wright, J.K. and Crank, J., On the Removal of Organic Vehicle from Moulded Ceramic Bodies, Proc. Roy. Soc. (London) A432 (1991) 321-40
4. Pinwill, I.E., Edirisinghe, M.J.and Bevis, M.J., Development of Temperature-Heating Rate Diagrams for the Pyrolytic Removal of Binder Used for Powder Injection Moulding", J. Mat. Sc., 27, (1992) 4381-88
5. Cima, M.J., Lewis, J.A. and Devoe, A.D., Binder Distribution in Ceramic Greenware During Thermolysis, J. Am. Ceram. Soc., 72(7) (1989) 1192-99
6. Wright, J.K. and Evans, J.R.G., Removal of Organic Vehicle from Moulded Ceramic Bodies by Capillary Action, Ceram. Int., 17 (1991) 79-87
7. Barone, M.R. and Ulicny, J.C., Liquid-Phase Transport During Removal of Organic Binders in Injection-Molded Ceramics, J. Am. Ceram. Soc., **73** (11) (1990) 3323-33
8. Matson, D.W., Smith, R.D., J. Am. Ceram. Soc., Vol. 72 (1989), p. 871
9. Nakashima, N., Nishikawa, E., Wakao, N., Proceedings of the 2nd International Symposium on Supercritical Fluids (May 20-22, 1991, Boston, USA.). Edited by M.McHugh, Butterworth publishers, Boston (1991), p. 357
10. Chartier, T., Ferrato, M., Baumard, J.F., J.Am.Ceram.Soc, Vol. 78 (1995), p. 1787
11. Marteau, Ph., Obriot, J. and Tufeu, R., "Experimental Determination of Vapor-Liquid Equilibria of CO_2+Limonene and CO_2+Citral Mixtures", J. of Supercritical Fluids, 8 (1995), 20-24
12. Chartier, T., Delhomme, E., Baumard, J.F., J. Phys. III, Vol. 7 (1997), p. 291
13. Chartier, T., Ferrato, M., Baumard, J.F., J.Eur.Ceram.Soc., Vol. 15 (1995), p. 899
14. Crank, J., The mathematics of diffusion, 2nd edition, Clarendon Press (1975), p. 49

CERAMIC OXIDE (MeO_2) SOLID SOLUTIONS OBTAINED BY MECHANICAL ALLOYING

Federica Bondioli, Paolo Veronesi, Cristina Leonelli and Tiziano Manfredini
Department of Chemistry, Faculty of Engineering,
University of Modena and Reggio Emilia, Modena, ITALY

1. Introduction

There has been a long-term trend toward the use of finer powders in ceramic practices. Increasing demands for improved performance of ceramics in loadbearing applications and in functional application have increased the importance of chemical purification and processing of the starting powders. Maximum control of a ceramic process begins with the starting powder and powder preparation. Consequently numerous "soft chemistry" synthesis methods, liable to lead to suitable powders, have been largely investigated. Only a few people are aware that milling, a technique which was and still is mainly used for reducing the size of particles of various materials, is a possible method for this purpose. For a period of about thirty years, grinding has indeed been a method of synthesis of advanced materials, intermetallic and alloy compounds[1-3], hence is called *mechanical alloying*. More recently, it was applied to prepare a few ceramic powders, including oxides such as ZrO_2[4], $PbTiO_3$[5] and $BaTiO_3$ [6].

The intrinsic advantage of this technique is that the solid-state reaction is activated via mechanical energy instead of temperature. High-energy ball-milling is a way of modifying the conditions in which chemical reactions usually take place either by changing the reactivity of as-milled solids (mechanical activation[7]) by increase of reaction rates, lowering of reaction temperatures of ground powders, or by inducing chemical reactions during milling (mechanochemistry[8-10]).

This work has to be regarded as a laboratory study concerning the synthesis and the application as ceramic pigments of powders obtained by mechanochemistry. In ceramic industry pigments, that must show thermal and chemical stability at high temperature and must be inert to the chemical action of the molten glaze, were traditionally obtained by calcination of raw materials. These heat treatment was carried out adding mineralizers (fluorides) to lower the high synthesis temperatures (1200-1400°C). The possibility to obtain them at room temperature and without mineralizers is of great interest: avoiding high calcination temperatures can cause drastic reduction in toxic gas emission and minimizes high temperature technological investments.

The two systems (ceria and zirconia powders containing 10 mol% praseodymium) were chosen because of their thermal and chemical properties and (from an industrial point of view) because of the interesting red and yellow tonalities of color that they can develop. In this study the thermal, chemical and optical properties of the as-obtained solid solution were analyzed to determine the effects of the synthesis conditions. To verify the applicability in the industrial ceramic bodies, color tests were performed on clay bodies loaded with 2-10 wt% ceria pigment and fired in semi-industrial kiln.

2. Experimental Procedure

High purity MeO_2 (Me = Ce, Zr) and praseodymium oxide (Pr_6O_{11}, Carlo Erba) powders were mixed in order to obtain the stoichiometric composition of $Me_{0.9}Pr_{0.1}O_2$. The powders were charged in a cylindrical agate vial of 40 mm in diameter together with agate balls of determined diameter (12 mm for the small type, 14 mm for the big type as milling media). All the experiments have been performed in a Fritsch "Pulverisette 7" planetary ball mill using a rotation speed of 200 rpm keeping constant the ball-to-powder weight ratio at 20:1. A complete description of the experimental procedure can be find elsewhere[11]. The optimized synthesis conditions are reported in Table 1.

Table 1. Synthesis conditions

System	Raw Materials		Milling Time (hrs)
$Ce_{0.9}Pr_{0.1}O_2$	CeO_2	Pr_6O_{11}	20
$Zr_{0.9}Pr_{0.1}O_2$	m-ZrO_2-BET (15 m^2/g)	Pr_6O_{11}	15
$Zr_{0.9}Pr_{0.1}O_2$	m-ZrO_2-BET(100 m^2/g)	Pr_6O_{11}	20

To determine the thermal stability of the as-obtained powders, thermogravimetric and differential thermal analysis were made in air at a heating rate of 20°C/min up to 1400°C, using a simultaneous TGA/DTA equipment (Netzsch STA409). To investigate the color of the pigments, the samples were analyzed with UV-Vis spectrophotometer with analytical software for color measurements (Perkin Elmer, Lamba 19) with CIE-Lab method. The method measures the adsorption intensity in the visible region to obtain the three parameters L^*, a^* and b^*, measuring brightness, red/green and yellow/blue color intensity, respectively.

Finally to evaluate their industrial usefulness as pigments for coloring fast-fired porcelain stoneware tiles, semiindustrial test according to the ceramic practice were done. The $Ce_{0.9}Pr_{0.1}O_2$ pigment were added to two typical compositions at percentages varying from 2 to 10 wt% of porcelain tile bodies powders, A and B (Table 2). The mixtures were homogenized, wet mixed (35 wt% of water) for 30 min in a blender mixer, and then dried. The dried powders were humidified to a water content of 6 wt%, then pressed at 40 MPa to produce disks (40 mm in diameter). Compacts were dried to constant weight at 110°C and fired in an industrial high-speed roller kiln using a typical industrial cycle (45-50 min up to 1220°C). The fired samples were characterized with a computer-assisted X-ray (Cukα) powder diffractometer (XRD, PW 3710 Philips). Apparent density of fired tiles was measured with the Archimedes method by immersion in mercury while linear shrinkage during firing and water absorption were determined according to the EN 99 standard. The chromophore role of the pigments was studied by ultraviolet and visible spectrophotometer (UV-Vis, Lamda 19 Perkin Elmer) with analytical software for color measurements (CIE-Lab).

Table 2. Chemical analysis of the two ceramic bodies used.

OXIDES	A	B	OXIDES	A	B
SiO_2	70.6	68.5	Al_2O_3	17.9	17.8
Na_2O	2.6	2.6	ZrO_2	/	3.4
K_2O	2.1	2.2	TiO_2	0.7	0.3
CaO	0.8	0.4	Fe_2O_3	0.5	0.5
MgO	0.8	1.1	I.L.	3.6	3.2

3. Result and Discussion

3.1 Pigment Characterization

The powders obtained were subjected to thermal analysis. No peaks in DTA curve were observed throughout the heating process, except, in the praseodymia-ceria system's curve, for a small endothermic peak, correlated to a weight loss resulting from the release of adsorbed water at 70-120°C (figure 1). Thus DTA shows a good thermal stability of the powders up to 1400°C.

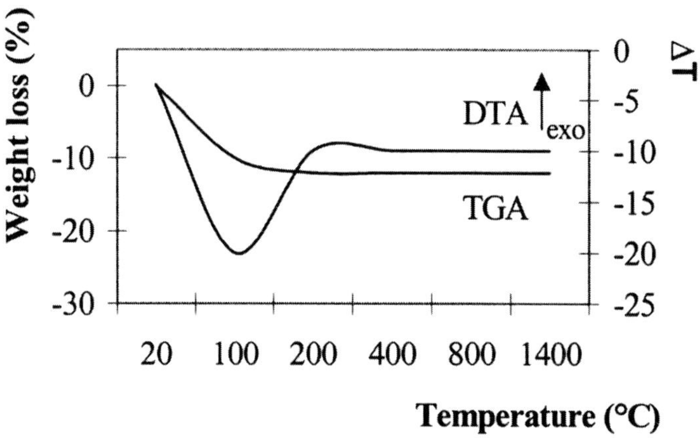

Figure 1. TGA/DTA curves of $(Ce,Pr)O_2$ sample after 15 hrs milling.

Table 3 shows the color of the optimized samples.

Table 3. Powders color

System		L^*	a^*	b^*
$Ce_{0.9}Pr_{0.1}O_2$		39.86	5.26	8.29
$Zr_{0.9}Pr_{0.1}O_2$	low BET	81.2	-4.2	11.98
$Zr_{0.9}Pr_{0.1}O_2$	high BET	85.3	-2.3	10.65

The decreasing in a^* and b^* values in the ceria and zirconia systems respectively, observed for higher milling time, is due to contamination with SiO_2, coming from the agate container and balls.

3.2 Ceramic Application

The dense microstructure of fired porcelain stoneware tiles is characterized by low porosity and by the presence of quartz, mullite, and a relevant amount of glassy phase. Additions of $Ce_{0.9}Pr_{0.1}O_2$ pigments in the 2-10 wt% amount do not seem to cause relevant changes in the reactivity of the pressed powders during firing, since the same phases, developed in fired tiles from base body, are present. Undissolved crystalline Pr-doped ceria in fired samples was observed. No evident changes in water absorption and in linear shrinkage, in mechanical and physical properties were observed for fired tiles containing different amounts of pigments. For these reasons coloration with $Ce_{0.9}Pr_{0.1}O_2$ pigment up to 10 wt%, can be done without modifying the composition of the body and the industrial process.

According to Jörgensen and Rittershaus, the red color of the praseodymium-doped powders is related to a charge transfer band due to electron transfer from the ligand orbitals to the praseodymium cation. The red color obtained corresponds to a strong absorption for $\lambda < 600$ nm. With regard to the color of the bodies, a continuous change is observed by comparing the L^*, a^*, b^* values as a function of the amount of pigment in the bodies. In particular, the values of the a^* measured on fired tiles reach the maximum value for addition of 10 wt% of pigments: the values obtained with B body, in which $ZrSiO_4$ was contained, are always higher than the ones measured in A body.

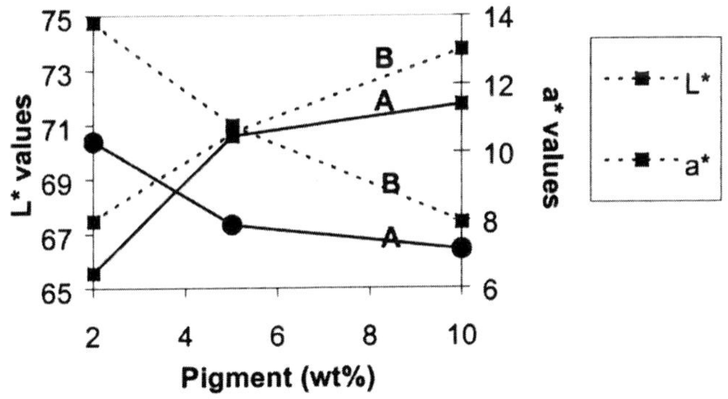

Figure 2. L* and a* values of the fired tiles

The measured values of the L^* parameter is considerably modified by the pigment: increasing percentages of added pigment cause a lowering in the L^* values.

4. Conclusion

The experiments demonstrated the capability of $Ce_{0.9}Pr_{0.1}O_2$ pigments obtained by mechanical alloying to efficiently act as inorganic pigment for porcelain stoneware tiles to obtain different shades of red and orange by varying the amount of powder. The addition of 2-10 wt% of pigment into the body did not cause major modification in the microstructure of the tiles, and the measured values corresponding to the main mechanical properties of porcelain stoneware tiles are consistent with the values prescribed by the European standards and with the real values of the products present on the market.

As far as the synthesis method is concerned, the SiO_2 impurities released by the agate vial and balls do not affect the sintering behavior of the ceramic tiles, neither influence their color. Considering that contamination could be reduced by using harder but more expensive materials for vial and milling media, the use of a relatively cheap material like agate can be considered satisfactory when applied to the production of low added value product as ceramic pigments are.

References

1) J.S. Benjamin, Sci. Am. 1976, 234, 40 - 48.
2) P.S. Gilman, J.S. Benjamin, Annu. Rev. Mater. Sci. 1983, 13, 279 - 300.
3) E. Gaffet, N. Malhouroux, M. Abdellaoui, J. Alloys Compd. 1993, 194, 339 - 360.
4) Y.L. Chen, D.Z. Yang, Scripta Metall. Mater. 1993, 29, 1349 - 1352.
5) K. Hamada, M. Senna, J. Mater. Sci. 1996, 31, 1725 - 1728.
6) O. Abe, Y. Suzuki, Mater. Sci. Forum. 1996, 225, 563 - 568.
7) J. Temuujin, K. Okada, K.J.D. MacKenzie, J. Eur. Cer. Soc. 1998, 18, 831 - 835.
8) A. Corrias, G. Paschina, P. Sirigu, J. Non-Cryst. Solids. 1998, 232-234, 358 – 363.
9) P. Matteazzi, G. Le Caër, J. Am. Ceram. Soc. 1992, 75[10], 2749 – 2755.
10) J. Hue, D. Wan, See-Ee Lee, J. Wang, J. Am. Ceram. Soc. 1999, 82[7], 1687 - 1692.
11) F. Bondioli, M. Romagnoli, L. Barbieri, T. Manfredini, Cer. Trans. Vol. 94, 1999, 295 -304.

Effects of Sintering Atmosphere and Sintering Time on Gas Emission Behavior of Alumina During Fracture under Ultrahigh Vacuum

M.Matsumoto, S.Kitaoka*, H.Matsubara*, S.Kuramoto and M.Kanno
Department of Materials Science, Graduate school of Engineering, The University of Tokyo, 7-3-1, Hongo, Bunkyo-Ku, Tokyo 113-8656, Japan
*Japan Fine Ceramics Center, 2-4-1, Mutsuno, Atsuta-Ku, Nagoya-shi 456-8587, Japan

1 Introduction

The present authors have been investigating gases emitted at the moment of fracture for ceramics such as Al_2O_3 and Si_3N_4 under ultrahigh vacuum using a mass spectrometer[1)2)3)]. In our previous study on Al_2O_3, it was found that various gases such as N_2 and Ar were emitted from polycrystalline specimens sintered in air, showing that it is an effective technique to analyze trace amount of gases entrapped in sintered materials. It has been reported that sintering atmosphere has a great influence on pore formation of sintered materials[4)5)]. However, the role of gases on pore formation is not fully understood because of difficulty in analyzing trace amount of gases in the material. In this paper, effects of sintering atmosphere and prolonged annealing on gas emission behavior of high purity Al_2O_3 were investigated to clarify the behavior of gases in the material and their influence on microstructure development.

2 Experimental Method

High purity Al_2O_3 powder (>99.99%) of 0.2μm average particle size (Taimei Chemicals Co., Ltd., TM-DAR) was used as starting powder. Powder compacts of 50x60x10mm were made by uniaxial pressing at 22MPa, followed by isostatic pressing at 306MPa. The green compacts were sintered at 1500°C for 0.5h under four kinds of atmospheres in a high purity Al_2O_3 tube to fabricate starting materials for subsequent annealing. The sintering atmospheres (pressures) of the starting materials were air (0.1MPa), Ar (0.1MPa), O_2 (0.1MPa), and vacuum (0.1Pa). Annealing of each starting material was also performed at 1500°C for 10h or 100h in order to investigate effects of annealing time on microstructure development. The other conditions of annealing (atmosphere, pressure and heating rate) were the same for these of each starting material. Densities of sintered bodies were determined according to Archimedes' principle. Microstructures of materials were observed by scanning electron microscopy (SEM) on polished sections after thermal etching.

Bending specimens for gas analysis were cut from sintered blocks and ground to dimensions of 3x4x20mm. A V-notch was introduced into the center of the tensile surface(4x20mm) of each specimen to make the area of fracture surface almost constant. The depth of the notch was about 200μm, and its radius of the curvature was about 10μm.

Three-point bending tests were carried out at ambient temperature at a crosshead speed of 0.5mm/min under background pressure of about $1x10^{-8}$Pa. Gases emitted from the samples during fracture were detected by a extracter gage (IE 514 Lybolt) and analyzed by a quadrupole mass spectorometer (MSQ-400 ULVAC). The response time of the extracter gage electronics was 200ms. The mass spectrometer enables us to analyze ten kinds of gas species having different *m/e* values at an interval of 10ms in a scan mode, where *m* is the mass number and *e* is the valence of gas species ionized in the spectrometer. The mass values (*m/e*) of examined gases (possible kinds of ions) were as follows: 2(H_2^+), 12(C^+), 14(CH_2^+,N^+), 15(CH_3^+), 16(CH_4^+,O^+), 18(H_2O^+), 28(N_2^+,CO^+), 32(O_2^+), 40(Ar^+), and 44(CO_2^+).

3 Results

Figure 1 presents the effect of annealing time on relative density of each sample. The starting materials show relative densities of about 99.7%, which is almost independent on sintering atmosphere. Whereas, during additional annealing at 1500°, relative densities of samples sintered in air, Ar and O_2 tend to decrease. In contrast, the vacuum sintered samples do not have such a negative tendency and the density is kept approximately constant.

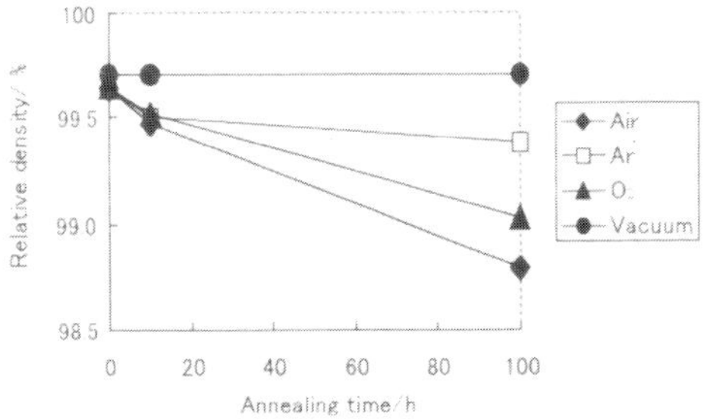

Figure 1. Effect of sintering atmosphere and annealing time on relative density.

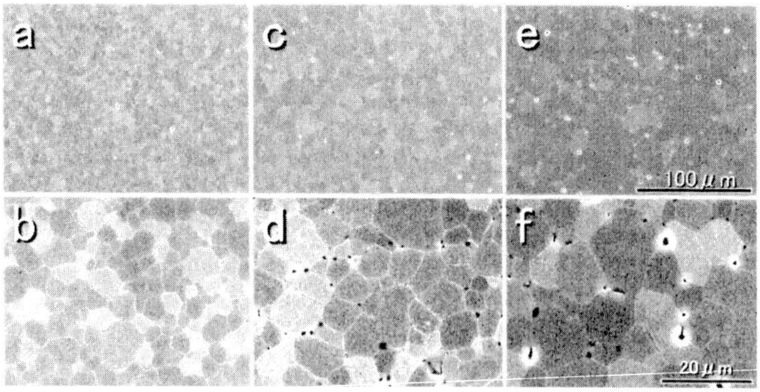

Figure 2. Microstructures of Al_2O_3 sintered in air at 1500□ for (a)(b) 0.5h and annealed for (c)(d) 10h and (e)(f) 100h.

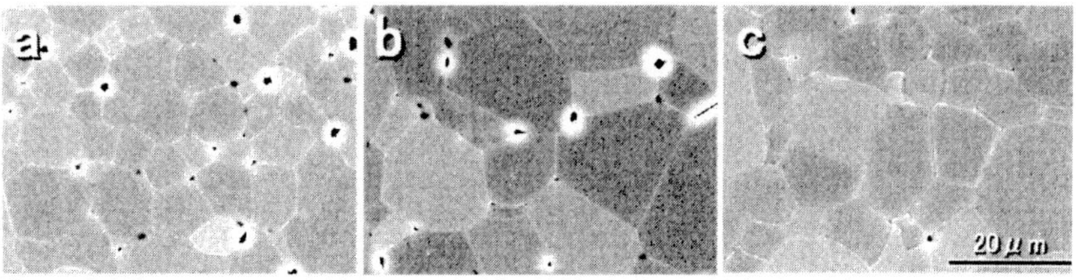

Figure 3. Microstructures of Al_2O_3 annealed for 100h; (a) in Ar (b) in O_2 (c) in vacuum.

Microstructures of samples sintered in air are shown in Fig.2. Pores are seldom observed for the starting material. However, as annealing time is prolonged, pores of a few μm radius appear mainly at grain boundary junctions. It is apparent that the pore formation and growth correspond to the density decrease. Microstructures of samples sintered in the other atmospheres with additional annealing for 100h can be compared in Fig.3. As is the case of air sintered samples, pores also appear at grain boundaries for those sintered in Ar and O_2 during annealing. In contrast, pores are scarcely observed when sintered in vacuum. These results represent that the entrapped gases in the sintered compacts lower density during prolonged annealing through pore formation.

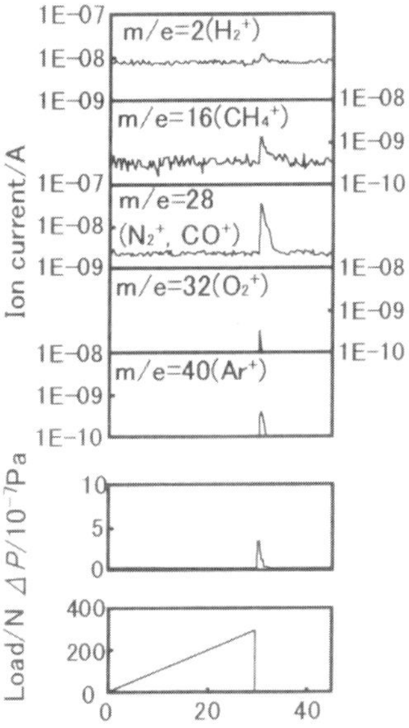

Figure 4. Typical results of gas emission behavior from a specimen sintered in air at 1500°C for 0.5h.

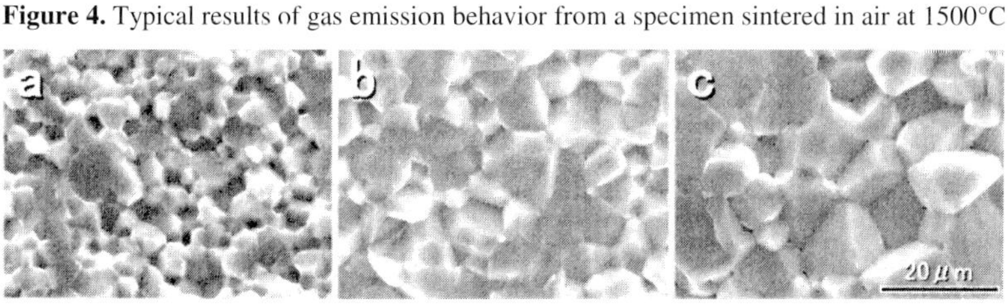

Figure 5. Fracture surfaces of Al_2O_3 sintered in air at 1500°C for (a) 0.5h and annealed for (b) 10h or (c)100h.

Figure 4 shows typical output signals of the mass spectrometer and the extracter gage when tested a specimen sintered in air for 0.5h (the starting material) and a corresponding load-displacement curve. At the time of fracture, a total pressure change of the chamber(ΔP) and mass signals increase suddenly. Mass peaks of *m/e* =2, 14, 15, 16, 28, 32, 40 were observed at the moment of fracture for all air-sintered specimens as shown in Fig.4, and these peaks were assigned as the emissions of H_2 (*m/e*=2), CH_4 (*m/e*=15, 16), N_2 (*m/e*=14, 28), O_2 (*m/e*=32) and Ar (*m/e*=40). Irrespective of annealing time, N_2 is the predominant gas emitted from specimens sintered in air. It can be said that N_2 is introduced from sintering atmosphere because the peaks of O_2 and Ar, which are the gases contained in air about 20% and 1%, respectively, are also observed. The emission of H_2 and CH_4 were confirmed to be artifacts such as adsorbants on surfaces of the specimen.

To investigate the origins of emitted gases, fracture surfaces of samples prepared in air are observed (Fig.5). Because of introducing V-notch, the macroscopic areas of fracture surfaces were nearly same among the specimens. The fracture mode are primary intergranular (about 70%) for each specimen and is independent on grain diameter. Pores were clearly seen on grain boundaries for all samples. The detected gases are likely to be released mainly from such pores accompanying intergranular fracture.

Figure 6 illustrates the total amount of emitted gas (*N*) calculated from the pressure change of the chamber at the time of fracture for all samples. Analysis with the mass spectrometer confirmed that O_2 and Ar were the predominant gases emitted from the specimens sintered in O_2 and Ar, respectively. In the case of sintering in vacuum, only adsorbants such as H_2 were found. Compared under the same annealing time, the amount of emitted gas is the largest for specimens sintered in Ar, followed by specimens sintered in air, O_2 and vacuum. These differences are considered to result from differences in the mobility of the gases in the material. In the case of specimens sintered in air and Ar, the amount of emitted gas increases with annealing time. In contrast, the total amount of emitted gas turns to decrease for O_2 fired samples in the course of annealing showing that a part of O_2 could move out of pores.

4 Discussion

Since pores located on grain boundaries were seen to grow, it is thought that gases in the sintered compacts concentrate on grain boundaries during annealing through some mechanisms to cause pore growth. This idea is supported by experimental results of gas emission behavior. Namely, the increase in the amounts of emitted gases with annealing time means that the concentration of gases on grain boundary increases during annealing.

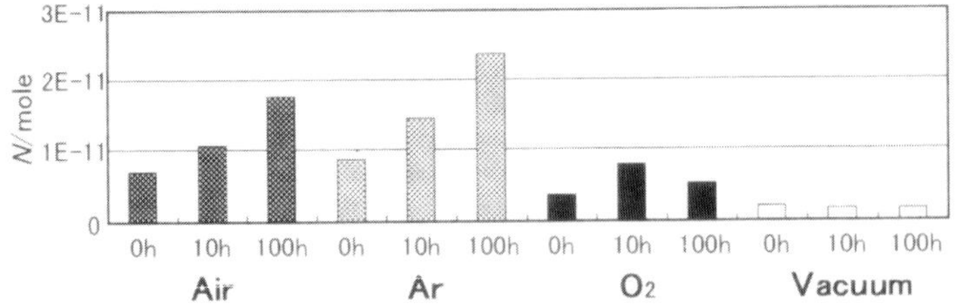

Figure 6. Effects of sintering atmosphere and annealing time of the total amount of emitted gas at the moment of fracture.

There are some expected mechanisms of the density decrease according to entrapment site of gases in the material. The possible entrapment sites of gases are intergranular and/or transgranular defects such as pores and vacancies.

First of all, discussion is made on the possibility of presence of gases in the interior of grains. In such a case, the amount of gas on grain boundary would increase during grain growth due to sweeping out of the gas-filled transgranular pores or interstitially dissolved gas by grain boundary migration[6], resulting in pore growth on grain boundaries. However, contribution of such a mechanism to pore growth was believed to be small in this study because transgranular pores were rarely found as shown in Fig. 2 and Fig.3. In addition, N_2 or Ar are thought to be too large to enter interstitial position[4]. Indeed, the amount of emitted gas was very small for single-crystal Al_2O_3 which was melted in Ar atmosphere[3], showing that Ar has very low solubility in Al_2O_3 matrix although the solubility of Ar has not been reported.

It is most likely that gases are mainly present on grain boundaries. Being assumed that gases are present in pores located on grain boundaries, the phenomenon can be explained in terms of pore coalescence[8]. The coalescences of gas-filled pores during grain growth would lead to the growth of intergranular pores and an increase in the total amount of emitted gases accompanying intergranular fracture.

Nevertheless, theoretical density was not achieved even if sintered in O_2, which was somewhat different from the previous findings[4)5]. The reason for not sintering to full density in O_2 could be attributed to rapid grain growth. In this study on high purity Al_2O_3, grain diameter of samples sintered in O_2 was much lager than that of the other materials sintered in air or Ar as shown in Fig.3, and the excess grain growth would bring about extensive pore coalescence and further decrease in density. The high mobility of O_2 is thought to accelerate grain growth by reducing pore size or pore number density, which would control grain growth[8]. It is speculated that MgO addition is effective for helping O_2 to move out by limiting grain growth and pore coalescence and that making it possible to sinter to theoretical density. On the contrary, the density decrease in Ar sintered sample is relatively small in comparison to the samples sintered in air or O_2 as shown in Fig.1. The reason for this is believed that the grain growth is highly inhibited by pores containing Ar which is believed to have extremely low mobility in the material.

The density decrease during prolonged sintering was reported for other sintered materials and thought to be caused by gases in the sintered body and pore coalescence[9)10]. The present results provide good support for these reports.

5 Conclusions

(1) Predominant gases emitted from samples sintered in air, Ar and O_2 were N_2, Ar and O_2, respectively, while gases were rarely detected when sintered in vacuum. It is clear that gases from the sintering atmosphere are the main gases entrapped in Al_2O_3. Further, the amounts of emitted gases for samples sintered in air and Ar were larger than those of O_2 sintered samples. The result shows that mobility of O_2 in the material is much higher than that of N_2 or Ar.

(2) Density decreased with increasing annealing time when sintered in air, Ar and O_2 . However, such a trend was not observed when sintered in vacuum. The density decrease was caused by formation and growth of intergranular pores. Corresponding to the density

decrease, the amounts of emitted gases for samples sintered in air and Ar tended to increase. The density decrease is concluded to be caused by entrapped gases in the sintered bodies through a pore coalescence process.

6 Acknowledgements

This work was partially supported by a Grant-in-Aid from the Ministry of Education, Science and Culture of Japan.

7 References

1. M.Kanno, H.Okada and G.Itoh, J. Japan. Inst.Metals, **59**, 296-302 (1995)
2. S.Kitaoka, H.Matsubara, H.Kawamoto, M.Matsumoto and M.Kanno, J. Ceram. Soc. Japan, **105**, 915-17 (1997)
3. M.Matsumoto, H.Okada, M.Kanno, S.Kitaoka and H.Matsubara, Scripta Mater., **37**, 2047-51(1997)
4. R.L.Coble, J. Am.Ceram. Soc., **45,** 123-27 (1962)
5. Y.K.Peak, K.Y.Eun and S.J.L.Kang, J. Am.Ceram. Soc., **71**, C380-82(1988)
6. R.J.Brook, J. Am.Ceram. Soc., **40**, 80-85(1957)
7. W.D.Kingry and B.Francois, J. Am.Ceram. Soc., **48**, 546-47(1965)
8. R.J.Brook, J. Am.Ceram. Soc., **52**, 56-57(1969)
9. F.A.Nichols, J. Appl. Phys., **37**, 4599-602 (1966)
10. C.Greskovich, J. Am. Ceram. Soc., **64**, 725-30 (1981)

Influence of Magnesia on Sintering of Transition Alumina

C. Legros, F. Herbst, C. Carry
LEMHE, UMR8647, Université de Paris-Sud, 91405 ORSAY Cedex, France
LTPCM, UMR5614, INPG/UJF, 38042 SAINT-MARTIN D'HERES Cedex, France

1 Introduction

Most nanocrystalline oxide ceramic powders are metastable and this metastability may have a critical influence on the sintering behaviour. It is particularly true for transition alumina powders which are currently produced with very high specific surface area and ultrafine crystallite sizes ; the transformations into the stable α-alumina are generally accompanied by the development of porous vermicular microstructures. The final stages of sintering then require very high temperatures to achieve high densities. Studies on producing nanocrystalline ceramics from fine powders (<100 nm) have highlighted the problem of achieving high densities without excessive grain growth. In recent literature on this subject, three main routes have been investigated to overcome this difficulty : α-alumina seeding of precursor gels, pressure sintering methods such as uniaxial hot-pressing or sinter-forging and doping routes. In a recent study [1], we have clearly highlighted the effects of several parameters such as green density, heating rate and α-alumina initial content on the pressureless sintering of a commercial transition alumina. An enhanced densification, over and above that expected for the γ to α phase transition, has been shown. This enhanced relative density change is brought about particle re-arrangement during the transformation, the degree of particle re-arrangement being influenced by the above mentionned parameters which also have an effect on the subsequent sintering of the α phase.

The ultimate goal of our study is the synthesis of α-Al_2O_3 nanoceramics starting from nanocrystalline γ-Al_2O_3, doped or un-doped but without pressure assisted sintering. Considerable controversy has surrounded the role of MgO in the sintering of α-Al_2O_3 and many various interpretations of the role of MgO have been proposed since the pioneering work in 1961 [2]. Nevertheless the most important effect of MgO doping is its apparent ability to significantly reduce the grain boundary mobility. The consequences of the action of MgO are the suppression of discontinuous or abnormal grain growth and an indirect enhancement of densification processes resulting in a nearly 100% dense material [3-4]. We present in this contribution the effects of magnesium as a doping element on the transformation-densification behavior of a γ-alumina, with the idea that a phase transformation could be used to enhance the densification of nanostructured alumina. In this study we use readily available commercial transition alumina powders. The effects of other dopants (yttrium and zirconium) will be reported elsewhere [5].

2 Experimental Procedures

Mg-doped powders were prepared from the same non-doped batch (n°1135R, named ND) of Baïkalox CR 125 commercial grade which is an alum derived γ-alumina with a specific area of $108m^2/g$ and α initial content less than 0.5%. This γ-alumina was doped with magnesia by the following method. γ-Al_2O_3 powder was dispersed in distilled water by a vibrating stirrer and a solution (1M) of $Mg(NO_3)_2$,$6H_2O$ in distilled water was prepared. For a given doping level, the corresponding amount of solution was poured into the alumina slurry and stirred for 2min. Then each slurry was dried and later calcined in an oxygen flow at 670°C for 24h. This procedure is expected to yield a uniform repartition of the dopant in alumina particles. Doped powders and samples are labelled by their cationic ratio (ppm/100) as reported in Table 1. A25Mg samples corresponds to a commercial CR125 γ-powder doped with magnesia at the alum stage ; its α content is rather high (6%) and this powder has to be considered as involontary α–seeded.

Table 1. Nominal compositions of the various doped samples

Batch	cationic ratio Mg/Al (ppm)	MgO content (ppm wt)	α content (%) (XRD)
ND	0	0	< 0.5
6Mg	633	500	< 0.5
25Mg	2531	2000	< 0.5
57Mg	5697	4500	< 0.5
75Mg	7593	6000	< 0.5
A25Mg	2531	2000	6

Green compacts were produced by cold isostatic pressing (250 MPa) of the as-received or doped powders. Cylindrical samples (diameters from 5-7mm) were cold machined from these compacts with lengths from 6 to 15mm depending on dilatometric experiments. The apparent shrinkage was followed by dilatometry (TMA92 Setaram France with 4mm maximum length variation and 2g constant load on the sample). All samples were heated to 800°C at 10°/min and thereafter under constant heating rate conditions 1 or 10°/min to 1550°C and rapidly cooled. All densification and sintering treatments were performed in air. Densities, ρ(T), and densification rate curves were computed from the recorded shrinkage data and from the final densities measured by the classical Archimede's method on cooled samples.

X-ray diffraction (XRD) measurements have been carried out with a Philips PW 3020 diffractomer with nickel-filtered CuK_α radiation. Microstructures and grain sizes were observed by scanning electron microscopy (SEM) on fractured surfaces.

3 Results

Figure 1 compares the densification behaviors of magnesium doped samples with the same cationic ratio Mg/Al (2531ppm), but with different α contents, sintered under constant heating rates (1°/min and 10°/min) up to 1550°C.

For A25Mg sample, the two typical regimes of densification of such transition alumina compact [1] are observed : R1, is a rapid apparent densification associated with phase transition of γ to the stable α in the 1100°C-1200°C temperature range, and R2, a slower densification of α alumina at higher temperatures. In all cases, the temperature of first densification rate peak, R1, is lowered on decreasing the heating rate. For all heating rates, the R1 temperature decreases with increasing α content. The second densification rate peaks, R2, are clearly observed for A25Mg samples while for the 25Mg compacts the R2 peak is shifted to higher temperatures (> 1550°C for 10°/min). The R2 temperature generally decreases with increasing α contents. The transformation relative density variation ($\Delta\rho_R$ as defined in [1]) is much higher than the increase of density from γ to α phase (10.75% with 3.6g.cm^{-3} for γ-phase or 17.25% with pratical density of the powder). Significant differences in $\Delta\rho_R$ values are observed and depend on powder batch and heating rate.

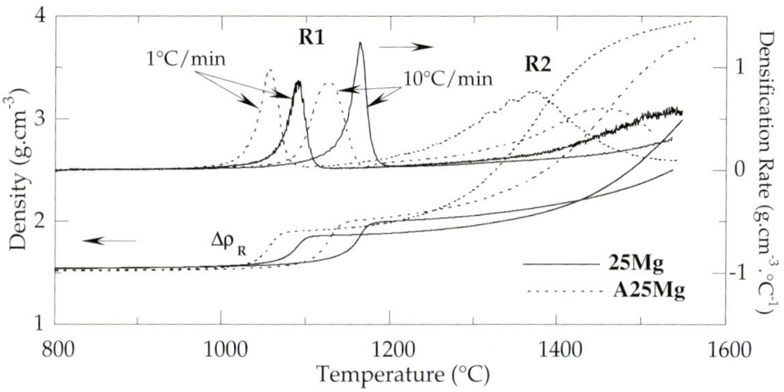

Figure 1. Densification behaviors of magnesium doped alumina compacts (250MPa) with the same cationic ratio (2531 ppm) but with different α-alumina contents : < 0.5% (25Mg) and 6 % (A25Mg)

Figure 2 presents the effects of magnesium contents on densification behavior of transition alumina compacts sintered at constant heating rate (1°/min) up to 1550°C. First of all, doping with Mg does not greatly influence the γ–α transformation, but general trends can be stated : in any case, the first densification regime starts at the same temperature, although the R1 temperature slightly increases for cationic ratio Mg/Al higher than 633ppm. For yttrium and zirconium doped samples [4], the same but more pronounced effect is observed with an increase of more than 50°C of the R1 temperature for samples with the higher yttrium contents.

The transformation relative density variation, $\Delta\rho_R$, increases with cationic ratio Mg/Al. The total shrinkage and final densities are not improved for magnesium doped compacts while for a ND sample processed via a different route (Attrition milled + cold isostatically Pressed) the R2 step leads to a much higher density. Processing effects on sintering behavior of this transition alumina powder will be reported elsewhere [6]. The second regime of densification is affected by the magnesium doping level, as can be seen in figure 3. This figure compares the densification rates of the ND sample and the doped samples in the 1200-1600°C temperature range. An increase of the cationic ratio Mg/Al up to 5697ppm enhances the second densification step, although for higher doping level, this R2 densification step is identical to the non-doped sample one's.

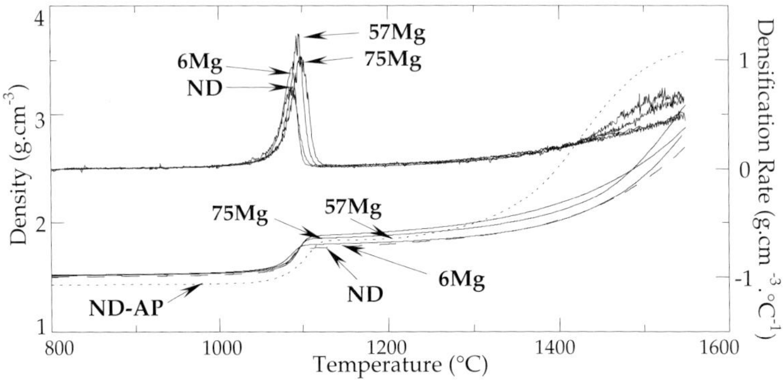

Figure 2. Effects of magnesium contents and processing routes (Non Doped - Attrited Pressed powder) on sintering behavior of the commercial transition alumina powder (1°/min heating rate to 1550°C).

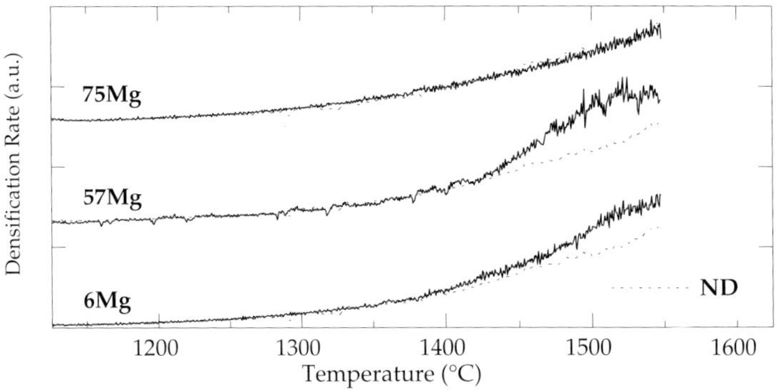

Figure 3. Densification rate curves of magnesium doped and non-doped (ND) compacts after the transformation step (1°/min to 1550°C).

To investigate the precipitation of $MgAl_2O_4$ spinel second phase, XRD analyses were performed on rapidly cooled samples from various temperatures of the second densification step (1°/min heating rate). For the 75Mg composition, $MgAl_2O_4$ was not detected at 1130°C, while corresponding XRD peaks are clearly identified for a sample cooled from 1200°C ; the α-alumina grain sizes were respectively about 40 and 50nm For 6Mg composition, this second phase is only detected for temperatures higher than 1400°C.

4 Interpretation, Discussion and Prospects

In this section we first focuse on the effect of doping on the first densification step which corresponds to the transformation of transition metastable alumina into the stable α phase. The formation of stable phase from metastable transition phase occurs by a nucleation and growth process during heating. The apparent transformation temperature, defined in this work as the temperature of the first densification rate peak (R1), depends on the kinetics of nucleation and of growth processes. The nucleation depends on the density of nuclei which results from : (i) the number per unit volume of α grains already present in the green body ;

(ii) the number of mechanical contacts between transition alumina particles which are potential nucleation sites, the higher is the green density the higher the number of such sites ; (iii) the heating rate. In principle the doping does not change any of these factors. Consequently, as seen in figure 2, the predominent factors in lowering R1 and R2 are the initial α content and heating rate. This result is in agreement with the behavior of other alumina precursor powders seeded with α-particles [7] and confirms our recent study [1].

However, we can deduce that the doping has an effect mainly on the growth step of the stable phase. Possible data on solubility of MgO in metastable alumina should be a priori questionable . For the α stable phase, literature reports that the solubility limit is typically on the order of a few hundred ppm or a thousand ppm at most [3]. Thus, with MgO contents less than 1000 ppm, the magnesium should be distributed in solution in the growing α phase during the transformation and does not modify the first step densification R1. This is in agreement with the densification behavior of 6Mg sample. For higher contents, the excess of the dopant has to diffuse towards the surfaces α-particles which grow at the expense of γ-grains, and this would delay the γ to α transformation. The resulting magnesium segregation at grain surfaces could have an influence on the particle re-arrangement mechanism associated with the phase transformation and consequently increase the apparent transformation density variation $\Delta\rho_R$, as observed in figure 2. But possible modifications of agglomeration green microstructures by the doping process could also affect this R1 densification regime.

Now, if we analyse the effect of Mg-doping on the second densification step which corresponds to an α alumina densification. In any case, final microstructures are porous and more or less vermicular (Figure 4). Magnesium seems enhance and shift at lower temperature the R2 densification regime for doping level lower than 7000ppm Mg/Al. During this regime grain growth or coarsening and spinel phase precipitation are observed. For a given magnesium content, we can consider than the excess of Mg above the solubility limit would segregates at grain surfaces or grain boundaries ; this segregation would increase with grain growth and a spinel precipitation would result of a magnesium grain interface saturation. Such apparent temporaly densification enhancement has been already observed for yttrium doped α–alumina and correlated to a grain boundary second phase precipitation [8]. The observed modification of the R2 regime by Mg doping could have the same origin. For the highest Mg doping level (75Mg) the spinel precipitation occurs at lower temperature and for a smaller grain size range (40-50 nm). A much finer dispersion of the spinel phase and a lower temperature range could explain the non-observation of this apparent densification enhancement and the finer final microstructures of this composition ; spinel phase precipitates slow down the grain growth and coarsening kinetics.

Figure 4. Microstructures of 6Mg (a) and 75Mg (b) samples heated up to 1550°C under constant heating rate (1°/min).

Works are in progress for a better understanding of the transformation-densification behavior of doped transition alumina powders. A particularly interesting research field is the production of doped green samples via colloïdal processing [6] : the influence of doping element should not be partialy biased, as probably in the present work, by the agglomeration phenomena which is the main obstacle to the putting in practice of nanometric oxyde powders to fabricate dense nanograined oxyde ceramics.

5 References

1. C. Legros, C. Carry, P. Bowen and H. Hofmann, J. Eur. Ceram. Soc., 1999, 19, 1967-1978.
2. R.L. Coble, J. App. Phys., 1961, 32, 793-799.
3. S.J. Bennison and M.P. Harmer in Sintering of Advanced Ceramics (Ed.: C.A. Handwerker et al.), American Ceramic Society, Westerville, OH, 1991, 13-45.
4. S.I. Bae and S. Baik, J. Amer. Ceram. Soc., 1994, 77, 2499-2504.
5. P. Bowen, C.Carry, F. Herbst, C. Legros, to be presented at SINTERING 99 (1-3 Nov 1999 Penn Stae College USA) and to be published in the conference proceedings
6. P. Bowen, H. Hofmann, M. Staiger, C.Carry, to be presented at SINTERING99 (1-3 Nov 1999 Penn State College USA) and to be published in the conference proceedings.
7. G. Messing, M. Kumagai, Am. Ceram. Soc. Bull., 1994, 73, 88-91.
8. E. Sato, C Carry, J. Amer. Ceram. Soc. ,1996, 79, 2156-2160.

High - Performance Composite Ceramics with AlN Matrix

Reiner Schober, Enrico Mählig

Fraunhofer Institute for Ceramic Technologies
and Sintering Materials, IKTS Dresden

1. Introduction

Ceramic - matrix composites consist of different components which are combined to produce properties not inherent in single components.
Tailor made high performance ceramics may be developed in this way for different applications.
Synthetic raw materials such as Si_3N_4, SiC, BN and AlN are used for manufacturing advanced refractory materials.
The Lanxide procedure is a widely used technique to produce aluminum/AlN composites by the directed nitridation of molten aluminum and particle – reinforced composites /1/.
The goal of the investigation in this paper focuses on the development of AlN/TiN/BN composite ceramics by reaction sintering and the use of TiB_2 as an active filler.
Aluminum nitride has a number of properties that make it an attractive material for use as a refractory material, in particular, the low thermal expansion and the high thermal conductivity. AlN is not wetted by melts of Ag, Cu, In, Sn, Bi and Pb showing optuse contact angles, and the interface between metals and AlN does not reveal the diffusion of any metal /2/.
TiB_2 is known as a material that withstands melt corrosion and has a low specific electrical resistance, which is important for use that as heating electrodes. The densification of TiB_2 requires a metallic or ceramic binder because of its low diffusion coefficient.

2. Experimental procedure

Composite ceramics were manufactured by preparing of powder mixtures comprising 40 vol.% TiB_2 and 60 vol.% AlN.
The AlN that was used is a commercially available powder (ESK, AlN -10 µm). Similarly, the TiB_2 powders that were employed are of Typ S – 325 mesh, ESK. The aluminum powder used (Alunova GmbH) has a particle distribution – 15 µm and 99.3 % Al.

The powders were mixed in a planetary mill, dried and compacted by uniaxial die pressing.

To manufacture the AlN matrix composites two routes have been investigated, on the one hand the nitridation and post - sintering of Al/TiB_2 or $Al/AlN/TiB_2$ precompacted powder mixtures and on the other hand the consolidation of precompacted powder mixtures of $AlN/TiB_2/CaO$ by liquid - phase sintering. The nitridation took place in nominally pure nitrogen up to 1300°C, followed by a post sintering - process at 1800°C.

3. Results and discussion

A typical microstructure is shown in Figure 1. The TiB_2 (light grains) and AlN (gray grains) are embedded in secondary formed AlN with some metal phase (mixed) with a very fine microstructure. The X-ray diffraction investigations reveal the main phases to be AlN, TiB_2 and Al. Decomposition of the TiB_2 phase was not found as long as metallic aluminum appeared in the microstructure.

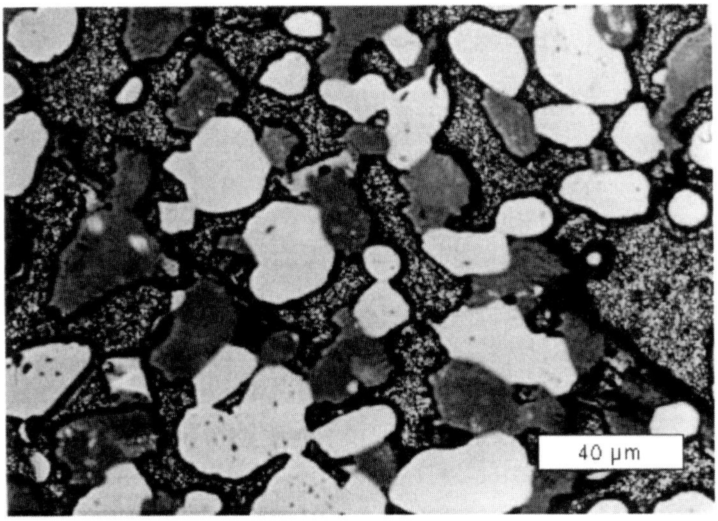

Figure 1: Optical photomicrograph of $AlN/Al/TiB_2$ ceramic, sintered at 1800°C,

Using $AlN/TiB_2/CaO$ as a starting powder mixture a composite ceramic consisting of AlN, TiN and BN is formed during liquid - phase sintering.

The filler component TiB_2 reacts in a N_2 sintering atmosphere according to the equation:

$$TiB_2 + 3/2\,N_2 \rightarrow TiN + 2\,BN \qquad (1)$$

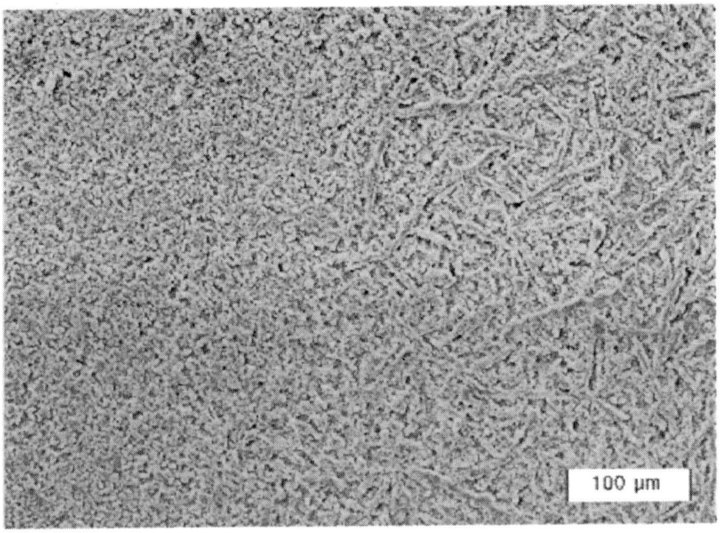

Figure 2: Composite ceramic AlN/TiN/BN with gradient structure

The microstructure and the degree of the phase transformation are influenced by the kinetics. Rapid heating and high sintering temperatures create a composite ceramic with a gradient structure, with equiaxial grains in the volume and columnar grains in the outer layer of the sample.

The kinetics of the phase transformation depend on:

- temperature – time regime and composition of the reaction gas,
- the particle size of the reactants, and
- the density of the green body.

The properties of the developed AlN matrix composites are summarized in comparison with a commercial AlN ceramic /3/ in Table 1.
The bulk density varies in dependence of the phase concentration. The sample 6.1a contains a coarser TiB_2 than the sample 6.1b, caused by a shorter milling time, which results in a lower rate of phase transformation.
The modulus of elasticity and the specific electrical resistance can be varied by the chemical composition of metallic aluminum and/or TiB_2 and TiN.

Table 1: Properties of AlN composites produced by nitridation and post - sintering (N + PS) and gas - pressure sintering (GPS)

Material	Techn.	Density g/cm3	Porosity %	MOR MPa	MOE GPa	Spec. el. resistance Ω cm
$AlN/Al/TiB_2$	N+PS	3,25	<5	300	190	0,008
AlN/Ti(B,N) 6.1a	GPS	3,18	<1	210	-	5
6.1b	GPS	2,9	<2	170	-	1
AlN/TiN/BN	GPS	3,0	<5	250	140	0,8
AlN	S	3,3	0	390	310	10^{19}

The results of the hot corrosion test at 1000°C in air are plotted in Figure 3. The weight gain of partially transformed AlN/Ti(B,N) composites 6.1a and 6.1b is plotted as a function of time in comparison with data of O'-sialon - bonded SiC and silicon – nitride - bonded SiC, taken from the literature /4/.

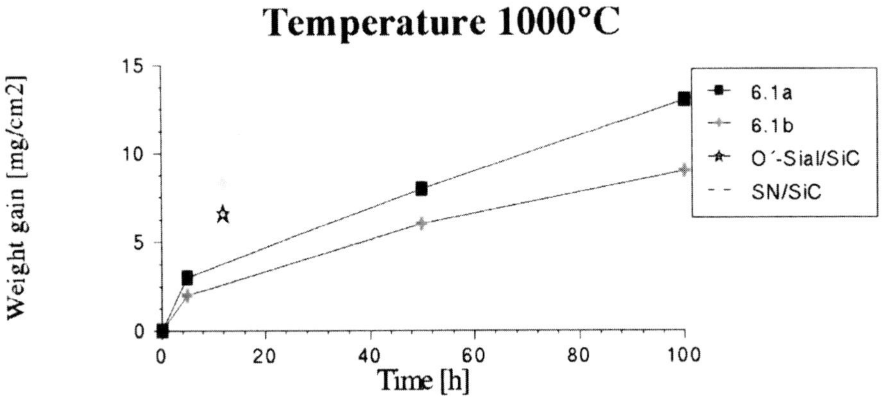

Figure 3: Comparison of measured weight gain as a function of time for the oxidation of AlN matrix composites and O'-sialon or silicon – nitrid - bonded SiC

The lower oxidation rate of the AlN composite ceramics is attributed to the higher open porosity and the active oxidation at 1000°C of the sialon - bonded SiC and/or the vaporization of B_2O_3 during the oxidation of TiB_2 or BN.

The data that refer to the oxidation in air at temperatures of 1000 °C can be fit by a parabolic rate law. The decrease of the bending strength after a storage time of 100 h at 1000 °C is significant.
The cross – section of the oxide layer in Figure 4 reveals that a gradated layer is formed during the hot corrosion process, with three distinct layers determined by energy dispersive X – ray analysis. Whereas the outer layer consists of Al_2O_3, the detected main phases in the middle layer are Al_2O_3 and TiOx. The region near the interface to the volume of the sample contains solid solutions of Al-Si and and a multiphase mixture of the system Al – B – Si – N. Deeper regions of the sample consist of TiB_2, TiN, AlN and AlON (Figure 4).

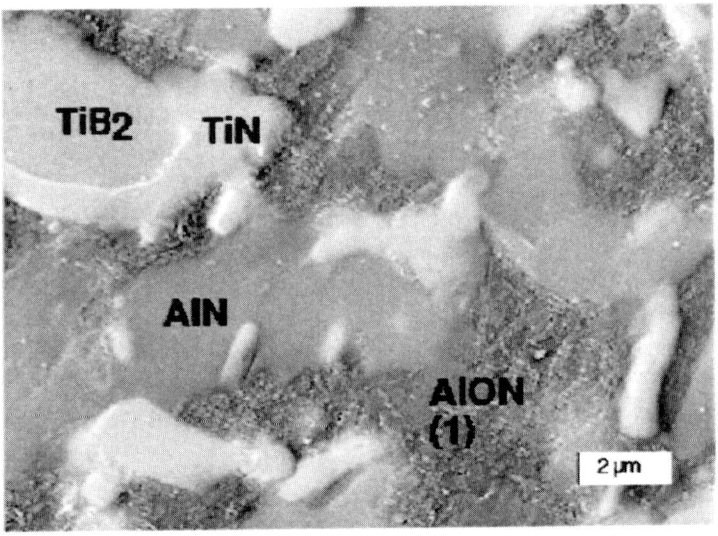

Figure 4: SEM photomicrograph showing TiB_2 - reinforced AlN matrix composite after oxidation at 1000 °C, 100 h with the formed AlON phase in the centre of the sample

Corrosion tests were carried out with liquid – phase - sintered samples of AlN/TiN/BN.
The wetting investigations by the sessile drop technique reveal no wetting by an FeSiB alloy. In the crucible test a passive layer is formed during the melting of the iron alloy at 1350 °C. There was no adhesion of the frozen melt on the wall of the crucible after removing the ingot (Fig. 5).

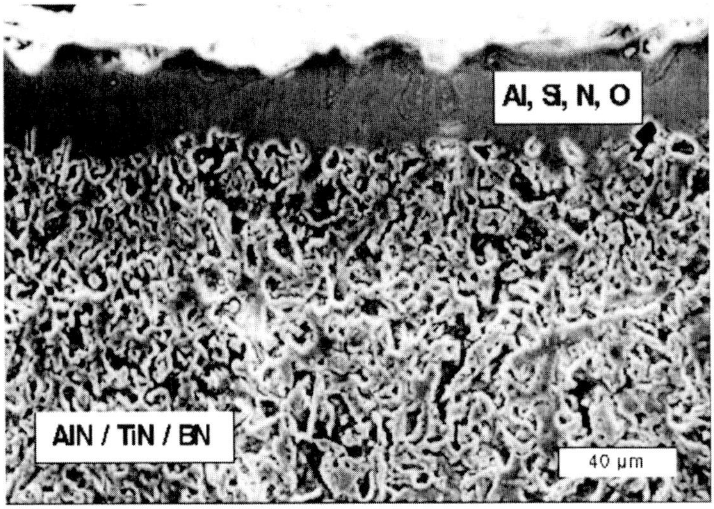

Figure 5: Composite ceramic AlN/TiN/BN corroded FeSiB/refractory interface with a protective layer (Si-Al-O-N)

No corrosion or infiltration could be detected for bars of AlN/TiN/BN which were stored under conditions of high-rate evaporation of aluminum at 1600 °C in an Ar atmosphere.

4. Summary

The manufacturing of CMCs based on AlN/TiB_2 was investigated by nitridation and post - sintering in comparison with gas - pressure sintering in a nitrogen atmosphere.
During the nitridation and post sintering the TiB_2 will be thermodynamic stable as long as metallic aluminum appears in the microstructure.
During sintering of AlN with CaO as a sintering additive, the TiB_2 has been transformed into TiN and BN with a gradient structure of equiaxial and columnar grains.
The kinetics of oxidation could be approximated by a parabolic rate law.
The AlN matrix composite is not wetted by iron and aluminum alloys. The mechanical and physical properties can be varied in dependence of the phase concentration.
Prospective application is expected for nozzles, crucibles and evaporation boats in the ferrous- and non-ferrous metallurgical fields..

The Use of Coalescence in Microstructural Design

Jens Adam, Rainer Telle
Department of Ceramics and Refractories, Institut für Gesteinshüttenkunde (Mineral Engineering), Rheinisch-Westfälische Technische Hochschule Aachen, Germany

1 Abstract

Alumina was liquid phase sintered with two different compositions of the liquid phase. It was found that a ZrO_2-containing liquid phase leads to an increased grain growth, what can be put down to an increased solid-liquid interface energy. To prevent grain boundary infiltration in melt corrosion, a high solid-liquid interface energy is desired. The effect of ZrO_2 is supposed to be another aspect of the corrosion mechanism of Zirconia containing refractories, for example in application in glass melt furnaces.

2 Introduction

Coalescence is a well-known effect from liquid phase sintering. It is understood as the direct contact and merger of grains. Thus, next to grain boundary migration and solution-reprecipitation it is a mechanism of grain coarsening.

To aim at a satisfying definition of coalescence, it is necessary to take different cases of interaction between polycrystals and liquid phases into account. Next to liquid phase sintering with its well-defined small amount of intergranular liquid phase, there is the case of melt corrosion where an excess amount of liquid phase attacks a refractory ceramic brick. A polycrystal under attack of a melt can be destroyed by three mechanisms: The first two mechanisms are chemical reactions such as decomposition or formation of new phases and the dissolution of the crystalline phase in the liquid. The third one is the penetration of grain boundaries followed by the disintegration of the microstructure. For the most part, in refractory applications the liquid is in motion, so that disintegrated grains are transported away from the solid material leading to higher rates of erosion than in the case of dissolution of the solid in the liquid. Infiltration of grain boundaries is energetically favorable if the energy of the two solid-liquid interfaces is lower than the energy of the grain boundary, i.e. if

$$2\gamma_{SL} < \gamma_{GB} \quad (1)$$

where γ_{SL} is the solid-liquid interface energy and γ_{GB} is the grain boundary energy (γ_{SL}, γ_{GB} in J/m^2). Considering both cases of interaction between polycrystals and liquid phases - liquid phase sintering and melt corrosion - coalescence should be defined as the opposite effect of grain boundary infiltration:

„Coalescence is the direct merger of two grains while they displace the liquid phase between them. It is caused by the energy balance $2\gamma_{SL} > \gamma_{GB}$."

This definition determines the given energy balance as the reason for contact formation. In the literature about sintering other reasons for contact formation are given, table 1. In these cases, other follow-up mechanisms complete the coalescence. The combination of contact formation and atomic transport leads to one new grain formed out of two. But if the process stops after contact formation, a remaining grain boundary would be infiltrated by a liquid if $2\gamma_{SL} > \gamma_{GB}$ is not fulfilled. Two coalesced grains with a remaining solid-solid grain boundary matching the energy balance could *not* be infiltrated. This difference is the reason for the given definition and the differentiation in table 1, that gives an overview of all causes for contact formation and the possible cases of following grain boundary development.

Table 1. Coalescence mechanisms

	Contact Formation due to	Grain Boundary
Own Definition	• $2\gamma_{SL} > \gamma_{GB}$ • grain contacts in as-pressed powders that randomly fulfil the balance and form low-energy grain boundaries at the sinter necks	• none, in case of coherent alignment of the crystal lattices • persisting, even under attack of a liquid with the same γ_{SL} • being eliminated by diffusion or particle rotation
More Comprehensive Definitions [1-6]	• thermal motion • settling • contacts in as-pressed powders • local forces and outer pressure	• remaining grain boundaries will be infiltrated • following elimination due to rotation • following elimination due to atomic transport mechanisms, as there are: - grain boundary migration - liquid film migration - solution-reprecipitation

The aim of this paper is to give first answers to the question if it is possible to use the coalescence effect in the development of materials with a high infiltration resistance in melt corrosion. The examined material combination is alumina and a silicate-based liquid phase. Most of the grain boundaries of the alumina polycrystals are infiltrated by silicate glass melts [7,8], because only a small fraction fulfils $2\gamma_{SL} > \gamma_{GB}$. These are the grain boundaries in special low energy arrangements like low angle tilt boundaries.

If one succeeds to increase γ_{SL} or to decrease γ_{GB}, the fraction of grain boundaries that can't be infiltrated should be increased. The value of γ_{GB} depends on the material and on the orientation of the two crystal lattices relative to each other. The variation of the crystalline phase, for example by adding Cr_2O_3 to Al_2O_3 in solid solution is not subject of this paper. But γ_{SL} is a function of the composition of both the liquid and the solid phases and of the crystallographic orientation of the plane facing the liquid, so that it can be influenced by varying the composition of the glassy phase and the habit of crystal growth. If a glass compound could be found, that increases the solid-liquid interface energy, a strategy for an improvement of refractories could be the addition of this kind of compound. This phase would be dissolved in the glass, increase γ_{SL} in the contact layers and causing more stable solid-solid grain boundaries which would protect against infiltration.

In own preliminary experiments ZrO_2 was found as a possible glass constituent increasing γ_{SL}. CAS-glass (CaO-Al_2O_3-SiO_2) and a similar ZrO_2-containing glass (CASZ) were used to infiltrate a pressed alumina body and to disperse the grains in the glass melt. Similar experiments were conducted by other authors [7,8] without testing the effect of ZrO_2. Reflected light microscopy photographs were examined by quantitative image analysis revealing a higher share of alumina solid state grain boundaries in the CASZ-glass. Since this kind of experiment resulted ino an unsatisfactory comparability of the samples, liquid phase sintering experiments have been carried out to test the effect of the two glasses. The experimental conditions in liquid phase sintering are more simple, because the local volume fraction of glass phase is constant in the entire sample.

The main mechanism of grain growth in liquid phase sintering is solution-reprecipitation. Alumina with a silicate-based liquid phase is known as a system with a grain growth controlled by interfacial reaction [9]. For this case the following equation for the grain growth G can be given [1]:

$$\overline{G}^2 = \overline{G}_0^2 + \frac{256 \cdot \gamma_{SL} \cdot C \cdot \Omega \cdot k_t \cdot t}{81 \cdot k \cdot T} \qquad (2)$$

where G_o is the initial grain size, C is the concentration of the solid in the liquid, Ω is the atomic volume of the solid, k_t is the interfacial reaction constant, t is the time, k is Boltzmann's constant and T is the absolute temperature.

3 Experimental Procedure

Mixtures of Al_2O_3 (Condea, Ceralox APA 0.5, without MgO-doping), SiO_2 (CE Minerals, Teco-Sil 44C, fused silica), $CaCO_3$ (Merck, p.A.) and $ZrSiO_4$ (Atochem, zircon, > 98 wt.-%) were melted at 1600°C in dense Alumina crucibles and quenched from 1500°C in water to gain glasses with the compositions given in table 2. Afterwards, the glasses were crushed and ball-milled to particle sizes < 5 μm. The true densities were measured by a gas-displacement pycnometer.

Table 2. Glass compositions (wt.-%)

	Al_2O_3	SiO_2	CaO	ZrO_2
CAS-glass	35,0	33,6	31,4	0
CASZ-glass	33,4	29,7	29,9	7,0

Different mixtures of alumina (same quality as above, D-50 value: 0,44 μm) and glass powders were ball-milled in the wet state for homogenization and to produce the samples by slip casting. The samples with the CAS-glass contain 1, 2, 3, 4 and 10 wt.-% of SiO_2 + CaO, respectively, and the samples with the CASZ-glass the same amounts of SiO_2 + CaO + ZrO_2. After drying the samples were sintered at 1650°C for 6 hours. The samples were polished for microstructural analysis after thermal etching at 1360°C for ½ hours. Theoretical densities were calculated from the true densities of the glass, the glass volume fractions in the microstructures and the measured bulk densities.

Photographs of the microstructures were taken by reflected light microscopy (Reichert Polyvar 2 Pol) and grain size measurements were conducted with a Cambridge Quantimet 570 analyzer.

4 Results

In the sintering experiments an average theoretical density of 98 % was achieved, the minimum value was 95 %. The apparent porosity is lower than 0,5 % for all samples.

Figure 1 gives the micrograph of the sample containing 1 wt.-% SiO_2 + CaO as an example. Most of the grains are elongated due to the lack of MgO that usually enhances an isotropic grain growth. The mean aspect ratio is 2,1 and 2,0 for the samples without and with ZrO_2-doping, respectively. Intergranular and intragranular porosity is visible. These microstructural aspects are similar for all samples, but the grain shape depends on the amount of liquid phase. With increasing content of liquid phase, the grain boundaries become more flat because in this case there is a better geometric condition for the grains to achieve their equilibrium shape according to Wulff´s construction.

The dependency of the grain size on the glass type and the amount of liquid phase is shown in figure 2. The main aspect is that the grain sizes of the CASZ-doped samples are clearly larger than those of the samples with CAS-glass addition.

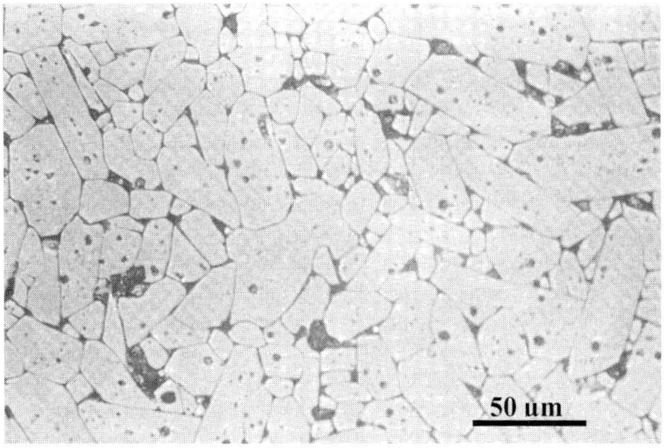

Figure 1. Microstructure of the sample containing 1 wt.-% SiO_2 + CaO + ZrO_2

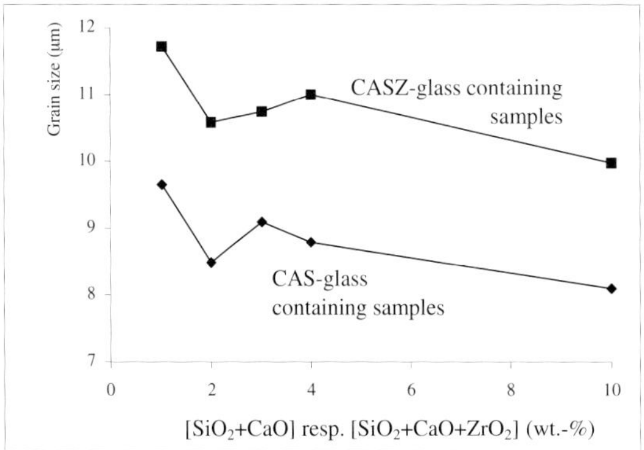

Figure 2. Grain sizes vs. dopant content

4 Discussion and Conclusions

While sintering the grain size of all samples increases more than one order of magnitude, the grain sizes of the CASZ-glass containing samples are more than 20 % larger than those of the CAS samples. Both mechanisms, solution-reprecipitation and coalescence are involved in the grain growth. The faster grain growth in the CASZ samples can be explained by one or both of the mechanisms. It can't be quantified which mechanism was favoured. *But for both mechanisms an increased* γ_{SL} *is precondition!* For solution-reprecipitation this can be shown by equation 2. Only γ_{SL} and the concentration of the solid in the liquid C can be influenced by the glass composition. EDS measurements proved, however, the same Al_2O_3-content dissolved in both glassy phases after sintering. Thus, only an increased γ_{SL} can be assumed responsible for the faster solution-reprecipitation. A stronger coalescence caused by $2\gamma_{SL} > \gamma_{GB}$ can only be explained by an increased γ_{SL}, too, because γ_{GB} is independent from the liquid phase composition.

The tendency of the grain coarsening is slightly decreasing with increasing dopant content. This can be explained with longer diffusion distances for the Al_2O_3 through the melt. A possible reason for the intermediate maxima could be that medium dopant amounts are the best condition for coalescence, but there are too little measurement points for a secure diagnosis of this effect.

It has been shown that the addition of a compound to a glass can clearly influence the solid-liquid interface energy. ZrO_2 increases γ_{SL} between alumina and a CAS-glass.

The practice in the use of alumina-based refractories for application under melt attack showed that small zirconia additions improve the corrosion resistance. The reasons for this improvement are not entirely clear as yet. One mechanism discussed is the dissolution of zirconia in the melt accompanied by an increase of the viscosity, which reduces flow-, creep- and diffusion-based corrosion mechanisms. From the result of this paper it can be concluded that the dissolution of the zirconia leads to a higher γ_{SL} as well. Due to this fact, a higher amount of grain boundaries are protected against infiltration by non-wetting.

Future research will focus on the selection of other phases that can be added to refractories having a stronger influence on the solid-liquid interface energy.

5 References

[1] R. M. German, Sintering Theory and Practice, 1st ed., John Wiley & Sons, New York, 1996
[2] W. Schatt, Sintervorgänge - Grundlagen, 1st ed., VDI-Verlag, Düsseldorf, 1992
[3] G. Petzow, W. J. Huppmann, Z. Metallkunde 1976, 67, 579 - 590
[4] W. J. Huppmann, G. Petzow, Ber. Bunsenges. Phys. Chem. 1978, 308 - 312
[5] W. J. Huppmann, Z. Metallkunde 1979, 70, 792 - 797
[6] S. Takayo, W.A. Kaysser, G. Petzow, Acta metall. 1984, 32, 107 - 113
[7] T. M. Shaw, P.R. Duncombe, J. Am. Ceram. Soc. 1991, 74, 2495 - 2505
[8] P. L. Flaitz, J. A. Pask, J. Am. Ceram. Soc. 1987, 70, 449 - 455
[9] O.-H. Kwon, G. L. Messing, J. Am. Ceram. Soc. 1990, 73, 275 - 281

Microstructure of boride ceramics in the ternary system TiB_2-WB_2-CrB_2

Clemens Schmalzried, Rainer Telle, Institut für Gesteinshüttenkunde, RWTH Aachen, Aachen, Germany

1 Introduction

Boride ceramics based on TiB_2 are an interesting material for application as cutting tools because they are wear resistant and oxidation resistant up to about 1200°C. Due to a rather low fracture toughness their is demand for a reinforcement of these materials. It is obvious that a process leading to particle reinforcement should lead to a homogeneous microstructure and a dense ceramic material, a request which can be best fulfilled by in-situ reinforcement of a TiB_2-based matrix. Early investigations on the system TiB_2-WB_2 showed the possibility of growing lamellar precipitates out of a supersaturated TiB_2-rich solid solution by annealing a homogeneous matrix in the two phase region of the phase diagram (1). The precipitation kinetics are rather slow in this system, so further investigation was carried out and concentrated on doping to increase reaction rates. CrB_2 which has a high solubility in both components fulfills the requirements in this system, leading to an increased precipitation- and growth rate and an increased solubility of WB_2 in TiB_2 (2).

The identification and evaluation of the occuring phases by means of atomic structure is the topic of the present work. There has been done a lot of work in the identification of layered structures on the atomic scale by HRTEM and EELS in the past years (3). This knowledge has been used in this work to identify layered boride structures and the occuring stacking faults in the investigated system.

2 Experimental

2.1 Sample preparation

The raw powders TiB_2, W_2B_5 and CrB_2 (H.C.Starck Grade A) were attrition milled with WC/Co balls in isopropanol for 2h. The dried powders were prepressed into cylinders of 18mm diameter and hot pressed for 0.5 h at 1800°C in graphite dies with a uniaxial load of 60 MPa under 1 atm Ar pressure. In a first series the samples were annealed at 1600°C for 2h. In a second series the hot pressed samples were annealed at 2000°C for 8h followed by precipitation annealing for 8h at 1600°C. The annealing treatments were all carried out without load under 1 atm Argon pressure in a graphite heated resistance furnace.

2.2 TEM and EELS investigations

For TEM investigations the specimen were cut and thinned by grinding and dimpling. For electron transparency they were thinned by conventional ion beam milling.

The microscopes CM30ST/STEM and CM300STEM/UTFEG were operated with 300KV and are both equipped with a post column Gatan imaging filter. They are equipped with a LaB_6-cathode and a field emission gun. The element specific images were calculated by the three window method using pre and post edge images of the EEL spectrum. It can be shown that the contrast is of pure chemical nature (3).

It should be mentioned that one reaches a resolution of 0.38 nm in element specific imaging of boron in the subsequent layers of boron and metal atoms.

3. Results

The samples annealed after hot pressing reveal a microstructure consisting of unreacted TiB_2 cores that are coated with a layer of $(Ti,W,Cr)B_2$ solid solution. The interface is

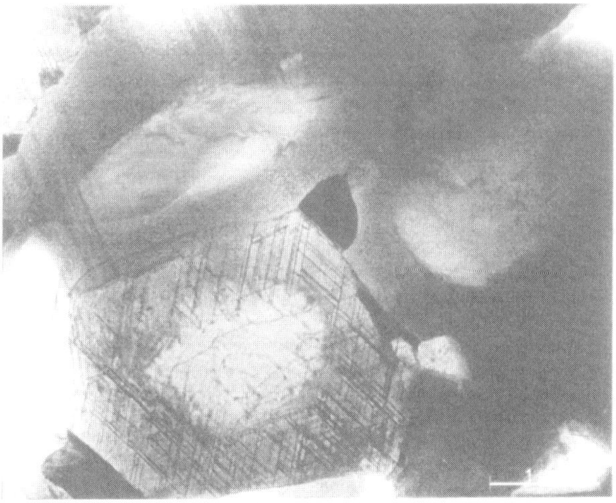

Figure 1. Unreacted TiB_2 cores with lamellar ß-WB precipitates

roughly sharp and is build up by a pile up of dislocations (fig.1). Within the solid solution fine planar defects occur intersecting at angles of 60° and 120° respectively. In addition one can identify a W-rich lamellar structure which could be adressed to spinodal decomposition (fig.2). The fine precipitates are only build inside the solid solution and stop immediately at the interface. The black phase can be identified as unreacted WB_2.

The samples annealed at two different temperatures show in addition to the spinodal decomposition and the fine lamellar precipitates stacking faults in the TiB_2- (fig.3) and W_2B_5-structure (fig.4). They can be identified as incorporation of layers of the W_2B_5-structure into TiB_2 and vice versa.

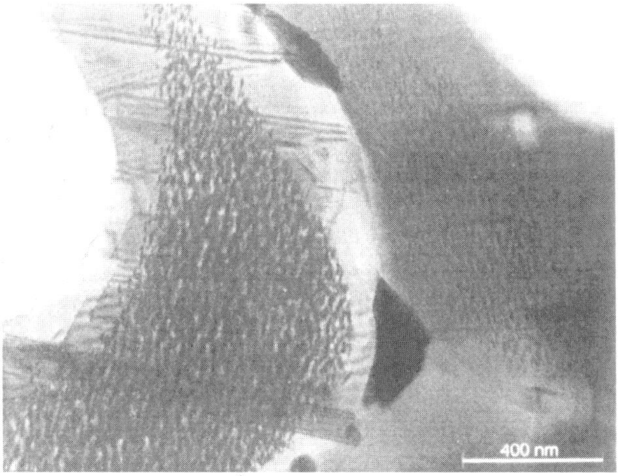

Figure 2. Spinodal decomposition in $(Ti,W,Cr)B_2$

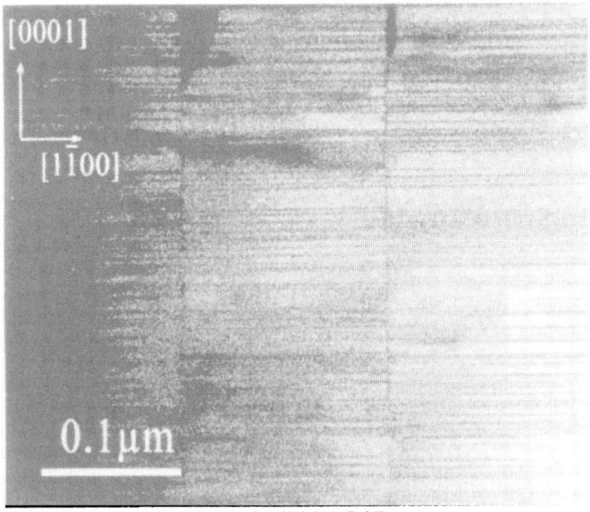

Figure 3. Stacking faults in $(Ti,W,Cr)B_2$

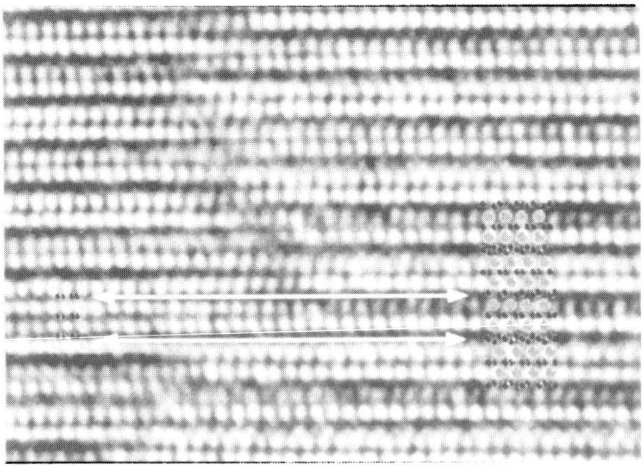

Figure 4. Stacking faults in W_2B_5

4 Discussion

Due to the mismatch in thermal expansion of TiB_2 and $(Ti,W,Cr)B_2$ solid solutions (2) dislocations pile up in the interface of the core-shell grain structure (fig.1). This phenomenon could be beneficial to the mechanical properties including creep resistance. In addition the fine precipitates could serve as obstacles against creep by dislocation motion.

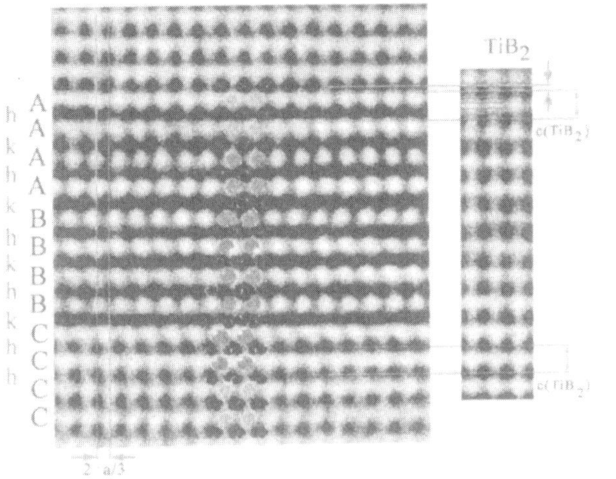

Figure 5. Modeling and HRTEM of stacking faults in $(Ti,W,Cr)B_2$

The structure of the stacking faults was solved by modeling (fig.5). The W_2B_5-structure reveals a shift of the metal layers in the B-position the puckered layers by multiples of a/3 depending on the number of blocks involved. This shift can be indeed measured with

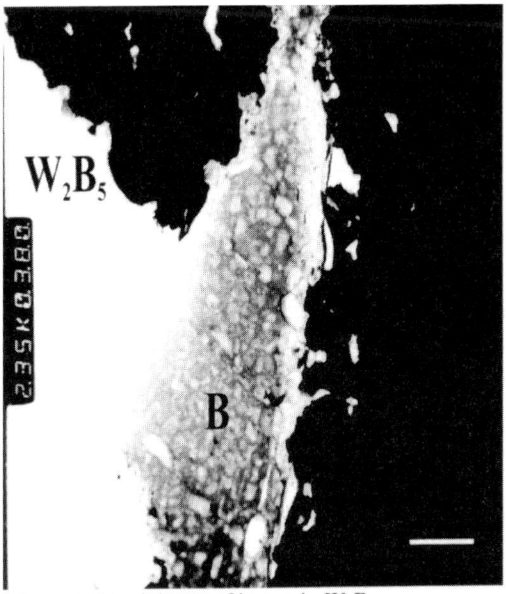

Figure 6. Precipitates of boron in W_2B_5

HRTEM. This is consistant with a shift in c-direction due to an increase in layer distance in the puckered blocks. This is shown on the right side of fig.5. It reveals a comparison of the atomic positions in the ideal TiB_2 structure and the modified structure. The breakdown of the puckered boron layers may originate from selective diffusion of boron out of these layers. In addition we could identify large areas of boron in contact with unreacted W_2B_5 (fig.6).

To proof the modeling of the fine precipitates we carried out element specific imaging with a good resolution of 0,055nm per pixel. It can be shown that the precipitates exist of two boron deficient layers with a distance of 0,38 nm, which is consistant with the ß-WB structure (fig.7).

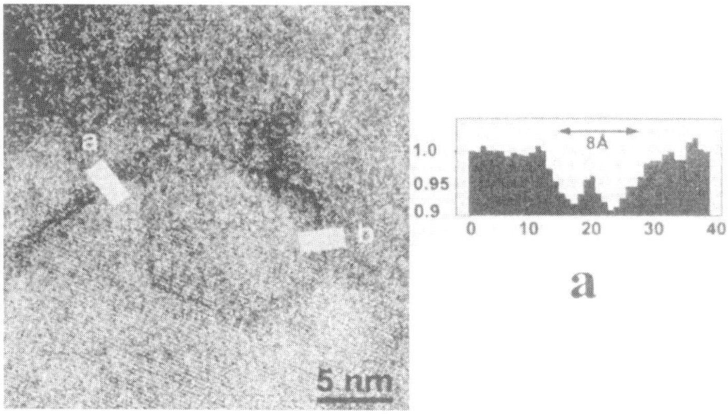

Figure 7. Element specific imaging of ß-WB

4 Acknowledgements

We greatfully acknowledge the German Volkswagen-Stiftung for financial support. We greatfully acknowledge Dr. Bert Freitag from the Institute of Inorganic Chemistry, Bonn, for the TEM and EELS investigations.

5 References

1. R. Telle, G. Petzow, Mat. Sci. Eng. 1988, A 105/106, 97

2. I. Mitra, R. Telle, Werkstoffwoche Stuttgart, 1996

3. B. Freitag, W. Mader, Journal of Microscopy 1999, 194, 42-57

Ceramic Composites from Bioorganic Derived Materials

Heino Sieber and Peter Greil
University of Erlangen-Nuremberg, Department of Materials Science (III) Glass and Ceramics, D-91058 Erlangen, Germany

1 Introduction

Conversion of biological materials such as organic fibres, paper or wood into ceramics represents a novel technology of composite processing and has recently attained particular interest [1-3]. Hierarchically built bioorganic structures exhibit interesting cellular and porous morphologies on the micro as well as on the macro scale. On the other hand, natural regenerating materials like wood are available on a commercial scale, less expensive and exhibit excellent shaping abilities. Using a high-temperature infiltration-reaction process [4], the bioorganic structures can be converted into ceramic composites within reasonable time, Fig. 1.

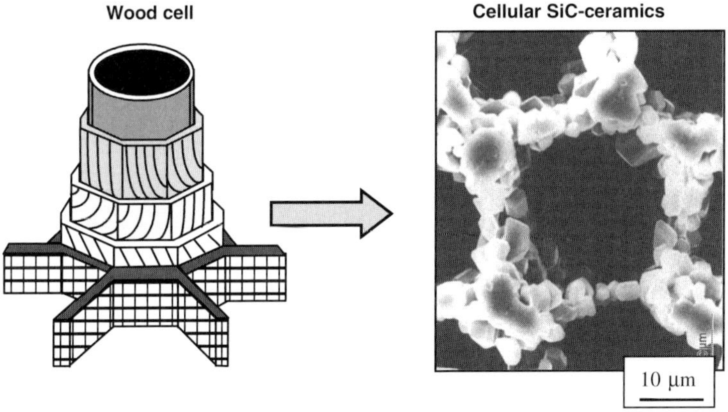

Figure 1: Basic principles of biotemplating: Conversion of bioorganic carbon structures into ceramic composites by high-temperature processing.

The inherent cellular and open porous morphology of biological structures is accessible for liquid or gaseous infiltrants. For example, Ota et al. [5] produced SiC with wood structure by infiltration of charcoal with TEOS (tetraethyl orthosilicate) and subsequent high-temperature pyrolysis. Greil et al. [6] converted different kinds of wood by infiltration of pyrolyzed wood preforms with liquid Si into cellular SiSiC composites that exhibit superior mechanical properties compared to conventional porous SiSiC materials. However the variety of natural structures and their easy workability enables a large field of technical applications of biostructural derived ceramic composites, Fig. 2.

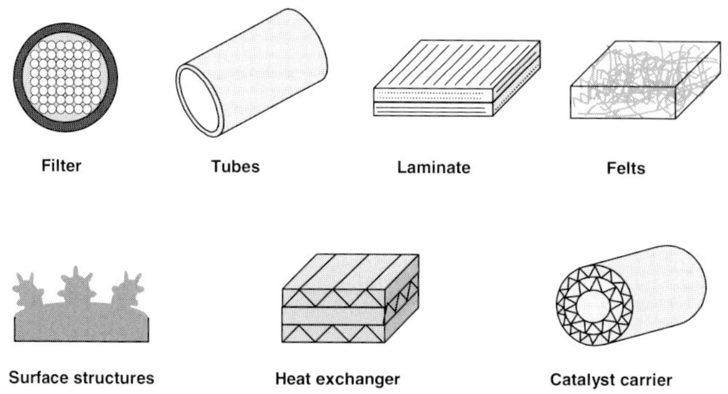

Figure 2: Conversion of biostructures into ceramics: possible applications.

In the present paper, the processing principles during the organic ➔ inorganic conversion of bioorganic derived cellulose materials like wood, paper or cardboard structures into ceramic composites will be discussed. Novel ceramic materials were prepared by infiltration of liquid and gaseous Si into carbonized wood preforms and subsequent reaction into Si-C or SiSiC ceramics. Low viscous preceramic polymer/filler suspensions were infiltrated into cellulose paper sheets and 3-dimensional paper structures and subsequently converted by filler controlled reaction pyrolysis into Si-C-Al-O composites. The resulting ceramic composites show low density, high stiffness and excellent high temperature stability.

2 Conversion of Wood Structures

Natural wood structures are characterized by different micro cellular and porous structures on the micrometer scale and can be used as biocarbon templates for conversion into SiC-ceramics. Different wood specimens were dried and subsequently pyrolyzed in N_2-atmosphere at temperatures up to 1800°C [5]. During pyrolysis, a slow heating rate of 1°C/min was applied up to 500°C where the organic components are completely decomposed to carbon, followed by a higher rate of 5°C/min up to the peak temperature. The biocarbon preforms were infiltrated with liquid or gaseous silicon at 1600°C without pressure, according to the conventional processing of SiSiC materials [7], and completely reacted into cellular β-SiC ceramic structures.

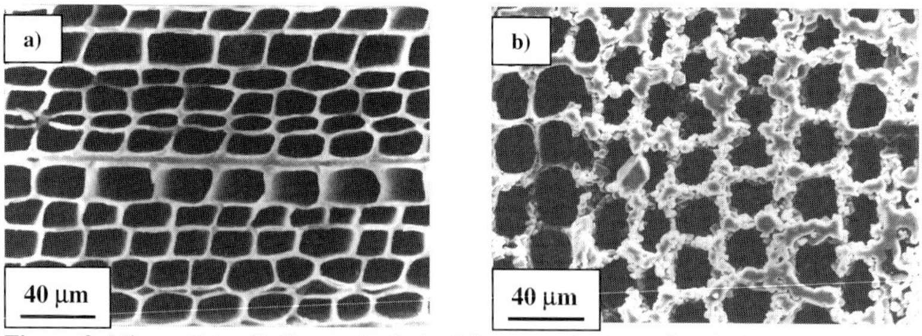

Figure 3: Microcellular Si-C ceramic derived from wood: a) pyrolyzed pine template, b) Si-gas infiltrated pine (pyrolysis and infiltration at 1600°C in Ar atmosphere).

The weight loss during pyrolysis of the wood preforms was found to be 70-80wt%. The pyrolysis involves an anisotropic shrinkage of about 30% depending on the orientation of the specimen and the kind of wood. After pyrolysis the total porosity of carbonized wood is 20...25% higher than the initial density of (dried) wood and exhibits a multimodal pore size distribution with pores between 1 to 200 μm. Despite the changes in porosity and anisotropic shrinkage the porous macro- and microstructure of the carbonized wood is retained after pyrolysis with high precision, see Fig. 3a. After Si-infiltration, the carbon structure is completely converted into β-SiC ceramics (Si-gas infiltration) or SiSiC material (Si-melt infiltration) and exhibits a micro cellular morphology. For Si-gas infiltration samples, the final porosity of the SiC ceramics is up to 40%, Fig. 3b. After Si-melt infiltration, small pores up to 30 μm are filled with residual Si yielding porosities between 6 and 25%. The mean particle size of the β-SiC grains is typically in the range of 5 μm.

3 Conversion of Laminated Papers

Conversion of preprocessed paper sheets into ceramic composites is of interest for laminated object manufacturing (LOM), for preparation of high-temperature structural materials as well as advanced processing of light weight macroscopic substrates and structures. Paper consisting of pure cellulose fibres (cleaning and filter paper) was used as organic preforms for preparing infiltrated paper laminates and conversion into ceramic composites [8]. The papers were infiltrated with a slurry of the preceramic polymer (poly(methylsiloxane), NH21, Chemiewerke Nuenchritz, Germany), Si as reactive and SiC as inert filler materials in different polymer/filler ratios. The poly(methylsiloxane) has an average molecular composition $[CH_3SiO_{1.5}]n$ ($n \approx 300\text{-}400$) and it is characterized by a high ceramic yield (SiCO) of more than 70wt% upon pyrolysis in inert atmosphere. After drying in air, different paper sheets were laminated together by hot pressing at 18 MPa and a temperature of 230°C for 20 min. The laminates were pyrolyzed at 1450°C for 4 h in argon atmosphere.

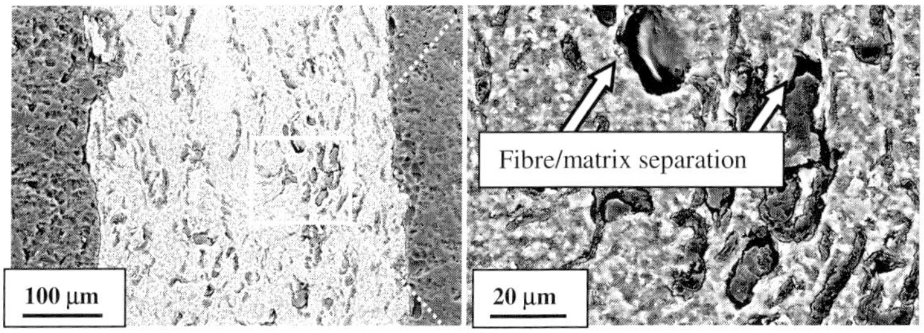

Figure 4: SEM/BSE micrographs of a laminated specimen initially consisting of 3 infiltrated papers after hot pressing and pyrolysis (1450°C in Ar atmosphere).

After pyrolysis, the morphology of the laminates depends on the polymer/filler ratio of the slurry. While for high polymer fraction (80vol% of polymer) a linear shrinkage of about 20% and a very low porosity of 3% were obtained, slurries containing only 30vol% polymer results in a low shrinkage (3%) but a porosity of more than 30%. The slurry, the initial cellulose fibres and the inert SiC filler were distributed homogeneously over the cross section of the specimen; the initial laminate interfaces were completely disappeared after pyrolysis, Fig. 4a. The initial cellulose fibres of the paper were completely reacted into SiC.

4 Conversion of Preformed Cardboard Structures

For conversion of preformed paper structures into ceramic composites, commercially available corrugated cardboard (secondary fibre paper, 190TL, Wellpappenwerk Bruchsal/Germany) with different porosity and impurity levels were used [8]. The cardboard has a mean cellular morphology with a cell diameter of 6 mm and a mean cell wall thickness of 100 μm. The specimens were dried (70°C, 1hr, vacuum) and infiltrated with a slurry containing 80 vol.% of elemental powder (40 vol.% Si - average grain size below 10 μm, 40 vol.% Al – average grain size 50 μm) and 20 vol.% of a preceramic poly(methylsiloxane) polymer. After infiltration, the cardboard structure was annealed at a temperature of 230°C (4 hs, vacuum) to crosslink the preceramic polymer via a polycondensation reaction of [Si-OH] groups. The infiltration and crosslinking procedure was repeated up to 3 times to get a high ratio of filler material/carbon and to ensure a complete reaction of the bioorganic carbon into carbide phases. After pyrolysis at 1450°C in inert atmosphere, an inorganic, porous SiC/Al/Si composite was obtained, subsequent annealing in air yields a Si-Al-O-C ceramic composite, Fig. 5.

Due to the large amount of filler in the slurry a linear shrinkage of less then 2.5 % could be obtained after processing. The cellular structure of the initial cardboard was well transformed into the ceramic composite. The final specimen was composed of a mixture of SiO_2, Al_2O_3 and $Al_6Si_2O_{13}$ (mullite) with minor amounts of SiC and residual Si.

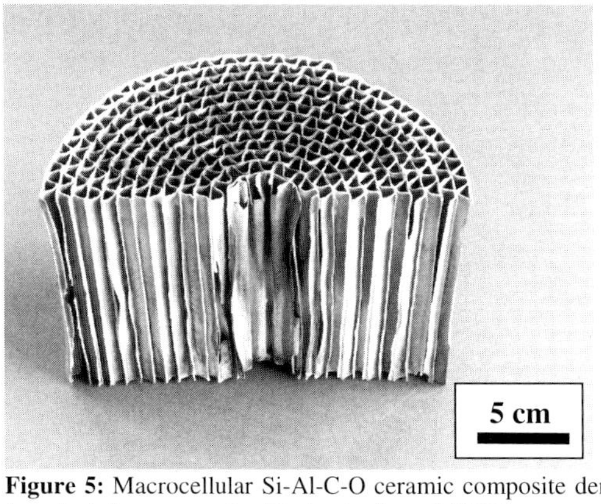

Figure 5: Macrocellular Si-Al-C-O ceramic composite derived from preformed corrugated cardboard structure (pyrolysis in argon and oxidation in air at 1400°C).

5 Conclusions

Different technological procedures for manufacturing of structural ceramic composite derived from bioorganic structures were presented. The conversion of wood structures yields SiC and SiSiC ceramics with different micro cellular morphologies. The slurry infiltration of preprocessed paper and cardboard structures offers a versatile processing scheme to produce light-weight, high-temperature ceramic composites with unique oriented cellular or porous structures. The resulting anisotropic and cellular ceramics exhibit a low density; they could be of particular interest for applications in heat insulation structures, as substrate, filter and catalysis carrier at high temperatures as well as for thermally and mechanically loaded light-weight structures.

6 Acknowledgement

Financial support from the Volkswagen Foundation under contract # I/73 043 is gratefully acknowledged.

7 References

1. I.A. Aksay, E. Baer, M. Sarikaya, D.A. Tirell (editors), Hierarchically Structured Materials, *Mat. Res. Soc. Symp. Proc. Vol.* 1992, *255*, Mat. Res. Soc. Pittsburg/PE.
2. M. Sarikaya, I.A. Aksay, *Biomimetics: Design and Processing of Materials*, AIP Series in Polymer and Complex Materials, Woodbury, NY, 1995.
3. A.H. Heuer, D.J. Fink, V.J. Laraia, J.L. Arias, P.D. Calvert, K. Kendali, G.L. Messing, J. Blackwell, P.C. Rieke, D.H. Thompson, A.P. Wheeler, A. Veis, A.I. Caplan, *Science* 1992, *255, 1098.*
4. Kaindl, Ph.-D. thesis, University of Erlangen-Nuremberg, 1999.
5. T. Ota, M. Takahashi, T. Hibi, M. Ozawa, S. Suzuki, Y. Hikichi and H. Suzuki, *J. Am. Ceram. Soc.* 1995, *78, 3409.*
6. P. Greil, T. Lifka, A. Kaindl, J. Europ. Ceram. Soc. 1998, *18, 1961,* and J. Europ. Ceram. Soc. 1998, *18, 1974.*
7. W.B. Hillig, J. Am. Ceram. Soc. 1988, *71, C 96.*
8. H. Sieber, D. Schwarze, H. Friedrich, A. Kaindl and P. Greil, *Proc. of the101st Annual Meeting of the American Ceramic Society, April 25-28,* 1999, Indianapolis/IN, Symposium N: Innovative Processing and Synthesis of Ceramics, Glasses and Composites.

Materials and Processing for Ceramic Flat Membranes

Jörg Adler, Reinhard Lenk and Hans-Jürgen Richter
Fraunhofer Institut für Keramische Technologien und Sinterwerkstoffe, Dresden

1 Introduction

Porous membranes are used for the filtration of liquids and gases in a lot of technical processes. Membranes with a pore diameter between 0,1 µm up to 10 µm are utilized for microfiltration. Ultrafiltration requires a pore diameter between 5 nm and 100 nm. Ceramic membranes have got some advantages compared with polymer membranes concerning the mechanical, thermal and chemical stability. However the application of ceramic membranes is still limited because of the high price per filter area. The prices of commercial ceramic membranes for micro- and ultrafiltration are in the range 3500 - 4000 DM per square meter filter area inclusive the modul [1].

At present most of the ceramic membranes are manufactured as a compact ceramic body with inner channels. The geometry is a tubular shape or a honeycomb structure. These multichannel elements are fabricated by extrusion of the support material. After sintering the support the membrane layer is prepared by dip-coating of the inner channels. The steps dip-coating, drying and sintering are repeated several times if more than one membrane layer is necessary or if the membrane shows defects after sintering. Flat membranes represent another type of ceramic membranes. The support of the flat membrane is prepared by tape casting, slip casting or calendering [2]. After the sintering of the support the membrane layer is made by the same procedure as described above (dip-coating, drying, sintering). The finer microfiltration and the ultrafiltration require a second membrane layer.

The purpose of this work is to develope a material and a fabrication technique for ceramic microfiltration membranes with the aim of low costs and high filter area. For a membrane with a flat shape the following solutions were worked out: the development of a ceramic material which allows the sintering of the support and the membrane layer in one step (= cofiring), the development of a green-in-green tape casting process for this material, fabrication of flat multichannel elements from the support/ membrane tapes.

2 Development of a Silicon Carbide Material for Cofiring

Two problems occur at the cofiring of several layers with different grain size. First the strength of the particles bond differs. The coarse particles of the support have a sufficient bond at high temperature. On the other hand the fine particles of the membrane layer sinter too strong at the same temperature. At low temperature the membrane layer shows an adequate strength, but the coarse particles of the support are insuffiently connected. Second the different shrinkage of the support and membrane layer often result in distortion and formation of cracks.

These problems are overcome if the powder particles do not sinter, but an additionally liquid phase bonds the ceramic particles.

With the aim of cofiring at a temperature below 1000 °C under air a special glass was selected as liquid phase additive. The silicon carbide/ glass system with maximum 15 mass per cent glass allows the cofiring of layers with coarse and fine SiC-particles. The mean particle diameter of the support SiC is 23 μm. The SiC-particles of the membrane layer have a mean diameter of 3,5 μm. Figure 1 shows the influence of the sintering temperature on the bending strength and the porosity of the support and membrane material. At a sintering temperature of 850 °C the porosity of the support and the membrane materials is high enough for a good permeability. The bending strength is in the range of 25 to 40 MPa and it guarantees a sufficient mechanical stability of the ceramic filter material. The difference in the shrinkage between the two materials is lower than 0,1 per cent. Therefore a cofiring is possible without cracks and distortion. The pore diameter of the support material is 10 μm. The membrane material has a pore diameter of 1 μm and it is suitable for the microfiltration. Experiments with a third (finer) silicon carbide powder show a principle way to prepare a second membrane layer with 0,2 μm pore diameter via the powder route.

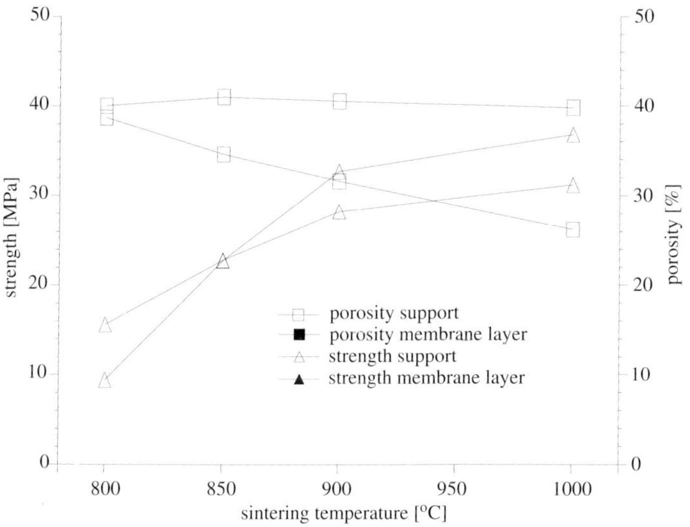

Figure 1. Strength (3-point bending strength) and porosity of the SiC/glass material in dependence on the sintering temperature, sintering under air

3 Tape Casting of the Support and the Membrane Layer

With the objective of low costs in combination with environmental and health aspects an aqueous tape casting process was chosen for the fabrication of the flat membrane. Compared with an organic slurry an aqueous slurry is more sensible to changes in composition, casting conditions and drying conditions [3]. Therefore all process parameters must be adjusted very exactly.

Aqueous slurries were developed for the support and the membrane layer. Figure 2 shows the zetapotential-pH dependence of glass powder and the fine SiC powder. The powders have a similar behaviour because in both cases the Si-OH groups determine the surface charge properties. The slurry pH should be between 8 and 10 to obtain a high electrostatic stabilizing of the particles. However the main component of the used binder system is polyvinylalcohol. Therefore the pH value of the tape casting slurries is lower than seven because of the acid groups of the polyvinylalcohol. Using a polyelectrolyte dispersant good dispersion and stabilizing of the glass powder and the fine SiC powder are reached. In the case of the coarse SiC powder for the support the electrostatic and steric stabilizing by the colloidal particle interactions play a secondary role. Rather the slurry viscosity must be adjusted so highly that the coarse particles do not sedimentate. Typical viscosity curves of both tape casting slurries are presented in figure 3. The developed binder system meets the usual demands regarding slurry stability and rheology, drying and handling of the tape. Furthermore the binder must also ensure the forming and the lamination of the green tape (see 4.). The volume concentration of the powder in the slurry and the compaction of the powder particles during the drying of the tape determine the green density and consequently the porosity of the sintered membrane.

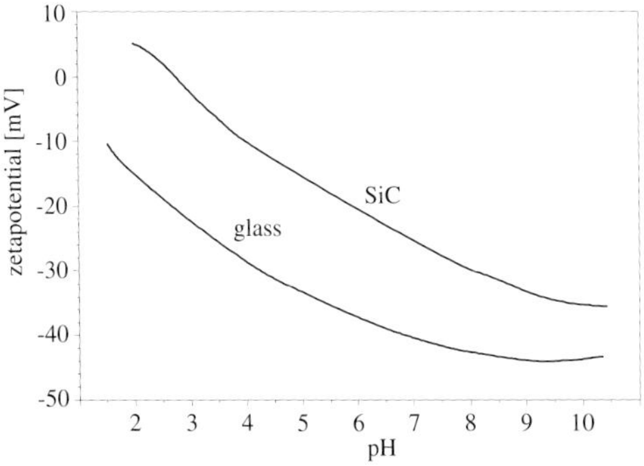

Figure 2. Zetapotential-pH dependence of SiC and glass powder in a dilute aqueous suspension

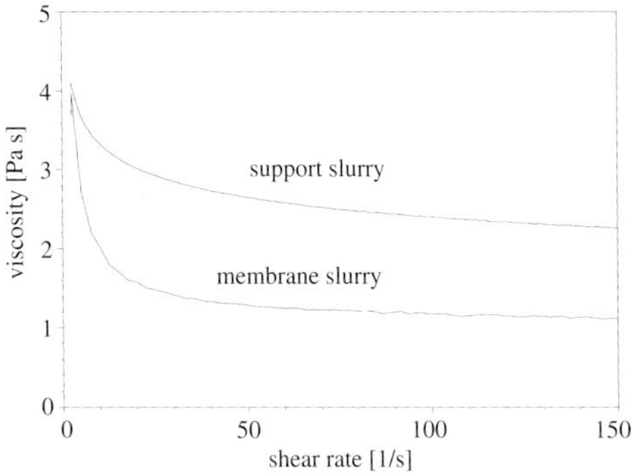

Figure 3. Rheology of the membrane and support slurries

The tape casting was carried out on a laboratory tape casting machine (Gebrüder Netzsch Maschinenfabrik, Germany). First the support is cast. In-situ measuremts of the tape thickness during the drying at 20 °C exhibit that the support thickness do not reduce any longer after 200 min. At this point the support tape is almost dry and the membrane can be cast on the support. The casting high (= high of the blade) for the support and the menbrane is justified so that the dry support has a thickness of 0,55 mm and the membrane layer thickness is about 50 µm. The dry support/ membrane double layer tape can be manufactured directly to a multichannel flat membrane.

4 Forming and Lamination to Flat Multichannel Elements

The support/membrane doublelayer tape is formed to a corrugated shape. This shape is obtained by coining or by folding. For example two rolls with a toothed fluting similar the manufacturing of corrugated board were used. The corrugated tape is laminated with the plane tape so that a flat multichannel element is obtained with the membrane layer on the outside (see figure 4).

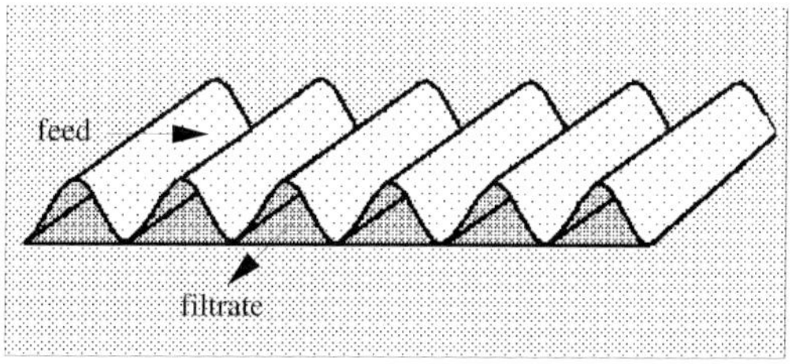

Figure 4. Principle structure of the multichannel flat membran

The lamination is performed by the use of a laboratory warm press. A strong bond between the corrugated support layer and the plane support layer originate during the sintering process. The multichannel element is debindered and sintered in one thermal process. It is verified by SEM studies that no defects occur at the laminate interfacc (see figure 5). Is is also obvious that the glass additive bonds the particles both within the support and the membrane layer as at the interface support/ membrane (see figure 6) .

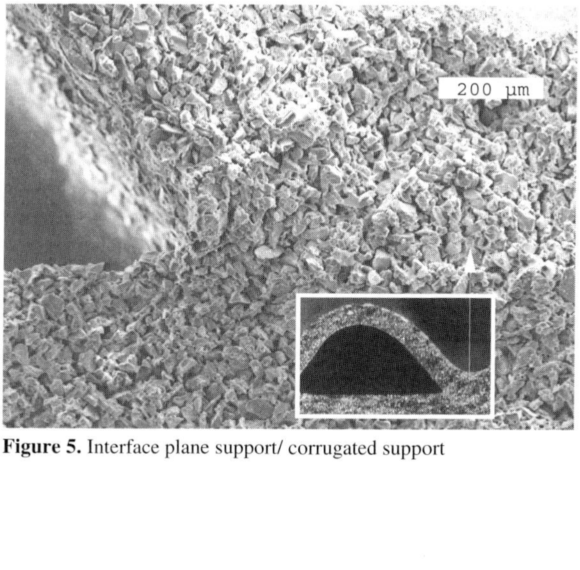

Figure 5. Interface plane support/ corrugated support

Figure 6. Interface support layer/ membrane layer

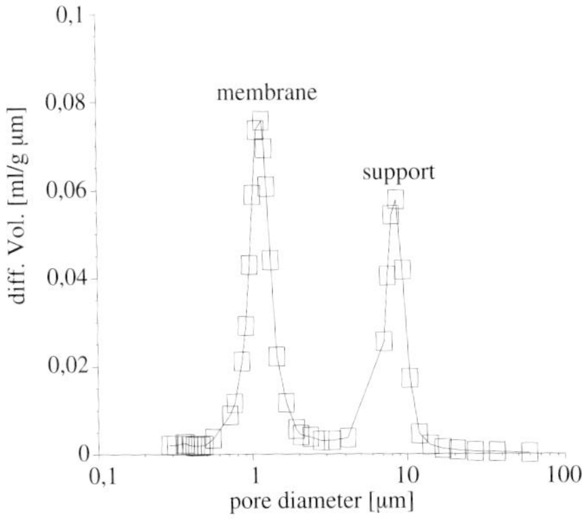

Figure 7. Pore diameter distribution (Hg-porosimetry) of the sintered multichannel flat membrane

The material properties of the multichannel flat membrane confirm with the properties of compact samples (see 2.). The bending strength (3-point, biaxial) is in the range 25 to 35 MPa depending on the porosity. The porosity varies from 40 to 45 per cent. The pore diameter distribution is shown in figure 7.

5 Conclusions

The developed SiC/ glass material system and the used technology allow a low cost fabrication of multichannel flat membranes. This membrane has a high filter area and its surface structure generates favourable streaming conditions. The water treatment is aimed as the main application of this ceramic flat membrane in the future. The fabrication by green-in-green tape casting makes possible the casting of a third layer too. Because the membrane layer is located on the outside of the multichannel element further layers can also be coated simply by dip-coating or by spray-coating.

6 References

[1] S. Tudyka, F. Aldinger, H. Brunner in Werkstoffwoche '98, Vol. VII (Ed. J. Heinrich, G. Ziegler, W. Hermel), Wiley-VCH, Weinheim 1999, pp. 745 - 751
[2] K. Lindquist, E. Liden, J. Europ. Ceram. Soc. 1997, 17, 359 - 366
[3] D. Hotza, P. Greil, Mat. Sci. Eng. A, 1995, A202, 206 - 217

Dependence of the Mechanical Properties on Low Nano-SiC Particle Content of Hot Pressed Alumina

Matthias Wilhelm / Werner Wruss
Institute of Chemical Technology of Inorganic Materials, University of Technology of Vienna

1 Abstract

The purpose of the study was to investigate the mechanical properties of hot pressed Al_2O_3 / SiC ceramics. As starting materials two alumina powders (I and II) with mean particle sizes <1µm are used. The Nano-SiC (mean particle size: 10nm) is used at low volume fractions (<1 vol%). Attrition milling and a dispersion agent is used to produce the composites. The mixtures are hot-pressed at 1500°C and 1650°C at 25 MPa. Using alumina I, a clear enhancement of the fracture strength with increasing SiC fraction is obtained (from 600 MPa for monolthic alumina up to 860 MPa for the composite containing 0.7vol% SiC nanopowder). The fracture toughness keeps almost constant at 6 $MPam^{1/2}$. In the case of using alumina II, the fracture strength is independent on the SiC fraction as well as on the sintering temperature and keeps at approximately 800 MPa. The toughness keeps constant at 6.3 $MPam^{1/2}$. The grain growth of alumina is dependent on the alumina quality and is clearly reduced with increasing SiC volume fraction. The addition of an dispersion agent can partially improve the homogeneous distribution of the nano SiC particles within the composites.

2 Introduction

Ceramic materials have many attractive properties which makes them appropriate for structural applications. Nanocomposite materials, where submicrometer secondary phase particles are incorporated into ceramic matrices, have become a growing interest. Significant improvements in strength were achieved in the ceramic system Al_2O_3 - nano SiC. Despite the clear increased strength only a much smaller enhancement of toughness was reported [1,2,3,4]. The observed increase of strength was explained by various mechanisms like formation of sub-grain boundaries, crack healing during annealing, reduction of the matrix grain-size by secondary phase particles with no abnormal large grains, reduced process flaw-size and distribution and compressive surface stresses introduced during grinding [1,2,3,4,5,6]. However, in all studies reported the amount of secondary phase (SiC particles) was chosen greater than 1 vol%. The incorporation of small quantities of nanopowders determines several problems regarding uniform dispersion of ultrafine particles within the ceramic matrix. Due to the nano particles are very expensive and for some industrial applications large amounts of nano-powders can not be used, low fractions of SiC particles are assumed to result in high mechanical properties.

The purpose of this study was to investigate the effects of low nano-SiC powder content (< 1vol%, mean grain-size 10nm) on the sintering behavior and mechanical properties of hot pressed alumina. In order to provide a better particle distribution within the alumina matrix a dispersing agent [7] was used

3 Experimental Procedure

To obtain the alumina – SiC nanocomposites two alumina powders supplied by ALCOA, Germany labeled CPR60 and Premalox 10 were used. These are referred to below as alumina I and alumina II. The alumina powders show differences in production route and purity. The SiC nanopowder was provided by ENEA, Italy and was produced by laser induced gas phase reaction. Detailed information about the starting materials are described in Table 1.

Table 1. Characteristics of the starting materials

	specific surface [m²/g]	mean grain size [nm]	cryst. phase	purity [w%]
Alumina I CPR60	5 – 6	400 ± 200	α-Al_2O_3	>99.9 Al_2O_3, 0.01 Na_2O, 0.02 SiO_2
Alumina II Premalox10	9 – 11	500 ± 150	α-Al_2O_3	>99.9 Al_2O_3, 0.05 Na_2O, 0.01 SiO_2
SiC 1	29	10 (own: 500 ± 200)	β-SiC	29 C, 1.7 O

Results of previous works have shown that to produce ceramics with high mechanical properties it is necessary to avoid agglomerates and, if they are present, to destroy them. For this purpose there are some possible ways: (1) adjust a right zeta-potential by pH-Value, (2) improved milling conditions i.e attriton milling, (3) adding an dispersing agent. The following mixing technique was elaborated to obtain a dry powder ready for hot pressing. In the case of using alumina I the SiC content was adjusted to 0.1, 0.4, 0.7 and 1.0 vol%; composites of alumina II contained 0.3, 0.7 and 1.0 vol% SiC nanopowder.

The starting alumina powder was suspended in deionized water and the pH-Value was adjusted to 10.5 by adding either NH_4OH (0.1m) or HNO_3 (0.1m). Subsequently about 10 drops (1.5 ml) of a dispersant (ET 85, Zschimmer & Schwarz, Germany) was added. The appropriate amounts of Nano SiC particles were separately suspended in deionized water and treated with U-Sonic for 30 min to promote deagglomeration. Both slurries were then mixed together and attrition milled for 30 min subsequently dried in a rotovapor and finally sieved through a 0,5mm plastic sieve. The powder mixture were hot-pressed applying a pressure of 25 MPa in flowing argon at 1500°C and 1650°C sintering temperature. After hot pressing specimens with 50mm diameter and a height of about 4.3mm were received.

After pressing the discs were cut to bars into 3mm x 4mm cross section and given length. The bars were ground flat and polished to a 3μm diamond finish on the side to be used later as the tensile side. Density measurements were carried out using the Archimedes technique using deionized water and wetting agent. To measure fracture toughness, a notch was sawed into the polished surface of the bars with a 0,1mm Diamond saw (SENB-S technique) [8]. The depth of the notch was determined by optical analysis. After notching the edges were bevelled to their length on a metal bonded Diamond wheel (10μm) in order to eliminate edge flaws.

Fracture toughness was determined using three-point loading and single-edge notched beam technique with a notch width of 100 μm, a span of 11 mm and a cross head speed of 0,5 mm/min at room temperature (9 specimens for every material). The flexural strength was determined on 18 specimens by 3-point test with a span of 11mm and a cross-head speed of 0,2mm/min at room temperature (18 bars for every material is sufficient for determining bending strength). The fracture strength was calculated using Weibull statistics. The broken specimens were etched with sodiumtetraboride at 800°C in air for 10 min, and examined using SEM.

4 Results and Discussion

Tables 2 and 3 gives a summary of the properties of all samples produced.

Table 2 Mechanical properties of the Al_2O_3 -SiC (alumina I) composites sintered at 1500°C and 1650°C

Content of SiC (10nm)/ [Vol%]	Weibull-modulus		Fracture toughness [MPa$\sqrt{m}$]		Fracture strength σ_{B50} [MPa]		Relative Density [%]	
	1500°C	1650°C	1500°C	1650°C	1500°C	1650°C	1500°C	1650°C
0	12 ± 0.7	11 ± 0.6	5.7 ± 0.2	4.9 ± 0.1	547 ± 45	597 ± 45	99.5 ± 0.2	99.9 ± 0.2
0.1	4 ± 0.5	6 ± 0.4	4.4 ± 0.7	5.4 ± 0.2	634 ± 36	723 ± 50	99.7 ± 0.1	100 ± 0.1
0.4	4 ± 0.4	5 ± 0.5	5.5 ± 0.6	6.0 ± 0.4	603 ± 67	763 ± 68	99.6 ± 0.2	99.6 ± 0.1
0.7	5 ± 0.3	8 ± 0.6	6.3 ± 0.7	6.4 ± 0.4	589 ± 55	864 ± 70	99.4 ± 0.2	99.6 ± 0.1
1.0	5 ± 0.2	6 ± 0.1	4.2 ± 0.4	4.1 ± 0.5	630 ± 69	711 ± 54	99 ± 0.2	99.7 ± 0.1

Content of SiC (10nm)/ [Vol%]	critical defect size [μm]		Grain-size of Alumina after sintering [μm]	
	1500°C	1650°C	1500°C	1650°C
0	85 ± 8	53 ± 6	1.0 ± 0.3	2.0 ± 0.1
0.1	51 ± 7	44 ± 7	0.9 ± 0.3	1.8 ± 0.2
0.4	65 ± 7	49 ± 5	0.8 ± 0.3	1.6 ± 0.2
0.7	90 ± 9	43 ± 8	0.7 ± 0,2	1.5 ± 0.1
1.0	35 ± 7	26 ± 6	1.0 ± 0.2	1.4 ± 0.1

Table 3 Mechanical properties of the Al_2O_3-SiC (alumina II) composites sintered at 1500°C and 1650°C

Content of SiC (10nm)/ [Vol%]	Weibull-modulus		Fracture toughness [MPa$\sqrt{m}$]		Fracture strength σ_{B50} [MPa]		Relative Density [%]	
	1500°C	1650°C	1500°C	1650°C	1500°C	1650°C	1500°C	1650°C
0	4 ± 0.2	8 ± 0.3	7.7 ± 0.2	6.0 ± 0.3	919 ± 65	747 ± 35	99.7 ± 0.1	99.5 ± 0.1
0.3	6 ± 0.3	9 ± 0.3	6.8 ± 0.7	6.3 ± 0.2	779 ± 35	768 ± 56	99.3 ± 0.2	99.9 ± 0.1
0.7	7 ± 0.2	11 ± 0.5	7.1 ± 0.9	6.4 ± 0.3	828 ± 55	826 ± 62	99.6 ± 0.2	99.5 ± 0.2
1.0	8 ± 0.4	6 ± 0.3	7.0 ± 0.8	6.9 ± 0.4	800 ± 76	838 ± 46	98.7 ± 0.1	99.9 ± 0.1

Content of SiC (10nm)/ [Vol%]	critical defect size [μm]		Grain-size of Alumina after sintering [μm]	
	1500°C	1650°C	1500°C	1650°C
0	55 ± 5	51 ± 6	1.3 ± 0.1	3.6 ± 0.2
0.3	60 ± 4	53 ± 8	0.7 ± 0.2	3.6 ± 0.3
0.7	58 ± 5	47 ± 7	0.9 ± 0.2	2.3 ± 0.3
1.0	60 ± 6	53 ± 6	0.7 ± 0.2	2.0 ± 0.2

The theoretical densities for each composite were calculated using the rule of mixtures. Considering both, the low SiC content of the composites as well as the low oxygen content of the SiC powder used, the amount of a silica-mullite phase generated during sintering process is very low and is assumed to have only less influence on the densification [9,10]. Up to 0.7 vol% SiC content relative densities higher than 99,5% were achieved and no dependence in sintering behavior on the starting alumina was seen. Only at 1 vol% SiC and 1500°C sintering temperature a beginning decrease of the relative density was obtained.

Contrary to densification a dependence of the grain growth (determined by linear intercept method in SEM images) on the alumina quality was noticed for the different sintering temperatures. At 1500°C sintering temperature only a small inhibition of alumina grain growth by increasing the amount of SiC powder was found. The grain size of alumina kept constant at approximately 1μm and is independent of the alumina used. At 1650°C sintering temperature the grain growth inhibition of alumina is dependent on the SiC content as well as the alumina quality. It has to be noticed that the grain growth of monolithic alumina II is aproximately 2 times higher (3.7μm) than that of alumina I (2μm) which clearly shows the higher sinter ability of alumina II compared to alumina I. With increasing SiC content the grain size of alumina clearly decreased. Adding 1 vol% SiC, the grain size of alumina II was halved (from 3,8 to 2μm) whereas the grainsize of alumina I decreased slightly from 2 to 1,5μm. The differences in the sinter abilities of the alumina used in this study may be explained by the different purities and crystallite sizes (higher specific surface is equal to higher surface energy). Although the amount of SiC powder used is very low a big retardation effect on grain growth of alumina was observed. This leads to the assumption that not only the content of a secondary phase is important, more emphasis must be put on the *number* of particles (not agglomerates) [9]. To reach high strength values (and probably toughness) it is assumed that only a few particles around and within each alumina grain must be enough to enhance mechanical properties due to grain growth inhibition by the particles through pinning effects. However, in several studies it was found that to enhance toughness the amount of SiC powder must be clearly higher than 1 vol% [11,12] due to the change of fracture mode from intergranular to transgraunular (grain boundary strengthening due to residual stresses)

The mechanical properties are summarized in Table 2 and 3. In the case of using alumina II the addition of SiC particles and sintering at 1500°C leads to a decrease of the strength from near 1000MPa for monolithic alumina II (Figure 1b) to 790MPa for the composite containing 1.0 vol% SiC. Sintering at 1650 °C caused only a slight increase of strength (from 730 MPa for monolithic to 830 MPa for 1 vol% sample). The toughness of alumina II composites remained constant at approximatley 6.3 $MPam^{1/2}$ (Figure 2b). In the case of using alumina I and 1500 °C sintering temperature the strength values kept constant at 600 MPa for all composites. Sintering at 1650°C showed that with increasing SiC content the observed strength clearly increased from 600 MPa for the monolithic alumina up to 864 MPa for the composite containing 0,7 vol% SiC (Figure 1a). This enhancement is due to the lowered grain size of alumina. The same behavior was found for the toughness. With increasing SiC content the toughness slightly increased (Figure 2a). Further enhancement of the SiC content up to 1 vol% brought a decrease of both, strength and toughness. As mentioned above it is assumed that too much particles within the microstructure enhance the stress distribution due to the different thermal expansion coefficients of SiC particle and alumina matrix. If these stresses are well dispersed (means the particles are well dispersed) and do not exceed a certain value microcracks may be formed either directly during cooling or if the microcracks do not form during cooling but form only in the presence of a stress field like that near a crack tip, these microcracks may increase toughness and strength due to absorbing energy. However, if the amount of particles is too high and/or the powders used are heavy agglomerated this lead to degradation of the mechanical properties. In good accordance with the mechanical properties found are the resulting critical defect sizes which were calculated applying Griffith criterion.

$$\sigma * Y * \sqrt{\pi * a} \equiv K \geq K_{Ic}$$

where σ is the load amplitude, a the crack length, Y a geometric factor, K the stress intensityfactor and K_{Ic} the fracture toughness. Using this equation the critical defect radius is $a_c = 1/\pi * (K_{Ic}/Y*\sigma_B)^2$ with σ_B as the fracture strength and $Y=2/\pi$ for a penny shaped short crack.

For the composites produced of alumina II the critical defect sizes remained more or less constant (53 µm) and are independent on the sintering temperature (Table 3). At 1650°C sintering temperature the critical defect size of the composites produced of alumina I decreased with increasing SiC content. The enhancement of strength and toughness may be explained by these reduced critical defect sizes as well as by compressive stress induced by grinding and polishing [2]. It has to be noticed that the time for preparing Al_2O_3 - nano SiC samples by grinding and polishing is clearly extended compared to that of monolithic alumina. However, at 1500°C the critical defect sizes are clearly higher that those sintered at 1650°C (Table 2). SEM investigation of all fractured surfaces confirmed the calculated sizes for the critical defects. They were found to be mainly processing failures i.e. pores.

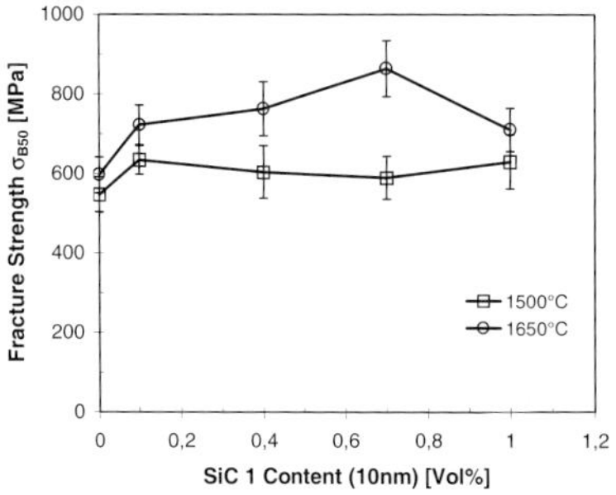

Figure 1 (a): Strength vs SiC content of alumina I composites

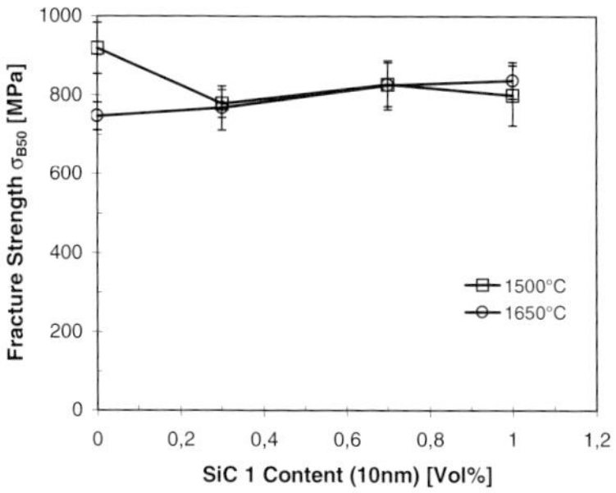

Figure 1 (b): Strength vs SiC content of alumina II composites

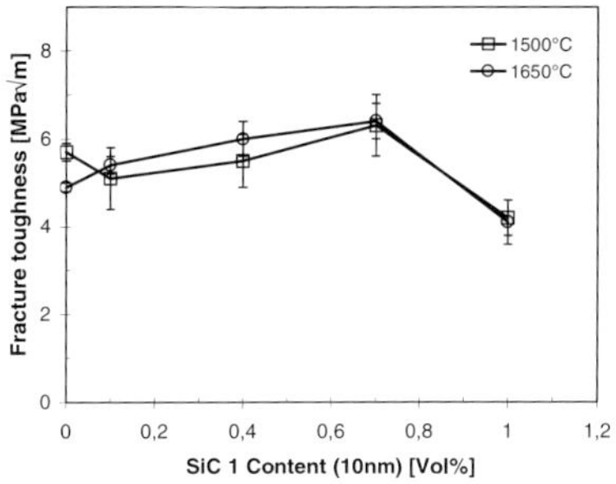

Figure 2 (a): Fracture toughness vs SiC content of alumina I composites

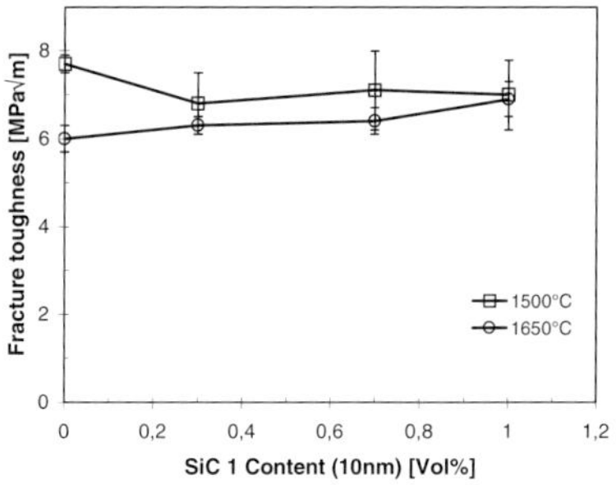

Figure 2 (b): Fracture toughness vs SiC content of alumina II composites

Figure 3 shows the microstructures of a sample made of alumina I with 0.3 vol% SiC (Figure 3) and the alumina II composite containing 0.7 vol% SiC particles, respectively (Figure 4). The grain size distribution of the sample made from alumina I at 1650°C is relatively non-uniform, which suggests abnormal grain growth has occurred (Figure 3). Also a crack propagation (trans- as well as intergranular) is shown in the image. Figure 4 reveals that the SiC particles (bright dots in the image indicated by arrows) are still agglomerated but less compared to the starting SiC powder. This shows that the use of a dsipersant can improve the distribution of the SiC particles within the matrix. The SiC Agglomerates are situated at the grain boundaries of the alumina grains and are partially within the alumina grains. This result is in accordance to those found in [11].

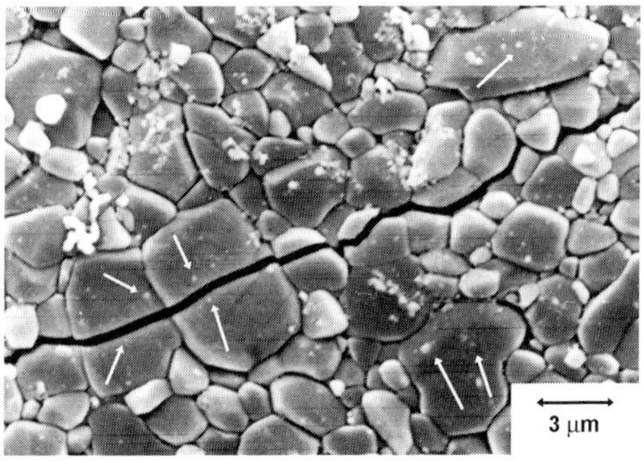

Figure 3. SEM Image of the etched surface of the composite produced of alumina I with 0.3 vol% SiC (1650°C), arrows indicate SiC particles

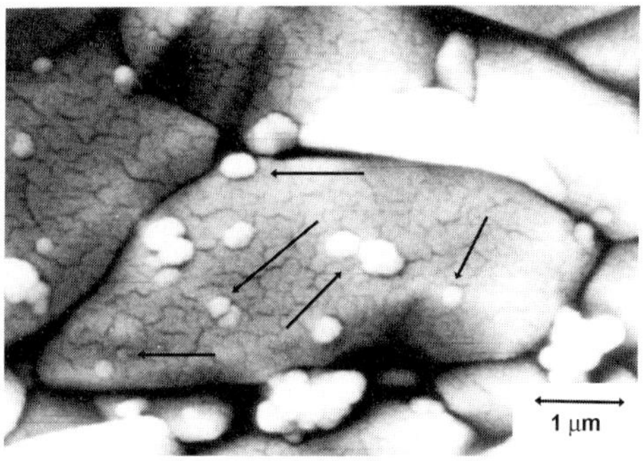

Figure 4. SEM Image of the etched surface of the composite produced of alumina II with 0.7 vol% SiC (1650°C), arrows indicate SiC particles

5 Conclusions

The purpose of the study was to investigate the mechanical properties of hot pressed Al_2O_3 / SiC ceramics. As starting materials two alumina powders (I and II) with mean particle sizes <1μm and differences in purity and production route are used. It was found that nano SiC particles (10nm) even at low volume fraction can improve strength of hot pressed alumina I composites. Alumina II composites showed high strength and toughness but only a slight improvement by adding nano-SiC particle was obtained. The chosen processing route (adjusting pH-value and adding a dispersant) could not destroy all SiC agglomerates. However, the grain growth of alumina is inhibited even at low SiC contents and is dependent on the alumina quality. Most of the SiC particles (agglomerates) are located at the grain boundaries, only few within the alumina grains. Due to the results obtained in this work it is assumed that even low nano SiC fractions may be used to obtain high strength and toughness if it succeeds to distribute the nano particles homogeneously within the matrix and use powders which show less or no agglomeration. The emphasis of future studies of nano-particle reinforced ceramic materials should be put in the number of nano-particles rather than in too high volume fractions.

6 References

1. K. Niihara, J. Ceram. Soc. Jpn. 1991, 99, 974
2. J. Zhao, L. Stearns, L.C. M.P. Harmer, H.M. Chan, G.A. Miller, R.E. Cook, J. Am. Ceram. Soc. 1993, 76, 503
3. L. Carrol, M. Sternitzke, B. Derby., Acta. Mater, 1996, 44, 4543
4. M. Hoffmann, M. Sternitzke, J. Rödel, R.J. Brook, Fracture Mechanics of Ceramics, R.C. Bradt, D.P.H. Hasselmann, D. Munz, M. Sakai and V.Y. Shevchenko, Vol.12, Plenum Press, New York, 1996, 176
5. Levin, W.D. Kaplan, D.G. Brandon, A.A. Layyous, J. Am. Ceram. Soc. 1995, 78, 254
6. M. Sternitzke, J. Eur. Ceram. Soc. 1997, 17, 1061
7. M. Burke, R. Greenwood, K. Kendall, J. Mater. Sci. 1998, 33, 5149
8. R. Damiani , R. Gstrein, R. Danzer, J. Eur. Ceram. Soc., 1996, 16, 695
9. M. Wilhelm, W. Wruss, cfi/DKG, 1998, 75, No. 9, 40
10. D. Sciti, D. Dalle-Fabbriche, A. Belosi, Key Eng. Mater., 1997, 132-136, 2001
11. Y. Xu, A. Zangvil, A. Kerber, J. Europ. Ceram. Soc., 1997, 17, 921
12. Z. Deng, J.L. Shi, Y. Zhang, D. Jiang, J. Guo, J. Europ. Ceram. Soc., 1998, 18, 501

Synthesis, Processing and Sintering of Nanosized Alpha Alumina Powder

G.R.Karagedov
Institute of Solid State Chemistry and Mechanochemistry, Novosibirsk, Russia

1 Introduction

To obtain nanocrystalline material of α-alumina is largely hampered by the difficulty in preparing its nanosized powder. The overwhelming majority of oxide synthesis procedures leads to the formation of metastable γ-modification. Peculiarities of $\gamma \rightarrow \alpha$ phase transition, with the rate determining step being the new phase nuclei formation [1], make it as proceeding at high temperatures. Such a process is bound to be accompanied by a considerable particle growth up to 100-200nm at best.

An intensive comminution of coarse grained α-Al_2O_3 might be considered as an alternative way to nanocrystalline powder unless it has been previously shown [2] that both dry and wet grinding reduces particle dimensions only till a certain limiting value. Beyond this limit, being equal to 400nm in case of α-Al_2O_3, particle size begins to grow as the time of grinding increased. Nevertheless, it is of importance to note, that the kinetics of variation of surface area in course of intensive alumina grinding does not correlate with concomitant change in crystallite size. As the specific surface area passes through the maximum, median crystallite size steadily reduces down to 25-30nm [3]. The latter observation opens the way to nanosized α-Al_2O_3 if one is able to separate crystallites from each other.

The present paper deals with an idea of introducing in a powder subjected to severe comminution a certain substance referred to as «grinding catalyst». It must be inert to aluminum oxide and provides mechanical barrier between crystallites, thus preventing them from coalescence. The elimination of this substance from the resulting powder should destroy the aggregates and lead to the powder composed of separate crystallites. The properties of the resulting powder with respect to green body formation and sintering into a dense material has been studied.

Additionally it has been shown that nonopowder produced can be used as a seed material to reduce $\gamma \rightarrow \alpha$ transformation temperature and to get pure and fine alumina.

2 Experimental procedure

Reaction grade alumina (pure for chemical analysis) has been preliminary calcined at 1150 °C for 4 h. Subsequent X-ray phase analysis indicated 100% α-Al_2O_3. The powder was ground in planetary mill AGO-2M 10 to 20 min followed by reflaxing in 3-5% HCl for the purpose of removing «grinding catalyst» and milling media (steel) wear. After having been dried at 70-120 °C, the powder was pressed uniaxially in a die at 145MPa and sintered in a dilatometer.

Impurities content was determined be laser ionization mass spectrometry and routine chemical analysis, surface area measured by BET method, and crystallite sizes were taken from X-ray line profile analysis. Particle size distribution curve was derived from the data of laser autocorreletion spectroscopy in slightly acidified water dispersion. Electron microscopy has been performed by JEM-2000FX2.

3 Results and Discussion

3.1 Preparation by intensive grinding and characterization

Previously it has been pointed out [4] that planetary milled with AGO-2M α-alumina presents highly agglomerated powder and that a degree of agglomeration rises with the time of mechanical treatment. Addition of «grinding catalyst» to the powder being ground does not affect this type of behavior. However, the removal of the «catalyst» from the ground powder drastically changes the relationship between powder specific surface area and time of treatment (fig.1) and, consequently, brings about powder deaggregation. Fig.1 presents also values of equivalent spherical diameter calculated from specific surface area and (in brackets) crystallite dimensions. One can focus attention on that the diameters given by two quite different methods coincide reasonably well which speaks for the powder as being, in fact, consisted of separated, presumably, single crystalline particles.

Particle size distribution curve of the powder ground for 20 min (taken as an example) indicates that more than 90wt.% of the sample assigned to the peak with average diameter of 20-25nm. The curve presents also few particles of 100nm but they are likely to be aggregates formed in the dispersion. Electron micrograph of ultrasonically dispersed suspension in alcohol shows separate particles of about 10-15nm and some polycristalline aggregates of them. Electron microdiffraction as well as X-ray pattern points to pure α-Al_2O_3 as the only phase present.

The main impurities detected for powders mechanically treated at most intensive and prolonged regimes were Fe-1.1%, Cr-0.4%, Si-0.4%, Mn-0.01%. The contaminants seems to build in crystal lattice of alumina and, hence, can not be dissolved by acid.

If the powder described above being dispersed in ethanol, the possibility arise to classify it with centrifuge (fig.2). The data are indicative of the ability to isolate fairly narrow in size fractions and of sufficiently large amount of a fraction with crystallite size equal to 18nm. The latter powder possess 80.6 m^2/g specific surface area that gives surface-based mean particle diameter as 18.6nm. Again, we observe an excellent consistency of findings of distinct methods. For all we know, it is the finest α-Al_2O_3 powder available at the moment.

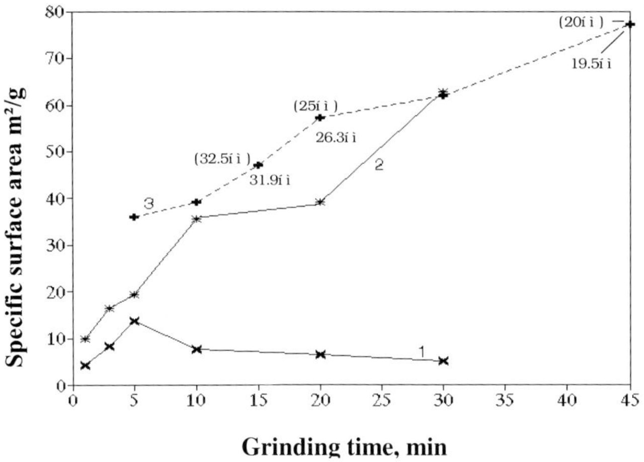

Figure 1 Specific surface area of powder ground at acceleration 20g with 10mm balls-1, then deaggregated-2, and powder ground at acceleration 40g with 5mm balls and deaggregated-3 versus time of grinding .

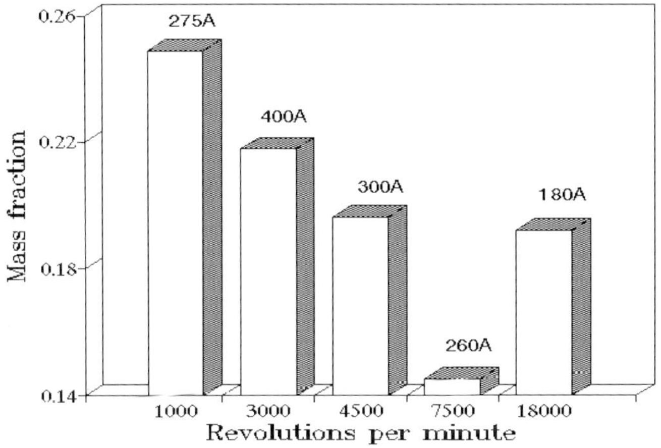

Figure 2 Mass fraction of particles isolated by centrifuging of a powder ground 20min at 40g acceleration and deaggregated Crystallite sizes also given above columns.

3.2 Compaction and Sintering

In spite of the fact that due to unfavorable gravitational/adhesive forces ratio for ultrafine particles they ordinarily show very poor formability by dry pressing, powders produced by the above procedure are completely devoid of this drawback. By way of example, the powder described above (20min grinding) may be pressed uniaxially at merely 145MPa to 60% of theoretical value (3.98g/cm^3) and isostatically at 280MPa to 65%.

Green body obtained has an enhanced sinterability at relatively low temperatures. As low temperature as 1270°C is requisite to attain 92-93% density while retaining uniform and fine microstructure with grain sizes ranged 100-200nm and not a single grain markedly exceeding this interval. Narrow in size fractions isolated from this powder by centrifuge sinter more active the smaller is the comprising crystallite diameter. The sinterability of the finest fraction of about 18 nm in size is shown in fig.3 as being higher than that of unclassified powder. This figure clearly indicates that mechanical treatment of alumina in planetary mill considerably improves its densification behavior comparative to starting material even without «grinding catalyst». The latter is very useful for the increase in density attained at low temperatures. In fact, this observation illustrates that powder deaggregation prevents from secondary pores generation originated from differential sintering of aggregated body[5,6].

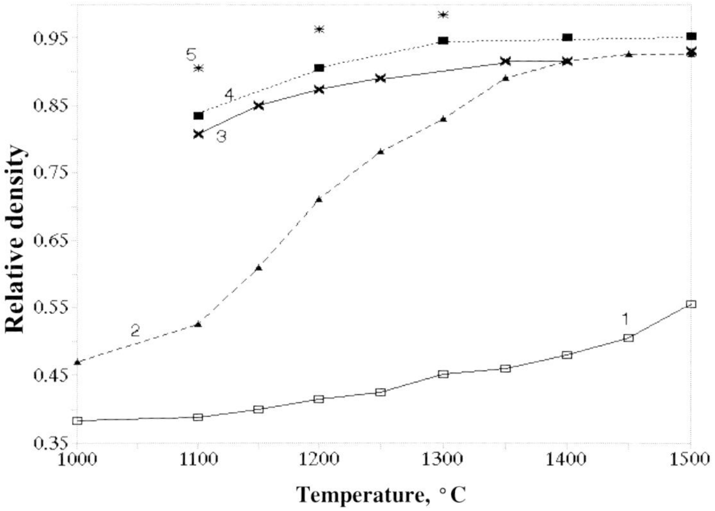

Figure 3 Sintered density as a function of a temperature of 1h isothermal exposure. 1-sarting powder, 2-ground for 20 min at 40g and 3-then deaggregated with the use of grinding catalyst, 4-fines of deaggregated powder separated by 7000 rpm centrifuging , 5-the same as 4 but separated from less contaminated powder with coarser particles..

Nevertheless the characteristic feature of compacted mechanically treated powders is that the porosity achieved at 1200-1270°C can not be reduced as the sintering temperature elevated even up to 1600°C (fig.3). Sometimes high temperature swelling took place.

This disadvantage may well be attributed to powder being contaminated with Si. As has been claimed by authors [7], 0.045 mass% of silicon worsen sintering behavior of alumina and with the level of impurity brought to 0.5% closed pores formation occurred resulting in no more than 90-92% density at 1600°C. This observation falls in line with our experimental evidence: 92-93% density at 0.4% Si content.

To clear the issue a batch of α-alumina powder has been subjected to comminution under reduced intensity of mechanical treatment. As a consequence, after eliminating «grinding catalyst» a powder coarser in size (mean crystallite dimension 40.5nm, specific surface area 39.75 m^2/g) but much less contaminated (Fe-0.5%, Si-0.05%) has been at our disposal. Leaving beyond the scope of our attention the green density which turned out to be unexpectedly high for dry pressing (68.5% at 145MPa) we would like to stress that final density achieved for 1hr at 1270°C exceeded 99%. The sinterability of the finest powder fraction isolated from this sample is presented in fig.5 (see curve5).

.Thus, it becomes evident that the development of a technique for ultrafine grinding of α-Al_2O_3 with a decreased level of contamination will lead to the material sintered to full density at even lower temperatures. This material is expected to be of much promise for sintering nanosized ceramics with enhanced mechanical properties.

3.3 Preparation and Sintering Gel-derived nanopowder

An additional advantage of thus prepared nanosized alumina powder is that it may be used as a seed facilitating a chemical reaction towards alpha alumina. Very similar idea has been applied to enhance boehmite gel transformation into α-Al_2O_3 [8] Authors reduced the temperature of $\gamma \rightarrow \alpha$ phase transition from common 1200 °C to 1025 °C by addition of 1.5wt% of 100nm α-alumina particles.

Being able to employ particles of much smaller dimension, we carried out the following process:

$Al(NO_3)_3$ + 1-3wt% of 25nm α-$Al_2O_3 \longrightarrow$ gel $\longrightarrow$ α-alumina

Dropwise addition of appropriate amount of ammonia to vigorously stirred water dispersion caused gel formation which, after having been aged and dried, gave rise to fully transformed α-Al_2O_3 within 15min at no more than 900 °C. The powder exhibits 40-60 m^2/g specific surface area depending on the conditions of heat treatment and consistent crystallite sizes. The level of contamination is determined by the amount of seed added, the latter being the main source of impurities, and does not exceed 0.02wt%.

Powder possess poor formability which is usually observed for ultrafine species and as a consequence sinters to porous material. Table 1 presents density data resulted from an hour of isothermal sintering for intensively ground powder with mean particle size of 45nm -sample1, and for gel derived powder of 47.5nm -sample2. In spite of the very similar particle dimension the behavior of mechanically ground sample is quite different. It comes to clear that intensive grinding involves a kind of mechanical activation together with mere size reduction and supplies the powder with features responsible for high compactability.

Table 1. Sintered density of compacted ground and gel derived powders

	25 °C	1100 °C	1200 °C	1300 °C
Sample1	2.6 g/cm^3	3.31 g/cm^3	3.59 g/cm^3	3.86 g/cm^3
Sample2	1.9 g/cm^3	2.45 g/cm^3	3.64 g/cm^3	3.4 g/cm^3
Sample3	2.42 g/cm^3	3.27 g/cm^3	3.59 g/cm^3	3.7 g/cm^3

The most striking peculiarity, however, is that after sample 2 has been subjected to grinding for only two minutes at relatively soft conditions it acquires ability to be pressed to a high density-(sample 3) and sinters fairly well. It seems that short-term grinding under suitable conditions , being quite insufficient to improve sintering behavior of course powder, when applied to nanosized powder can readily adjust size distribution function and drastically increase green density.

4 Conclusions

Highly intensive mechanical grinding in planetary mill with the use of so called «grinding catalyst» makes it possible to produce nanosized α-Al_2O_3 powders possessing mean particle diameters down to 20-25nm.

Powder fractions with mean particle diameter even less than 20nm may by isolated by centrifuging.

Samples dry pressed of mechanically comminuted powders exhibit excellent formability and enhanced sintering kinetics yielding dense material at but 1200 °C.

Pure nanosized α-Al_2O_3 powder can be produced by enhanced $\gamma \rightarrow \alpha$ transformation when starting reagent is seeded by mechanically ground fines.

Applying short duration intensive mechanical treatment to ultrafine powder considerably increases green density attained.

5 References

1. F.W.Dynys, J.W.Halloran, J.Amer.Ceram.Soc., 1982, 65, 442-448.
2. G.Gimbo et al., Proc.2nd World Cong. Particle Technology, Kyoto, Japan, 1990, v.2, 305-312
3. E.G.Avvakumov, G.R.Karagedov, and N.Z.Lyakhov, Proc. IUPAC CHEMRAWN 9 World Cong., 1996, 398-403.
4. G.R.Karagedov, N.Z.Lyakhov, J. Mater. Synthesis and Processing., 1996, 4, 291-297.
5. W.H.Rhodes, J.Amer.Ceram.Soc., 1981, 64, 19-22.
6. F.W.Dynys, J.W.Halloran, J.Amer.Ceram.Soc., 1984, 67, 596-601.
7. O.Abe, , Kona, 1995, 13, 159-166.
8. M.Kumagai, G.L.Messing, J.Amer.Ceram.Soc., 1985, 68, 500-505.

Stereolithography for Ceramic Part Manufacturing

F. Doreau, C.Chaput
Centre de Transfert de Technologies Céramique, Limoges.
T. Chartier
SPCTS, UMR CNRS 6638, Ecole Nationale Supérieure de Céramiques Industrielles, Limoges.

1 Introduction

Ceramics offer a wide range of properties in term of mechanical strength, thermal stability, hardness, chemical durability, electrical/optical/magnetic properties, and/or thermal conductivity. Processing of ceramic generally starts with raw materials, proceeds through batch preparation and shaping, and concludes with firing and sometimes surface finishing, as required by the application. Among the different classical ceramic shaping techniques, most of them require a mould, particularly when complex shapes have to be produced. It makes this step difficult and costly. In this context, rapid prototyping presents a great opportunity to fabricate new ceramic prototypes especially when complex shapes are required.

The first part of this paper present a short review of rapid prototyping of ceramics. The following parts deals with the principle of our machine and its main advantages. Before concluding, a last part presents example of a ceramic parts made by stereolithography.

2 Overview of Rapid Prototyping of Ceramics

Rapid prototyping and ceramics are two broad terms which include, for the first one, numerous technologies and for the second one, a lot of materials and applications. It is not possible to make here an exaustive review of all the research done in the field of rapid prototyping of ceramics. Therefore, only the main technologies are presented in this paper.

Among the different rapid prototyping technologies, Solid Freeform Fabrication (SFF) are the most suitable for ceramic applications. These technologies could be devided in 5 classes:

- Photolithography
- Laser fusion
- Lamination
- Extrusion
- Ink-jet printing

Numerous rapid prototyping devices based on these 5 technologies have been developed throughout the world (principally in US, Europe and Japan). At the bigining, they were developed for polymer applications but rapidely the need of functionnal materials led research teams to work on metal and ceramic applications.

2.1 Photolithography

Photolithography systems build 3D objects using light to selectively solidify photocurable resins either by laser beam scaning or by UV radiation exposure through masks generated for each layer.

Several teams work on the development of photocurable ceramic suspensions, i.e. photocurable resin loaded with ceramic powders[1-3] to fabricate fully dense materials. In order to obtain an homogeneous green part and to promote binder removal and sintering, high concentrated resin are suitable.

2.2 Laser fusion

Laser fusion systems use the energy of a laser beam to locally fuse the building material or a phase of this material, spread as a layer of powder (technology named SLS).

This technology is used for ceramic applications in the field of molds or cores for casting and inverstment casting in foundry. In these applications, a binder added to ceramic particles is fused by the laser an give the cohesion to the part[4, 5]. Post treatments like thermal treatments and/or infiltration are often necessary to improve the properties of the pieces because the green parts are very porous.

Some investigations are conducted on high power laser systems for directly sinter metals or ceramics.

2.3 Lamination

Lamination systems build parts stacking sheets of material the one onto the other and promoting adhesion between each layer by a lamination process (pressing, rolling...). Different techniques are used to cut individual sheets at the right shape (laser, knife...).

In the field of ceramics, works have been conducted to perform structural ceramics or composites[6-11]. A great quantity of ceramic sheets are necessary as feedstock. They are mainly obtained by tape casting which finally leads to a complex shaping process from raw material to sintered part.

2.4 Extrusion

In the extrusion systems, a heated nozzle is moved over the working table in the required geometry and deposits a thin bead of extruded plastic to form each layer. The plastic hardens immediately after being squirted from the nozzle and bonds to the layer below.

Works are conducted in the field of ceramics to use comercialy available apparatus or more experimentally devices with loaded thermoplastics in order to build ceramic green preforms similar to those obtained by injection molding[12-15].

2.5 Ink-jet printing

Ink-jet printing could be devided in two sub-classes: indirect or direct ink-jet printing. Indirect ink-jet printing was the first developped technology. An ink-jet printing head scans a layer of powder of the material to be used and selectively injects droplets of binder into the powder. In the direct processes, the material is directly deposited via the ink-jet printing head. The material could be a suspension of powder or a fused low melting point system.

Both indirect and direct techniques are used for several ceramic applications[16-18].

Several other rapid prototyping technologies are presented in the litterature. Some of them are tested with ceramic materials but these works are often at an early stage.

3 Rapid Prototyping of Ceramic at the CTTC

3.1 Stereolithography device

CTTC is equiped with a new machine developed by the French company OPTOFORM. This machine operates under the classical principle of building an object in a layer-by-layer fashion. It is based on the photolitographic technique. Contrary to classical stereolithography, where a fluid monomer is required, this machine uses a paste with a high viscosity and a specific rheological behaviour. The adapted rheological behaviour does not require any vat. The paste layers are spreaded by means of a specific scraper patented by OPTOFORM. The schematic principle and a view of this machine are presented on figure 1.

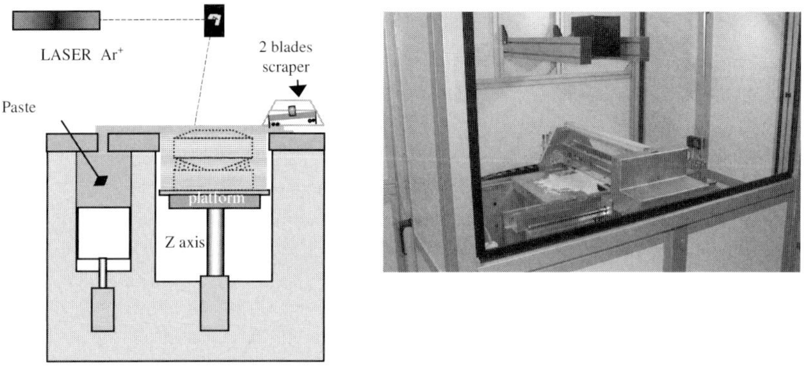

Figure 1 : Schematic principle and view of the Optoform stereolithography apparatus

The main advantage of this technique is speed. Indeed, the design of the machine allows a high speed to spread the layers and, owing to the rheological behaviour of the paste, no relaxation time is necessary for the menisci. As a consequence, the UV exposure (via a laser and galvanometric mirror) is possible immediately after deposition of a layer. There is no dead time in the piece building. Another advantage, linked to the use of a paste, is the possibility to use high concentrated suspension. It is suitable for the following steps of the process (debinding and sintering) and for the stability of the suspension. High viscosity prevents settling and allows long time storage.

The size of the building volume is 250 x 250 x 250 mm but the design and the use of paste allow to work with smaller volumes on a smaller building plateform. This is very suitable for material development, and economical when expensive materials are used.

3.2 Ceramic paste preparation

Suspension preparation is of prime importance to obtain homogeneous and dense ceramic parts. A rheological behaviour adapted to the machine is required as well as a high level of homogeneity. Indeed, the paste has firstly to be very homogeneous to be properly spread in thin layers. Secondly, the homogeneity of the paste is required to ensure the continuity of the ceramic media in the bulk polymerised part to promote a good debinding and sintering without any cracks.

Different components are mixed together to obtain the paste with the required properties. They are at least four and some times five :

- photocurable resin
- photo initiator
- thickner
- dispersant
- ceramic powder

To reach the objective of a homogeneous paste, the various components are mixed using an appropriate sequence of addition and an energetic mixing technique.

Beside the homogeneity, the other feature of the ceramic paste is the rheological behaviour. Because of the principle of the machine, the paste has to exhibit a viscosity with a high yield value to avoid the need of vat. A shear thinning behaviour, with a lower viscosity at the considered shear rate imposed by the machine, facilitates the spread of the layer and allows to use a higher speed of deposition. A typical rheological behaviour is presented on figure 2.

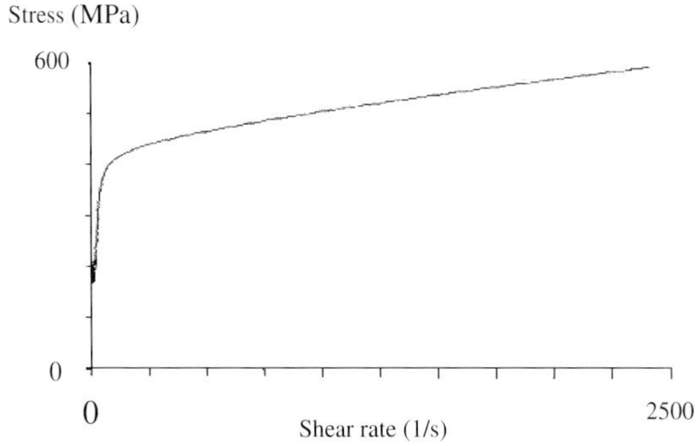

Figure 2 : Suitable rheological behaviour of ceramic paste for rapid prototyping

3.3 Example of ceramic parts fabrication

We develop photocurable pastes, based on ceramic powders and acrylic photocurable resins. Load up to 65 vol% of ceramic powders are filled in the resin. Alumina parts have been developed and zirconia is now studied.

Parts are classicaly built from a STL file with a layer thickness of 100 or 50 μm. After debinding up to 500°C, the piece was sintered at a temperature linked to the material used.. Figure 3 presents alumina pieces. Few hours are necessary to obtain a green ceramic body from the CAD file.

Sintered alumina parts fabricated by stereolithography exhibit similar mechanical properties than those obtained using classical shaping techniques.

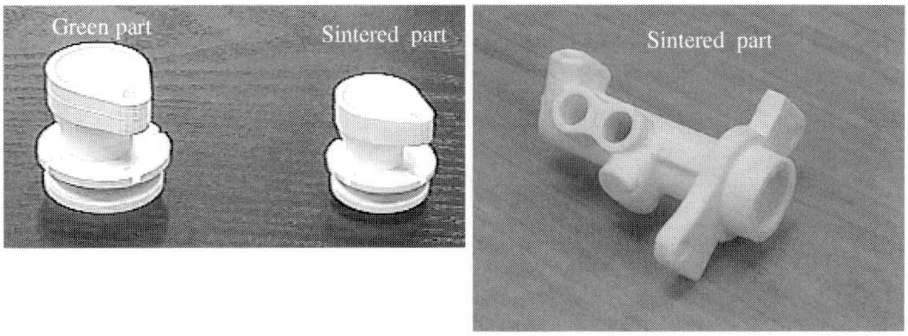

Figure 3 : Examples of alumina parts

4 Conclusion

The major drawback of the fabrication of complex shape ceramic parts is the cost of the mould. Engineers often avoid the use of ceramic pieces because of the high cost linked to the development and validation of these parts. Rapid prototyping offers an unique opportunity to test ceramic solutions in a short time and for limited costs. On an other hand, it is probably the unique possibility for fabricating parts with very complex shapes.

Ceramic pastes with a rheological behaviour adapted to a new rapid prototyping machine have been developed. These pastes, composed of ceramic particles dispersed in a photocurable resin reach solid lodings up to 65 vol%. The process was validated with the fabrication of several alumina pieces which exhibit similar flexural stregth than alumina parts made by classical techniques like pressing.

5 Acknowledgements

We want to thank the French organisation for the research valorisation, ANVAR, for its financial support.

6 References

1. M.L. Griffith, J.W. Halloran, J. Am. Ceram. Soc,1995, 82, 1653-1658
2. T. Himmer, T. Nakagawa, H. Noguchi, Proc. Of the SFF Symposium (Ed.: D.L. Bourell, J.J. Beaman , R.H. Crawford, H.L. Marcus and J.W. Barlow), 1997, p.363
3. C. Hinczewski, S. Corbel, T. Chartier, J. Europ. Ceram. Soc., 1998, 18, 583-590.
4. P.K. Subramanian, J.W. Barlow, H.L. Marcus, Proc. Of the SFF Symposium, 1995, p.346
5. C. Griffin, S. McMillin, Proc. Of the SFF Symposium, 1995, p.25
6. D. Klosterman, R. Chartoff, N. Osborne, G. Graves, A. Lightman, Proc. Of the SFF Symposium (Ed.: D.L. Bourell, J.J. Beaman , R.H. Crawford, H.L. Marcus and J.W. Barlow), 1997, p.247
7. D. Klosterman, R. Chartoff, N. Osborne, G. Graves, Proc. Of the SFF Symposium (Ed.: D.L. Bourell, J.J. Beaman , R.H. Crawford, H.L. Marcus and J.W. Barlow), 1997, p.537

8. M.J. Pope, M. Patterson, W. Zimbeck, M. Fehrenbacher, Proc. Of the SFF Symposium (Ed.: D.L. Bourell, J.J. Beaman , R.H. Crawford, H.L. Marcus and J.W. Barlow), 1997, p.529
9. Z.E. Liu, T.C. Ku, J. Best, J.D. Cawley, A.H. Heuer, Proc. Of the SFF Symposium (Ed.: D.L. Bourell, J.J. Beaman , R.H. Crawford, H.L. Marcus and J.W. Barlow), 1997, p.379
10. D. Klosterman, R. Chartoff, N. Osborne, G. Graves, A. Lightman, G.han, Proc. of the 7th International Conference on Rapid Prototyping, 1997
11. D.M. Kupp, H. Eifert, M. Greul, M. Künstner, Proc. Of the SFF Symposium (Ed.: D.L. Bourell, J.J. Beaman , R.H. Crawford, H.L. Marcus and J.W. Barlow), 1997, p.203
12. Safari, S.C. Danforth, Proc of the 7th European Conference on Rapid Prototyping and Manufacturing, (Ed.: R. I. Campbell), 1998, p. 259
13. R. Clancy, V. Jamalaba, P. Whalen, P. Bhargava, C. Dai, S. Rangarajan, S. Wu, S. Danforth, N. Langrana, A. Safari, Proc. Of the SFF Symposium (Ed.: D.L. Bourell, J.J. Beaman , R.H. Crawford, H.L. Marcus and J.W. Barlow), 1997, p.185
14. Dai, G. Qi, S. Rangarajan, S. Wu, N. Langrana, A. Safari, S.C. Danforth Proc. Of the SFF Symposium (Ed.: D.L. Bourell, J.J. Beaman , R.H. Crawford, H.L. Marcus and J.W. Barlow), 1997, p.411
15. E. Sachs, M. Cima, J. Cornie, Proc. of the Second International Conference on Rapid Prototyping, 1991
16. M.J. Cima, J. Yoo, S. Khanuya, M. Rynerson, D. Nammour, B. Giriliglu, J. Grau, E.M. Sachs, Proc. Of the SFF Symposium, 1995, p.479
17. J. Grau, J. Moon, S. Uhland, M. Cima, E. Sachs, Proc. Of the SFF Symposium (Ed.: D.L. Bourell, J.J. Beaman , R.H. Crawford, H.L. Marcus and J.W. Barlow), 1997, p.371
18. M.Mott, J-H. Song, J.R.G. Evans, J. Am. Ceram. Soc., 1999, 82, 1653-1658

Rapid Manufacturing of Ceramic Microcomponents

Regina Knitter, Werner Bauer, Britta Linner-Krčmar, Edgar Hansjosten*
Forschungszentrum Karlsruhe GmbH
Institut für Materialforschung III, *Hauptabteilung Versuchstechnik

1 Introduction

The increasing applications of microsystems and the development of new microcomponents create the need for different materials. The variety of available materials, however, is often limited by the use of micropatterning techniques that are mostly based on Si-technologies or the LIGA technique [1]. If high strength, high corrosion or thermal resistance, or special physical properties, e.g. piezo- or dielectricity are required, polymers or metals are not always capable to meet the demands, and ceramics are often the only materials to fill this gap. Because of their high hardness and brittleness, forming in the densified state is excluded or at least very cost-intensive. As long as the direct shaping of ceramics with generative methods like 3D-printing, laser sintering or laminated object manufacturing does not reach the precision or surface finishes demanded for microsystems, ceramics have to be shaped by replication techniques.

Ceramic microcomponents like almost all ceramics are formed in the green, unfired state by consolidating the ceramic powder with the help of more or less additives into the desired shape. In the past years a variety of processes were developed or known techniques adapted to the shaping of ceramic components with inner dimensions in the micrometer range. These techniques differ in the design restrictions, the attainable aspect ratio and the economic lot size, but have in common that a micropatterned mold is needed [2]. By methods like slip pressing [3, 4] or centrifugal casting [5], a ceramic slip is cast into a polymer or wax mold which is used as a lost mold. Methods with higher requirements made on the molding die, like ceramic injection molding [6] or tape casting & stamping [7], use a metallic mold, from which the green part has to be demolded before firing. In contrast to the high-pressure injection molding of ceramics, where the plastification of the feedstock is achieved by thermoplastic plastics of high viscosity, the hot molding or low-pressure injection molding of ceramics uses low-melting paraffins [8]. Due to the significantly reduced pressure and temperature conditions and the lowered demands made on the die, it is possible to use non-metallic molds. But even if the final step of the replication technique does not need a metallic mold, a metallic die is used in one step of the replication in almost all cases, e. g. in the fabrication of the polymer molds by injection molding. Depending on the demands in terms of geometry and aspect ratio, the metallic molds, however, have to be made by electrical discharge machinery, metallic microfabrication [9] or electrodeposition.

Within the last fifteen years, a number of processes were developed with the goal to reduce the time for the fabrication of prototypes and, thus, to shorten the time of product design and to enable rapid manufacturing. The first rapid prototyping process patented by C. W. Hull in 1986 was the stereolithography, a technique for the preparation of three-dimensional prototypes made from polymers [10]. The growing wish to use the prototypes not only for design studies but also as functional test models led to the development of a variety of generative techniques for the preparation of prototypes made not only from polymer, waxes or paper, but also from metallic and ceramic materials [11-13]. On the other hand, it was aimed at developing rapid prototyping processes with higher precision and resolution for the use in microtechnologies [14, 15]. Stereolithography was the first commercialized prototyping process [16] and up to now is the one with the best resolution [17]. In this process the desired component is first generated as a 3D-CAD model. These data are used to control a laser which selectively cures a layer of liquid resin on a platform in a vat. The platform is then lowered by a certain degree into the liquid before the next layer is built. In first approximation the possible resolution depends on the layer thickness, the spot size of the laser and the beam intensity which affects the interaction volume of the monomer. Polymer parts made by Rapid Micro Product Development (RMPD), a micro-modification of stereolithography, which enables multiple arrangement production and offers a resolution of a few micrometers have already been made commercially available [18]. The combination of micro-stereolithography which is used here as a rapid tooling process and the hot molding of ceramics now opens up the possibility for the rapid manufacturing of ceramic microcomponents.

2 Experimental

For the rapid manufacturing of ceramic microcomponents presented here, the designs were first generated as a 3D model via CAD and then transferred to polymer molds by stereolithography. The polymer parts were copied into silicone molds that were directly used as tools for the hot molding process. The final ceramic microcomponents were obtained after debinding and sintering (Fig. 1).

2.1 Preparation of the mold

The preparation of the primary molds was carried out either on a stereolithography machine (FS-REALIZER, Fockele & Schwarze) at FZK or in RMPD technique at microTEC, Duisburg. The former has a positioning accuracy of 10 µm and works with a solid-state laser with a spot size of 200 µm. For the selective curing of the monomer the support platform was lowered in 100 µm steps. Some of the RMPD microparts were not selectively cured, but fabricated by mask technology, i. e. one layer is cured simultaneously through a mask [19]. For the given geometry a step size of 25 and 50 µm, respectively was sufficient, although this technique also allows to build objects in small steps of approx. 1 µm. The negative molds for the hot molding of ceramics were prepared by casting silicone into the polymer parts. The silicone was evacuated to eliminate air inclu-

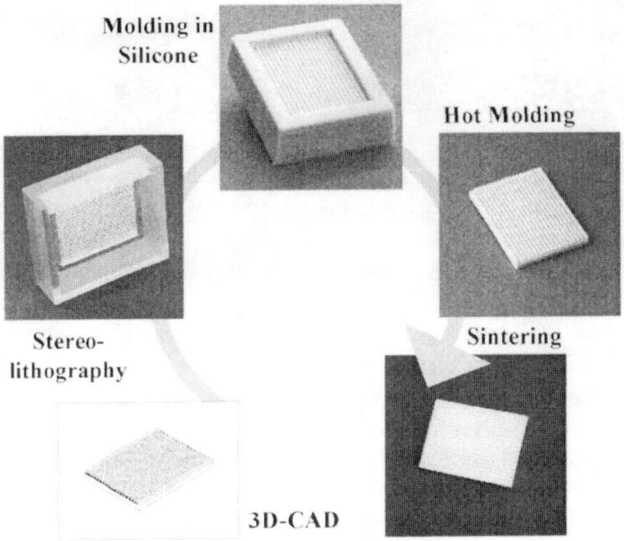

Figure 1. The process steps of stereolithography and hot molding.

sions and after curing it was easily removed from the polymer part. The process of curing can be reduced to a few hours by heating the silicone up to 80°C.

2.2 Low-pressure injection molding

For a good filling of the mold the feedstock must be optimized with regard to its viscosity. In order to increase the green density, however, and thus to minimize shrinkage during sintering, the amount of binder should be as low as possible. The feedstock was prepared in a blade kneader which was evacuated and heated to 80°C by melting 11 wt% of paraffin and about 1 wt% of an emulsifying agent. In this mixture 88 wt% of the ceramic powder were dispersed and subsequently kneaded for about 4 hours. The used alumina powder has an average particle size of 1.2 µm (MR 52, Martinswerk).

The filling of the silicon molds can be done either manually in case of small feedstock amounts or in a hot molding machine (Peltsman, Minneapolis, USA) where the silicone mold is placed inside an evacuatable tool. If the molding is done by hand, the mold has to be heated and evacuated to minimize inclusions of air. In both cases it was not necessary to use a mold release spray. After demolding, the paraffin was removed from the green parts by a slow heating process up to 500°C. For sintering the parts were heated up to 1700°C. The heating rates for debinding and sintering depend on the geometry and the cross-sections of the parts, i. e. large parts with large or varying cross-sections are difficult to debind and sinter without introducing defects which may cause cracking. The debinding process was optimized by thermoanalytical investigations that gave information about the temperature and intensity of wax melting and decomposition. All additives were removed when a constant weight was observed at about 500°C.

3 Results and Discussion

After sintering 94 % of theoretical density was typically achieved in the ceramic parts. Starting with a green density of 62 %, the linear shrinkage was about 13 %. Measurements of the micropattern showed a precision of the patterning with a deviation in the range of only 0.2 %. In all cases the reproduction of the details was very high. Even small defects caused by insufficient cleaning of the polymer mold were copied into the silicone mold and subsequently into the ceramic part. Fig. 2 shows a detail of a micropatterned part with sloping surfaces, replicated of a polymer part produced with conventional stereolithography. The steps that result from the layerwise curing of the primary mold are clearly visible at the surface and at the walls. With the used stereolithography equipment the patterning in the horizontal plane is limited to about 250 μm. An improvement of the resolution can be realized if the primary mold is fabricated by RMPD technique. Nevertheless, the smoothness of the walls is affected by the layer structure of the primary mold (Fig. 3). Using RMPD-Mask technology for the preparation of the primary mold, it is possible to reduce the roughness of the vertical surfaces. However, in this case the design is restricted within relatively narrow limits by the otherwise soon increasing number of necessary masks.

For the fabrication of macroscopic parts the successful combination of conventional stereolithography and hot molding has already been demonstrated. The polymer tools were either directly used as molds [20] or copied into polymer or metallic molds [21]. For the direct molding with stereolithographic parts it was necessary to use a mold release spray before each injection. With the molding by silicone molds as presented here, however, there is no need for a mold release agent and demolding is significantly simplified due to the elasticity and the superior release characteristics of the silicone. Even dimensions of a few micrometers are easily removed from the mold without being damaged. The replication of the polymer part in silicone is fast and inexpensive and can be

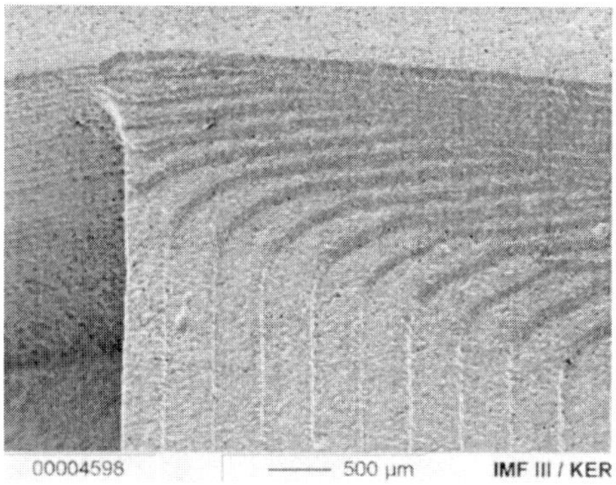

Figure 2. Detail of a micropatterned alumina component replicated from a polymer part made by conventional stereolithography.

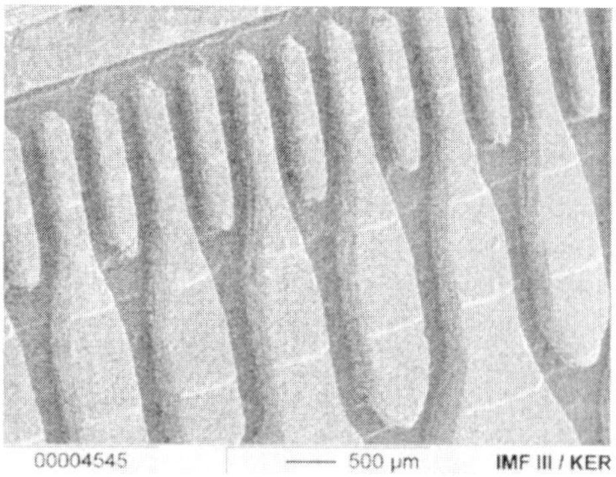

Figure 3. Detail of a micropatterned alumina component replicated from a polymer part made by RMPD technique.

repeated several times to increase the number of available molds. The use of wax parts made by 3D-printing for the replication into silicone appeared to be less advantageous, as the micropatterning of the primary mold was destroyed to some extent. Nevertheless, it should be pointed out that the use of silicone molds may be limited to the injection of small parts, as the mold may be deformed by the weight of larger parts due to the softness of the material.

It was already proved that ceramic microcomponents with high aspect ratios and precision can be realized by hot molding into silicone molds [22], and it seems that the achievable resolution is limited by the grain size of the powder rather than by the molding process. In the past it was necessary to use expensive primary molds made by LIGA or mechanical microfabrication, whereas now in combination with stereolithographic techniques – used as rapid tooling – hot molding is suitable for the rapid manufacturing of ceramic components. Even product designs can be realized that would be impossible to produce by mechanical microfabrication or too uneconomic if the LIGA technique is used for tooling. The fast and flexible tooling enables the hot molding process to produce functional test models or to modify the design in a few days. As a great variety of suitable ceramic powders is commercially available, the process combination is applicable to produce microcomponents which may meet various demands of future microsystems. Experience has already been gained for a variety of other materials, like ZrO_2, Al_2O_3/TiN, $BaTiO_3$, and PZT.

4 Conclusions

Stereolithography techniques can be used as rapid tooling in combination with low-pressure injection molding for the rapid manufacturing of ceramic components. Thus, the time for the preparation of functional models and the product development cycle can be significantly reduced. By combining these processes, it is possible to fabricate prototypes or to modify the design within one week. Due to the rapid tooling, changes in the feedstock or in the material can be realized at low costs and within a short period of time. As the hot molding may be extended up to small-lot productions with high precision and reproducibility, the problems of up-scaling are eliminated, which are typically related with the transition from prototyping to the fabrication of larger series. Newly developed techniques for the production of polymer microparts like RMPD, open up the possibility to extend the application of this promising combination in order to fabricate ceramic parts even with inner dimensions down to the 10 micron range.

Acknowledgements

The authors gratefully acknowledge the active support of the work by Dr. D. Göhring, Mr. W. Laub and Mr. B. Dörzapf, and appreciate the cooperation with Mr. R. Götzen, microTEC, Duisburg.

References

[1] E. W. Becker, W. Ehrfeld, Phys. Bl. 1988, 44 [6] 166-170.
[2] W. Bauer, R. Knitter, Nachrichten Forschungszentrum Karlsruhe 1998, 30 [3-4] 243-250.
[3] F. Nöker, E. Beyer, Keram. Z. 1992, 44 [10] 677-681.
[4] W. Bauer, H.-J. Ritzhaupt-Kleissl, J. Haußelt, Ceram. Int. 1999, 25, 201-205.
[5] W. Bauer, H.-J. Ritzhaupt-Kleissl, J. Haußelt, Keram. Z. 1998, 50 [6] 411-415.
[6] B. C. Mutsuddy, R. G. Ford, Ceramic Injection Molding, Materials Technology Series, Chapman & Hall, 1995.
[7] R. Knitter, E. Günther, U. Maciejewski, C. Odemer, cfi/Ber. DKG 1994, 71 [9] 549-556.
[8] R. Lenk, cfi/Ber. DKG 1995, 72 [10] 636-642.
[9] T. Schaller, W. Bier, G. Linder, K. Schubert, F&M 1994, 102 [5-6] 274-278.
[10] C. Hull, U.S. Patent 4,575,330, 1986.
[11] M. Greulich, M. Greul, T. Pintat, Rapid Prototyping Journal 1995, 1 [1] 20-25.
[12] M. L. Griffith, J. W. Halloran, J. Am. Ceram. Soc. 1996, 79 [10] 2601-608.
[13] J. G. Heinrich, cfi/Ber. DKG 1999, 76 [5] 29-35.
[14] T. Cord, V. Schillen, Proc. Micro Engineering '98, Symposium Fertigung miniaturisierter Teile, Sep. 15-16, 1998, Stuttgart, 39-44.
[15] D. Kochan, Proc. Micro Engineering '98, Symposium Fertigung miniaturisierter Teile, Sep. 15-16, 1998, Stuttgart, 45-50.
[16] P. F. Jacobs: Rapid Prototyping & Manufacturing - Fundamentals of Stereo-Lithography. Publ. by Society of Manufacturing Engineers, Dearborn, MI, 1992.
[17] V. Schillen, T. Cord, P. Levi, Proc. 31st Isata, Automotive Mechatronics Design & Engineering, Düsseldorf, June 1998.
[18] R. Götzen, F&M 1998, 106 [3] 134-136.
[19] R. Götzen, Proc. Micro Engineering '98, Symposium Fertigung miniaturisierter Teile, Sep. 15-16, 1998, Stuttgart,17-21.
[20] E. A. Judson, T. L. Starr, Annual Meeting Am. Ceram. Soc., May 3-6, 1998, Cincinnati, Ohio.
[21] R. Lenk, B. Alm, C. Richter, cfi/Ber. DKG 1999, 76 [4] D12-D14.
[22] T. Benzler, W. Bauer in Technische Keramische Werkstoffe (Ed.: J. Kriegesmann) Deutscher Wirtschaftsdienst, 51. Erg.-Lfg.(1999), Kapitel 3.4.8.3.

Pressure Casting of SiC Suspensions

Holger von Both[1] / Matthias Thommes[2] / Rainer Oberacker[1] / Michael J. Hoffmann[1]

[1] Institut für Keramik im Maschinenbau, Universität Karlsruhe (TH), Haid-und-Neu-Str. 7, D-76131 Karlsruhe, Germany

[2] QUANTACHROME GmbH, Rudolf Diesel Str. 12, D-85235 Odelzhausen, Germany

1 Introduction

Aqueous ceramic suspensions are used to prepare ceramic components by pressure casting. The interparticle forces between the submicron particles, which can be attractive or repulsive, are necessary to get a stable and dispersed powder slurry [1-3]. The role of these forces on the pressure casting kinetics and resulting filter cake properties was analyzed. A correlation between the filter cake structure, as well as compact properties like compressibility, porosity, pore size distribution and permeability will be discussed on the basis of these experiments. Finally, the properties will be related to the structural permeability model of Carman and Kozeny.

2 Experimental Procedure

An aqueous, electrostatically stabilized SiC suspension was used for the present experiments. This slurry has a solids loading of 32.5 vol% of SiC particles (mean particle size of 0.8 µm) plus B_4C and nanoscaled carbon particles as sintering additives [4]. The zeta-potential of the diluted SiC suspensions was measured in the pH interval from 3 to 10 by a Malvern Zetasizer 3000 (Malvern Instruments, Herrenberg, Germany). The titration was carried out by stepwise addition of 0.001 mol/l KCl solution, starting at a pH of 10. Suspensions with strongly and weakly agglomerated, as well as completely dispersed structures were prepared from these slurries by changing the repulsive potential by pH variation from 4 to 10 [5]. Pressure casting experiments of the suspensions with the above described conditions were carried out in a laboratory pressure filtration equipment under constant pressure of 1 MPa. The filtration pressure, filtration time, filtrate volume and mass flow were recorded during the filtration experiments.

The consolidated, liquid-saturated filter cakes were dried in a controlled atmosphere in a climatic chamber at 20 °C and 95 % relative humidity. The samples were subsequently dried in an oven at 60 °C under vacuum to remove the residual water. Dimensional shrinkage and weight loss of the filter cakes were measured during drying. The green body porosity was measured by Archimedes method after infiltrating the pores with paraffin. The porosity of the filter cakes was calculated from the porosity of the green bodies, taking into account the drying shrinkage. The porosity was used for the calculation of the permeability from the filtration kinetics [6]. The pore size distribution of the green bodies was analyzed by in- and extrusion mercury porosimetry with a Quantachrome Poremaster 60 (Quantachrome, Odelzhausen, Germany).

3 Results and Discussion

3.1 Zeta-Potential of SiC Suspension

The zeta-potential of particles in the suspension correlates with the surface charge of the particles and directly influences the interparticle forces [1-3]. Fig. 1 shows the zeta-potential of the diluted SiC suspensions as a function of pH.

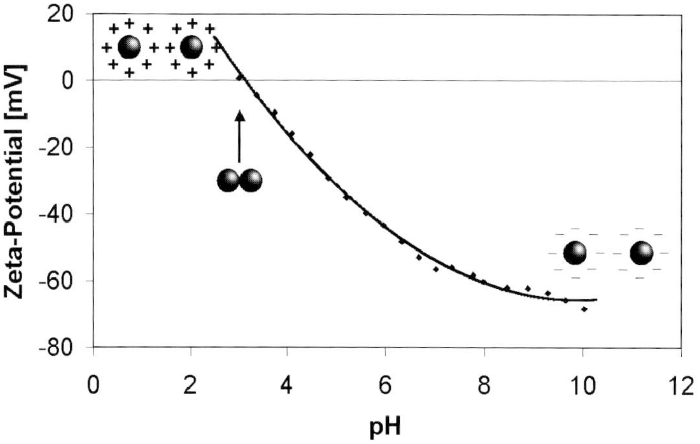

Figure 1: Zeta-Potential versus pH of the diluted SiC suspension.

For a pH below 3, the particles are positively charged. The isoelectrical point (IEP) is at a pH of 3. Above this neutral point, the suspensions show a negative zeta-potential. Between the IEP and a pH of 7, a steep decrease from 0 to -60 mV is observed. In the region above pH 7, a slight decrease from -60 to -70 mV is determined. The change in zeta-potential indicates strong differences in interparticle forces, from adhesive at the IEP to strong repulsive at low and high pH values. The resulting suspension structure is at a pH of 3 strongly agglomerated, at 5 weakly agglomerated and at 7 well dispersed [5].

3.2 Pressure Casting Kinetics

During the consolidation the filtration process can be analyzed by the kinetic plot in Fig. 2 which shows the recorded pressure casting data as time over filtrate volume (t/V_l) versus filtrate volume V_l. The shape of the curves characterizes the filtration behavior and the slope correlates with the reciprocal of the filter cake permeability [6].

The slope of the curves in Fig. 2 is very sensitive to the pH and the particle arrangement, respectively, and increases strongly from pH 4 to 10. The dispersed suspension at a pH of 7 shows a linear relationship between (t/V_l) and V_l and confirms an incompressible filter cake behavior [6]. The kinetics of the partially agglomerated suspension with a pH of 5 shows a slightly progressive curve. The physical understanding of this curvature behavior is an increase of the filter cake flow resistance with increasing cake thickness. The structural explanation is compressibility of the filter cake due to a rearrangement or fracturing of the agglomerates during the pressure casting process. The filtration process of a suspension with pH 4 is very fast and therefore only a few filtration data can be recorded so that the expected progressive curve behavior is not detected.

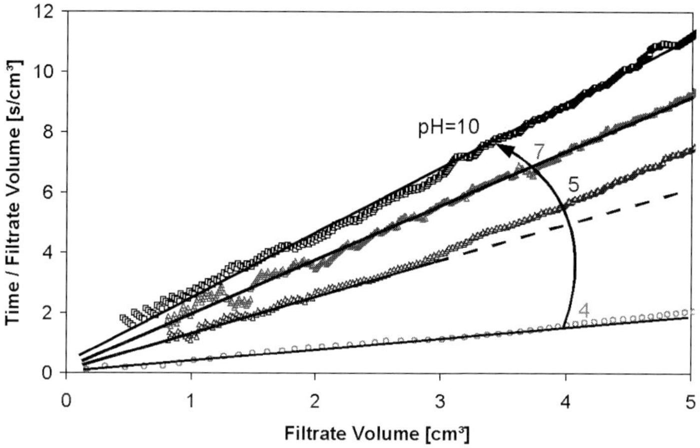

Figure 2: Pressure casting plot (t/V_l versus V_l) of SiC suspensions with varying pH from 4 to 10.

3.3 Characterization of the Filter Cake

The porosity of the filter cake as a function of pH is shown in Fig. 3. The permeability is evaluated from the slope in Fig. 2 using the equations given in [7]. With increasing pH the filter cake porosity decreases from 57 % at a pH of 4 to a minimum of 47 % at a pH of 7.8. With higher pH, the porosity slightly increases to 48 % at pH 10. The permeability of the filter cakes shows a similar pH dependency as for the porosity. A minimum of permeability ($40 \cdot 10^{-18}$ m²) is obtained at a pH of 7.4. By changing the pH to the acidic region, the permeability increases steeply and much stronger as it would be expected from the porosity data. This behavior suggest that the pore size of the dried compacts is influenced by the agglomerate size which is determined by the interparticle forces. In order to get more detailed information about the filter cake structure, the pore size distribution has been analyzed by intrusion mercury porosity of green bodies.

Fig. 4 shows a mean pore diameter of 0.13 µm for compacts consolidated with suspensions of pH 7; the pore diameter frequency distribution is relatively narrow. A green body obtained from suspensions with pH 4 shows a wider distribution with a mean pore diameter of 0.24 µm. The compacts prepared with suspensions at pH 5 and 6 confirm that an increase in pH increases the interparticle forces and decreases the agglomerate size so that the frequency distribution is shifted to lower pore diameters.

The schematic model structures in Fig. 4 illustrates a clear evidence for a correlation between porosity, pore size distribution and permeability of the consolidated compacts. A high porosity, permeability and large pore diameter in Fig. 3 and 4 can be attributed to agglomerates, which already exist in the suspension [5]. A well dispersed suspension generates a homogeneous particle packing and results in a lower porosity, permeability and smaller pore diameters.

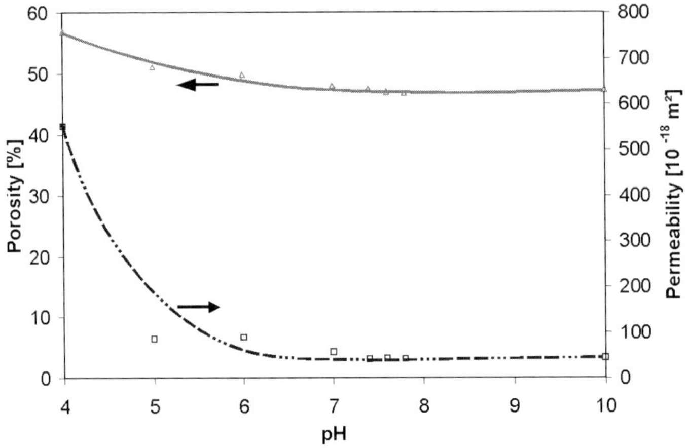

Figure 3: Permeability and porosity of the SiC filter cake as a function of pH.

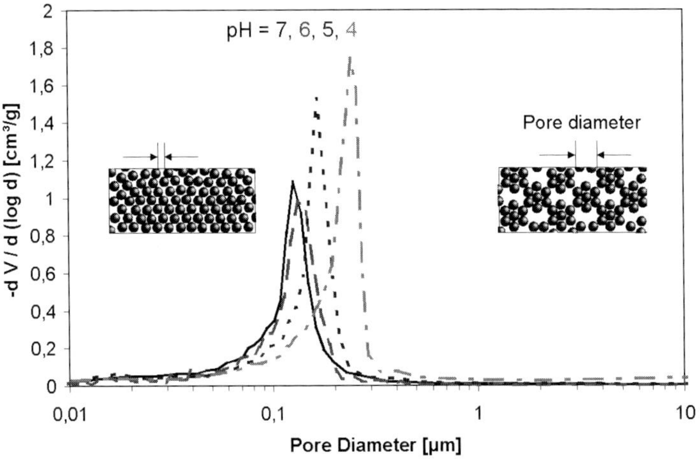

Figure 4:. Pore size distribution of SiC compacts with varying pH determined by mercury porosimetry.

3.4 Capillaric Permeability Model of Carman and Kozeny

A common model for the description of a filter cake structure is the Carman-Kozeny approach, which relates the compact porosity ε_{FC}, permeability k_{FC} and the hydraulic pore diameter D_H with a shape factor a_0 and the tortuosity factor (L_e/L) of the filtration channels [7].

$$a_0 \cdot \left(\frac{L_e}{L}\right)^2 = \frac{\varepsilon_{FC} \cdot D_H^2}{16 \cdot k_{FC}} \tag{1}$$

The structural information of the filter cake is obtained from the left side of equation (1), which gives the so-called Carman-Kozeny constant. This constant is assumed in many cases to have a value of approximately 5 [7]. Using the drying shrinkage and the assumption, that the pore diameter of the mercury intrusion porosimetry is similar to the hydraulic pore diameter, the Carman-Kozeny constant can be calculated. Fig. 5 shows that the Kozeny constant increases from 4 to 8 (from pH 4 to 7) with increasing interparticle forces.

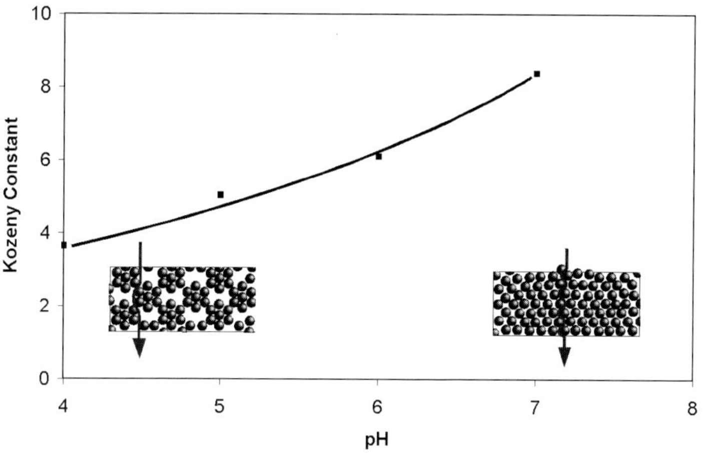

Figure 5:. Kozeny constant as a function of pH.

This indicates that the path of the filtrate flow through the filter cake is longer and therefore more tortuous for a well dispersed filter cake structure at pH 7 compared to a agglomerated one at pH 4, which is schematically demonstrated in Fig. 5.

4 Conclusions

The present results demonstrate the influence of the interparticle forces of a SiC suspension on the consolidation behavior and properties of a green compact. The aqueous SiC slurries were electrostatically stabilized and the structure properties have been systematically varied by changing the pH-value.

The pressure casting experiments indicate, that the suspension structure directly influences the filtration kinetics and filter cake properties. An agglomerated suspension condition leads to an agglomerated filter cake with high porosity and large pore diameters. For well dispersed slurries a lower porosity and a smaller pore diameter with a narrow pore size distribution is obtained. The permeability of the compacts shows a correlation between porosity and pore diameter which is combined with the tortuosity of the filtration channels estimated by the Carman-Kozeny model.

5 References

[1] Horn, R. G., Surface Forces and Their Action in Ceramic Materials, Journal of the American Ceramic Society 73 (1990), 1117-1135

[2] Fries, R., Rand, B., Slip-Casting and Filter-Pressing, in: Cahn, R. W.: Material Science and Technology, Vol. 17a (Vol.- Ed. Brook, R. J. , Processing of Ceramics), VCH Weinheim/New York/Basel/Cambridge/Tokyo (1996), 155-185

[3] Russel, W.B., Saville, D.A., Schowalter, W.R., Colloidal Dispersions, Cambridge University Press (1989), 258-309

[4] Aslan, M., Herstellung großvolumiger SiC-Bauteile über kolloidale Formgebungsverfahren, in: Ziegler, G, Werkstoffwoche 1996, Symposium 6, Werkstoff- und Verfahrenstechnik, DGM - Informationsgesellschaft (1997), 543-549

[5] von Both, H., Structural Analysis of Ceramic Suspensions by CRYO-SEM Investigations, in: to be published in the Proceedings of Euromat 1999, Symposium F 1.4, Munich, Germany

[6] Gasper, H., Handbuch der industriellen Fest/Flüssig-Filtration, Hüthig Buch Verlag, Heidelberg (1990), 33-45

[7] Dullien, F.A.L, Porous Media - Fluid Transport and Pore Structure, Second Edition, Academic Press, San Diego/New York/Boston/London/Sydney/Tokyo/ Toronto (1992), 237-317

Material and Technology Development for Microwave Sintering of High Performance Ceramics

M. Willert-Porada*
University of Bayreuth, W. Bartusch, GERO Hochtemperaturöfen GmbH
G. Dhupia
Industrie-Ofenbau Rudolf Brands GmbH
G. Müller
University of Stuttgart,
A. Nagel
Ceram Tec AG, G. Wötting, cfi GmbH

1 Introduction

In modern materials processing, extreme effort is directed to the "tailoring" of material properties as well as to an increase of process reliability at the same time. With respect to sintering, new processing methods are required, capable of precise energy deposition not only within the individual part to be sintered but also within a large batch of parts inside a sintering furnace. The "quasi optical" behavior of electromagnetic waves from the microwave frequency range led to the interest in application of microwave heating for sintering purposes, because by careful control of the interference pattern of the microwave radiation inside a furnace a homogeneous illumination of each sintering part and therefore a homogeneous heating should be achievable [1]. Furthermore, changes in the microstructure of ceramics sintered under the influence of microwave radiation were observed [2], as well as a significant decrease in sintering temperature and time [3]. Therefore, microwave sintering of advanced ceramics was expected to open up new opportunities for cost effective, large scale sintering without application of high pressure, which usually is required to obtain the desired mechanical properties in non-oxide ceramics, e.g., Si_3N_4-valves for future automotive applications or in dispersion ceramics, e.g., Al_2O_3-TiC, already used as cutting tools.

Despite quite an extended work on the development of microwave sintering of ceramics at university level since the mid-80 [4], scale up and implementation of this technology into industrial processes was not successful, mainly because of the lack of an appropriate furnace technology as well as an insufficient understanding of the process parameters governing microwave sintering. The present paper gives a brief summary of some of the results on sintering equipment development and material selection for successful microwave sintering of industrially relevant ceramics, based on the collaboration between two material producers, two furnace manufacturers and two university institutes, involved in microwave sintering for several years. The potential of microwave sintering as an industrially viable technique will be discussed for oxide and nitride ceramics.

2 Experimental

2.1 Materials selection

Commercial oxide ceramics of 94% and 99.7% α-Al_2O_3-grade were employed as well as Si_3N_4-ceramics with Al_2O_3-Y_2O_3-MgO additives. Furthermore Al_2O_3-TiC and Si_3N_4-ceramics for cutting tool applications were investigated. In order to relate the heating ability of the different ceramics by dissipation of microwave radiation to the phase composition of the ceramics, electrically conductive materials, like e.g., Si, SiC, TiC and TiN as well as ionically conductive materials, like e.g., ZrO_2 were added to the basic mixtures.

2.2 Microwave frequency selection and applicator concept

In order to investigate the dependence of heating behavior as well as microstructure evolution of a ceramic material on the frequency of radiation, besides the commercially available and industrially quite well adopted microwave frequency of 2.45 GHz a higher frequency, namely 28 GHz, was used within the frame of the project, too.

The analysis of the energy consumption of large high temperature furnaces, the need to compensate heat loss and the necessity to equilibrate the inverse temperature profile of a sintered batch heated by microwaves led to the development of multi-mode, hybrid heated, hot wall [5,6] furnaces. The multi mode cavities are made from Molybdenum or Graphite in case of inert gas sintering furnaces or from steel in case of a microwave only, suszeptor hybrid heated microwave furnace, which is used in air atmosphere up to 1700°C. In Figure 1 the industrial equipment developed within the project and in Figure 2 the hot wall furnace developed at an university is shown.

Manufacturer: GERO **Manufacturer: Brands Industrieofenbau**

Figure 1. Microwave sintering furnaces developed by industrial furnace manufacturers; left: 28 GHz/2.45 GHz microwave frequency + graphite resistance heating elements (40 kW 2.45 GHz with a 6% frequency sweep, 10 kW CW 28 GHz, 120 kW resistant heating), graphite hot wall applicator; right: 2.45 GHz (18 kW) microwave frequency

3 Results and Discussion

3.1 Microwave Sintering Technology

One major goal of microwave application for sintering purposes is the benefit from a volumetric heating to densify large ceramic parts as well as parts with a geometry prone to uneven shrinkage. As shown in Figure 2, parts of very different geometry can be sintered within one sintering cycle, providing the thermal insulation accommodates the heat loss at T>1000°C very homogeneously and the microwave field distribution is quite even. The scatter of density, found within a 49-Si_3N_4-valves load reflects mainly the inhomogeneous thermal insulation, rather than a scatter of the microwave field distribution within the hot-wall Mo-wall microwave cavity, as shown in Figure 3 from a partially sintered load.

3.2 Material selection for microwave sintering

Opposite to conventional sintering, microwave sintering adds a new parameter - the amount of microwave power - to the parameter set of a sintering process. For gas-assisted microwave sintering, according to reports in literature, the amount of microwave power is kept on a 10-20% level of total power needed for sintering of silicate ceramics [7]. For high temperature sintering, e.g., at 1600°C and above, the amount of microwave power applied may vary during the sintering cycle, depending upon the dissipation behavior of the material to be sintered, as shown in Figure 4. The temperature distribution within a furnace can therefore also be varied within quite a range, starting from microwave sintering only, which imposes a strong inverse temperature gradient – hot center, cold periphery - to the load up to a homogeneous temperature distribution, when nearly all heat loss at the surface of the load is accommodated for by conventional heating.

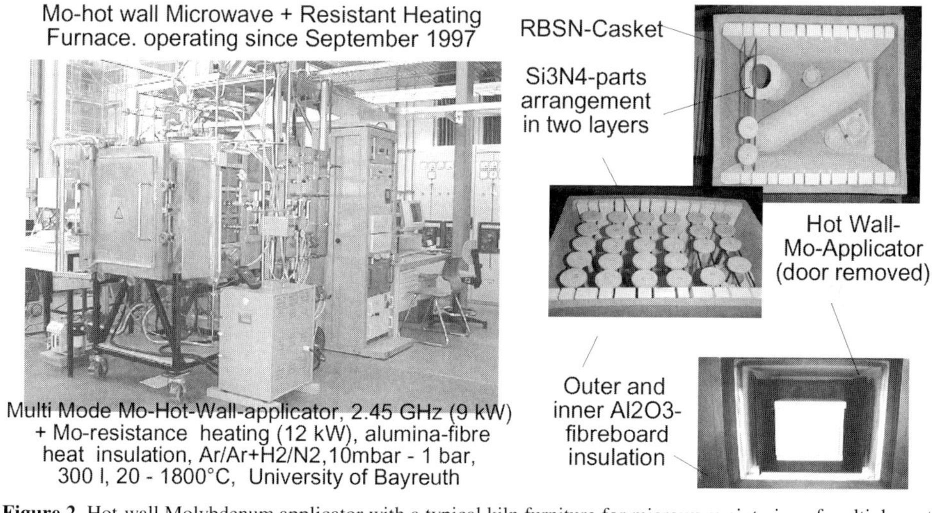

Figure 2. Hot-wall Molybdenum applicator with a typical kiln furniture for microwave sintering of multiple parts with different geometry

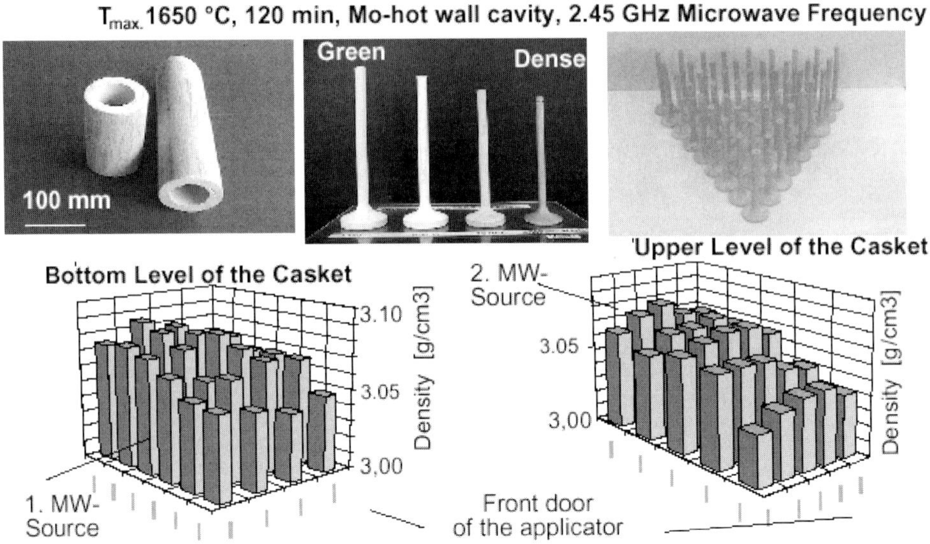

Figure 3. Results on Si_3N_4-parts densification from a partial sintering process, reflecting the heat distribution within a hot wall microwave sintering furnace, shown in Figure 2.

An example of a strong temperature gradient imposed on the sintered load is shown in Figure 4 for microwave sintering of Si_3N_4- at 2.45 GHz.

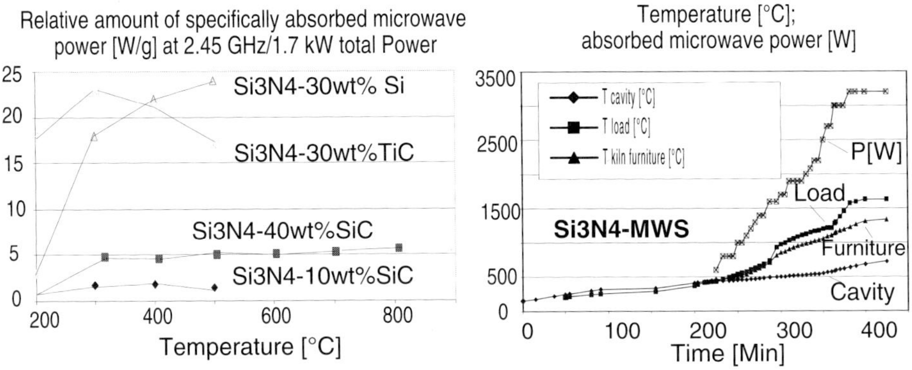

Figure 4. Relation between heating ability and phase composition for Si_3N_4-ceramics with electrically conductive additives (left); Temperature profiles during microwave sintering with a very low (750°C) amount of conventional heating, therefore with inverse temperature gradient

Besides heating a ceramic material up to very high temperatures, microwave radiation can significantly influence the microstructure and the pore size distribution as well as the overall porosity during sintering, as shown in Figure 5 and Figure 6 for alumina ceramics doped with small amounts of ZrO_2 or with silicate additives (94% pure commercial alumina).

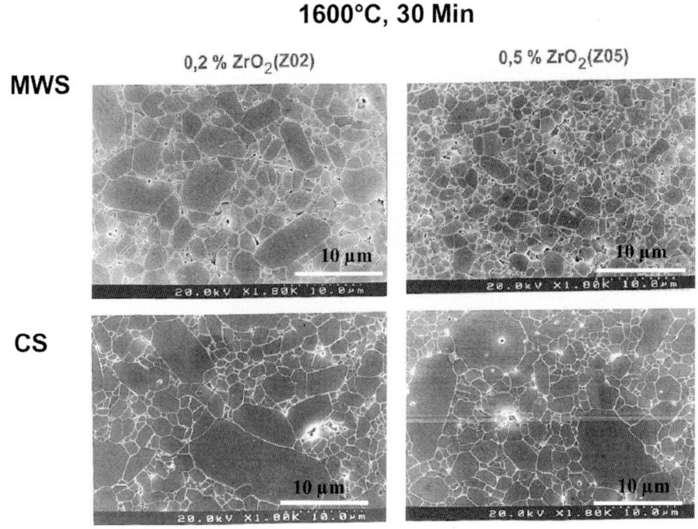

Figure 5. Microstructural difference in alumina ceramics doped with an ionic conductive additive (ZrO_2) upon (upper row) microwave sintering (2.45 GHz) as compared to (lower row) conventional sintering at nominally identical sintering schedules.

Depending upon the material to be sintered, the amount and the sequence of microwave power addition may vary during the sintering cycle in order to facilitate the development of particular microstructure features. Furthermore, the penetration depth for microwave radiation is not constant during a densification cycle, adding a further unique processing parameter to the development of microstructure within a material sintered upon application of a microwave field [2]. In case of cemented carbides this effect is extremely pronounced [8], but even in less conductive materials with rather low values of DK, like e.g. Si_3N_4-ceramics, a difference of the near surface composition as compared to the volume composition is detected, as shown in Figure 7.

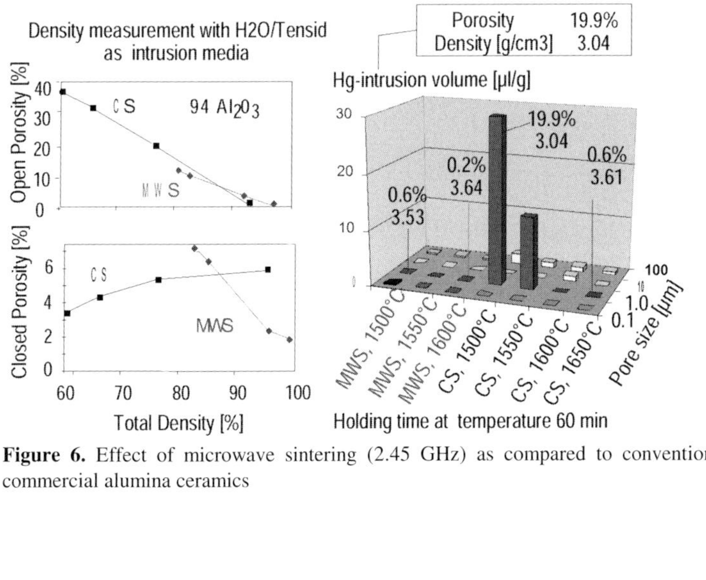

Figure 6. Effect of microwave sintering (2.45 GHz) as compared to conventional sintering on porosity in commercial alumina ceramics

In the case of 2.45 GHz microwave sintering as compared to conventional sintering differences were found in the $\alpha \rightarrow \beta$ phase transformation process and in the surface phase composition, as shown in Figure 7.

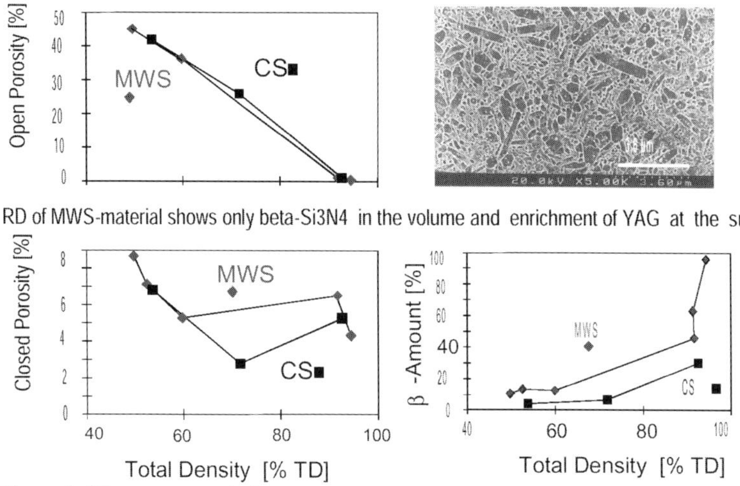

Figure 7: Microstructure of microwave (2.45 GHz) sintered Si_3N_4-ceramic and microstructural features of MWS as compared to CS

However, the conventional experiments were performed in an industrial furnace.

Until now, no *major* difference in microstructure evolution was found for microwave sintering with 28 GHz radiation as compared to conventional sintering, providing both sintering processes are performed in the same furnace. Some results are given in Figure 8.

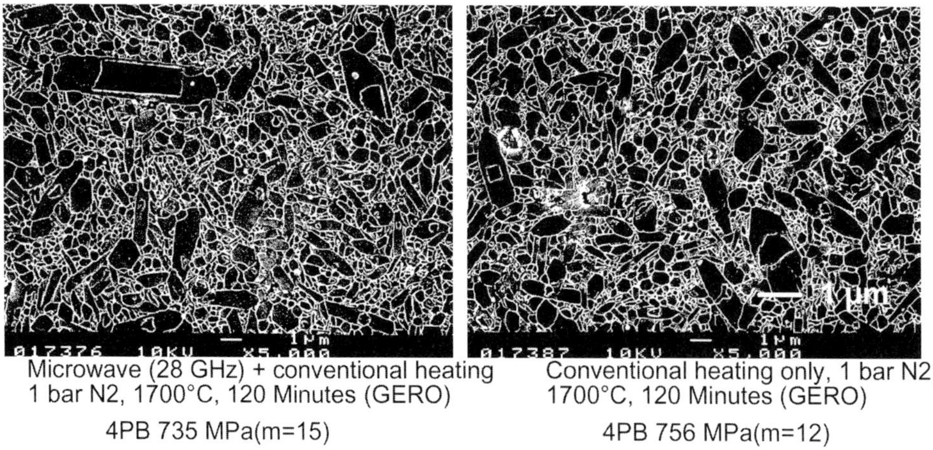

Figure 8. Microstructure and bending strength of Si_3N_4 ceramics sintered at 28 GHz and conventionally in the same furnace

4 Conclusion

New sintering equipment and the systematic study of microwave sintering on commercial high performance ceramics show this new technology to be mature enough for entering into the competition with conventional sintering processes, with respect to reproducibility and reliability criteria. In certain materials microstructural differences are observed between microwave and conventional sintering, not only as volume but also as surface effect. No limitation on part geometry and heating ability was found for larger sintering batches at the commercial microwave frequency of 2.45 GHz. Besides hot-wall electrical+microwave heating furnaces a microwave only furnace with SiC-suscepting kiln furniture has been developed as a low cost furnace for oxidizing sintering atmospheres.

5 Acknowledgment

Financial support of the German Federal Ministry of Education, Science, Research and Technology is gratefully acknowledged (Project No. 03N5010 within the Material R&D Programme 1997-1999).

6 References

[1] T. Gerdes, M. Willert-Porada, K. Rödiger, K. Dreyer; *Microwave Reaction Sintering of Cemented Carbides*, MRS. Vol. 430, 175-180 (1996)

[2] M. Willert-Porada, *A Microstructural Approach to the Origin of Microwave Effects in Sintering of Ceramics and Composites*, Ceram. Trans. Vol. *80*, 153-163 (1997)

[3] M.A. Janney, H.D. Kimrey, *Diffusion Controlled Processes in Microwave-Fired Oxide Ceramics*, MRS Vol. 189, 215-227 (1991)

[4] see Ed. W.H. Sutton, M.H. Brooks. I.J. Chabinsky, *Microwave Processing of Materials*, MRS Vol. 124, 1988 and Lit. Reported there

[5] T. Gerdes, M. Willert-Porada, *Technology Development for Microwave Sintering of Silicon Nitride Components in a Hot-Wall Applicator*, 9thCimtec 98, *Ceramics : Getting into the 2000's, [C]*

[6] P. Mehringer, G.A. Müller, W. Bartusch, *Mikrowellensintern: vom Labor zum Industriemaßstab*, cfi/Ber. DKG 76 [5], D9 (1999)

[7] F.C.R. Wroe, *Transferring Microwave Assisted Technology to the Ceramic Industry*, MRS. Vol. 347, 47-56 (1994)

Microstructure of Sinters Produced from Al-Ni-Al_2O_3 Composite Powders

Anita Olszówka – Myalska, Bolesław Formanek, Adolf Maciejny, Krzysztof Szopiński
Silesian University of Technology ul. Krasińskiego 8, 40-019 Katowice, Poland

1 Introduction

The microstructure of composites of ceramic particles - metallic matrix type forms both during the process of direct joining of components e.g. casting and sintering and during additional technological treatments, such as plastic working or heat treatment. Properties of a composite lower than theoretical ones, resulting from the mixing rule, are connected among others with the reinforcement distribution and the kind of a ceramic – metal bonding. Widely used technologies of producing this group of materials are founding technologies, however their main defect is the fact that particles tend to form bands and clusters that are difficult to be eliminated both during mixing of particles - liquid metal system, and during plastic working, whereas this is where mechanical degradation of composites is initiated and from where it spreads. The application of sintering technologies, although they require bigger expenditures, limits the above mentioned reinforcement segregation and even can eliminate it.

A separate problem is the selection of components. A classical system of ceramic particles e.g. Al_2O_3, SiC with an aluminium matrix, which is the subject of many publications, can be supplemented with new elements of the structure, i.e. intermetallic phases. Additional strengthening of the composites with intermetallic phases allows to increase the maximum working temperature, improves tribological properties and expands the range of their application [1].

One of the technological solutions which facilitate obtaining of the kind of composite discussed is joining two processes together, i.e. producing of composite powder by means of MA method (mechanical alloying) and then its sintering. Mechanical alloying allows to initially form particles distribution in the matrix, particle – matrix connection, and to generate new phases. During sintering under pressure with a liquid phase, the composite powder undergoes structural changes. Reactions take place on interfaces. As a result of the reactions between nickel and liquid aluminium particles, intermetallic phases will form from Ni-Al system, the chemical constitution and morphology of which will depend on the size of particles and sintering parameters. Thus, joining of the MA technology and sintering allows to obtain material often enriched with new phases in comparison with the introduced components, characterised by a uniform distribution of ceramic phases in the metallic matrix.

The main objective of the presented research was to define the structure of composite sinters produced by means of MA method with an aluminium matrix, the sinters being strengthened with aluminium oxide and intermetallic phases.

2 Materials and Experimental Procedure

The range of the research presented in the article comprised:

- production of Al-Ni-Al_2O_3 composite powders by means of MA method,
- defining the structure of the powders produced,
- development of powders sintering parameters and producing of a composite,
- defining the structure of composite sinters and its analysis.

For composite powders production the following initial powders were used:

- atomised aluminium powder of 71 μm granulation,
- nickel powder of 22-45 μm granulation,
- aluminium oxide powder of 25 μm granulation.

In the adopted material conception it was assumed that the fraction of the strengthening oxide phase amounts to 20-30 vol.% and the fraction of intermetallic phases formed during sintering amounts to 10-20 wt.%. It should be pointed out that the volumetric fraction of the oxide phase in the composite is bigger in comparison to the amount introduced as a component to produce composite powder because during the MA process fine-dispersive oxides are also formed.

Al-Ni-A_2O_3 composite powders were produced in a rotary-vibration mill in 6 hours at a changeable amplitude, frequency and degree of chamber filling. The method of obtaining composite powders with a dispersively strengthened matrix is presented in detail in publication [2]. The structure of powders obtained and used for further research was analysed by means of light microscopy methods and by means of scanning electron microscopy with the use of FE-SEM Hitachi 4200 equipped with EDS Voyager 4500.

Composite sinters were produced from agglomerates of up to 160 μm granulation in a Deguss high temperature press. Sintering processes with a liquid phase were carried out in graphite mould in vacuum of 10^{-5}tr, under pressure of up to 150 MPa, at the temperature 600-640°C in variable time of 15-45 min. It was assumed that with a diversified size of nickel particles and at variable time of the sintering process one can obtain intermetallic phases from the Ni-Al system, in which the chemical constitution of the intermetallic phases will be determined by the phenomena of solving nickel in liquid aluminium and by reactive diffusion. For structural research of composite sinters, as in the case of powders, methods of light and scanning electron microscopy were applied. Samples both in the form of microsections and fractures were analysed by means of SEM methods. The phase composition was determined by means of TEM by making thin foil followed by electron diffractions.

3 Results and Discussion

3.1 Composite Powders

Agglomerates in a form resembling spheres or discs (Fig. 1a) obtained by means of a rotary-vibration mill featured uniform distribution of Al_2O_3 and nickel particles introduced in aluminium. Between Al_2O_3 and Al ceramic particles as well as Ni and Al particles a mechanical – adhesive bonding was formed (Fig 1b). No important changes of Al_2O_3 or Ni particles were noticed, however the effects of the ceramic phase breaking up and nickel plastic strain were found, which was the result of MA nature.

Microscope research also showed presence of oxidizing during the formation of agglomerates. As a result numerous fine-dispersive aluminium oxides were formed. The way of formation of this new structural element is presented in the drawing (Fig. 1c). It consists in local growth of oxide particles on the surface of much developed aluminium grains followed by their detachment and dislocation within the agglomerate during MA. The presence of nickel oxide in agglomerates cannot be excluded, although no identification of this phase was made.

3.2 Composite Sinters

The composite sinters obtained were characterized by a compact structure without pores. Simultaneously, there were no cracks of the ceramic phase, which shows that the pressure and sintering time had been properly adjusted. No particles segregation in the aluminium matrix was found (Fig. 2a).

In the composite matrix there were Al_2O_3 primary particles, fine-dispersive Al_2O_3 particles formed during the MA process and new particles of a shape similar to spherical; a part of them had laminar structure where the thickness of individual layers was diversified (Fig 2b). Such diversification results from the methodology applied, i.e. observations are carried out on flat sections, as well as from differences in shapes and sizes of nickel particles themselves. The structure of a new element of the composite formed as a result of solving nickel in liquid aluminium and diffusion is also influenced by close distribution of aluminium oxide particles which limit these processes.

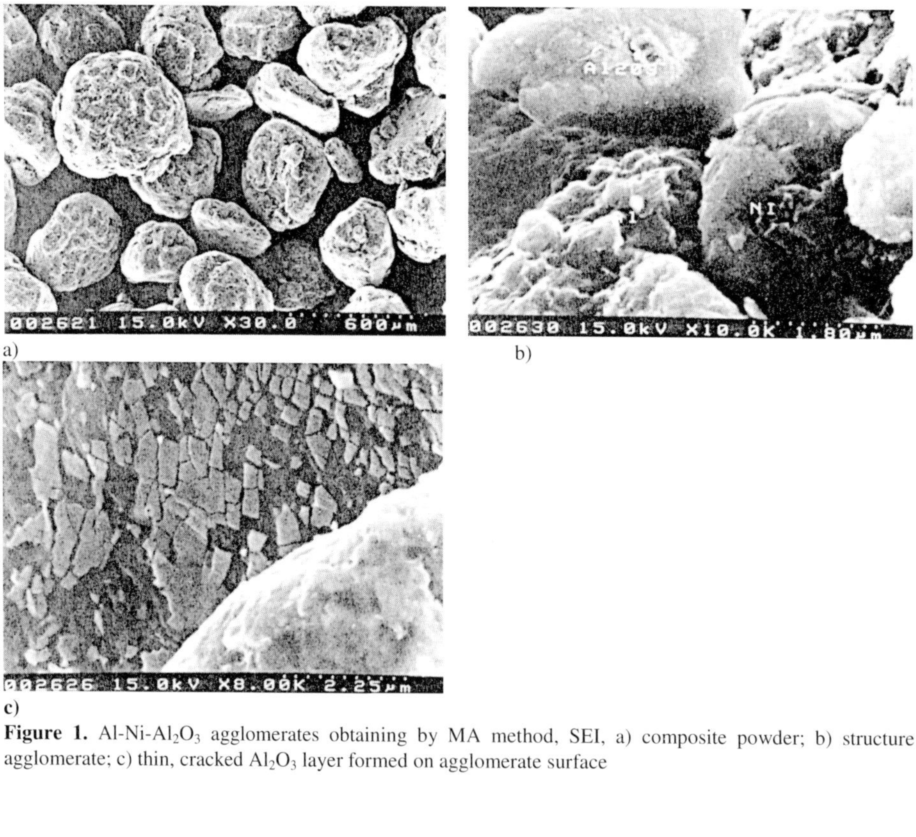

c)

Figure 1. Al-Ni-Al_2O_3 agglomerates obtaining by MA method, SEI, a) composite powder; b) structure of agglomerate; c) thin, cracked Al_2O_3 layer formed on agglomerate surface

The analysis of chemical constitution by means of EDX in microareas of the newly formed particles of homogeneous structure showed approximately 25% at. of Ni and approximately 75% at.of Al, which shows that $NiAl_3$ intermetallic phase was formed. In the case of particles of laminar structure it were observed microareas with variable concentration of Ni and Al. In I - inner zone 98.4-99.8% at. of Ni occurred. In some microareas within this zone up to 1.6% at. of Al was also detected. The results of EDX and TEM investigations (Fig. 3) showed presence of $AlNi_3$ and Al_3Ni_5 in II layer, AlNi and Al_3Ni in III - outer layer.

The laminar structure of the particles formed as a result of mutual interaction between nickel and aluminium also appeared on the composite fractures (Fig. 2c). I zone – nickel has a ductile fracture, II zone – intermetallic phase have a brittle fracture of very fine grains, and III zone – intermetallic phases also have a brittle fracture with definitely bigger grains.

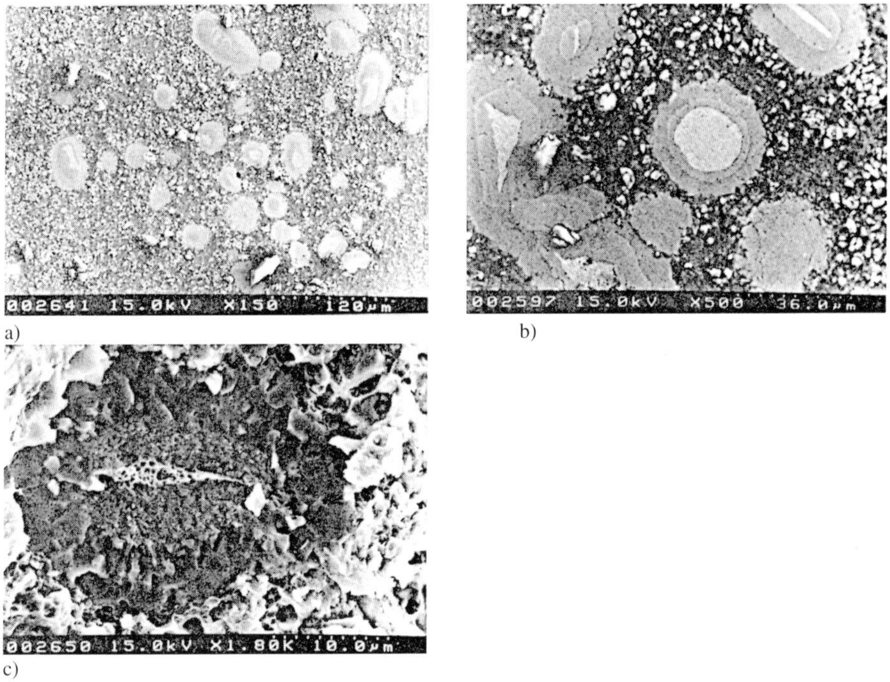

Figure 2. Structure of $Al\text{-}Ni\text{-}Al_2O_3$ sinter, SEI, a;b) polished composite; c) fracture of laminar particle

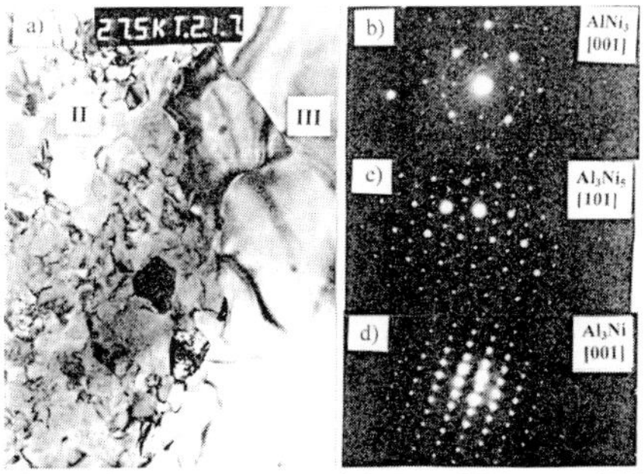

Figure 3. The laminar particle II and III zones, TEM thin foil, a)structure; b-d) diffractions

4 Conclusions

The research done showed that one can form the structure of Al-(Ni-Al)-Al_2O_3 composite sinter as regards its phase composition and the distribution of structural elements by selecting the chemical constitution and the structure of composite powders produced in MA and by selecting the temperature and sintering time.

Combining of the MA processes and sintering in a technological cycle allows to produce composite sinters with an aluminium matrix, strengthened with oxides and intermetallic phases of different dispersity.

By changing the size of nickel particles at the assumed parameters of the technology of composite formation a complex phase composition of intermetallic particles can be obtained, and it is possible to obtain particles of laminar structure. The morphology of intermetallic particles is influenced by the presence of ceramic particles (Al_2O_3) in the aluminium matrix.

5 References

1 ASM Handbook, Vol. 7, Powder Metallurgy, 1997, USA
2 B. Formanek, et al., Inżynieria Materiałowa, 1999, 110-111, 137-143

The research was financed within the project of Polish State Committee for Scientific Research (KBN) (Contract No. 7 T08C 001 16)

Processing and Properties of Fiber Reinforced Reaction Bonded Silicon Nitride (RBSN)

Andreas Herzog, Gerhard Woetting
CFI Ceramics for Industry GmbH & Co.KG, Roedental

1 Abstract

The paper describes the achievement of fiber-reinforced laminates by filament winding technique using a silicon/siloxan/solvent slurry. Correlation between slurry and laminate properties is shown. Conclusions to optimize the slurry referring to laminate values are drawn. Reactions during the RBSN-process between matrix constituents and fibers as well as properties of the received composites are discussed.

2 Introduction

Because of the outstanding high-temperature strength and the oxidation resistance RBSN is expected for use in high-temperature structural components. The limiting factor for use in engineering applications is the lack of crack resistance resp. its brittle behavior and consequently a catastrophic failure while overloading.

During the last two decades the introduction of second phases like fibers, whiskers, platelets into ceramics has been approached to achieve a quasi ductile behavior. As it is known the processing of ceramic matrix composites (CMC) is mostly demanding a firing step with temperatures above 1400°C to obtain a dense matrix or convert the matrix into a ceramic. The high-temperature step often causes serious damage in ceramic fibers connected with a loss of mechanical properties. To avoid fiber damaging temperatures above 1400°C, RBSN is a promising CMC matrix candidate as formation of RBSN can be completed between 1200-1400°C, though there remains some residual porosity.

3 Experimental Procedure

For filament winding technique applied, primarily a stable slurry with high solid load is necessary. The slurry was prepared by ball milling of a silicon powder (Bayer) in different organic solvents. To avoid impurities balls made from Si_3N_4 were used. Afterwards siloxan was added which offers the application of standard filament winding techniques normally used for fiber-reinforced plastics to prepare ceramic pepregs. Viscosity was specified by efflux method (DIN 53211) and fixed to a certain level by addition of solvent.

Different types of continuous fiber tows (Toray 800, UBE Tyranno ZMI, $SiBN_3C$ Bayer fiber) were conducted through a slurry bath, infiltrated by the slurry and laid up over a rotating cylindrical mandrel. After evaporation of the solvent the dried prepregs are held together by the polymeric binder. The prepregs were cut and piled up in 0°/90° fiber orientation to achieve a more isotropic material. Laminates were achieved by consolidation in an autoclave at 230°C. At this temperature the polymer lowers viscosity by melting and causes interconnection of the prepreg as well as the curing of the siloxan. For more detailed description of this process refer to [1].

The nitridation of the silicon based matrix was carried out in N_2 atmosphere at 1400°C. During the firing step the siloxan is converted to a SiOC-glass and interaction between filler and volatizing species of the siloxan are possible [2, 3].

The damage of C-fibers was assessed by determination of total and free carbon content of the nitrided laminates. The free carbon content was determined by burning off the carbon as described by DIN 51075. At 1800°C total carbon content was determined by oxidation in the automatic analyzer CSA 2003.

Strength degradation of SiC-fibers were examined by SEM inspection of the fracture mirrors after tensile testing. Fracture mirror show two different areas - a smooth and a scattered fracture profile, following to eq. 1 [4]

$$\sigma = 3.5(E_f \Gamma_f / r)^{0.5} \qquad (1)$$

The radius r of the smooth area, Young's moduls of the fiber E_f and the fibre fracture energy ($\Gamma_f = 20\ J/m^2$) were used to calculate the in-situ fiber strength. Fibers with no fracture mirror or several fracture mirrors have not been taken into account [5].

Mechanical testing of bending strength and interlaminar shear strength (ILS) were done according to ENV 658-3 and 5 in a 3-point bending test.

To lower the porosity of the CMC they were vacuum-infiltrated by siloxan/organic solvent mixture. Afterwards CMC's were dried and in a following firing step up to 1300°C the siloxan was converted to a SiOC-glass.

Cyclic oxidation tests were carried out at 1200°C in air. The samples were removed from the furnace after 6min, 44min, 4h, 24h, 144h and cooled down very rapidly in air. After measurement of weight change they were replaced in the furnace.

4 Results

4.1 Slurry and laminate properties

To achieve prepregs with low porosity it is desirable to have slurries with high solid content and low viscosity. To fulfill this aim different solvents were tested for slurry optimization. Two different groups of solvents were found which allow high (cyclohexan, xylol) and low (isopropanol, butanol, ethoxyethanol [EOE]) solid contents at the same efflux time.

The transfer of slurry properties to laminate properties can be seen in *Fig. 1*. For two butanol slurry based laminates the deviation of porosity and fiber volume content can be estimated to 5vol%. When fiber volume goes up porosity arises. This reversal tendency makes it difficult to assess improvements in laminate properties, as high fiber volume content and low porosity is demanded. To overcome this problem and to compare different laminates the ratio of fiber volume content to porosity was calculated. The higher this value the better the laminate. The deviation of this value is 0.05 for the same slurry and winding parameters. Best properties were achieved by xylol slurry. Referring to the deviation of the fiber volume/porosity ratio improvement by xylol is quite reasonable. However, as xylol is harmful to humans it was decided to use butanol slurries in further development.

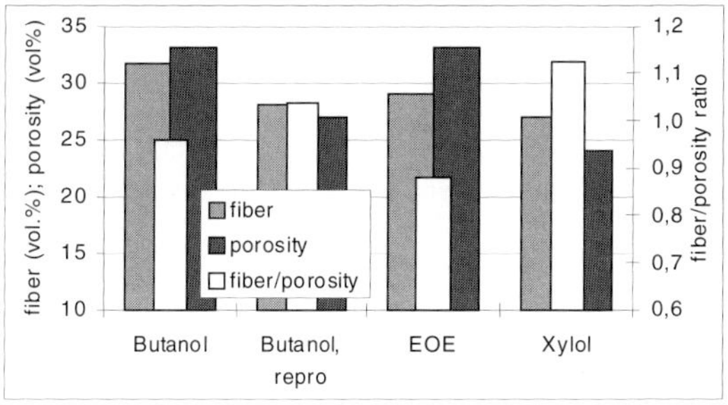

Fig.1 Influence of different solvents on laminate properties (same efflux time)

The variation of efflux time has no or little influence to laminate properties, *Fig.2*. There is a slight but unsteady tendency of lower fiber content and porosity with rising viscosity of the slurry.

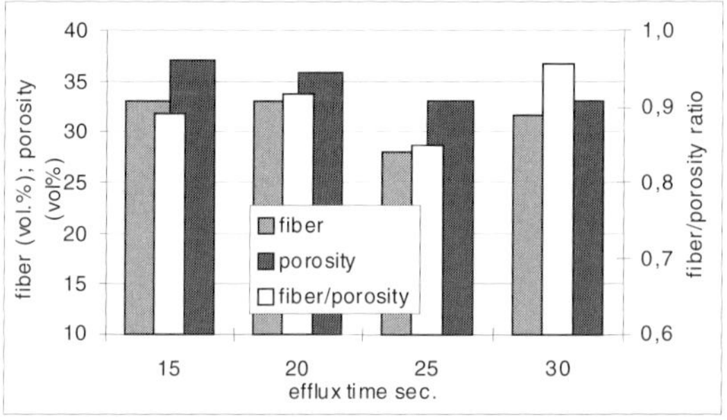

Fig.2 Influence of different efflux time on laminate properties (all butanol slurry)

Changes in slurry composition by partially substitution of silicon by siloxan is shown in *Fig 3*. When siloxan content is pushed up porosity decreases rapidly. Unfortunately, high siloxan content leads to high volume shrinkage of the siloxan during pyrolysis which causes porosity and decreases reaction kinetics during silicon nitridation caused by glass formation.

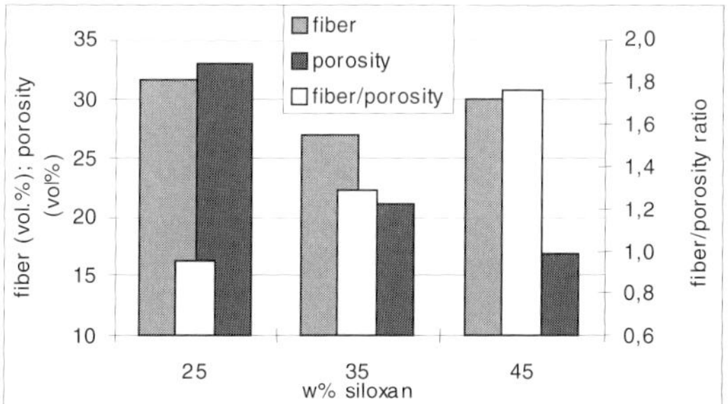

Fig.3 Influence of siloxan content on laminate properties (all butanol slurry)

Results of optimization of C-fiber laminates transferred to SiC and $SiBN_3C$ fibers and the role of interface can be seen in *Fig.4a*. As fiber surface is changed from pyC coating to epoxy sizing the fiber volume content rises while porosity remains. The fiber volume/porosity ratio is improved. An even better increase can be recognized for SiC fiber laminates. Improved wettability of non-coated fibers seems to be responsible for the increase. Furtheron, pyC coated fibers sometimes stick together and cannot be properly infiltrated by slurry. This can be seen in micrograph A, *Fig. 4b*, where the larger pores are caused by not perfectly separated pyC-coated fibers. Laminates B and C non pyC-coated fibers were used and consequently large pores disappeared.

Comparison between C-fiber and SiC-fiber laminates indicates a further step forward in laminate optimization. Porosity of SiC laminates is as low as 22% and fiber volume fraction as high as 52%. First experiments with the new Bayer $SiBN_3C$ fiber show laminate values in the range of C-fiber laminates.

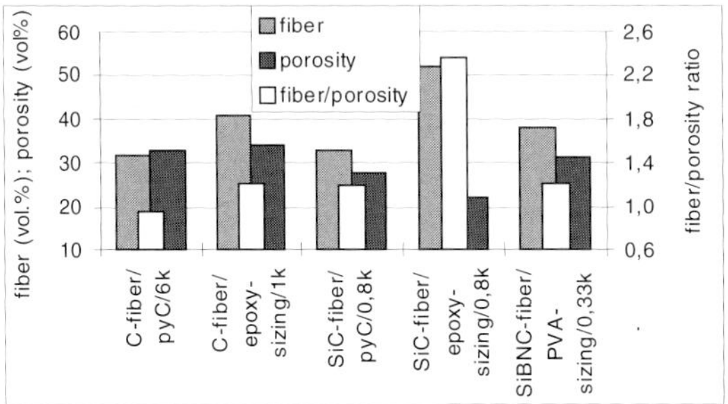

Fig.4a Influence of different interfaces and fibers on laminate properties (all butanol slurry), notice different number of filaments in a fiber tow

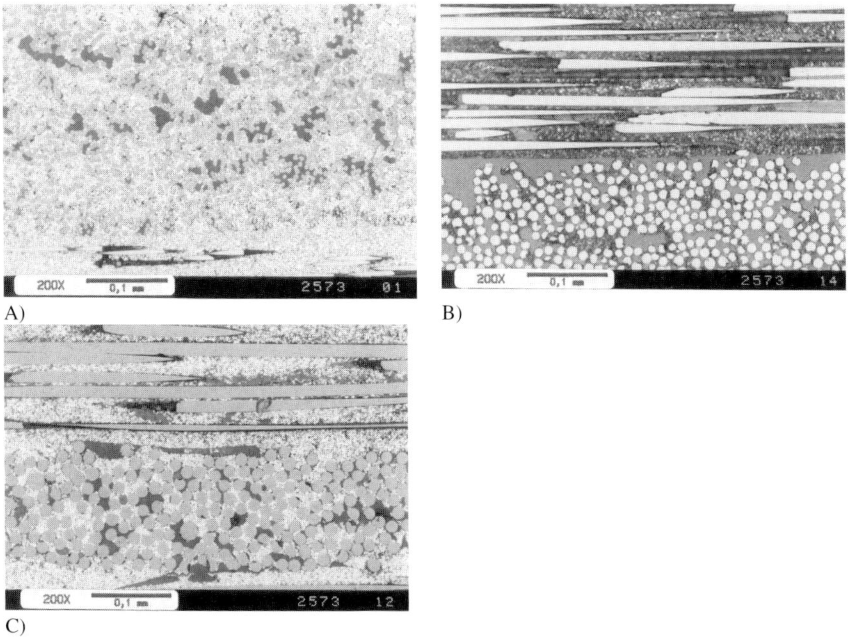

Fig. 4b A: C-fiber laminate B: SiC-fiber laminate C: $SiBN_3C$-fiber laminate

4.2 Nitridation

Thermodynamic calculation of the SiOCN system indicates a tendency of SiC formation above 1200°C, *Fig.5*. Phases Si_2ON_2, Si_3N_4 and carbon are suppressed. To the real system this means an attack of C-fibers and formation of SiC.

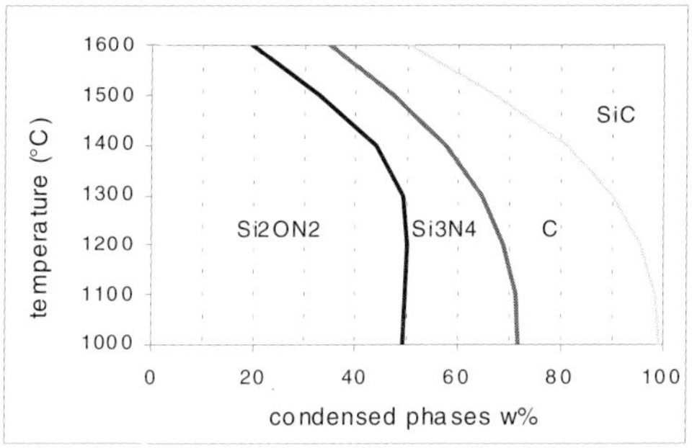

Fig. 5 Thermodynamic calculation of SiOCN-system vs temperature

In *Fig.6* the degree of reaction, free and total carbon content of C-fiber and SiC-fiber laminates after standard and improved nitridation can be seen. Degree of reaction and free carbon content after standard nitridation is much lower than after improved nitridation. Difference between free and total carbon after standard nitridation indicates the formation of SiC as predicted by thermodynamic calculation. Even pyC-coating cannot protect C-fibers from SiC formation, as can be seen by the difference of total and free carbon and the SEM picture. The SiC coating around the fiber is much thicker than the originally pyC coating of 170nm. After improved nitridation there is no evidence of SiC formation in C-fiber composites with and without pyC-coating as free and total carbon content are equal.

For both nitridation programs a large difference to 100% total carbon was measured. This difference can be explained by volatizing species like N_2 from PAN carbon fibers and CO by carbothermal reduction of SiO_2 from the SiOC-glass.

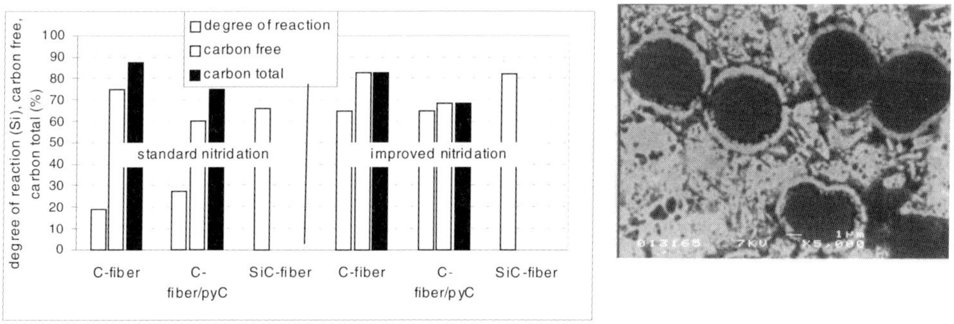

Fig.6 Degree of reaction, free and total carbon after standard and improved nitridation programs, SiC-shell around C-fibre after standard nitridation.

The monitoring of the reaction bonding process by XRD, *Fig. 7*, proves SiC formation during standard nitridation. At 1300°C first SiC peaks can be detected but no Si-N-phases have been established yet. During standard nitridation these peaks are growing and getting sharper. After improved nitridation SiC peaks are smaller and seem to refer only to the SiC formation in the SiOC-glass.

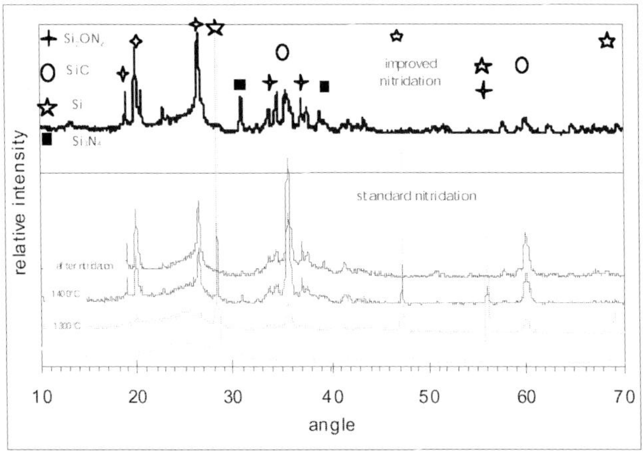

Fig.7 XRD-files after different nitridation temperatures and programms

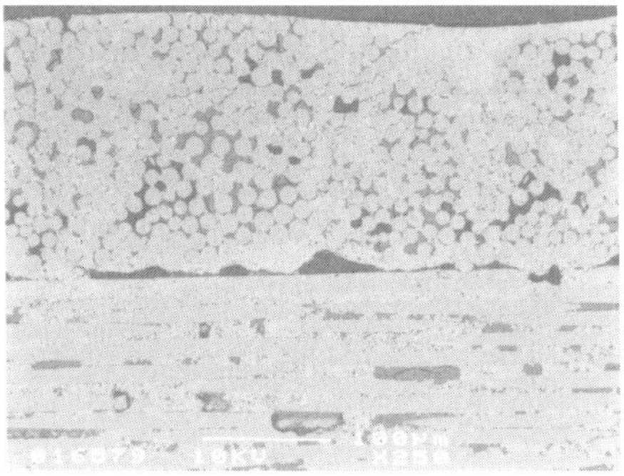

Fig. 8 SiC-fiber laminate

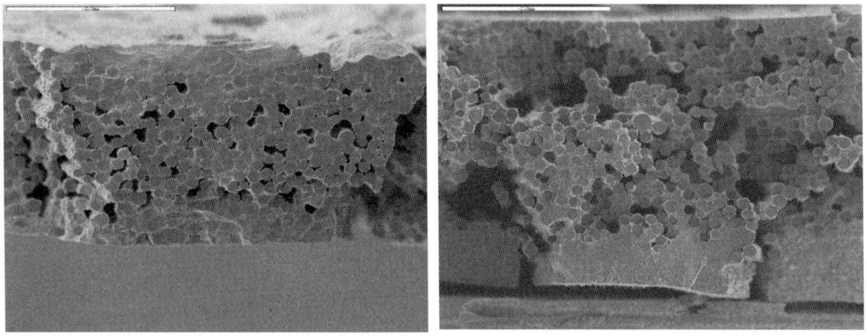

A) B)

Fig. 9a/b SiC-fiber laminate fracture profile after infiltration step 1. A: no coating, B: pyC-coating

4.3 Properties

Mechanical properties of SiC-fiber laminates can be seen in Tab. 1. As porosity and strength values were not satisfactory after nitridation siloxan infiltration was done. Depending on number of infiltration steps the porosity decreases, *Fig. 8*. Bending strength and ILS show a slight rising tendency as porosity is lowerd. Only laminates with a pyC fiber coating show fiber pull out, *Fig. 9*.

Table 1 Properties of SiC-fiber laminates, improved nitridation

Infiltration step	Porosity (%)	Y-Modulus (GPa)	Strain (%)	Bending strength (MPa)	ILS (MPa)
0	25	89	0,08	70	4
1	23			90	5,5
3	13			111	6,5

Inspection of the fiber mirror reveal a slight degradation of fibers referring to Weibull modulus. Between 900°C and finish of improved nitridation fiber strength remains constant, *Fig. 10*. After standard nitridation fiber strength could not be measured and indicates either fiber weakening or different reasons as suggestet by [5].

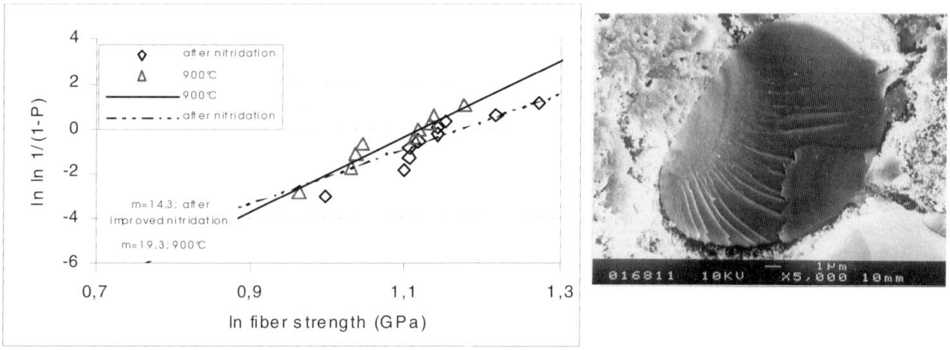

Fig. 10 Fiber fracture mirror and strength distribution at different temperatures

Oxidation stability of as nitrided SiC-fiber laminates is primarly poor compared to RBSN, *Fig. 11*. Siloxan infiltration improves oxidation stability but changes oxidation behavior. Infiltrated laminates show at first a weight loss after exposure to temperature. Laminates and RBSN show parabolic oxidation behavior. While RBSN has a very strong asymtotic behavior, laminates show a constant low weight gain within the oxidation time.

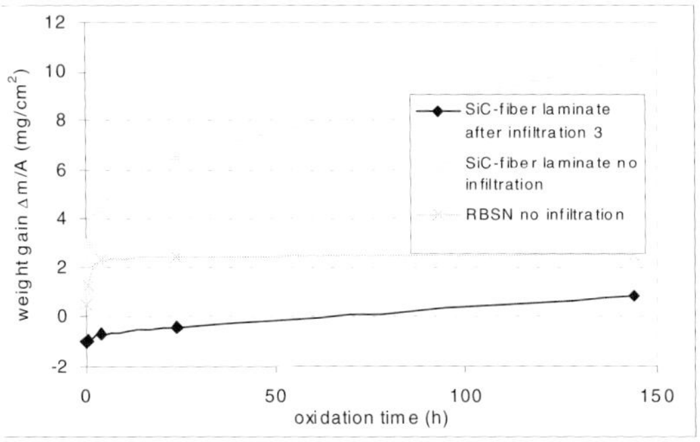

Fig. 11 Oxidation behavior of SiC-fiber/RBSN laminates at 1200°C, air

5 Discussion

It is demonstrated that standard filament winding technique can be used to produce continuos fiber reinforced RBSN-matrix composites. Specification of the slurries has to occur with respect to the demand of boiling point of 130°C as well as the harmful nature of the solvents. Wettability of the fibers is a further factor to improve laminate properties and is mainly influenced by fiber finish. This factor rules the transfer of the results of C-fiber laminates to SiC- and $SiBN_3C$-fiber laminates.

Laminate fracture profiles demonstrate the absolute necessity to apply a pyC-interface to achieve fiber pullout and quasi-ductile mechanics. This means to compromise between wettability and pyC-coating as pyC-coated fibers are not as properly infiltrated as epoxy sized fibers [6].

As solvent has to be removed after filament winding, a certain porosity always remains and probably cannot be pushed below 20vol% by this technique. To fill this porosity several infiltration steps might be applied, but even after 3 infiltration steps porosity is still above 10vol%. More infiltration steps are of decreasing effect as well as cost and time intensive. Another promising opportunity is to fill porosity by more expansive active fillers (AFCOP) i.e. boron [7, 8].

To protect C-fibers or pyC coating from SiC-formation alteration in nitiridation program and matrix composition proved to be necessary and realized. This alteration also reduces damage of the SiC-fibers as observed before by [9, 10].

As fiber volume fraction is reasonably high, low matrix strength caused by high porosity is responsible for low laminate strength. However, achieved SiC-fiber laminate strength is in the range of comparable bidirectional low fiber radius RBSN matrix composites [6]. Infiltration closes pores and therefore improves oxidation resistance but cannot improve matrix strength considerably.

We gratefully acknowledge financial support of this project by BMBF under contract MaTech 03N3038A1. Special thanks to our project partners, especially DaimlerChrysler Dornier GmbH.

6 References

[1] Lehman, R. et al (1995). "Handbook on continuous fiber reinforced CMC." Ceramics information analysis center and American Ceramic Society (ISBN 0-931682-43-5).

[2] Renlund, P., et al (1991). "Silicon oxycarbid glasses, I+II." J. Mater. Res.(6,12):p. 2716ff.

[3] Seibold, M. (1993). "Füllergesteuerte Reaktionspyrolyse polymerer Precursoren- ein neues Verfahren zur Herstellung keramischer Verbundwerkstoffe." Diss. Universität Hamburg-Harburg; VDI Verlag 1993

[4] Evans, A., Zok, F. (1994). "Review, The physics and mechanics of fibre reinforced brittle matrix composites." J. Mater. Sci.(29): p. 3857ff.

[5] Jamet, F., et al (1984). "Characterization of mechanical behavior and fractographic observations on Compglas SiC/LAS composites." Ceram. Eng. Sci. Proc.:p. 625ff.

[6] Brandt, J. et al (1992), "SiC continous fiber reinforced Si_3N_4 by infiltration and reaction bonding" Ceram. Eng. Sci. Proc. 13 p. 622-631

[7] Greil,P. (1995) "Active filler controlled pyrolysis of preceramic polymers", J. Am. Ceram. Soc. 78,4 p. 835-848

[8] Suttor,D. et al (1997). "Fiber reinforced CMC with a polysiloxane/boron derived matrix." J. Am. Ceram. Soc.(80, 7): 1831ff.

[9] Wroblewska, G.H et al (1995) . "Reaction bonded Si_3N_4 reinforced with continous SiC fibers" Processing and interface characteristics, Proceedings of High temperature CMC II, St. Barbara 1995 p. 131-136

[10] Hyuga, O. et al (1997). "Mechanical and thermal properties on the Si_3N_4 composite with SiC fibers." Int. symp.on ceramic materials and components for engines 1997: p.551-5.

A Novel CMC Material for Train Brake Systems

Zbigniew S. Rak
Netherlands Energy Research Foundation ECN, Petten

1 Abstract

A continuous carbon fiber/silicon nitride matrix composite material has been produced by an inexpensive method. According to this method, the space between 3D carbon fiber preform is filled with a S_3N_4 powder by a pressure infiltration method. High particle packing densities are achieved within the fiber preform in this way. The compact body is heat treated to form a porous framework without shrinkage, which is then strenghtened with an inorganic matrix synthezed from a liquid preceramic polymer. The densification degree, microstructure, thermal and mechanical properties of the composite material are characterized. The C/Si_3N_4 composite material pyrolyzed at 1300°C is considered to be a very promising material for low temperature applications such as brake discs for rapid train systems.

2 Introduction

The conventional brake equipment of a usual bogie of a high speed railway coach consists of four brake discs plus the necessary calipers, hand brake, and electromagnetic track brake of total weight app. 460 kg, made from a grey iron, which is about of 20% of the total bogie mass [1]. The weight reduction of this system was the main issue of the last years leading to the development of brake discs made from light alloys and metal matrix composites (MMC). The weight was reduced so far as to 65 kg for the aluminum based MMC discs. However, despite their low weight and their superior thermal properties, the reduced ductility elevated crack growth behavior and relatively high price shifted the attention of researchers to ceramic matrix composites (CMC). Materials of interest are the C/SiC, SiC/SiC, C/Si_3N_4, SiC/Si_3N_4 composites. Although primarily developed for high temperature structural engineering applications CMCs are potential candidates for low temperature applications such as brake discs in transportation systems, pump sealings in chemical engineering and so on [4]. Several processes, including tape casting, chemical vapor infiltration, solid and liquid infiltration followed by hot pressing or hot isostatic pressing have been employed for the fabrication of this new class of materials [2-5]. Complex shape C/SiC composites are already being manufactured by the infiltration of C preforms using molten silicon, followed by transformation by a reactive sintering process into the dense C/SiC material [4,6]. Another interesting method is based on organometallic polymeric SiC/Si_3N_4 precursors such as silazanes or carbosilazanes [3-5, 7]. Hoewever, the high number of infiltration/pyrolysis steps employed for manufacturing a relatively dense material is both expensive and time consuming.

The aim of the present work is to describe the densification degree, microstructure, thermal and mechanical properties of C/Si_3N_4 composites manufactured by an inexpensive method, consisting of a ceramic slurry infiltration and liquid polymer infiltration followed by a pyrolysis step, and to compare these properties to the properties of CMC materials manufactured by other routes.

3 Experimental Procedure

3.1 Starting Materials

The 2D ceramic fiber plain weave supplied by Amoc P25 Carbon FMI Composites Ltd, Scotland, was used in the preparation of the fiber preform. Silicon nitride powder, UBE SN-10, from UBE Europe GmbH, Germany, with an average particle size of 0.2 μm was used as the major component of the slurry for the pressure impregnation process. Yttrium oxide powder, grade Fine, from H.C. Starck and alumina powder Alcan RA 207LS from Alcan Chemicals Europe, UK, were used as sintering additives and Darvan C from R.T. Vanderbilt, USA, as a dispersing agent. From the large variety of possible organometallic polymers, CERASET™ SN preceramic polymer, made by Lanxide Performance Materials, USA, was selected due to a high ceramic yield and formation of SiC/Si_3N_4 compounds after pyrolysis. The solvent free polymer has a low viscosity, can be processed in an ambient air environment and, what is equally important, it is available in large quantities at a relatively low cost [8].

3.2 Composite Processing

Stacks of 10 sheets of 2D woven fibers in the shape of disc 80 mm in diameter were pressed together in a steel mould and infiltrated with a Si_3N_4 slurry under a pressure of 20 bar for 60-75 min. After the infiltration process the green body was dried for 24 h at ambient temperature and calcinated at 400°C for 5 hours. Then the calcinated material was infiltrated with a liquid preceramic polymer containing 0.1 wt% dicumyl peroxide initiator at a temperature of 80°C for 1 hour. The plate disc was removed from the preceramic polymer, wiped free of excess resin, and then cured at 160°C for 1 hour. The consolidated material had a remaining open porosity of 0.1%. Pyrolysis of samples was first performed in a resistance heated laboratory furnace at a temperature of 800°C for 2 hours while the samples were kept under a pressure of 1 MPa to prevent plate delamination and then at a temperature of 1000°C or higher under a flowing nitrogen atmosphere for a further 2 h.

3.3 Characterization

The density of CERASET™ SN preceramic polymer was measured after the curing process, the phase composition measured after pyrolysis at 1200 and 1600°C, and the "char yield" in N_2, Ar and air measured by thermal gravimetric analysis (TGA) in the temperature range of 20-1000°C. The manufactured C/Si_3N_4 composites were characterized according to their density and residual porosity, microstructure, phase identification, strength in bending, fracture toughness, elasticity, thermal expansion and thermal conductivity.

4 Results and Discussion

4.1 Characterization of preceramic polymer

CERASET^{TM} SN inorganic polymer is a versatile liquid thermosetting polyureasilazane-based resin of the composition $(Si_{1.0}N_{1.0}C_{1.4}H_{5.4})_x$ [7]. The polymer is easily thermoset to a coherent solid through a vinyl crosslinking mechanism by heating with a small quantity of a free radical generator. Dicumyl peroxide of the composition $[C_6H_5C(CH_3)_2]O_2$ is commonly used for thermosetting of CERASET^{TM} SN. Thermal curing of liquid polymer to a solid can occur in times from 1 to 90 minutes and at temperatures ranging from 90 to 190°C. After the curing process the rigid solid is non-melting and does not flow or slump. The properties of CERASET^{TM} SN polymer according to the supplier are as follows [9]:

density	0.96 g/cm^3
viscosity at 25°C	5.0 cPa.s
viscosity at 75°C	2.0 cPa.s

The polymer converts to Si_3N_4 or SiC ceramics upon pyrolysis at elevated temperatures. The solid is characterized by the density of 1.12 g/cm^3. The polysilazane CERASET^{TM} SN cured at 160°C and later pyrolysed at temperatures up to 1400°C in nitrogen was mainly amorphous. The samples fired at 1500 and 1600°C crystallized to both alpha and beta forms of silicon nitride. The "char yield", an indicator of mass conversion to ceramic material, measured by TGA was about 70% in nitrogen and argon and about 88% in air. These values are slightly lower than the values given by the supplier, 75 and 95%, respectively [9].

4.2 The C/Si_3N_4 composite infiltrated with preceramic polymer

The preparation method of the C/Si_3N_4 composite material from the carbon preform by a pressure impregnation with a Si_3N_4 slurry is described in [10]. High particle packing densities can be achieved within the fiber preform using this method. Filling the interstices with a powder increases the composite density and also limits the size of crack-like voids within the matrix. The porous material after calcination at 400°C was infiltrated only once with CERASET^{TM} SN polymer. The up-take of the polymer into the open structure of calcinated C/Si_3N_4 composite was up to 10 wt %. The density of the green composite body cured at 160°C was 2.10 g/cm^3. The density and porosity of the infiltrated composite body was changed during the pyrolysis process (Table 1).

The density decreased slightly after pyrolysis at 1000°C due to the evaporation of gaseous decomposition products at temperatures between 400-700°C. An open channel structure was formed which resulted in an open porosity between 8-9 %. After the pyrolysis at 1200-1400°C the density increased again to 2.13 g/cm^3, due to the rearrangement in the Si_3N_4 structure and interaction of the matrix material with the nitrogen present in the furnace. The open porosity has mostly disappeared, leaving the closed porosity. The SEM investigation showed that some places in the composite body were not completely filled with preceramic polymer and residual pores are still visible in the composite body pyrolyzed at 1400°C. The infiltration process should be improved by applying a vacuum for de-airing before the infiltration process and by infiltration under a high pressure.

Table 1. Density and porosity changes of the C/Si_3N_4 composite infiltrated with preceramic liquid polymer after pyrolysis process.

Temperature of heat treatment, °C	Density, g/cm^3	Residual porosity, %
160	2.10	0.0
500	2.10	0.0
1000	2.05	8.6
1200	2.13	8.7
1400	2.13	9.2
1600	2.25	15.7

The carbon fibres are very well embedded in the Si_3N_4-amorphous phase based matrix and it was difficult to distinguish the silicon nitride powder particles from the surrounding amorphous phase in the composite matrix, even under a very high magnification. The crystallization of the polymer-delivered amorphous phase to nanocrystalline S_3N_4 started rapidly above 1400°C, resulting in a high shrinkage of the ceramic matrix and formation of a large number of cracks. The density of the composite increased to 2.25 g/cm^3 after pyrolysis at 1600°C for 2 h, and the open porosity also, up to 15.7%. The whole composite matrix cracked into a large number of blocks due to the stiff structure of the carbon preform. Also the separation of the matrix from the fiber is visible. More cracks were observed in the plane direction than in the direction perpendicular to the fiber plane. To avoid any influence of the excessive comprehensive stresses formed in the composite body during the crystallization of the amorphous phase to silicon nitride and intensive formation of cracks in the composite body, the composite material pyrolyzed at 1300°C was selected for testing mechanical and thermal properties. In the composites pyrolyzed at temperatures higher than 1400°C, the majority of cracks run perpendicular to the carbon fiber direction which means that the shrinkage of the matrix is bigger in the direction parallel to the C plane.

Physical, mechanical and thermal properties of the C/Si_3N_4 composite are summarized in Table 2. The density of the C/Si_3N_4 composite infiltrated with preceramic polymer was significantly lower than the density of hot-pressed composite material, 2.13 compared to 2.75 g/cm^3, respectively [10]. The presence of the amorphous phase in the matrix and the residual porosity have had the major influence on this parameter. The existence of microcracks and the residual porosity also result in a low thermal conductivity. The measured thermal conductivity value of the developed composite was significantly lower than the values reported for separate components of the composite. The values of 22 W/m.K for carbon fibers Amoco P25 and 24 W/m.K for silicon nitride monolithic ceramics are reported in the open literature [2]. As expected the coefficient of thermal expansion (CTE) of the tested composite depends on the direction of layered fibers. The CTE measured for the C/Si_3N_4 composite perpendicular to the fiber orientation was twice as high as that measured for parallel direction. This is due to the marked anisotropy of the CTE of the carbon fibers reported by the manufacturer. The room temperature flexural strength of the composite material, approximately 200 MPa, is comparable to the values of the mechanical strength reported for the C/SiC composites made by the multiple infiltration process with preceramic polysilazane polymers and silicon infiltration [4, 6, 7]. The fracture toughness, relatively high, 8.9 $MPa.m^{1/2}$, may be still improved by a more careful processing or by applying carbon fibers coated with a BN nanolayer.

Table 2. Properties of the preceramic polymer infiltrated C/S_3N_4 composite (fiber content - 26 vol.%, single infiltration, pyrolysis at 1300°C).

Parameter	Unit	Value
Bulk density	g/cm^3	2.13
Residual porosity	%	8.7
Coefficient of thermal expansion	10^{-6} K^{-1}	∥ 3.0
		⊥ 4.5
Thermal conductivity	W/m.K	⊥ 4.9
Bending strength	MPa	198 ± 10.5
Weibull modulus		18.1
Fracture toughness	$MPa.m^{1/2}$	8.9
Young's modulus	GPa	121.0 ± 2.5

The SEM investigation of the sample used for the measurement of bending strength demonstrated qualitatively the damage-tolerant behavior of the composite, showing pull-out lengths of single fibers from matrix between 20-80 μm. The crack deflection is generally achieved by a "weak" interface between carbon preform and loaded Si_3N_4 powder bonded with pyrolyzed polymer residue, which allows for the crack to be deflected between the fiber and the matrix. However, processing composites to produce weak interfaces is a major limitation in achieving composites which are strong and which are stable at high temperatures. Much higher results for the mechanical strength for the C/SiC composite manufactured by the multiple polymer infiltration process using CERASET ™SN as a preceramic polymer, even up to 575 MPa, were reported by Miller et al [8]. It is clear that it would be possible to reach much higher values of thermal and mechanical properties of the developed composites by a further optimization of the manufacturing process. A pyrolysis temperature of 1300°C was also found to be the optimum temperature for the CMC composites manufactured by the polymer infiltration process. Temperatures higher than 1400°C are not recommended for pyrolysis due to the formation of a porous SiC layer at the interfaces and crack formation. The Weibull modulus of the tested composite was quite high, 18.1. The value of Young's modulus was also quite good, above 120 GPa, much higher than the values reported for the C/SiC composites manufactured by other methods [4]. The manufacturing method developed for CMC composites, which is still far from full optimization, proved that it is possible to produce the C fibers-ceramic matrix composites in an inexpensive way, avoiding the expensive hot pressing step as well as the time and energy required by multiple infiltration process in the manufacturing procedure for C/Si_3N_4 composites.

5 Conclusions

This research, although preliminary, has demonstrated several important features about processing of CMC materials:

- According to this method, the space between the fibers is first filled with a ceramic powder, which is subsequently strengthened with an inorganic amorphous phase synthesized from an infiltrated preceramic polymer.
- The preceramic polymer viscosity of approximately 2.0 mPa.s at 80°C is sufficient for the infiltration of porous composite materials, however, a further improvement of the

infiltration process is necessary to eliminate totally the residual porosity still present in the pyrolyzed composite body.

- The application of the developed composites is restricted mainly to temperatures not exceeding 1300°C due to the intensive crystallization process of the ceramic matrix above 1400°C and hence a high shrinkage of the ceramic matrix followed by an intensive cracking of the whole composite body.
- The method of manufacturing the C/Si_3N_4 composites by a single infiltration technique with a polysilazane polymer on a C/Si_3N_4 porous preform allows for an economical production of the CMC components.

6 References

1. T. Zeuner, P. Stojanov, P.R. Sahm, H. Ruppert, and A. Engels, Developing trends in disc brake technology for rail application, Materials Science and Technology, 1998, 14, 857-863.
2. K. K. Chawla, Ceramic Matrix Composites, Chapman & Hall, London, 1993
3. G. Ziegler, Advanced ceramic developments trends, cfi/Ber, DKG, 1991, 68, 399-404.
4. Mühlratzer, Production, properties and applications of ceramic matrix composites, cfi/Ber. DKG, 1999, 74, 4, 30-35.
5. F.F. Lange, Fiber and laminar composites with strong interfaces, Silicates Industries, 1998, 63, 3-4 33-43.
6. S. Simmer, B. Derby, The processing of novel reaction bonded SiC ceramics using alloyed silicon infiltrates, Fourth Euro-Ceramics, Ed. by A. Bellosi, 1995, 4, 393-400.
7. J. M. Schwark, M. J. Sullivan, Isocyanate-modified polysilazanes conversion to ceramic materials, Mat. Res. Soc. Symp. Proceed., 1992, 271, 807-812.
8. D.V. Miller, D.L Pommell, G.H. Schiroky, Fabrication and properties of SiC/SiC composites delivered from Ceraset™SN Preceramic Polymer, Ceram.Eng.& Science Proc., 1997, 18.
9. CERASET™ SN Inorganic Polymer, Information from Lanxide Performance Materials Inc, February 1997.
10. Z.S. Rak, Fiber reinforced CMC with organometallic polymers, Br. Ceramic Proceedings, 1999, 60, 1, 393-394.

Reactive Spray Forming as a Novel Process for the Production of Ceramic Components

Meinhard Kuntz, Stephan Dierkes and Georg Grathwohl
Universität Bremen, FG Keramik, Sonderforschungsbereich Sprühkompaktieren

1 Introduction

Spray Forming is a new process of increasing importance for manufacturing metallic components with a very fine grained and homogeneous microstructure. The process enables manifold opportunities for an efficient production of composites or alloys. Today, spray formed products made of steel, copper and aluminum are in use - the latter requires higher expenditures due to its reactivity which is nevertheless successfully applied even in large-scale facilities [1].

In this study, a new approach is proposed in which the advantageous features of the spray forming process are exploited in order to produce ceramic materials. Molten reactive metal alloys together with coreactant particle injection in an oxygen-rich atmosphere are sprayed and react in-situ to a ceramic material. In a first approach, Al-Mg alloys are used together with particle injection of MgO, which yields a ceramic compound with a high amount of spinel ($MgAl_2O_4$) as the dominating reaction product.

Related procedures for the transformation of metallic precursors to ceramic components are Reaction Bonding of Aluminum Oxide (RBAO) [2,3], Reactive Metal Penetration (RMP) [4] and Directed Metal Oxidation (DIMOX) [5,6]. At the RBAO-process aluminum is mixed with alumina powder and refined under inert conditions to particle sizes smaller than 1μm. This is the critical particle size in order to overcome the passivation of Al. The mixture reacts in oxidizing atmosphere at temperatures of 500°C - 1100°C with an reaction yield of more than 95%. The volume increase during oxidation compensates sintering shrinkage.

In the RMP-process the aluminum is mixed with an „oxygen supplier", e.g. TiO_2. The resulting compound consists of Al_2O_3, Al and an intermetallic phase of TiAl. At this process the amount of new formed ceramic depends on the oxygen supplier but is always much lower than 100%.

DIMOX is a further well known process for the manufacturing of ceramic components. Using an Al-Mg-alloy a double layer of spinel and magnesia is formed and enables oxygen diffusion to the reaction surface. By this mechanism the passivation of the Al is completely surmounted thus enabling complete reaction even of large bulk components. There is an incubation time until the process is started, whereas by using MgO-powder as a dopand the process is started immediately [7]. Thus, a high amount of MgO is favorable for the reaction kinetics and also the amount of spinel formed.

Spinel is a material with an attractive combination of properties. High melting point, hardness, chemical inertness and high thermal shock resistance due to a low coefficient of thermal expansion and its cubic crystal lattice. It is produced e.g. by mixing and sintering of magnesia and alumina powders [8,9] or by hydrolysis and calcining of Mg-Al-precursors at relatively low temperatures (400-600°C) [10,11]. The high sintering shrinkage, which is unavoidable by these processes, is overcome by liquid-Mg-infiltration of porous alumina bodies in oxidizing atmosphere [12].

In the following, the new concept of reactive spray forming is explained and the results of experiments are presented. It is shown that spray conditions are suitable for ending up in ceramic components with attractive properties.

2 Principle of the Process

Figure 1 shows the experimental setup. The metal is supplied in wire form, where pure Al, Al-Mg-alloys or particle reinforced composite wires are used. The assembly is mounted in a water-cooled, vacuum-tight chamber which is equipped with facilities for observation and additional components. Before each experiment the chamber is evacuated and sweeped with the process gas in order to adjust the intended oxygen content. The end of the wire is molten by a laser beam of up to 2 kW and immediately atomized by the gas jet. The gas is a mixture of Ar and O_2 with arbitrary O_2 content. Even 100% O_2 can be supplied. The particle injection is placed in a way that the particle streams coincide during the time of flight. The spray cone is focused to a diameter of appr. 30mm.

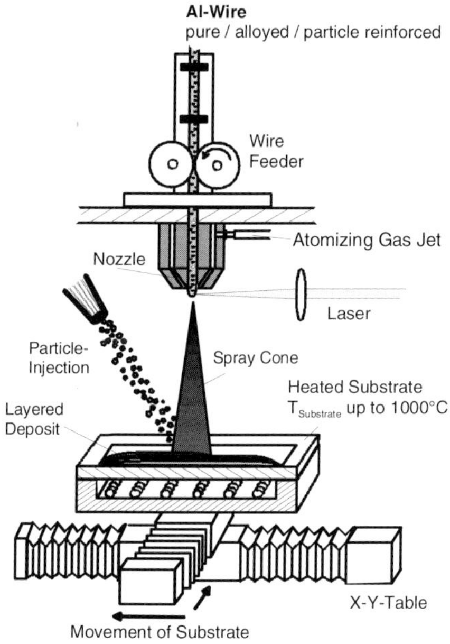

Fig. 1: Spray forming setup

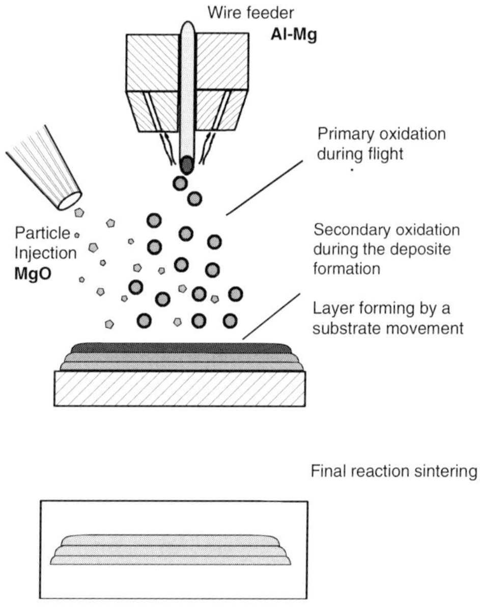

Fig. 2: Schematics of the reactive spray forming process

The substrate is heated by resistance heaters up to 1000°C. A computer-driven xy-table enables arbitrary lateral motions. An area of $100 \times 100 mm^2$ is sprayed in rows with a speed of 5-20mm/sec. By one complete cycle the deposit grows about 0.5-1mm depending on table speed and mass flow. The thickness of the deposit is not limited and usually chosen to 3-5mm. There is only little overspray of appr. 20%.

The conversion of the reactants (Al-Mg, MgO, O_2) performs in 3 steps, as it is schematically explained in fig. 2. Primary reaction is assigned to the time of flight of the metallic particles by reaction with the process gas. The intensity of the primary reaction depends on the amount of oxygen (p_{O2}) and the pressure of the process gas, the mass flow of Al-Mg, the distance to the substrate and on overheating. Oxidation of Al is exothermal, which assists the process by „heating" the particle during the time of flight, reducing the viscosity and thus improving the deposition conditions. The average particle size is about 50-100μm - thus, a complete oxidation during the flight is not possible.

The highest amount of the oxidation rate is reached during the secondary reaction, which is assigned to the in-situ oxidation of the freshly deposed layer while the process is running. The transformation to spinel utilizes the remaining oxygen content of the process gas, which is most effective at the surface of the layer. It is thus advantageous to keep the thickness of a single layer as low as possible in order to increase the reaction kinetics. The freshly deposed particles are at a high level of reactiveness and are therefore expected to be transformed within a short time (several minutes). The reaction rate of the secondary reaction can be adjusted by the speed and temperature of the substrate and the remaining oxygen content.

If full transformation to a ceramic material can not be reached by the secondary reaction, a final step of reaction sintering in a conventional furnace under oxidizing conditions can be applied.

3 Results

Tests have been performed in order to find out appropriate material combinations, process conditions and reaction kinetics. The reaction rate, i.e. the amount of Al which is transformed to a ceramic phase, depends on the amount of MgO in the mixture. In fig. 3 the reaction rate of pressed powder pellets after annealing at 6h, 975°C is plotted as a function of MgO-content. If it is assumed that the complete material is transformed to spinel it is easily calculated that a content of 43% would yield 100% spinel. Thus, a mixture of 43% MgO and 57% Al is referred to as the stoichiometric mixture. The reaction rate steadily increases with increasing amount of MgO. However, as it is obvious in fig. 3, the increase of reaction yield is higher below the stoichiometric mixture, whereas the increase is much less pronounced at higher MgO-contents.

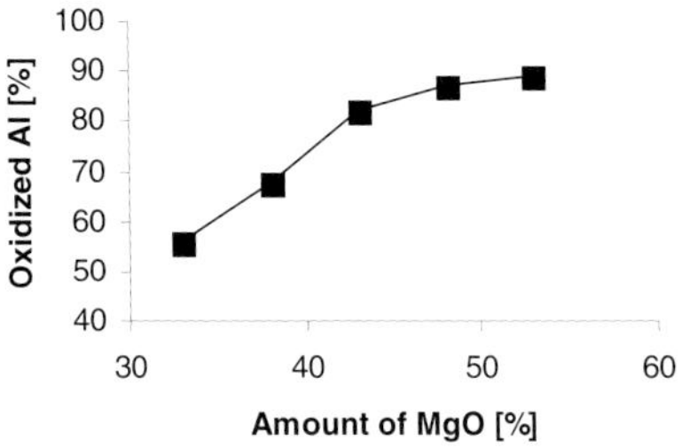

Fig. 3: Oxidation rate vs. MgO content (975°C, 6h)

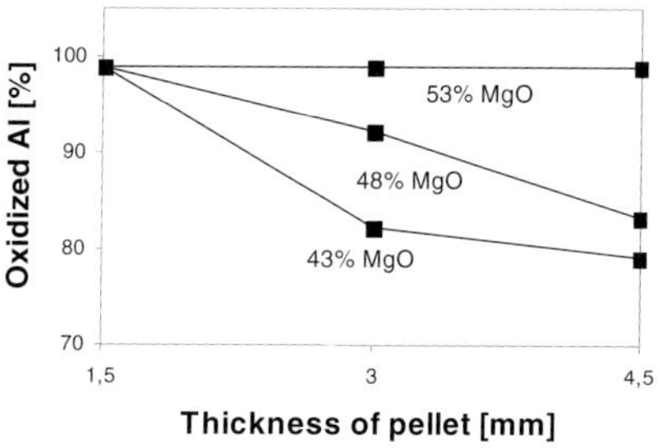

Fig. 4: Oxidation rate v. pellet thickness and MgO content (975°C, 6h + 1400°C, 12h)

As it was pointed out, an increase of the reaction rate is expected for thin layers. Therefore, pressed pellets of variable thickness and mixture were prepared (fig. 4) The specimens were first annealed at 975°C which is the starting temperature for the formation of spinel. To reach a stable condition, the specimen were finally annealed at 1400°C, 12h. At these conditions, all thin specimen reach a full ceramic state. In contrary to this, thick pellets can only be completely transformed when the MgO-content is 10% above the stochiometric mixture, whereas at lower MgO-contents a complete reaction can not be obtained.

This results were proved by X-ray diffraction. As shown in fig. 5 all peaks can be assigned to the 3 ceramic phases, i.e. spinel ($MgAl_2O_4$), corundum (α-Al_2O_3) and magnesia (MgO). Quantitative evaluation of the spectrum clearly indicates the dominating spinel-phase, whereas the amount of corundum is negligible.

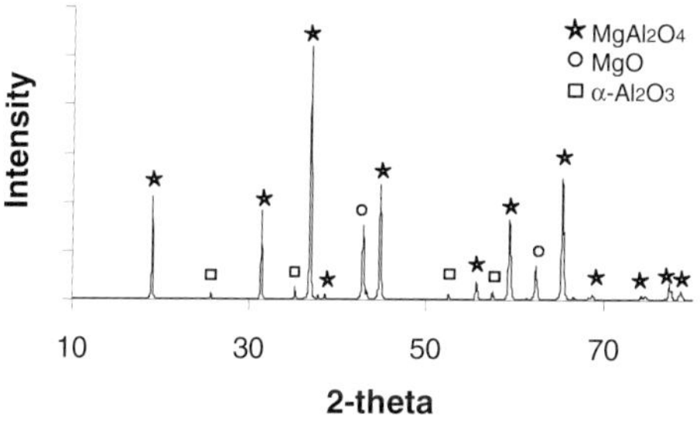

Fig. 5: X-Ray diffraction of anneald pellets (53% MgO) after 100% transformation

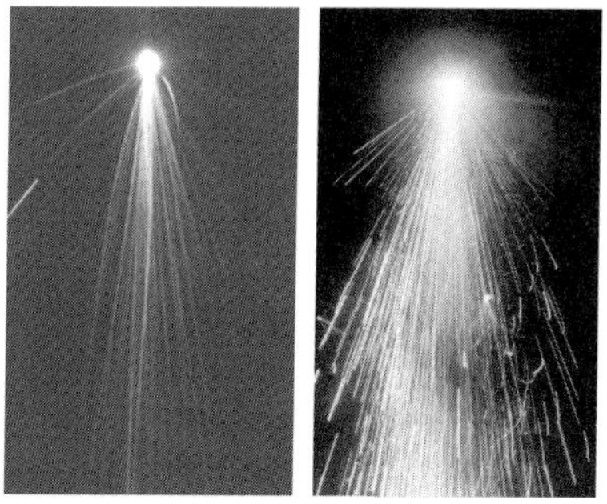

Fig. 6: Spraying of Al-Mg-Alloy by Ar+3%O_2 (left) and 100% O_2 (right)

The spray experiments in the chamber have been performed in order to set up and optimize the various facilities (laser, wire feeder, particle injection, process gas mixture, xy-table, substrate heating). Fig. 6 shows the particle jet at different process gas mixtures of 3% (left) or 100% (right) oxygen content. This figure gives an impression of the amount of primary reaction which is already present at low oxygen content but extremely high at pure oxygen. It should be noted that at a very high amount of primary oxidation the compaction rate (i.e. the ratio of the compacted to the sprayed mass of metal) decreases, because the particles are already in a high ceramic grade and thus unable to deform plastically which is a precondition for the compaction process. Thus, an optimum of the primary reaction should be reached which is as high as possible but still allows breakage of the ceramic skin.

A controlled deposition of layered deposites with MgO-content up to 30% is reached without substrate heating. A further increase of the MgO-content will be obtained when substrate heating is activated and smaller MgO-particles are used. Fig. 7 shows a typical micrograph of a spray formed compound with a MgO-content of 15-20%.

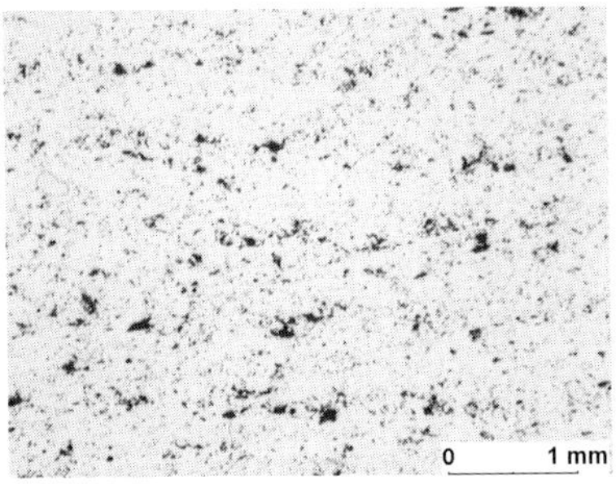

Fig. 7: Polished cross-section of a reactive spray formed compact

As a characteristic of the spray forming process a fine porosity is obtained, in particular in combination with particle injection. This will be advantageous for the final reaction step of the body. A further benefit is expected to originate from the lagered microstructure of the ceramic parts since crack deflection is found to be caused by this feature. In this case, enhanced fracture toughness of the body is obtained.

4 Summary

Reactive spray forming is a novel process for manufacturing large 2-dimensional ceramic composites. In this first study, a material based on spinel as the dominating phase is prefered because it is obtained cost-effectively and promises attractive material properties. It has been shown that transformation to a full ceramic grade material is possible even in the case of coarse grained precursors. A homogenous distribution of MgO in the Al-matrix is necessary. Thin layers promote reaction kinetics. The experimental reactive spray process as proposed here is extensively instrumented in order to enable parametric studies for the in-situ formation of ceramic components. First tests of the facilities have been successfully performed. It is now the task to systematically find out optimal spray conditions relating to mass flow, alloying, process gas mixture and pressure, substrate temperature and motion. The new process enables the production of components of extended geometries which can not be obtained by conventional processing. Moreover, the characteristics of the in-situ-produced spinel and the laminar microstructure are expected to yield increased thermal shock resistance and fracture toughness.

5 Literature

[1] H. Liu, X. Zeng, E. J. Lavernia: Processing Maps for Reactive Atomization and Deposition Processing, Scripta Metallurgica et Materialia, Vol. 29, No. 10 (1995) p. 1341-1344

[2] D. Holz: Herstellung und Charakterisierung von reaktionsgebundenen Al2O3-Keramiken (RBAO-Verfahren) am Beispiel des Systems Al2O3/ZrO2; Reihe 5: Grund- und Werkstoffe Nr.367, VDI-Verlag, Düsseldorf 1994

[3] N. Claussen, S. Wu, D. Holz: Reaction Bonding of Aluminum Oxide (RBAO) Composites: Processing, Reaction Mechanisms and Properties, J. Eur. Ceram. Soc., 14 (1994) p. 97-109

[4] R. E. Loehman, K. Ewsuk: Synthesis of Al2O3-Al Composites by Reactive Metal Penetration, J. Am. Ceram. Soc., Vol. 79, No. 1 (1996), p. 27-32

[5] K. C. Vlach, O. Salas, H. Ni, V. Jayaram, C.G. Levi, R. Mehrabian: A Thermogravimetric Study of the Oxidative Growth of Al2O3/Al Alloy Composites, Vol. 6, No. 9 (1991) p. 1982-1995

[6] M. Sindel, N. A. Travitzky, N. Claussen: Influence of Magnesium-Aluminum Spinel of the Directed Oxidation of Molten Aluminum Alloys, J. Am. Ceram. Soc., Vol. 73, No. 9 (1990) p. 2615-2618

[7] P. Xiao, B.Derby: Alumina/Aluminum Composites Formed by the Directed Oxidation of Aluminum Using Magnesia as a Surface Dopant, J. Am. Ceram. Soc., Vol. 77, No. 7 (1994) p. 1761-1770

[8] J. T. Bailey, R. Russell: Sintered Spinel Ceramics, Ceramic Bulletin, Vol. 47, No. 11 (1968) p. 1025-1029

[9] U. I. Isenmann: Spinele im System MgO-Al2O3 – Eigenschaften und Anwendungen für feuerfeste Werkstoffe, Dissertation, TU Clausthal (1996)

[10] R. J. Bratton: Charakterization and Sintering of Reactive MgAl2O4 Spinel, Ceramic Bulletin, Vol. 48, No. 11 (1969) p. 1069-1075

[11] K. B. Pflanz: Herstellung und Charakterisierung einer Spinel-Ultrafiltrationsmembran, Reihe 5: Grund- und Werkstoffe Nr.323, VDI-Verlag

[12] P. Kumar, K. H. Sandhage: The Fabrication of Near Net-Shaped Spinel Bodies by the Oxidative Transformation of Mg/Al2O3 Precursors, J. Mater. Res., Vol. 13, No. 12 (1998) p. 3423-3435

Thermal Analysis of Si-C-N Ceramics Derived from Polysilazanes

Jianqiang Peng, Hans Jürgen Seifert, and Fritz Aldinger
Max-Planck-Institut für Metallforschung and Institut für Nichtmetallische Anorganische Materialien, Universität Stuttgart, Pulvermetallurgisches Laboratorium, Heisenbergstr. 5, 70569 Stuttgart, Germany

1 Abstract

The phase reactions, the crystallisation behaviour and the accompanied thermal degradation of two Si-C-N precursor ceramics, which were derived from polyvinylsilazane and polyhydridomethylsilazane respectively, were studied by means of thermodynamic calculation of phase diagrams, DTA/TG , XRD and SEM. The CALPHAD type simulations were confirmed by the experimental results. The thermal degradation of both ceramics have been characterised quantitatively by taking into account the endothermic phase reactions $Si_3N_4+3C = 3SiC+2N_2$ and $Si_3N_4 = 3Si+2N_2$.

Keywords: precursor derived Si-C-N ceramics, phase reactions, DTA/TG, CALPHAD

2 Introduction

The synthesis of Si-C-N ceramics from precursor polymers is a new way to control the material structure and properties on an atomic scale [1-6]. According to this route polysilazane is crosslinked and then pyrolysed into the amorphous Si-C-N ceramics with a completely homogeneous distribution of the elements on an atomic scale and remaining amorphous up to 1400°C in argon or nitrogen and 1500°C in air [4, 5]. At temperatures higher than 1400°C in argon or nitrogen this amorphous state is transformed into polycrystalline ceramics with thermodynamic stable phases, such as Si_3N_4, SiC and graphite. Since the pyrolysis products can be used as both amorphous and crystalline materials, the understanding of the high temperature behaviour of the ceramics and the phase reactions, the crystallisation behaviour and the accompanied materials thermal degradation are of great importance with respect to the physico – chemical properties of the polycrystalline ceramics and to the maximum application temperature of the amorphous ceramics. Therefore, in the present work, two typical precursor derived Si-C-N ceramics VT50 and NCP200, which were derived from polyvinylsilazane (VT50, Hoechst AG, Germany) and polyhydridomethylsilazane (NCP200, Nichimen Corp., Tokyo, Japan) separately, were investigated by means of differential thermal analysis (DTA), thermogravimetry (TG) , X-ray diffraction (XRD) and scanning electron microscopy (SEM). In order to predict and understand the phase reactions, the crystallisation behaviour and the accompanied thermal degradation, the phase diagrams, phase fraction diagrams and enthalpy temperature diagram of the two ceramics were calculated by computer.

3 Results of Thermodynamic Calculations

CALPHAD (CALculation of PHAse Diagram) can be used to understand phase reactions and the crystallisation behaviour of precursor ceramics [7, 8, 9]. The phase diagrams were calculated using the Lukas software [10], ThermoCalc [11] and the database provided by Kasper and Seifert [12, 13].

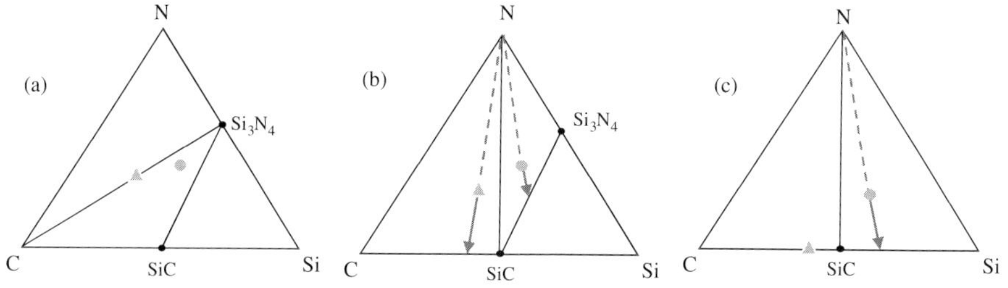

Fig. 1. The phase diagram of the ternary system Si-C-N for the temperature range (a) T< 1484°C, (b)1484°C<T<1841°C, (c) T>1841°C with indicated compositions of the amorphous ceramic VT50 (Δ) and NCP200 (O), respectively.

Fig. 1 shows the calculated isothermal sections of the ternary system Si-C-N at a total pressure of 1 bar for the temperature ranges (a) T<1484°C, (b) 1484°C<T<1841°C, (c) T>1841°C. The compositions of the as-pyrolysed amorphous ceramics VT50 (Δ), NCP200 (O) and reaction paths (arrows) are indicated. At temperatures below 1484°C (1757K) Si_3N_4 is stable in the presence of free carbon as shown in Fig. 1a. At T>1484°C (1757K) Si_3N_4 reacts with carbon to form SiC and gas phase according to

$$Si_3N_4 + 3C = 3SiC + 2N_2 \uparrow \quad (1)$$

The ceramic VT50 (Δ) with a ratio Si:C < 1 produces nitrogen until the reaction path ends on the SiC-C tie line (see Fig. 1b). The ceramic NCP200 (O) with a ratio Si:C > 1 gives off nitrogen and changes the composition to the SiC-Si_3N_4 tie line. At temperature above 1841°C (2114K) this ceramic will again give off nitrogen because the residual Si_3N_4 dissociates into liquid silicon and nitrogen gas according to

$$Si_3N_4 = 3Si(l) + 2N_2 \uparrow \quad (2)$$

Quantitative mass balances and the enthalpies of phase reactions for the ceramics can be derived from the related phase fraction diagrams (Fig. 2) and the enthalpy-temperature diagram (Fig. 3). Fig. 2a shows the calculated phase fraction diagram at a total pressure of 1 bar for the ceramic VT50 where at the aforementioned temperature of 1484°C (1757 K) 71% mass% Si_3N_4 and 29% mass% graphite (negligible amount of SiC) form 61 mass% of SiC and 28 mass% of nitrogen. The predicted sample mass loss due to nitrogen evolution is 28%. Excess 11 mass% graphite remain. The calculated enthalpy-temperature diagram shows an endothermic heat effect of +52kJ/mol for ceramic VT50. For ceramic NCP200 with a ratio Si:C > 1 all carbon is consumed and there is mass loss of 13% during reaction (1) and after the reaction excess Si_3N_4 remains. At temperatures higher than 1841°C (2114K) this residual Si_3N_4 decomposes according to reaction (2) causing a further mass loss of 15%. The enthalpy of phase reaction (1) for the ceramic NCP200 is +28kJ/mol and phase reaction (2) is +49kJ/mol. Both phase reactions are endothermic.

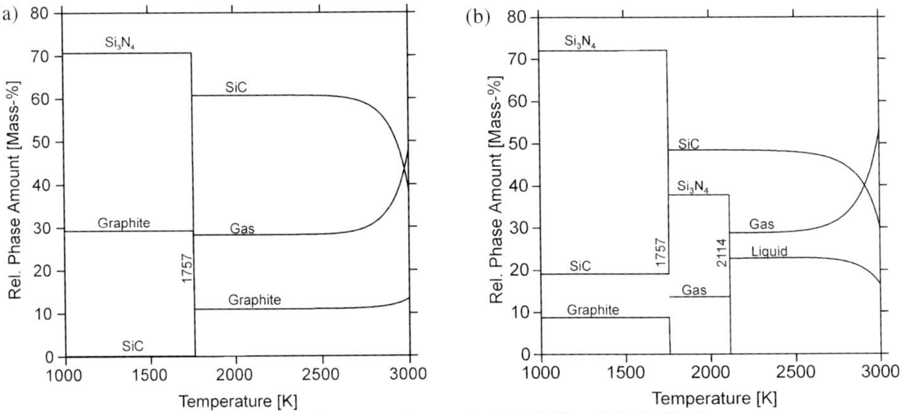

Fig. 2. The calculated phase fraction diagrams of ceramics (a) VT50 and (b) NCP200.

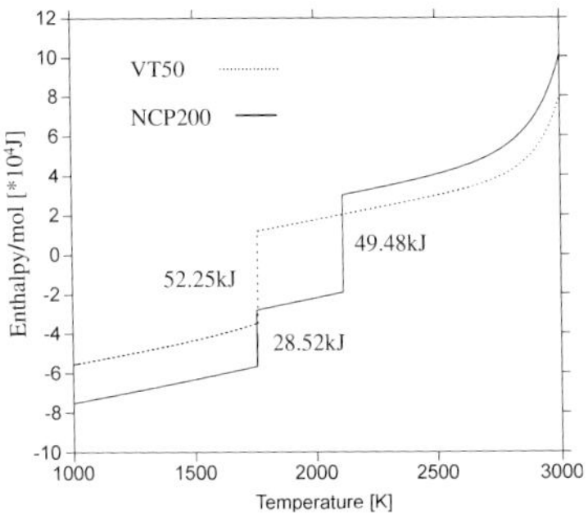

Fig. 3. The calculated enthalpy-temperature diagram of the ceramics VT50 and NCP200.

4 Experimental Procedure

To confirm the calculated high temperature behaviour of Si-C-N ceramics, the two amorphous Si-C-N ceramics VT50 and NCP200 were derived separately from the polyvinylsilazane (PVS) and polyhydridomethylsilazane (PHMS), as described in detail elsewhere [14, 15]. The polysilazanes were crosslinked between 200°C and 400°C under vacuum and Ar atmosphere, and then pyrolysed at a temperature of 1050°C under Ar atmosphere into the amorphous ceramics. The products were ball milled to powders with grain size of about 10μm and investigated by differential thermal analysis (DTA) and thermogravimetry (TG) carried out simultaneously in N_2 atmosphere (Bähr STA501). The heating rate was 10K/min up to 1000°C, 5K/min up to 1800°C and 2000°C, respectively. The phase compositions and structures were analysed by X-ray diffractometry (XRD) (Siemens Diffraktometer D5000/Kristalloflex, Cu-Kα_1 radiation). The microstructure and the phase composition were investigated by scanning electron microscopy (SEM) (Zeiss DSM982 GEMINI) coupled with EDX (Oxford – Instrument ISIS 300).

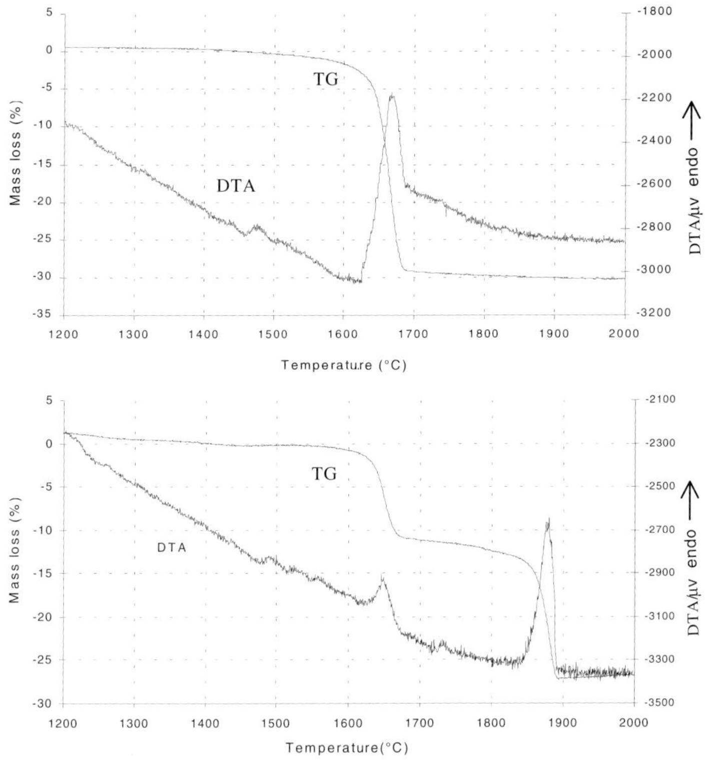

Fig. 4. The DTA/TG curves of the ceramics VT50 (a) and NCP200 (b), in N_2 atmosphere, 5K/min up to 2000°C.

5 Experimental Results and Discussion

Fig. 4 shows the DTA/TG curves of the ceramic VT50 and NCP200, respectively. The TG analysis of the ceramic VT50 shows a mass loss of 29% between 1610°C and 1690°C. Simultaneously an endothermic reaction peak was found by DTA. This mass loss and endothermic reaction peak can be explained by reaction (1) and agrees with the calculated results (Figs. 2a, 3). In the case of the ceramic NCP200, the two step mass loss accompanied endothermic reactions as predicted by thermodynamic calculation (Figs. 2b, 3) could be detected. In the temperature range from 1610°C to 1680°C 12% mass loss and a endothermic reaction peak was detected, and between 1840°C and 1900°C 15% mass loss and a endothermic reaction peak was determined. This first endothermic peak can be attributed to the reaction (1) and the second endothermic peak to residual Si_3N_4 decomposition according to reaction (2). The two step mass loss values agree also quantitatively with the calculated results. The ratio of the three experimentally derived enthalpy values (the area of the endothermic peaks) is similar to the ratio of the calculated enthalpies (Fig. 3). No direct measurements of the temperature of reaction (1) are documented. Our experiments showed significant influence of kinetics and a shift of the reaction temperatures from 1484°C (calculated) to 1610°C (detected). The temperature of reaction (2) is in agreement with the calculated result.

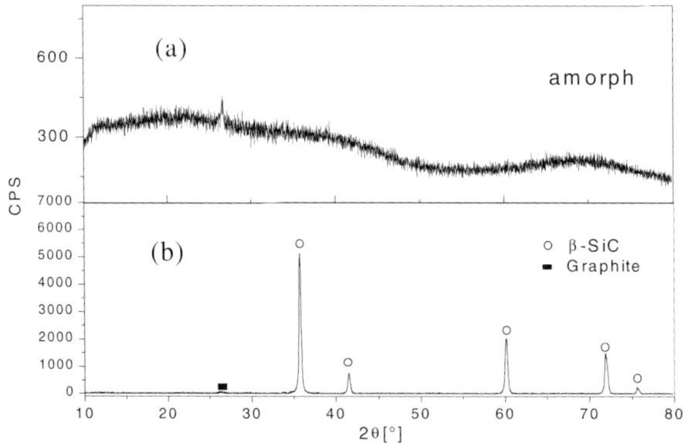

Fig. 5. The XRD patterns of the ceramic VT50, (a) before DTA/TG, (b) after DTA/TG up to 2000°C.

XRD and SEM were used to confirm the results obtained by DTA/TG and the thermodynamic calculations. Fig. 5 shows the XRD patterns of the ceramic VT50. The structure of the ceramic VT50 before DTA/TG is typically amorphous as shown in Fig. 5a. After DTA/TG with the highest temperature of 2000°C the material consists of β-SiC and graphite (Fig. 5b), which is in accordance with the calculated results (Figs. 1, 2). In case of the ceramic NCP200, after DTA/TG up to 1800°C the phase changes to β-SiC/α-Si_3N_4, but to β-SiC/Si after DTA/TG up to 2000°C, as shown in Fig. 6a-c, a result which is also correctly simulated by the calculation.

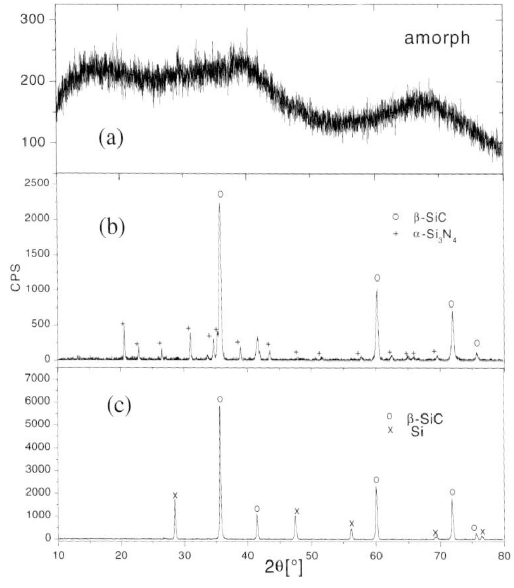

Fig. 6. The XRD patterns of the ceramic NCP200, (a) before DTA/TG, (b) after DTA/TG up to 1800°C, (c) after DTA/TG up to 2000°C.

The SEM diagrams of the ceramic NCP200 are shown in Fig. 7. It is explicitly shown that the phase assemblage is β-SiC/α-Si_3N_4 after DTA/TG up to 1800°C (Fig. 7a) and β-SiC/Si after DTA/TG up to 2000°C (Fig. 7b).

Fig. 7. The SEM diagrams of the ceramic NCP200, (a) after DTA/TG up to 1800°C, (b) after DTA/TG up to 2000°C.

6 Conclusion

The phase reactions, the crystallisation behaviour and the thermal degradation of the two Si-C-N ceramics VT50 and NCP200 were studied by means of CALPHAD type thermodynamic calculation, DTA / TG , XRD and SEM. The computer calculated simulations were confirmed by the experiments. The thermal degradation behaviours of these two ceramics are different. The endothermic phase reaction ($Si_3N_4 + 3C = 3SiC + 2N_2$) proceeds during the thermal degradation of both ceramics. The phase reaction ($Si_3N_4 = 3Si + 2N_2$) occurs only during the thermal degradation of the ceramic NCP200. Both phase reactions have been quantitatively characterised. The studies of DTA /TG, X-ray diffraction and SEM are in good agreement with the calculated results.

7 Acknowledgement

We thank the Deutsche Forschungsgemeinschaft (DFG) for financial support. For providing the samples and for helpful discussions we are grateful to J. Bill, P. Gerstel, S. Prinz and J. Seitz. The authors thank M. Thomas, H. Kummer and H. Labitzke for support in experimental analysis.

8 References

1. J. Bill, F. Aldinger, Advanced Materials, **7** (1995) 775-787.
2. F. Aldinger, M. Weinmann, J. Bill, Pure & Appl. Chem., **70** (1998) 439-448.
3. J. Bill, F. Aldinger, Z. Metallkd. **87** (1996) 827-840.
4. R. Riedel, H.-J. Kleebe, H. Schönfelder, and F. Aldinger, Nature, **374** (1995) 526-528.
5. R. Riedel, G. Passing, H. Schönfelder, and R. J. Brook, Nature, **355** (1992) 714-716.
6. H.-J. Kleebe, D. Suttor, H. Müller, and G. Ziegler, J. Am. Ceram. Soc., **81** (1998) 2971-77.
7. H. J. Seifert, J. Peng, F. Aldinger, "Die Konstitution von Si-B-C-N Keramiken" in: J. Heinrich, G. Ziegler, W. Hermel, H. Riedel (edits.), Werkstoffwoche '98, 12.-15.10. 1998, München, Vol. VII, Keramik/Simulation Keramik, Wiley VCH, Weinheim New York (1999) 339-343.
8. H. J. Seifert, H. L. Lukas, F. Aldinger, Ber. Bunsenges. Phys. Chem. **9** (1998) 1309-13.
9. H. J. Seifert, F. Aldinger, Z. Metallkd. **87** (1996) 841.
10. H. L. Lukas, S. G. Fries, J. Phase Equilibria **13** (1992) 532-541.
11. B. Sundman, B. Jansson, J.-O. Anderson, Calphad **9** (1985) 153-190.
12. B. Kasper, H. J. Seifert, A. Kußmaul, H. L. Lukas, F. Aldinger, "Entwicklung eines thermodynamischen Datensatzes für das System B-C-N-Si-O" in F. Aldinger, H. Mughrabi (edits), Werkstoffwoche'96, 28.-31.05.1996, Stuttgart, Symposium 7, Materialwissenschaftliche Grundlagen, DGM, (1996) 623-628.
13. B. Kasper, Phasengleichgewichte im System B-C-N-Si, PH. D. thesis, University of Stuttgart, Germany (1996).
14. J. Seitz, Polymer-Pyrolyse-Keramik auf Si/C/N-Basis, PH. D. Thesis, University of Stuttgart, Germany, (1996).
15. J. Seitz, J. Bill, J. Mater. Sci. Let., **15** (1996) 391-393.

Surface Modification of Mica with TiO_2 Powders Prepared by Alkoxide Hydrolysis Method

Sang–Pil Han, Sang-Hoon Lee *and Sung-Churl Choi
Departmrnt of Inorganic Materials Eng., Hanyang Univ., Seoul, Korea
*Technology & Research Institute of KORES, Seoul, Korea

1 Abstract

TiO_2 powder was adsorbed on the surface of mica using the heterocoagulation method in water. TiO_2 powder was prepared from hydrolysis of titanium-iso propoxide in a mixed solvent of anhydrous ethanol and water. When the molar ratio of water to titanium iso-propoxide was 0.25, monodispersed spherical TiO_2 particles were obtained. The prepared TiO_2 powder showed anatase phase after heat treatment at 500 °C for 2h and then fully transformed to rutile phase after heat treatment at 1000 °C for 2h. The iso-electric point of TiO_2 and mica were pH 3.9 and pH 3.25, respectively, which were measured by Z-potential analysis in water base. The maximum Z-potential difference between two powders was observed in the range of pH 3.6~3.7. TiO_2 powder was adsorbed on the surface of mica by heterocoagulation method in pH 3.6~3.7. The properties of prepared TiO_2 powder was characterized by XRD, SEM, etc. The color change and SPF (Sun Protection Factor) indices of TiO_2-adsorbed mica were examined.

2 Introduction

The Earth's ozon layer protects all life from the sun's harmful radiation, but human activities have damaged this shield using the CFCs for refrigerants, solvents etc. The UV-light can be divided into the UVA (320~340nm), UVB (290~320nm) and UVC (200~290nm) according to wavelengths. In these wavelengths, the UVB has been linked to many harmful effects, including various types of skin cancer, some crops, plastics and other materials. Since ozon layer filters out harmful UVB radiation, less ozone means higher UVB levels at the surface. Many materials such as PABA, PABA ester, cinoxate and TiO_2, talc, MgO, ZnO have been introduced as a physical or chemical sunscreen agents.

In those materials, TiO_2 powder is widely used as an anti-reflecting coating film, white pigment materials, paint materials and cosmetics because of its high refractive index, good stability for UV-light, inert chemical and whiteness properties. Recently, different fabrication methods, such as the sol-gel process, emulsion, pyrolysis and hydrothemal, have been employed to prepare monodispersed spherical TiO_2 powers.[2,3]

In this experiment, the surface of mica was modified with adsorbing TiO_2 powder to enhance the UV-light screening effect. TiO_2 powder was prepared from hydrolysis of titanium iso-propoxide in a mixed solvent of anhydrous ethanol and water. And the effect of concentration of starting materials, temperature and stirring rate on the morphology of prepared TiO_2 powder was investigated. And then TiO_2 powder was adsorbed on the surface of mica using the heterocoagulation method.[4)] The color change and SPF indices of TiO_2-adsorbed mica were examined.

3 Experimental Procedure

Titanium iso-propoxide(Titanium(IV)iso-propoxide, 98+%, ACROS), anhydrous ethanol and distilled water were used as a starting materials. For preparing monosized TiO_2 powders, our experiments were carried out under the different way. One was mixing the water and alkoxide at the same time, the other was dropping the water into the alkoxide soultion with a various time. The concentrations of alkoxide and water throughout this experiment ranged from 0.025 to 0.1mol and from 0.1 to 0.2mol, respectively. In hydrolysis process, the stirring rate and reaction temperature were varied from 150r.p.m to 320r.p.m and from 0 °C to 25 °C. The concentrations of starting materials and stirring conditions are shown in Table 1. After hydrolysis process was finished, mixed solution was filtered with 0.22μm glass filter and then washed with anhydrous ethanol until the precipitates of alkoxide(not reacted with water) were disappeared. And the prepared TiO_2 powder was fired with a various time for 2h to characterize the crystalline phase after dried at 80 °C for one day. To adsorb TiO_2 powder on the mica surface, surface potential of each powders were evaluated using the Z-potential analyzer (Zeta Potential Anlyzer, Brrokhaven Instruments, Co., U.S.A). Two slurries (mica and TiO_2 slurry) were adjusted to the pH range, which showed the maximum electrostatic force between two powders. The prepared TiO_2 powder was adsorbed on the mica surface with various TiO_2 amounts (0.3wt%, 5wt%, 10wt%, 15wt%, 50wt%). TiO_2-adsorbed mica were heated at 1000 °C for 2h and its properties were examined with XRD (Rigaku, Co., Japan), EDS, SEM (Scanning Electron Microscopy, JEOL JXA-840A, Japan). The color change and SPF (Sun Protection Factor) indices of TiO_2-adsorbed mica was measured using the Colorimeter and SPF analyzer, respectively.

Table 1. The concentrations of starting materials and starting conditions

	Titanium Alkoxide	H_2O	Temp.(°C)	stirring rate(r.p.m)
S1	0.1 mol	0.1 mol	0	150
S2	0.1 mol	0.1 mol	25	150
S3	0.1 mol	0.2 mol	25	150
S4	0.1 mol	0.2 mol	25	320
S5	0.025 mol	0.1 mol	25	320

4 Results and Discussion

Fig.1 shows SEM photographs of TiO_2 powder prepared from various conditions. Fig.1 (a) and (b) shows microstructure when 0.1mol distilled water is added to 0.1mol titanium iso-propoxide solution at 0 °C and 25 °C with a various time. There was no difference between reaction at 0 °C and 25 °C for the morphology and size of prepared particles. From these results, we knew that the reaction temperature did not affect the morphology of powders in this experiment. But Fig.1 (e) shows narrow-sized and spherical TiO_2 powder prepared from hydrolysis of 0.025mol titanium iso-propoxide with 0.1mol distilled water at the same time.

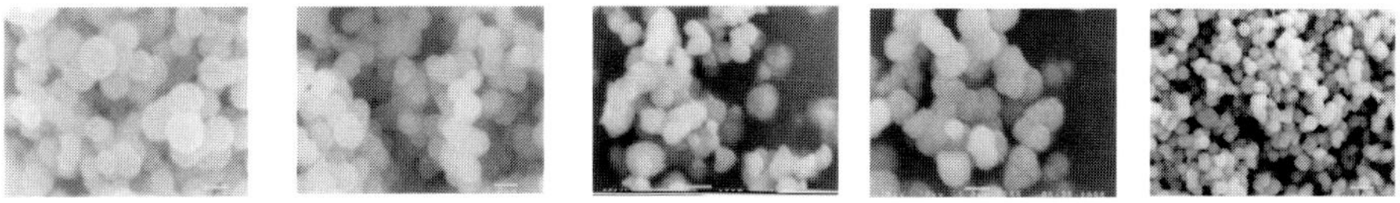

a) b) c) d) e)

Fig. 1 SEM micrographs for TiO_2 hydrates ; a) S1 b) S2 c) S3 d) S4 e)S5

In SEM photographs of Fig.1 (a)~(c), agglomerates and neck among the particles were observed. We supposed that these results were attributed to inhomogeneous nucleation and growth, which was because of various water amounts. And the portion of agglomerates among the particles was decreased as the stirring rate was increased more and more.

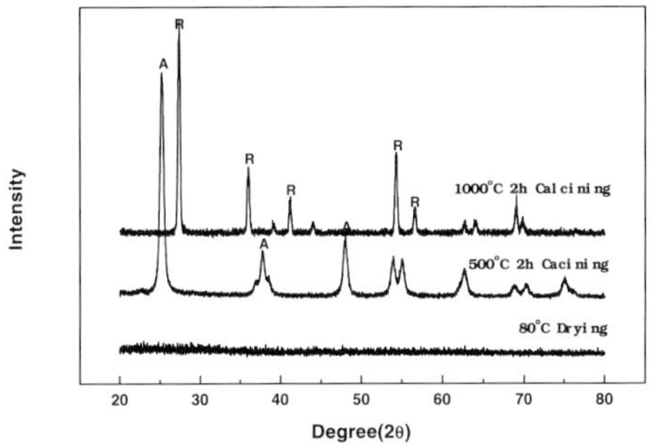

Fig. 2 XRD patterns for TiO_2 powders

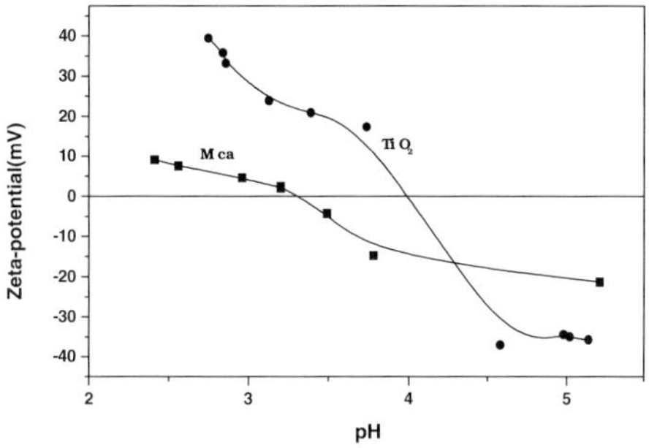

Fig. 3 Z-potential of TiO_2 and mica as a function of pH

Fig.2 shows the XRD diffraction results for prepared TiO_2 powder calcined at various temperatures. The prepared TiO_2 powder showed an amorphous phase after dried at 80 °C. And it showed an anatase phase after heat treatment at 500 °C for 2h and then fully transformed to rutile phase after heat treatment at 1000 °C for 2h. Fig.3 shows Z-potential analysis for prepared TiO_2 powders and mica. In this result, the iso-electric points of TiO_2 powder and mica were found in pH 3.8, pH 3.25, respectively and the maximum electrostatic force between TiO_2 and mica was observed in pH 3.6~3.7.

a) b)

Fig. 4 SEM micrographs for TiO_2-adsorbed mica ; a) Before b)After

Fig.4 shows SEM photographs of TiO_2 powder adsorbed on the surface of mica using heterocoagulation method in pH 3.6~3.7. But the Z-potential of TiO_2 powder at pH 3.6~3.7 was approximately 20mV, it was not enough to separate powders effectively. According to

Reed,[6] at 20 °C a repulsion barrier corresponding to a zeta potential of about 25mV is required to minimize coagulation. As a result, the coagulated TiO_2 powder was found in Fig. 4-b.

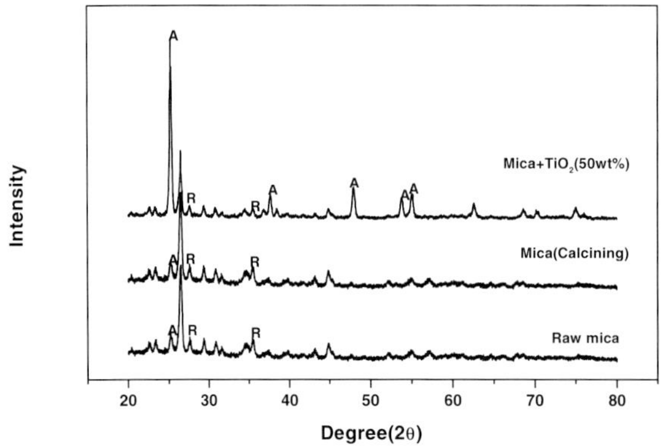

Fig. 5 XRD patterns for TiO_2-adsorbed mica

Fig. 5 shows XRD diffraction results for calcined TiO_2-adsorbed mica, calcined mica and raw mica. It showed a different result from that of prepared TiO_2 powder. (Fig.2) When the prepared TiO_2 powder was calcined at 1000 °C for 2h, an anatase phase was fully transformed to rutile phase in Fig.2. But there is no change of rutile peaks compare with that of calcined mica. On the contrary, the intensity of anatase peaks increased as the temperature was shifted from 500 °C to 1000 °C. It was attributed that the ionic radius of Ti^{4+} is 0.61 Å, while that of Si^{4+} is 0.40 Å, small enough to enter the titania lattice interstitially and inhibit the transformation of anatase to rutile.[7] These results could be also explained from the EDS profiles for raw mica and TiO_2-adsorbed mica. Fig.6 shows the EDS profile for raw mica. In this result, Ti and Si ion were found in raw mica and we knew that Ti ion was coexisted with anatase and rutile form from XRD diffraction result. And the intensity of Ti ion after TiO_2 powder was adsorbed on the surface of mica increased compared with that of raw mica. From these results, we supposed Si^{4+} ion inhibited the transformation of anatase to rutile.

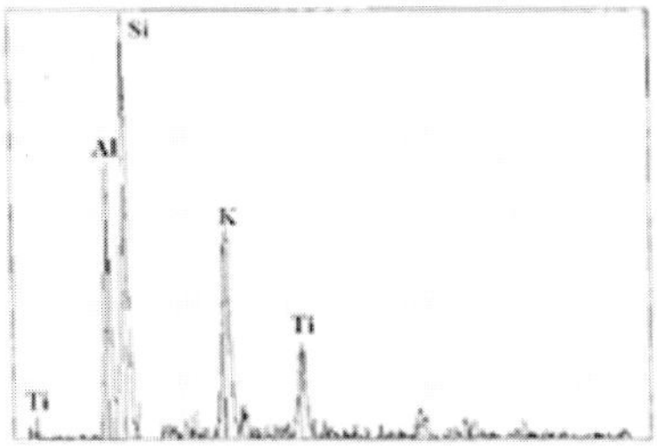

Fig. 6 EDS profile for raw mica

The results of color measurement for TiO_2-adsorbed mica were shown in Fig. 7 and Fig. 8. All the analysis was measured with a reflectance color and the measuring wavelength range was from 360nm to 740nm used with a D_{65} for light source. And the purity and dominant wavelength were calculated according to JIS Z8701. Fig.7 shows the purity result and Fig. 8

shows the dominant wavelength result for TiO_2-adsorbed mica. The purity and dominant wavelength for TiO_2-adsorbed mica were decreased as the TiO_2 amounts were increased. It means that the surface of mica is modified with TiO_2 powder effectively and the whiteness of mica can be increased also. From the purity and wavelength results, we could have a comprehensive color change result. Fig.9 shows the relationship between hue and x-y for TiO_2-adsorbed mica with combining purity and dominant wavelength result. It shows the whiteness of mica is increasing as an increase of TiO_2 powder amounts.

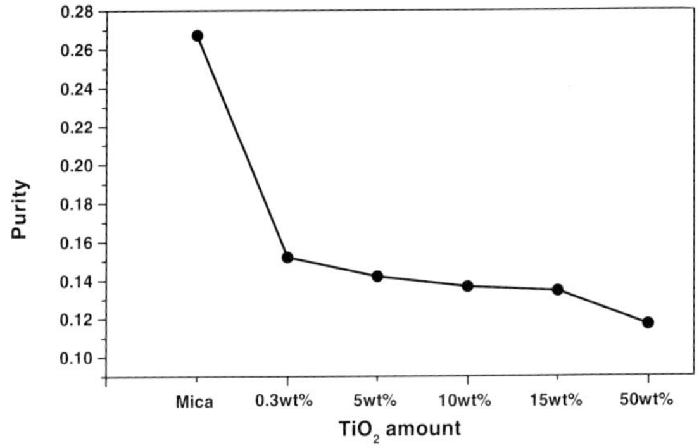

Fig. 7 Purity results for TiO_2-adsorbed mica

Fig. 10 shows the change of SPF(Sun Protection Factor) ratio of TiO_2 amount to mica. This is an in-vitro test result calculated with a SPF analyzer in the 290~400nm wavelength range. In-vitro SPF test has been used to attempt to predict the in-vivo efficacy of sunscreening products. Because there are no written regulations about evaluating the SPF number among nations, we showed only the change of SPF ratio result instead of SPF number. When 50wt% TiO_2 powder was added to mica, its SPF number showed over 4 times increase than that of mica. We supposed it was attributed to absorption and scattering properties against UV-light of TiO_2 powder.

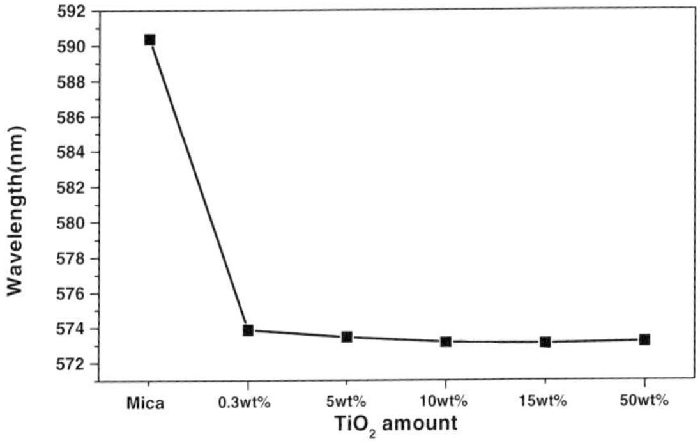

Fig. 8 Wavelength result for TiO_2-adsorbed mica

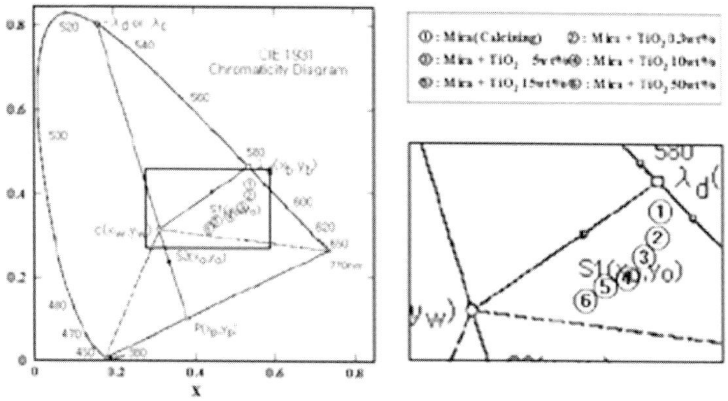

Fig. 9 The relationship between hue and x-y for TiO_2-adsorbed mica

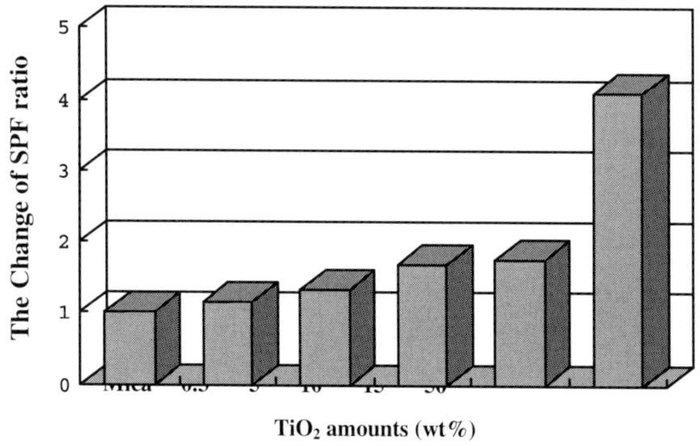

Fig. 10 The Change of SPF ratio of TiO_2 amount to mica

5 Conclusion

When the molar ratio of water to titanium iso-propoxide was 0.25, monodispersed spherical TiO_2 powder was obtained. TiO_2 powder was adsorbed on the surface of mica using the heterocoagulation method in the range of pH 3.6~3.7. After heat treatment at 1000 °C for 2h, TiO_2 powder did not transformed to rutile because of Si^{4+} ion in mica. TiO_2-adsorbed mica showed an increase of whiteness and SPF number compared with that of mica. As a result, we suppose that mica modified with a TiO_2 powder may have a possibility to prevent skin and various types of materials from UV-light.

6 References

1. Y. Takahashi, J.Chisaki.," Titania Coating of Alumina Powders," *J. Ceram. Soc. Jpn.,* 96(3), 240-246 (1988).
2. T. Ogihara, T. Yanagawa, N. Ogata, K. Yoshida, M. Lguchi, N. Nagata and K. Ogawa., " Preparation of Monodispersed $MTIO_3$(M=Ba, Sr, Pb) Fine Powders by Hydrolysis of Metal Alkoxide," *J. Soc. Powder Technol. Jpn.,* 31(11), 21-28 (1994).
3. J.H Jean, T.A Ring., " Processing Monosized TiO_2 Powders Generated with HPC Dispersant," *Am. Ceram. Soc. Bull.,* 65(12) (1986).
4. E. Liden, L. Bergstrom, M. Persson and R. Carlsson., "Surface Modification and Dispersion of Silicon Nitride and Silicon Carbide Powders," *J. Eur. Ceram. Soc.*, 7, 361-368 (1991).
5. J.L Look and C.F Zukoski., "Alkoxide-Derived Titania Particles : Use of Electrolytes to Control Size and Agglomeration Levels," *J. Am. Ceram. Soc.*, 7(6), 1587-1595 (1992).
6. J.S Reed., Principle of Ceramics Processing, PP. 159-167, JOHN WILEY & SONS, Inc., New York, 1995.
7. S. Vemury and S.E Pratsinis., "Dopants in Flame Synthesis of Titania," *J. Am. Ceram. Soc.*, 78(11), 2984-2992 (1995).

Indirect Characterization of Modified Silicon Nitride Powders by Chromatography

Siegfried Vieth*, Elke Mitzner, Dietmar Linke
Brandenburgische Technische Universität Cottbus, Lehrstuhl für Anorganische Chemie, Karl-Marx-Str. 17, 03044 Cottbus, Germany

1 Introduction

Powder processing of silicon nitride ceramics is not possible without organic additives, e.g. as binders or lubricants. Typically, the additives used are chosen by empirical methods. Usually, they are characterized by their influence (e.g., on the rheologic behavior of suspensions, on the density of green compact [1]). Systematic investigations of the physical and chemical processes between powder and additives are rare [2].

The change to modified surfaces of Si_3N_4 powders makes it necessary to get new insights into the interaction of powders and additives. The chromatographic process can be regarded as a mimic of the preparation of green compacts. In both processes organic molecules move along a surface and the adsorption behavior is of great importance. Furthermore, chromatographic methods are well-developed and a deep understanding of the processes exists [3]. Thus, we developed a simple column-chromatography method which is based on test compounds carrying typical functional groups and a chromophor (for UV detection) on one hand and modified silica on the other. Due to the small size of the particles column chromatography cannot be performed with modified Si_3N_4 powders itself. It is known that the surface of Si_3N_4 powders consists mainly of silica due to hydrolytic processes [4]. For that reason the effect of modification should be nearly the same for silica and silicon nitride powders. The eluents (mobile phases) used represent a broad area of polarity and consist of solvents carrying also typical functional groups known from additives (e.g. polyacrylates and polyglycoles).

To the best of our knowledge column chromatography was not used to investigate the interaction of ceramic powder surfaces and organic additives. Thin layer chromatography (TLC) was used to characterize the adsorption behavior of additives on clay [6].

2 Materials and Methods

The modified silica powders **1,2** were prepared from silica (40-63 μm, Merck [5]) and the appropriate silane by standard procedures [3] according fig. 1. Hydrocarbon-modified silicas **2a** were further treated with hexamethylsilazane (HMDS) to diminish the amount of remaining OH groups (endcapping [3]).

Solvents (HPLC quality) for eluent preparation (see table 1) and test substances (see fig. 3) were used as purchased. Solutions of test substances were prepared prior use - solvent: mobile phase A, except **6** and **8** which were dissolved in a 1:1 mixture of dimethoxyethane and n-hexane; concentration: 1 mg/ml except **3** (0.1 mg/ml) and **4** (10 mg/ml). The column chro-

matography was run with a usual equipment including pump (Waters 501 HPLC pump), injection valve (Rheodyne), column and UV detector (Varian 9050) and recorder/integrator (Hewlett Packard HP3394A). The column (steel, d_i = 3.5 mm, h = 24.5 cm) was filled with dry silica powder. The air in the column was displaced with a strong flush of mobile phase (7.5 ml/min) as long no gas bubbles were leaving the column. Chromatography was run at 0.5 ml/min.

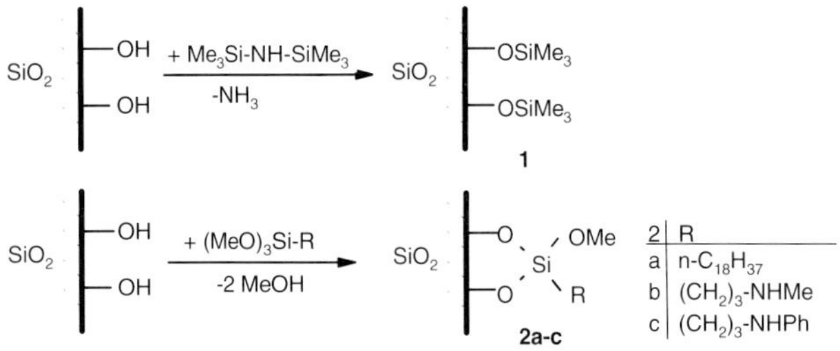

Figure 1: Synthesis of different types of modified silica

3 Data and Observations

Using column chromatography we could observe different processes. First to mention is the physical („normal") adsorption. In this case the compounds show well-shaped peaks from which retention times and k' values can be estimated (see table 1). The k' values - a direct measure of the adsorption strength (see fig. 2) - can be influenced by the polarity of the solvent (see 4).

$$\frac{t_R-t_0}{t_0} = k' = \frac{V_S}{V_M} K_A$$

t_R: retention time
t_0: time of passage (dead time)
k': capacity factor
V_S: volume of stationary phase
V_M: volume of mobile phase (in the column)
K_A: adsorption constant

Figure 2: Important parameters of column chromatography

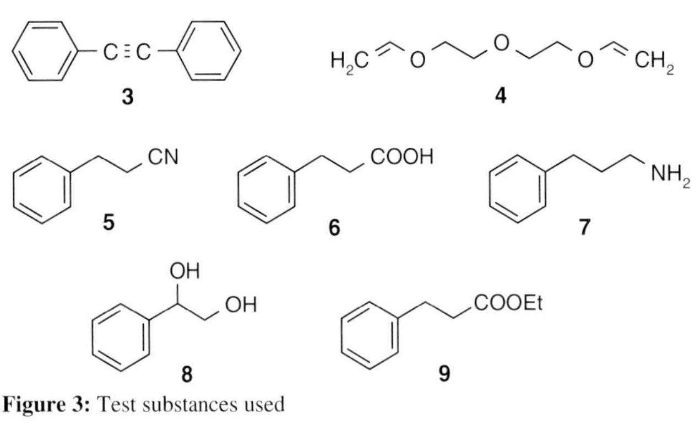

Figure 3: Test substances used

Table 1: Silica and modified silicas **1, 2** - k' values and observations

silica, eluent(a)	k'-values of test compounds **3**	**4**	**5**	**6**	**7**	**8**	**9**
silica A	0.40	0.31	1.64	(d)	(d)	(d)	0.62
B	0.19	0.21	0.23	(d)	(d)	0.56	0.38
C	0.17	0.17	0.17	(d)	(d)	0.54	0.31
1 A	0.22	0.25	0.30	0.65	(d)	0.58	0.26
B	0.40	0.38	0.42	0.60	1.2(c)	0.42	0.18
C	0.45	0.30	0.20	0.30(e)	(d)	0.20	0.28
2a A	0.23	0.23(b)	0.30	0.20	(d)	0.20(b)	0.28
B	0.23	0.23(b)	0.10(b)	0.13(b)	(d)	0.13	0.24
C	1.00	1.00	0.23	0.23	(d)	0.13	0.66
2b A	0.49	0.31	(f)	(d)	(d)	(d)	0.49
D	0.20	0.29	(f)	(d)	(d)	0.49	0.23
2c A	0.35	0.14	(f)	(d)	(d)	1.95	0.37
D	0.25	0.28	0.32	(d)	(d)	0.23(b)	0.22

a) A: n-heptane/dimethoxyethane 4:1; B: dimethoxyethane/acetonitrile 9:1;
C: acetonitrile/methanol 9:1; D: dimethoxyethane/methanol 85:15

b) values obtained after saturation of the „hot spots" by several injections (see also text)

c) peaks are broad and show a tailing

d) strong chemisorption

e) peaks are small and are followed by a broad peak

f) nitrile groups react with the immobilized amino groups, estimation of t_R-values is impossible

Very often chemisorption or other strong adsorption processes occur. These irreversible processes cause different phenomena in chromatographic experiments: The test compounds show, e.g., no peak, a broad peak, a tailing, or more than one peak (see footnotes under table 1). In some cases well-shaped peaks can be obtained after several injections (after all reactive centers on the surface have been saturated).

Also chemical reactions can be detected. If nitriles (eluents B and C, compound **5**) get in contact with amino modified silicas **2b,c**, no regular peaks can be observed. It is impossible to get a base line due to permanent flow of reaction products (which show a strong UV absorption)). Furthermore, the silica becomes yellow during the treatment with nitriles. Type of reaction was not investigated. We assume a self-condensation of nitriles to triazines catalyzed by the modified silica.

4 Relation to the Ceramic Processing of Silicon Nitride Powder

4.1 Amino-modified powders 2b and 2c

Column-chromatography experiments give information on the polarity of the stationary phase - in our case the modified silica surface [3]. Polar, so called normal phases show a strong adsorption of polar compounds. Nonpolar compounds show a weaker adsorption and therefore smaller k' values than polar compounds. Increasing polarity of the eluents causes a decrease in

k' values, independent of the polarity of the test compound. Silica, of course, is a normal phase. Hydrocarbon **3** and ether **4** show only a weak retention, whereas the strong polar substances **6-8** are irreversibly adsorbed even in the presence of the polar eluent C (see table).

The methylamino-modified silica **2b** behaves mainly like silica. Obviously, after the modification reaction a considerable number of OH groups is still present on the silica surface (see 4.2) and dominates the adsorption properties. This is confirmed by our results in the ceramic processing [8] Thus, the densities of green bodies from NHMe-modified Si_3N_4 powders are only slightly higher than those from unmodified powders.

The anilinogroups carrying **2c** is less polar, as seen from the smaller k' values compared with **2b**. The surface of **2c** is carrying hot spots (see below), as indicated by the strong adsorption of the amine **7**. Unfortunately, **2c** is not stable during storage. For this reason anilino-modified Si_3N_4 powders are ruled out for green-body preparation.

4.2 Hydrophobic Silicas 1 and 2a

Octadecyl(OD)-modified silicas are typical *reversed phase materials* (RP18, OD silica) which are widely used in analytical HPLC [3]. Also our ODC-modified silica **2b** shows this behavior. Increasing the polarity of the eluent, **3** shows increasing k' values, indicating that the hydrocarbon has a stronger affinity to the OCD-modified surface than the polar solvent molecules. Using a polar eluent, the much more polar diol **8** shows a lower retention than hydrocarbon **3**. Thus, **2b** has opposite (reversed) adsorption [7] properties compared with simple silica.

The surface of OD silica still carries free OH groups - the so-called hot spots [3]. In general, the concentration of hydroxy groups on a silica surface is about 8 $\mu mol/m^2$. Reactions with silanes are never quantitative by steric reasons, leaving intact about half the amount of hydroxy groups. These remaining OH groups still allow the irreversible adsorption of polar compounds even the powder is hydrophobic. Only after saturation of the hot spots normal peaks can be obtained (see table 1).

The time of passage t_0 is about 25 % shorter for OD silica **2a** compared with the other silicas investigated. On the other hand, OD Si_3N_4 has a smaller density (2,8 g/cm^3) than otherwise or nonmodified powders (3...3.1 g/cm^3). Both observations can be explained by the formation of closed micropores by the long alkyl chains. This thesis is further supported by the bad compaction behavior of OD Si_3N_4.

The TMS-modified silica **1** shows a "reversed phase behavior" similar to **2a**. It also carries hot spots on the surface (seen by the irreversible adsorption of **7**) but they are of less importance compared with **2a**. The smaller hydrophobic groups and the high reactivity of HMDS allow a higher extent of OH-group modification.

Our studies of the TMS-modified Si_3N_4 powders confirm the observations of chromatographic experiments. Such powders allow the preparation of green bodies also with nonpolar binders whereas we failed using common silicon nitride powders [8]. Due to the hot spots also TMS-modified Si_3N_4 powders can interact with polar functional groups. Thus, stearic acid improves also the compaction behavior of this nonpolar powder.

5 Conclusions

Column chromatography is a versatile tool for the investigation of the interaction between modified surfaces and organic additives which are interesting for ceramic processing. Dependence of retention time from polarity of eluents or test compounds allows conclusions on the polarity of the surface. Effects unwelcome in analytical or preparative chromatography, like strong absorption and chemical reactions, can be easily observed and give further information. The results obtained in chromatography fit very well with the observations of the compaction behavior of silicon nitride powders modified identically. Thus, column chromatography supports the choice of powder modification as well as of organic additives for ceramic processing.

6 Acknowledgments

Thanks are due to Dr. Bärbel Vieth and Dr. Lutz Alder (Bundesinstitut für gesundheitlichen Verbraucherschutz und Veterinärmedizin) for helpful discussions and for leaving of equipment. We thank DFG (Deutsche Forschungsgemeinschaft) for a research fellowship (E.M.).

7 References and Notes

1. P. Hidber, T. Graule, L. Gauckler in Technische Keramische Werkstoffe (Ed.: J. Kriegsmann.), 39. Ergänzungs-Lieferung, Deutscher Wirtschaftsdienst, Köln 1997, p. 1-21. b) V.A. Hackley, R. Premachandran, S.G. Malghan, Key Engineering Materials 1994, 89-91, 679-82. c) H. Gorter, Mermsen, S. Ayar, J. Gerretsen, Third Euro-Ceramics (Madrid 1993), Vol.1, p. 473-78. d) G. Riedel, Fortschrittsber. Dt. Keram. Ges. 1996,11 137-42. e) G. Riedel, S. Krieger, Keramische Zeitschrift 1996, 48 192-9.
2. D. Bortzmeyer in Material science and Technology (Eds.: R.W. Cahn, P. Haasen, E.J. Kramer), Vol.17a, VCH, Weinheim 1996, p. 147-52. b) Gmelin Handbook of Inorganic Chemistry, vol. 15, Supplement B5d1, Silicon nitride: Electrochemical behavior, colloidal chemistry and chemical reactions, 1995. c) Gmelin Handbook of Inorganic Chemistry, vol. 15, Supplement B5d2, Silicon nitride: chemical reactions (continued), 1995
3. Some examples of the plethora of literature: a) V.R. Meyer, Praxis der Hochleistungs-Flüssigchromatographie, Verlage Diesterweg und Sauerländer, Frankfurt a.M., 1988. b) K. Ballschmiter, M. Wößner, GIT Laborfachzeitschrift 1997, 593-600
4. G. Busca, V. Lorenzelli, G. Procile, M. I. Baraton, P. Quintard, R. Marchand, Mater. Chem. Phys. 1986, 14, 123-40. b) T. A: Dang, R. J. Gnanasekaran, J. Vac. Sci. Technol. 1991, 9A, 1406-9.
5. This silica is much easier to handle than the finer HPLC silica. As long no separation of mixtures is necessary, the silica used fulfills all needs.
6. F. Genel Ricciardello, S. Volpe, Ceramica Informazione, 1995, 353, 473-75.
7. If reversed stationary phases are used the process is often regarded as distribution chromatography [3]. Thus, absorption could be the better word.
8. S. Vieth, E. Mitzner, D. Stelmach, D. Linke, Jahrestagung der Deutschen Keramischen Gesellschaft, Freiberg 1999

STUDIES UPON AUTOMATED MANUFACTURING OF TUBES WITH THICK SUPERALUMINA CERAMIC WALLS DESTINED FOR ELECTRONICS AND ELECTROTECHNIQUE INDUSTRIES

Constantin S. Dumitru*,
Mihai E. Mangra*, Horatiu I. Niciu**

* Dept. of Mechanics, University of Craiova
** "MATPUR" Manufacturing Enterprise, Bucuresti

1. INTRODUCTION

The freqvent products chaning as well as ever greater requirements of parts manufacturing with minimum comsumption of material lead to number and typodimensions increasing of powder parts. The achievement of the parts from translucent polycrystalline superaluminizing of high purity (99 % $Al_2 O_3$) through isostatic pressing became important for the commercial point of view at the beginning of 1970 when there were achieved the first discharge tubes for sodium discharge lamp of high pressure. The were manufactured tubes of a great measurement precision (the well thickness 0,6-1mm, ratio H/D big 4:1) of a idealized structure with Al_2O_3 hexagonal grains, dense and fine and the surface ranghness is R_a=0,2-0,4 μm, accordind to the ruling norms.

In order to modernize the process of forming and compression of ceramic powder parts it has acted on the designing of intensive technological processes (the operations concentration) , of using of stamps with a great technological flexibility (group and nodule), but especially on mechanization and automate of auxiliary operations (semi-products feeding, transport between operations, parts exhausting, storage) /1 /.

2. THE ESTABLISHMENT OF CONSTRUCTIVE AND FUNCTIONAL PARAMETERS OF AUTOMATIC ISOSTATIC PRESS

Isostatic automatic presses for manufacturing of ceramics powder product work on principle of isostatic pressing. The grain material in the shape of powder measured in an elastic cover is compressed in this way under a high identical pressure achieving a body of requested structure and resistance as well as of uniform density in all sections of press body. The development of manufacturing equipments for isostatic pressing is based on

one post presses on two columns presses or on the principle of rotary presses. Automatic isostatic presses can be built for a closing pressure of about 2000 kN and maximum isostatic pressure of 200 MPa, in this way the application range being increased /4/. The hydraulic actuating using is recommended for compression process. The essential two criteria of a press are: the maximum closing force and the maximum isostatic pressure that can be developed. Both criteria directly depend on tool designing. The closing force shows the maximum power with which one or many tool holes can be closed during isostatic pressing. The isostatic pressure represents the guiding parameter of compressing pressure developed by the pressure fluid upon flexible die. Its maximum value is limited by the press designing on the one hand, and by the tool.

In order to allow the automatic loading the presses must be equipped with a device for material filling up and measuring. Mainly, the choosing of filling up method depends on the part to be pressed and on the required tool position to be pressed. At the designing of isostatic automatic presses must take also in the consideration the fact that compressed parts have to be automatically ejected and transferred. The used method depends on part geometric form and its resistance after its compression.

The whole process: the material measuring, the die filling up, the die closing stroke, the hydraulic pressure increasing, the die opening, the outside dumping throw of the pressed part are automatically achieved.

3. METHODS AND MEANS OF RESEARCH

For determining of technological parameters influence upon compressed parts characteristics, the paper has in view the procedural research method which consists in establishing of a strategy concerning an operation in predetermined situation. The procedural model for achieving a part through automatic isostatic compression includes several instructions upon the part shape and dimensions, working pressure, time of decompression. From value parameters analyze of the working regime used for the researches upon powders pressing with translucide polycrystalline structure in the reference /4/ for effective compressions under the condition previously mentioned were recommended the followings: 1) for high working pressure inside the recommended values are in the range 70 - 110 MPa; 2) maximum testing pressure is 120 MPa; 3) closing force, the range of values is F_p 300 KN; 4) decompression in pressure steps. The research testing on automatic isostatic press PIA 1000 type for manufacturing of discharge tubes used for sodium discharge lamps in big series were achieved at "MATPUR" SA Bucharest. Functional, the pieces studied, figure 1, are destined to some products wich constitute the ceramics-metal ensemble, fact wich involve a certain precision of execution and a certain surface roughness.

The ceramic materials that belong to alumina family α-Al_2O_3 with translucent polycrystalline structure used in experimental researches were as powers with particles of 1- 30 μm and residue humidity less than 1,5 %. The material allows the achieving of a

higher transparency degree. For a thickness of 0.5 mm in a visible range, it offers a total transmission of 90 % /2,3/. In order to prevent an exaggerated increasing of grains during sintering and a high transparency degree to fine alumina are added small quantities of MgO (0,05-0,06).

The experimental results which will be statistically processed are presented in table 1.

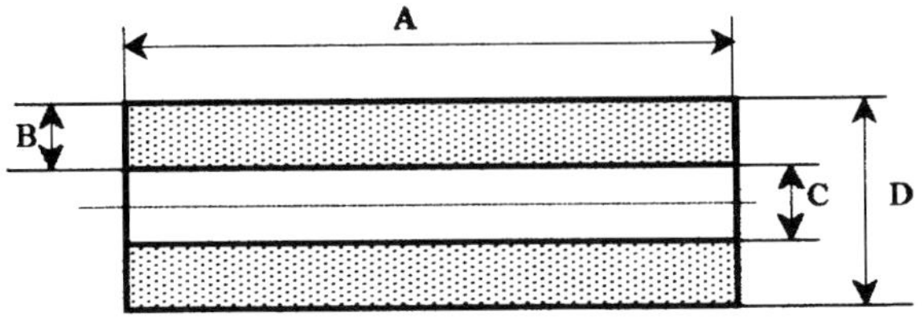

Power [W]	A [mm]	B [mm]	C [mm]
70	58 ±0.5	3,3 ±0,1	0,6 ±0,1
150	60 ±0.5	60,0 ±0,2	0,6 ± 0,1
250	90 ±0.5	7,2 ±0,2	0,75 ±0,1
400	114 ±0.5	8,0 ±0,1	0,75 ±0,1

Figure 1. The form and the dimensions of the ceramics tubes of translucent polycrystalline alumina

4. EXPERIMENTAL RESULTS

In order to achieve results as well as possible for automatic isostatic pressing it is necessary a manufacturing control as follows: control of powders preparing; control of technological flow operations; control of parts quality. The preparing of powders mixture for pressing is achieved according to existing norms. Moreover, for automatic isostatic pressing it is recommended to control groaning and particles humidity which influence the process of automatic dosage. The full automation of the process allows the following and the control of all manufacturing flow stages. The switching on of these cycles is achieved by confirming the kits position of the machine.

The external diameter (D = 9,5 ± 1) as well as the interior diameter of parts (d = 8 ±0,1) were framed in the required tolerances. Owing to elastic stamps the medium tolerances achieved by isostatic by isostatic pressing are 0,1 mm. For less values, the grinding of the parts is required after pressing.

Table 1. The obtained results of polycrystalline alumina isostatic pressing

Compression presure p_c MPa	Parts density measured ρ_m g/cm^3	Porosity after compression P_c %	Parts density sintering ρ_m g/cm^3	Porosity after sintering P_c %
70	1,87	53,10	3.44	13,57
80	2,00	49,86	3.52	11,56
90	2,10	47,34	3.60	9,55
100	2,15	46,00	3.45	9,86
110	2,17	45,52	3.76	8,56

In order to determine the probability that the parts density might be bigger with 1,90 g/cm^3 and the trust range for average m of parts density belonging to the respective lot it was used a statistic calculation. 84,31% represents the probability of a part to be chosen at random and it may have a density greater than 1.90 g/cm^3. That is 83,13 % of all the parts has the density greater than 1,90 g/cm^3 . With a probability of 95 %, the density of the parts belonging to the same lot there is within the trust range 12,255 and 12,299. Analysis the experimental results obtained upon the manners in achieving of working parameters of automatic isostatic pressing equipment and their relation with parts properties it is found out a distinct influence of pressing factors (densification pressure, decompression speed) and of particularities of this method upon structure and mechanic properties. The densities knowing of the parts in a crude state is useful for further research of their behavior during handing over and sintering under the conditions of some dynamic loads. There were performed tests for different working pressures at the ambient temperature. The pressure values of 70 -110 MPa assured the compact of pressed parts. The curve that represents the density increasing according to the density is of the following type:

$$\rho_c = \rho_n + k \log \rho \qquad (1)$$

where: apparent density of compressed part, g/cm^3; fill in density of the powder, g/cm^3 (0,88 g/cm^3); p = compact pressure, daN/cm^2; k =0.42 - constant, compress indicator depending on powders properties. From figure 1 achieved with the data from table 1 it may be seen a concordance good enough between the density experimentally determined and the estimated density. For any grain composition there is a compression pressure above which the density increasing is neglected. For a translucent polycrystalline grain alumina of 1-30μm isostatic pressed at a compression pressure of 100 MPa it was obtained a limit density of 2,15 g/cm^3. The level of maximum pressure and the time of decompression between 10 - 20 sec. actually did not influence upon pressed parts densities. Experimentally, to avoid stresses and part cracks, decompression was achieved in two pressure steps at different lapses of time. Constructively, the decompression in steps is likely, by introducing into hydraulic scheme of a machine of a proportional valves. Also, experimentally it has been established the correlation between density in nonsintered state and density of sintered parts of the translucent polycrystalline.

This correlation can be expressed from relation /5/ :

$$k = \left[\frac{1}{(D_U - D_o)^{\frac{1}{3}}} - \frac{1}{(D_F - D_o)^{\frac{1}{3}}} \right] \cdot \left[\frac{1}{(D_U - D_o)^{\frac{1}{3}}} - \frac{1}{(D_T - D_o)^{\frac{1}{3}}} \right] \tag{2}$$

where k is a constant which is calculated for different sintering temperatures; D_U is the density in the nonsintered pressed state; D_f is the density of the sintered parts; D_T is the theoretical density of the crystalline α - Al_2O_3 (3,98 g/cm^3); D_0 is a constant which represents the minimum density of which oxide aluminum powder has enough touch between the particles to maintain a semiestabshliment structure.

The relation has a different importance in admixtures determination in of alumina sintering by density determination after sintering of superaluminium products with and without admixtures, obtained at the some density of pressing, figure 2.

At the establishment of the technological scheme it referes on the one hand the achievements in fact of the author and on the other hand the operations for wich there was not the technological possibilities of making (final sintering 1850-1950°C).

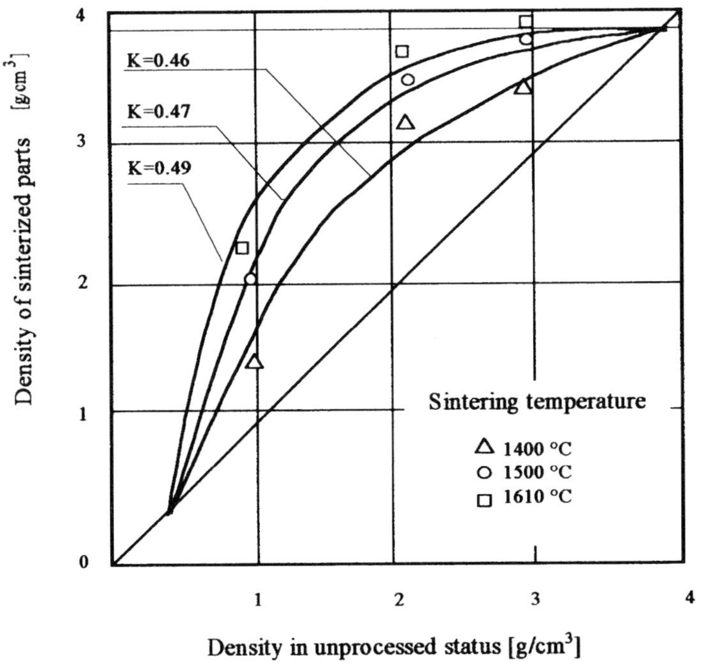

Figure 2. The experimental correlation between the density in non burnt state and the sintered parts from translucent polycrystalline alumina density

7. CONCLUSION

The optimal working pressure gives ensures the shape of the final outline of the parts, again accuracy of mechanic processing of a dies, the smooth and shiny surfaces, uniformity of the thinchness of the walls on length of the part and absence the of deformations. For to avoid nonuniform compacting, fissuring of the parts, in a correlation between the in creasing speed of a force, the time of pressure maintenance and decreasing speed of force is necessary. So decreasing speed as well as decompression speed have generally low spwwds.

On the results obtained from the laboratory and industrial experimental research there were laid the foundations of automated technologic flow for achieving the superaluminium ceramic powders. There were pressed parts from ceramic powder of thin walls of 0,7 -1 mm and big H/D ratio (4:1). There were obtained parts with precise dimension contour, with a dense and thin structure and the surface roughness is R_a =0,2 – 0,4 m according to existing norms. Concerning the structure quality of the parts for uniform densities, it was hexagonal of $Al_2 O_3$ grains and a glassy phase at interstices (triangular zones).

8. REFERENCES

1. C. Dumitru, S. Sontea, D. Petrisor - Cercetari tehnologice privind producerea corpurilor din alumina policristalina translucida. A III-a Conferinta Nationala de "Tehnologii moderne in constructia de masini", Craiova, Sept. 1994, p. 445-450;
2. Z. Aiexander -Sodium Vapour Lamp, U.S. Patent 3.806.747. April 23, 1974;
3. I. Takeo - Development press of transparency by vacuum sintering of alumina, Jaurn of Cer. Soc. of Japan 18, 8-9, 19;
4. E. Labusca - Studiu documentar privind caracteristicile materialelor superaluminoase pentru tuburi de descarcare pentu lampi cu vapori de sodiu, I.F.A. , Bucuresti, 1972;
5. T.Surdeanu, R. Dinescu - Ceramica din oxid de aluminiu, ET, Bucuresti, 1979;
6. C. H. Jackson – Isostatic Presing of Powderend Materials. New Methodes for the Consolidation of Metal Powders. New York, Plenum Press, 1967, pag. 13-27;
7. H. W. Gitzen – Alumina as a ceramic material. Ohio (SUA), The American Ceramic Society, 1970;
8. C. S. Carnighia – Jour. Amer. Ceram. Soc. nr 12, 1973.

Rapid Prototyping of Complex-Shaped Parts of Si/SiC-Ceramics by Laser Sintering

Wolfgang Löschau *, Reinhard Lenk **

* Fraunhofer-Institut für Werkstoff- und Strahltechnik IWS, Dresden

** Fraunhofer-Institut für Keramische Technologien und Sinterwerkstoffe IKTS, Dresden

1 Abstract

Results are presented on generating very complex parts and tools with undercuts, thin walls and internal voids by layer-wise selective laser sintering of pure SiC-powder without additives. The sintered parts are then infiltrated with Si in a vacuum furnace without distortions. The sintering process is brought about by the transient formation of different Si-base phases and by partial disintegration and interlocking of the SiC grains. The final porosity of the Si/SiC-parts is below 1%. Their mechanical and thermal properties (as density, bending strength, Young's modulus, thermal conductivity, thermal expansion coefficient) can be controlled via Si content. They are slightly below those of conventional Si/SiC-ceramics.

It is expected that by means of this technique, rigid light-weight mirrors and wear and temperature resistant dies and tools, for example, can be generated.

2 Introduction

Selective Laser Sintering (SLS) is distinguished by its advantages of layer-by-layer build-up of complex and delicate shapes with voids and undercuts, realized with minimal distortion owing to low sintering temperature.

At Fraunhofer IWS Dresden, the technology of direct generating complex-shaped prototypes from high-melting metal powders, mainly structural and tool steel, has been recently developed. It involves short-time laser liquid phase sintering of compound powders, consisting of a high-melting structural component (as steel) and a low-melting binder (as Cu or Sn-bronze) [1..2]. A liquid phase is necessary with SLS because of the extremely short heating time of $< 0.1 s$.

There is a demand of complex-shaped prototypes of structural ceramics, as light-weight mirrors, turbine and motor components, but there are not many attempts at laser sintering of ceramics reported in the literature. This can be ascribed to the high sintering temperatures, the low wettability by binders, and the notorious brittleness of the material. Experiments are known on indirect laser sintering of Al_2O_3 and SiC with low-melting binder: High volume fractions of binder (about 40%) had been used, often thermoplastics as in Croning sand sintering for casting dies which had to be removed afterwards by a troublesome process [3..5].

It has been the aim of the present work to check the feasibility of direct generation of complex-shaped parts from SiC-powder by laser sintering. SiC was chosen because of the well-known advantages of conventional Si-infiltrated SiC, known as Si/SiC, whose mechanical and thermal properties can be varied within wide bounds by varying the Si content. Conventional

Si/SiC is distinguished by low mass density, high hardness, high strength, high thermal and chemical resistance.

3 Solution of the Problem

The basic idea consisted in generating the object in two steps:

1. Layer-by-layer generation of green parts by laser liquid phase sintering involving partial oxidation of SiC during laser irradiation which gives rise to a vitreous binder phase.
2. Infiltration of the porous green parts with molten Si in a vacuum furnace at moderate temperature (just above the melting point of Si) for increasing strength and density, essentially without distortion.

4 Experimental Set-Up

The principle of SLS of powders is shown in Figure 1. A complex shaped part is build up layerwise, while loose powder is deposited by a blade with typical layer thicknesses of 50 ...200 μm. Each layer is irratiated by a scanned laser beam over that area, which has to be sintered. After irratiation the support sinks by the amount of layer thickness and the process starts anew.

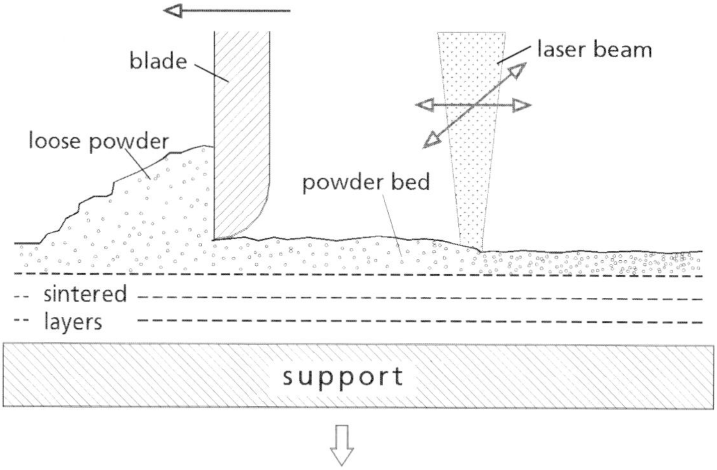

Figure 1. Principle of the laser powder sintering (SLS)

Commercially available pure SiC powder of 20-30 μm grain size was used. For reasons of heat conduction, the smoothly running but rather gritty powder had to be deposited in 50 μm layers. The sintering of the layers was done by means of a metal sintering machine with x/y-scanner, EOSINT M160 of the EOS company, which makes use of a focussed 100 W beam of a pulsed CO_2-laser. With optimized parameters and properly designed support, the whole build-up process can be run automatically. The process chamber of the device was flushed with an argon /air mixture or with plain air.

Preliminary test runs for process optimization were done with prismatic shapes. The results of the test runs were judged according to sticking to the support, adhesion between layers, crack formation, shape accuracy, and microstructure.

Optimum parameter ranges for generation of large parts were found to be 45 ... 52 W and 200 ... 300 mm/s. Infiltration with Si was done in a vacuum furnace at 1600 °C.

5 Results

The generated samples deviated from the CAD data by no more than 0.2 mm. The microstructure was investigated mainly on cross sections by light microscopy, SEM+EDX, and micro-Raman spectrometry. In contrast to the liquid phase sintering of metals, the strengthening mechanisms of SiC ceramics turned out to be much more complicated than expected. The layers are mutually interlocked. This is brought about by partial disintegration of the SiC-particles in the upper part of each layer, which is more or less intense depending on the energy density (Fig. 2). There is virtually no agglomeration of particles as in metal powder sintering. However, vitreous seams with high oxygen content (probably SiO_2) were observed on top of each layer, which glues the disintegrated SiC-particles together. Spongy structures on top of these seams have been found to consist of redeposited SiC. Elementary Si has been detected between the primary SiC grains.

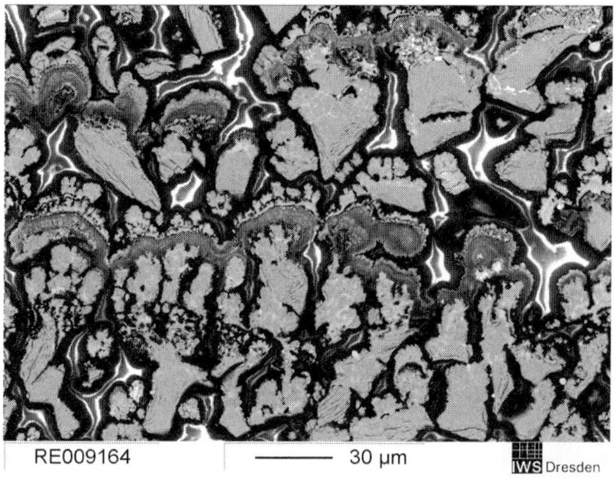

Figure 2. Cross section of a laser sintered SiC rod, SEM (BSE)

The density of the green parts was 42 ...52 %. The prisms and hollow objects had got enough strength to be handled but did not stand much loading.The infiltration of the green parts was flawless. They showed a distinct laminated structure (Fig.3) with an open porosity of about 0.5%. Their content of elemental Si is higher than that of conventional Si/SiC. As a result of infiltration, the transient binder phases have transformed into constituents of the final material. Prior to Si infiltration, some of the green parts were infiltrated with a carbon compound in order to obtain two useful effects: Easier handling of the green parts due to higher strength, and formation of secondary SiC after pyrolysis during Si infiltration. Higher SiC:Si ratios make higher thermal and chemical resistance.

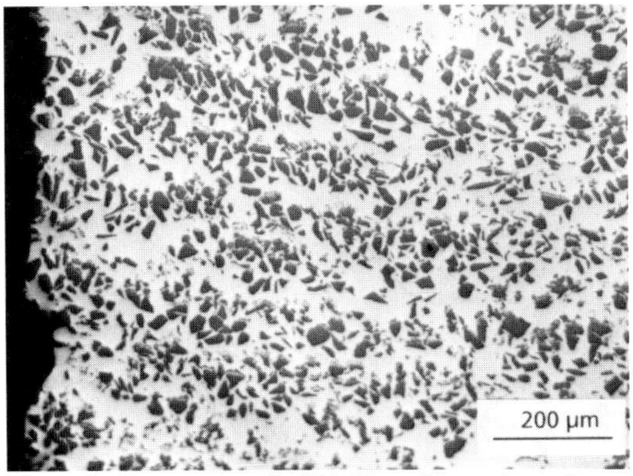

Figure 3. Cross section of a Si-infiltrated laser sintered sample, optical micrograph

High surface quality of the infiltrated parts was obtained by grinding and lapping (see below). Mechanical and thermal properties have been measured on laser sintered, infiltrated and ground bending bars for comparison with conventional Si/SiC (Table 1). Density, bending strength, and Young's modulus of the laser sintered material are slightly lower, which is expected from the lower SiC content. Remarkably, thermal conductivity is high so that it is comparable with that of pure iron, and thermal expansion is low. The results are well reproducable [6].

Table 1. Properties of infiltrated laser sintered Si/SiC samples

		Standard - Si/SiC	Laser - Si/SiC
Mass density	g/cm^3	3.05	2.65
Volume fraction	%	> 99	> 99
Bending strength (3p, rt)	MPa	350	195
Young´s modulus	GPa	320	225
Thermal expansion coefficient	10^{-7} K^{-1}	42	40
Thermal conductivity (rt)	W/mK	100	70

After optimization of the process, several model parts were generated:

1. Model of an extremely light-weight mirror support for use in outer space, honeycomb structure with 1 mm wall thickness and 120 mm diameter (Fig. 4).
2. Several laser light-weight mirrors, 55 mm in length, with coolant ducts inside, for use in articulated arms of laser beam guides (Fig. 5).

Figure 4. Laser sintered model mirror (green part, bottom), wall thickness 1 mm

The sintered parts were remarkably true to nominal size. Deformation due to infiltration was found to be as low as 0.1%. After lapping, the roughness R_a of the mirror faces was 20 ...30 nm. Subsequently, a 150 nm gold layer was deposited in the faces for test runs.

There are reasons to expect further improvements of the Si/SiC materials in this way since several parameters can influence both structure and properties: powder density of primary SiC, carbon containing additives, formation of secondary SiC, fraction of infiltrated Si. Larger and more complex parts can be generated by this process if several green parts are jointed together before infiltration.

Figure 5. Laser beam mirror made of Si/SiC-ceramic, left: raw part after laser sintering and infiltration, right: ground and lapped

5 References

1) W. Löschau, K.-P. Wieters, S. Nowotny, B. Kieback in Proceedings 2. ICRPD, International Conference on Rapid Product Development , Stuttgart, 10.-11.Juni 1996
2) W. Löschau, K.-P. Wieters, S. Nowotny, B. Kieback in Proceedings ECLAT 96, 6th European Conference on Laser Treatment of Materials, Stuttgart, 16.-18. September 1996, vol. 2, 919-926
3) D.L. Bourell, H.L. Marcus, J.W. Barlow, J.J. Beaman in The International Journal of Powder Metallurgy, vol. 28, No. 4, 1992, 369-381
4) H.L. Marcus, U. Lakshminaryan, US Patent 5,147,587, Sept. 15, 1992 [5]) A. Manthiram, H.L. Marcus, D.L. Bourell, US Patent 5,431,967, Jul. 11, 1995
5) W. Löschau in Fraunhofer IWS-Jahresbericht 1997, Dresden 1998, 46-47

Large Alumina Parts from an Aqueous Suspension: Hydrolysis Assisted Solidification (HAS)

Saša Novak, Tomaž Kosmač
Jožef Stefan Institute, Ljubljana

1 Abstract

Hydrolysis assisted solidification (HAS) is a new method for net-shaping ceramic green parts from aqueous suspensions. The process exploits thermally activated and/or accelerated hydrolysis of aluminum nitride powder added to highly loaded ceramic suspensions. Several simultaneously occurring effects accompany the hydrolysis of AlN: an increase in the effective solids content of the suspension due to internal water consumption, a decreasing zeta potential due to ammonia formation, an increase in the surface area of the solid phase, etc. The overall result is a rapid increase of the suspension's viscosity and hence the setting of the low viscous suspension to form a rigid solid part. As a result, green parts can be prepared by slip-casting or injection molding within an impermeable mould.

The present paper presents the basic principle of the hydrolysis-assisted solidification process and in addition, the preparation technique for forming alumina ceramic parts is described and the properties of the green and sintered products presented. The green parts of different size, formed by slip-casting or injection molding, show homogeneous microstructures and high green strength. The sintered alumina parts have high density, homogeneous microstructure and good mechanical properties.

2 Introduction

During the last few decades, systems containing gelling agents which cause setting of the water based ceramic feedstock with temperature change, have been intensively studied and new ceramic forming processes have emerged. Due to the ease of debinding, the environmental benefits and the lack of fire and health hazards, the application of an aqueous suspension for forming ceramics is becoming popular. The earliest emerging aqueous ceramic forming principles where those based on gelling methyl-cellulose during heating [1] and gelling agar on cooling [2], gel-casting [3] used polymerization of acrylamide monomer on heating and direct coagulation casting (DCC) [4] an enzymatic decomposition of urea for setting the water based feedstock. In addition, temperature induced forming (TIF) has been suggested as a novel method of ceramic forming based on alumina flocculation [5]. Kosmač and co-workers presented a method of ceramic forming based on setting due to hydrolysis of AlN powder, admixed in a small quantity into the aqueous ceramic suspension [6]. It should be pointed out that each of these setting principles could be applied for casting in an impermeable mold. Solidification of the green body is not affected by externally applied pressure or water absorption by an absorptive mold but only exploits thermally activated chemical processes in the suspensions. After opening the mold, the removed solid green parts have to be dried and during this process stage defects still readily appear and cause a decreased reliability of the sintered ceramics. For this reason, further studies to improve the processes are still needed.

As mentioned above, hydrolysis-assisted-solidification (HAS) exploits the hydrolysis of the AlN powder added as a setting agent to an aqueous suspension of ceramic powders. The aluminium nitride reacts with water and aluminium hydroxide forms, after sintering in the open air this reacts to form aluminum oxide. The process can therefore be applied for those ceramic materials that contain alumina at least as a minor phase, for example: all grades of alumina ceramics, alumina toughened zirconia, sialones [5,7] etc. As described in [8], the increase in viscosity and finally solidification of the suspension is predominantly a consequence of the cumulative effect of (1) flocculation due to a hange of zeta-potential caused by ammonia formation, (2) water consumption during the reaction and (3) the increase in surface area of the solid phase due to newly precipitated alumina hydroxide. The solidification rate depends on many parameters such as particle size of the setting agent (AlN), temperature, volume of the heated suspension etc. Knowledge of these relationships is important to accommodate the process parameters to a particular product.

The HAS forming principle has been used in slip casting as well as in the injection molding variety. Further, the applicability of the process for forming larger parts, that can only be pressed by difficulties due to the high pressures required or slip cast due to void formation were verified and is presented in the present paper.

3 Experimental Procedure

Aqueous alumina suspensions (>60 vol. % solids) were prepared by de-agglomeration and homogenization of commercial alumina powder in water containing a suitable amount of dispersant and 1 wt. % of AlN powder (see table 1). The suspension was further de-aired by slow mixing in vacuum chamber (p = 300 mbar) for 5 minutes.

Table 1. The materials used for the suspension preparation

Material		Producer
Alumina	CT 530SG	Alcoa
Liquid medium	Distilled water	
Dispersant	Dolapix CE 64	Zschimmer & Schwarz
Setting agent	AlN, grade B	Herman Starck

The viscosity of slips during heating were followed using a rotational viscometer (Haake, measuring system NV, D = 100 s^{-1}) during progressive heating from room temperature up to 80°C (heating rate ≈ 1°C per minute) or at a constant temperature. The time needed for the solidification of suspensions was examined visually or by measuring viscosity.

The strength of green and sintered parts was measured by a three-point bend test. The samples were rectangular bars of dimension 4 x 4 x 50 mm, formed by injecting the suspension into a preheated corrosion protected metal mold.

Larger ceramic parts were formed by casting the suspension into plastic or corrosion protected metal molds to produce parts of the following geometry: (a) disk ϕ34 mm x 18 mm,

(b) plate 300 mm x 200 mm x 10 mm and (c) cylinder with a diameter of 45 mm and height of 60 mm. The green parts were formed by heating the closed molds for 15 minutes to 24 hours at temperatures of 25 to 80°C. After removal from the molds the parts were carefully dried in air till their weight was constant and then sintered at 1640°C.

4 Results and Discussion

Fig. 1 shows the thermally activated viscosity increase for a suspension containing 80 wt. % of alumina powder and 1 wt.% of AlN in water. For a comparison, the temperature effect on the viscosity of a suspension without AlN was also analyzed. The viscosity of the both suspensions slightly decreases with increasing temperature. For the AlN containing suspension the viscosity is somewhat lower (solid squares), which is a consequence of the favorable combination of electrokinetic properties of Al_2O_3 and AlN powders [8]. At a temperature of about 65°C, the viscosity of the AlN containing suspension starts to increase steeply and within a few minutes reaches values above the measuring range. Since during the measurement a definite time was needed to heat the measuring system, the next experiment was performed at a constant (preset) temperature of 55°C. Fig. 2 shows that reaction also proceeds at lower temperatures and when initiated, causes a rapid increase in viscosity.

The AlN containing suspension was than cast into plastic containers (ϕ 34 mm, 18 mm height), closed by covers and heated at different temperatures for different times. The consistency of the cast was visually controlled in regular time intervals in order to define the time needed for solidification of the suspension to form a solid green part. The results collected in Table 2 show that solidification time decreases significantly with increasing preset temperature, while the density of sintered parts slightly decreases. Microstructural observation of the sample's cross-section at low magnification revealed that lower density resulted from the presence of individual large spherical pores in the material, while no additional micropores were observed at higher magnification (Fig. 3). This suggests that the observed pores were most probably the consequence of air bubbles remaining in the suspension after homogenization or vapor bubbles created during heating at higher temperatures.

Although the suspension used to form larger ceramic parts contained 87 wt. % (62.7 vol. % of alumina powder), its viscosity was still rather low. As presented in Table 3, after the addition of 1 wt. % AlN powder, the viscosity remained practically unchanged. In order to avoid air bubbles, the suspension was evacuated to 300 mbar during slow mixing. The increase in density of the suspension confirmed that most of the air was eliminated.

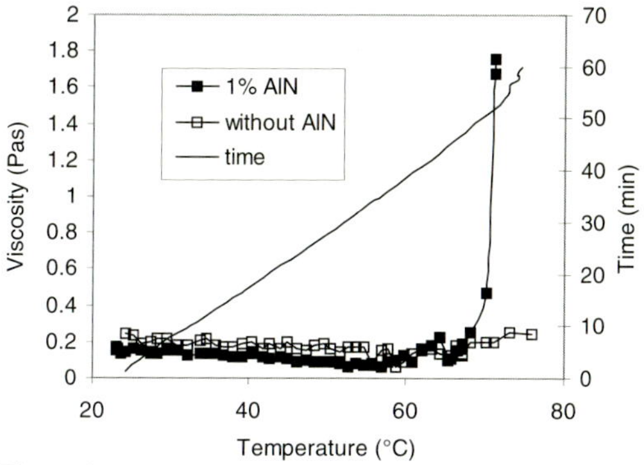

Figure 1: Viscosity of alumina suspension with or without AlN as a function of temperature.

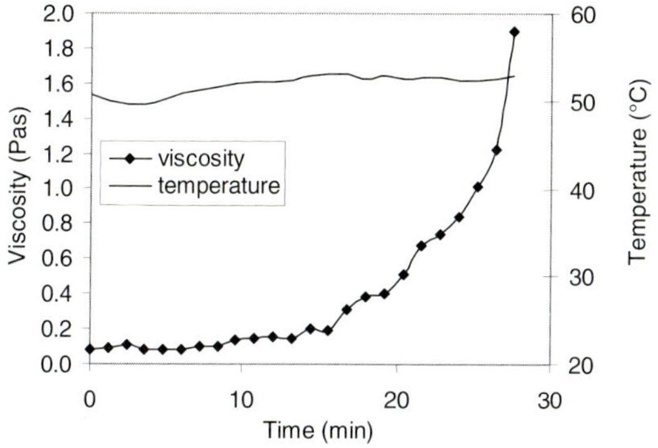

Figure 2: Viscosity of the alumina suspension as a function of time at a constant temperature.

Table 2. The effect of preset temperature on solidification time and density of alumina parts (ϕ = 34 mm, height 18 mm)

Preset temperature (°C)	Time for solidification	Density of sintered parts (% t.d.)
R.t.	> 24 h	98.6
40	6 h	98.3
60	1 h	97.6
80	10 min	96.7

Table 3. Properties of the alumina suspension and of the HAS formed parts

Suspension with 87 wt. % solids	
Viscosity after de-agglomeration, mPas	215
Viscosity after AlN addition, mPas	220
Density of the homogenized suspension, g/cm^3	2.67
Density of the suspension for molding, g/cm^3	2.80
Theoretical suspension density	2.87
Green strength of the solid part, MPa	10.6
Density of the sintered parts, g cm^{-3}	3.91
σ_{TRS}, MPa	390

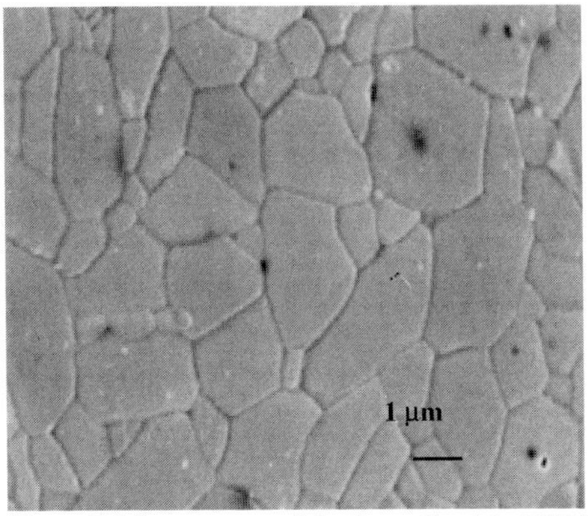

Figure 3: Microstructure of the thermally etched HAS formed alumina (setting temperature = 60°C)

Larger ceramic parts of dimensions 300 x 200 x 10 mm and cylinders with a diameter of 45 mm and height of 60 mm were formed by pouring the suspension into corrosion protected metal and plastic molds, respectively, and heated at 60°C for 2 hours. A longer time was chosen due to a larger volume of the suspension to be heated and solidified. After cooling to room temperature, the molds were open and the solid green parts were dried in air at controlled humidity. Dry parts were sintered at 1640°C for 2 hours. Microstructural analysis of the cross-sections revealed homogeneous microstructure over the whole profile of the products. The green strength was determined to be rather high (see Table 3), which can be ascribed to an aluminum hydroxide "gel" formed by the reaction that binds alumina particles into a stiff network. Finally, the flexural strength of the sintered parts also suggests that the materials properties were not affected by the reaction used for solidification.

5 Conclusions

The Hydrolysis Assisted Solidification (HAS) process was studied with the aim of establishing temperature and time effects on the solidification rate of an alumina aqueous suspension. The setting time decreases with increased temperature, however, the density decreases due to an increased probability of gas formation. This suggests that full attention should be paid to the gas presence. Considering this, the HAS exhibits a great potential for net-shape forming of large alumina parts.

6 Acknowledgements

The financial support of the Ministry of Science and Technology of Slovenia and Comet d.o.o., Zreče, Slovenia is acknowledged. The authors also wish to thank Mr. D. Eterovič and Mrs. M. Vrčon for their help in the experimental work.

7 References

1. R.D. Rivers, US patent No.4113480, 1978.
2. A.J. Fanelli, R.D. Silvers, W.S. Frei, J.V. Burlew, G.B. Marsh, New aqueous injection molding process for ceramic powders, J. Am. Ceram. Soc., 1989, 72, 1833.
3. M.A Janney, O.O. Omatete, Method for molding ceramic powders using a water –based gel casting, US Patent 5028362, 1991.
4. L.J. Gauckler, T. Graule, Process for producing ceramic bodies, WO 94/02429, PCT/CH93/00192.
5. N. S. Bell, L. Wang, W.M. Sigmund, F. Aldinger, Temperature induced forming: Application of bridging flocculation to near-net shape production of ceramic parts, Z. Metallkd., 90, 1999, 6.
6. T. Kosmač, S. Novak, M. Sajko, D. Eterovič, Patent No. EP 0-813-508 B1, 1999.
7. T. Kosmač, K. Krnel, Microstructure and mechanical properties of HAS formed Al_2O_3, Y-TZP and Si_3N_4 ceramics, Proc. 6th Int. Symp. on Ceramic Materials and Components for Engines, Arita, Japan, ed. K. Niihara, Tokyo, 1998, pp. 663.
8. S. Novak, T. Kosmač, Interactions in aqueous Al_2O_3-AlN suspensions during the HAS process, Mat.Sci.Eng. A256, 1998, 237.

Micro Powder Injection Molding

Volker Piotter, Tobias Benzler, Thomas Gietzelt, Robert Ruprecht, Jürgen Haußelt
Forschungszentrum Karlsruhe GmbH, Institut fuer Materialforschung III, P.O. Box 3640, D-76021 Karlsruhe

1 Introduction

Manufacturing of microsystems products is often designated as one of the leading technologies for the next century with a respectable market potential [Wec97]. An important condition for the economic breakthrough is the availability of manufacturing processes suitable for medium and large scale production. For parts made of silicon the wellknown processes from microelectronic production can be adapted while for polymeric materials the different types of micro injection molding are suitable methods. For example, first plastic products manufactured by micro injection molding have successfully entered the market.

But what about metal or ceramic micro parts? Their are also a lot of processes each one characterized through certain advantages but in most cases they have no sufficient qualities overall. An exception is micro powder injection molding because it combines the possibility for large scale production with a wide range of materials, so that it exhibits a rather great economic potential [Pio98]. Therefore, the specialities and possibilities of micro powder injection molding will be described in the following.

2 Powder Injection Molding Experiments

Powder injection molding is an established and economically viable process for manufacturing complex shaped metal or ceramic parts in large scale series.

For this purpose, metal or ceramic powders are mixed with a binder system to generate a thermoplastically processible material, usually called feedstock, which is injected into the mold. In order to achieve a good filling of the mold inserts, feedstocks with low viscosity would be preferable while for safe demolding of microstructures a relatively high mechanical stability of the binder system is necessary. The molded parts are processed further in a furnace to remove the binder system. Subsequently, it is sintered under defined atmosphere to achieve a solid part.

In order to apply this process to microsystems technology, the so-called micro powder injection molding process is currently being developed at the Forschungszentrum Karlsruhe.

The experiments started with a screening of established powders, binders and feedstock systems as well as debinding methods to test their compatibility to micro injection molding [Pio98].

As an example for metals, carbonyl iron powder with a mean particle size of 4-5µm has been chosen and for ceramic materials, most investigations have been carried out with aluminum oxide powder (0.6µm mean particle size). In certain cases, yttrium oxide stabilised zirconium oxide with a mean particle size of 0.3-0.4µm has been used.

For the investigations, two commercially available binder systems were selected: polyolefin/wax compounds and a polyacetal based system. Most experiments have been carried out using both systems for comparison [Pio98].

In the first experiments, test structures were injection molded by MIM with 260µm lateral width and an aspect ratio (comparable to the flow length to wall thickness ratio) of nearly 5 while for further investigations LIGA mold inserts with aspect ratios of more than 10 have been applied. In more than five hundred injection molding cycles so far, mostly with automated machine operation, no wear phenomena have been detected in the nickel mold inserts.

For debinding, three basically different methods are currently under investigation:

- Debinding by thermal elimination of the organic components.
- A catalytical debinding process as commonly used for polyacetal based systems.
- Debinding by supercritical carbon dioxide. This technology uses carbon dioxide under supercritical conditions in an autoclave which normally operates at temperatures of more than 430K and pressures of approx. 300bar.

For debinding of the polyolefin/wax structures, both methods thermal elimination and debinding by supercritical carbon dioxide were examined. Although supercritical debinding works sufficiently, additional modifications of the process parameters are necessary. Thus, a final assessment of supercritical debinding in micro powder injection molding is not possible yet. Therefore, most micro components molded with feedstocks based on polyolefines have been debindered by thermal elimination.

The sintering procedure was carried out in a tube furnace with a diameter of 88mm and a heating zone length of 600mm. While the ceramic micro parts were sintered under air supply, a reducing N_2/H_2-atmosphere was necessary for the metal microstructures to avoid oxidation of metal.

3 Examples of Molded Micro Components

As a typical example of micro components for micromechanical applications stepped gear wheel structures were manufactured using mold inserts made by the LIGA process. For demonstration, carbonyl iron, stainless steel (316L), aluminum oxide and zirconium oxide powders have been chosen for these structures, but processing of other materials should also be possible. Up to now minimal dimensions after sintering are 50µm lateral width and a structural height of approx. 480µm depending on the used microstructured mold inserts.

A second application for powder injection molding of micro components would be the manufacturing of nozzle structures using ceramic materials to obtain longer lifetimes and reduce standstill periods. First patterns with minimal dimensions in the range of 20µm were injection molded in zirconium oxide.

Usually, micro optical devices are made of PMMA or PC resulting in a relatively large thermal expansion limiting there usage to small temperature changes only. Therefore, the application of ceramics which exhibit a much lower coefficient of thermal expansion would be advantageous. Interesting examples are components for optical constructions like e.g. microoptical benches which can be used for communications systems in outdoor environment carrying functional units like microlenses or micro beam splitters. The smallest structural details of these parts are in the range of only 10μm.

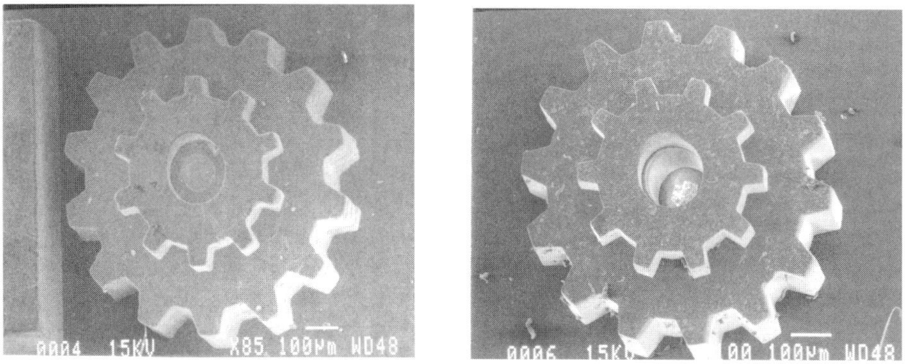

Fig.1,2: Stepped LIGA gear wheels made of stainless steel 316L (left) and zirconium oxide (right). Structure height is about 480 μm and minimum tooth width is about 50 μm; both values measured after sintering.

4 Results and Outlook

One important question during the investigations on micro powder injection molding was, up to what geometries replication could be carried out with the available feedstock systems successfully. These critical dimensions gained in the experiments are listed in table 1.

Table 1: Critical Dimensions in Micro Powder Injection Molding presently.

Material	d_{50} [μm]	Min. lateral dimensions [μm]	Min. structural-details [μm]	Max. aspect ratios
Carbonyl-Fe	4,5	50	50	14
Carbonyl-Fe	1,5	50	50	14
316LS	4,5	50	50	10
WC-Co	k.A.	50	50	10
Al_2O_3	0,4-0,6	50	10	14
ZrO_2	0,2-0,4	50	10	14

The densities of the final parts achieved after sintering are shown in table 2.

Table 2: Achieved Densities after sintering.

Material	d_{50} [μm]	Density [g/cm³]	Theoretical density [g/cm³]	In % of theo. density
Carbonyl-Fe	4,5	7,53	7,87	95,6
Carbonyl-Fe	1,5	7,53	7,87	95,6
316LS	4,5	7,78	8,05	96,7
Al_2O_3	0,4-0,6	3,85	3,98	96,7
ZrO_2	0,2-0,4	6,05	6,1	99,2

Liefern an:

Buchhandlung Buecherland CB 11
Bäckergasse 32
86150 Augsburg
Deutschland

Bestellnummer: 302-9412517-0559546

Vielen Dank für Ihren Einkauf bei BerlinParis2020 auf Amazon.de Marketplace.

Lieferanschrift:	Bestellt am:	Di., 22. Sept. 2026
Buchhandlung Buecherland	Versandart:	Standard
CB 11	Name des Käufers:	Irina
Bäckergasse 32	Name des Verkäufers:	BerlinParis2020
86150 Augsburg		
Deutschland		

Menge	Produktdetails	Preis pro Einheit	Gesamtbestellsumme
1	**Ceramics: Processing, Reliability, Tribology and Wear (Euromat 99) [Gebundene Ausgabe] [2000] Müller, G** **SKU:** AX-41VC-E1C5 **ASIN:** 3527301941 **Zustand:** Gebraucht - Sehr gut	60,00 €	Enthaltene Umsatzsteuer **Zwischensumme Artikel** 60,00 € 0,00 € **Gesamtbetrag Versand** 3,00 € 0,00 € **Gesamtbetrag Artikel** **63,00 €** 0,00 €

Vielen Dank für Ihren Einkauf bei Amazon Marketplace! Unter www.amazon.de/feedback können Sie Ihre Bewertung für diesen Verkäufer abgeben. Wenn Sie den Verkäufer kontaktieren möchten, gehen Sie über "Mein Konto" auf "Meine Bestellungen". Suchen Sie die Bestellung und klicken Sie dort bitte auf die Schaltfläche "Verkäufer kontaktieren" um zum Kontaktformular zu gelangen.

The lowest weights of the parts after separation and finishing were only 0,25mg for LIGA gear wheels made of aluminium oxide. This is much lower than the values mentioned in literature [Ger98]. Otherwise, because typical injection molded arrays of micro parts are connected with the runner systems as well as with auxiliary structures which are necessary to guarantee secure demolding, the total shot weights extend to more than 20g currently.

The experiments described above demonstrate, that the mean particle diameter have a significant influence on the accuracy and surface quality of the replicated structures [Rup98, Pio99a/b]. Special metallic powders with mean particle dimensions in the range of 1-5µm seem to be sufficient for many applications but best results have been achieved by using ceramic powders with mean particle diameters of 0.6µm or even smaller.

The linear shrinkage ranged from 15-22% depending on the composition of the feedstocks. In all cases, nominal sizes in the micrometer range have been obtained within a standard deviation of 0.3-0.7%. Due to the smaller particle size of the ceramic powders, the manufactured micro structures have been performed in better surface quality (Ra=0.02µm) compared to the metal microstructures (Ra=0.5-0.8µm).

After these promising experiments, development of micro powder injection molding continues in order to achieve an industrially feasible process [Pio99a/b, Rup99]. Further materials like alloyed steels, non oxide ceramics which are interesting because of their mechanical strength and relatively high thermal conductivities are currently under investigation. Another topic will be to identify binder systems with higher mechanical strength and stiffness for a better demolding behaviour which is important to obtain better process securities and higher aspect ratios.

5 Literature

[Ger98] German, R.M.: A Rationalization of the Powder Injection Molded Process for Stainless Steels based on Component Features; Proceedings of 1998 International Conference on Powder Metallurgy & Particulate Materials; Las Vegas NV; 1998.

[Pio98] Piotter, V.; Benzler, T.; Ruprecht, R.; Hausselt, J.: Manufacturing of Micro Sized Structures by MIM and CIM, Proceedings of 1998 International Conference on Powder Metallurgy & Particulate Materials; Las Vegas NV; 1998; pp. 5-161 – 5-169.

[Pio99a] Piotter, V. et al.: Innovative molding technologies for the fabrication of components for microsystems; Proceedings of SPIE Conference on Design, Test and Microfabrication of MEMS and MOEMS; SPIE Vol. 3580; Paris; 1999; pp. 456-463.

[Pio99b] Piotter, V.; Benzler, T.; Hausselt, J.: PIM looks for role in the micro world; Metal Powder Report; Elsevier Science Ltd.; 4; June 1999.

[Rot98] Rota, A.; Duong, V.; Hartwig, T.: Micro Powder Metallurgy for Micropart Production; Micro System Technologies 98, VDE-Verlag GmbH, Berlin; 1998; pp. 539-544.

[Rup98] Ruprecht, R.; Piotter, V.; Benzler, T.; Haußelt, J.: Spritzgießen von Mikroteilen aus Kunststoffen, Metallen und Keramiken; Tagungsband 3. Statuskolloquium Mikrosystemtechnik; FZKA-Bericht 6080 Forschungszentrum Karlsruhe; 1998; pp. 83-88.

[Rup99] Ruprecht, R.; Piotter, V.: Abformen von Mikroteilen aus Kunststoffen, Metallen und Keramiken; Jahrbuch Oberflächentechnik 1999; Hüthig Verlag; 1999; pp. 95-108.
[Shi99] Shimizu, T.; Murakoshi, Y.; Wang, Z.; Maeda, R.; Sano, T.: Microfabrication technique for thick structure of metals and PZT; Proceedings of SPIE Conference on Design, Test, and Microfabrication of MEMS/MOEMS; Paris; 1999; pp. 472-477.
[Wec97] R. Wechsung, N. Ünal, J.-C. Eloy, H. Wicht: "Market Analysis for Microsystems" – an Interim Report from the NEXUS TASK FORCE; Proceedings of MICRO SYSTEM Technologies 98; Potsdam; 1998; pp. 275-282.

6 Acknowledgment

We gratefully acknowledge financial support of the Deutsche Forschungsgemeinschaft (DFG) and the supply of materials by the Schunk Sintermetalltechnik GmbH, BASF AG, Hoechst AG, and Degussa AG. We would also like to thank all our colleagues, especially Mr. P. Holzer and Mr. K. Mueller, at the Forschungszentrum Karlsruhe for their helpful support.

Study of Sintering of Oxide Ceramics at Constant Rate of Heating and by means of Rate-Controlled Sintering Method

Karel Maca, Hynek Hadraba, Jaroslav Cihlar
Brno University of Technology

1 Introduction

The sintering process is very important in a wide range of material technologies, with the most important areas including powder metallurgy and ceramic materials. The most current method of densifying ceramic green bodies is high-temperature sintering at normal furnace pressure. Although we are concerned here with an experimentally simple technique (mere heating of bodies for a certain period of time), this process is theoretically very complex and to date there is no definite theory of the sintering process, not even for relatively simple cases. Even when a number of hitherto unresolved problems have been successfully overcome, a mathematical description of the process will be very complex and more demanding in time than a simple laboratory experiment, which has an irreplaceable role in the study of sintering.

Using a high-temperature dilatometer in the study of sintering process has some advantages, the most important of which is the possibility of monitoring longitudinal changes in the sintered body during the whole sintering process [1] and the possibility of applying the rate-controlled sintering method (subsequently referred to as RCS) [2]. In its philosophy the RCS method differs from the widely used method of sintering at a constant rate of heating with a holding time at the temperature of sintering (CRH in the following). With the RSC method, the relation between relative density and time is chosen and the result of experiment is the time dependence of the temperature that enables the given development of relative density. The chosen course of sintering is in fact an empirical optimization and, according to [3], for suitably chosen models equally high final relative densities can be obtained as with the CRH method, with the difference that the increase in grain-size of powder material is smaller.

Materials based on alumina and zirconia are of much promise for the area of medical, electroceramic and engineering applications. The aim of the present paper is a study of the sintering process of advanced ceramic materials based on alumina and zirconia, using a high-temperature dilatometer and both the CRH and the RCS methods.

2 Experimental

The following types of ceramic powder were used to prepare the ceramic bodies:

- Alumina, type RC-HP DBM (Reynolds Chemicals), mean particle size 0.6 μm
- Mixture of 85 wt% alumina + 15 wt% zirconia, type ZTA 85 (Daichii Kigenso), mean particle size 0.7 μm

- Mixture of 80 wt% zirconia + 20 wt% alumina, type ATZ 80 (Daichii Kigenso), mean particle size 1.45 μm
- Zirconia stabilized by 5.15 wt% Y_2O_3, type TZ-3YS (Tosoh), particle size 0.7μm.

Test samples were prepared by injection molding [4]. The samples used were in the form of cylinders and prisms of 10 to 16 mm in length and a maximum transverse dimension of 5.6 mm. The relative density of samples which had been presintered at 800°C/1h was established by measuring the sample mass and dimensions. The relative density of the ceramic materials under study ranged between 61.4% and 63.8%.

Sintering took place in type L70/1700 high-temperature dilatometer (Linseis, Germany) with vertically placed sample. The measuring system of alumina and the sample were placed in free atmosphere; they were insulated from a graphite furnace working with slight nitrogen overpressure by a closed alumina tube.

For CRH method the rates of heating the sample from 800°C to 1500°C (zirconia and ATZ samples) or to 1530°C (alumina and ZTA samples) were from +2°C/min up to +20°C/min. At the maximum temperature there was a holding time of 2 h. For sintering by the RCS method sintering curves were used that were chosen from experience with the sintering of alumina [3], i.e. there were two areas (up to ρ_{rel} = 75%, and up to ρ_{rel} = 85%) with relative density growing linearly with time, and the last area above ρ_{rel} = 85% with the rate of densification decreasing linearly with time. The values of relative sample density were calculated from the values of relative sample shrinkage ε[%] by the following relation, which holds on the assumption of isotropic sample shrinkage

$$\rho_{rel}=\rho_{rel0}(100/(100+\varepsilon))^3 \ [\%], \qquad (1)$$

where ρ_{rel0} [%] is the relative sample density prior to sintering.

The sintered specimens were cut with a diamond disk saw, polished and thermally etched at 1400°C for a period of 5 min. The microstructure was studied on a Philips XL30 electron microscope, and the grain size was assessed from the photographs by the linear intercept method [5].

3 Results

3.1 Sintering at a constant rate of heating

The dependence of relative density of individual types material on temperature with a constant rate of heating of +10°C/min is evident from Figure 1. It follows from the graph that the materials based on zirconia (zirconia and ATZ) sintered at lower temperatures and reached higher relative densities than those based on alumina (alumina and ZTA). It is also evident from the graph that in the region of the holding time at maximum temperature the specimen prepared of zirconia did not sinter at all.

Figure 2 gives the dependence of relative density of injection molded zirconia on temperature and rate of heating. It follows from the graph that using the +2°C/min rate of heating led to higher final relative densities, which held for all the materials studied.

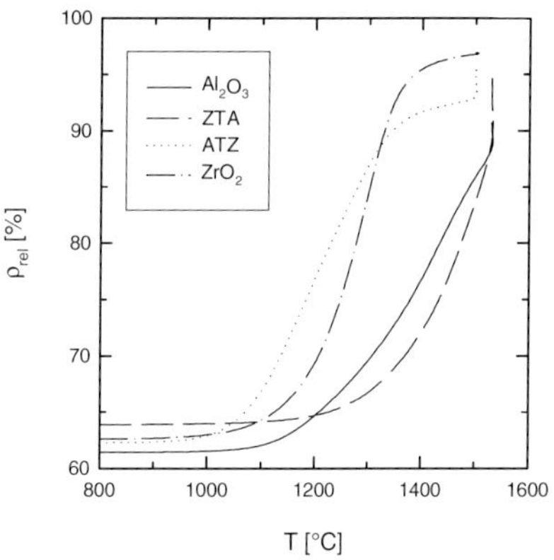

Figure 1. The dependence of relative density on temperature and type of material (heating rate +10°C/min)

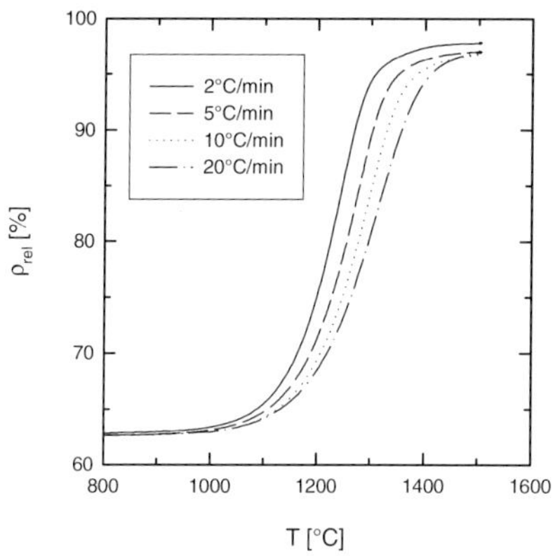

Figure 2. The dependence of relative density of zirconia on temperature and heating rate

3.2 Sintering by the rate-controlled sintering method, and comparison with the constant rate-of-heating method

Figure 3 shows the dependence of relative density of alumina and temperature on time as obtained with the RCS method, and for comparison the same dependence for a specimen sintered at a constant rate of heating of +5°C/min. Similar dependence values for zirconia are given in Figure 4. In both cases the two methods yielded similar resultant densities in comparable time. From the graphs we can see the difference between the two density – time plots: with the RCS method there was after the first rapid phase (from $\rho_{rel} = 75\%$) a decrease in the rate of sintering, which also found correspondence in the very slow increase in temperature. The temperature then started increasing more rapidly at relative densities of over 90%. By contrast, with the CRH method the maximum rate of densification was reached with the relative density ranging between 75 and 90%, after which it dropped rapidly and in the region of the two-hour holding time the specimen density increased very slowly or did not increase at all.

The different course of sintering cycle was also reflected in the microstructure of specimens. Figure 5 gives the microstructure of thermally etched alumina sintered by the RCS method while Figure 6 gives the microstructure of the same material sintered by the CRH method at a rate of heating of +5°C/min and two hours' holding time at 1530°C. A similar comparison of microstructures is given in Figures 7 and 8 but for zirconia. The application of RCS method resulted in a smaller size of grains and their narrower size distribution than was the case with the CRH method. The average grain size as established by the linear intercept method was as follows: alumina – 0.96 μm (RCS) and 1.6 μm (CRH); zirconia – 0.28 μm (RCS) and 0.40 μm (CRH).

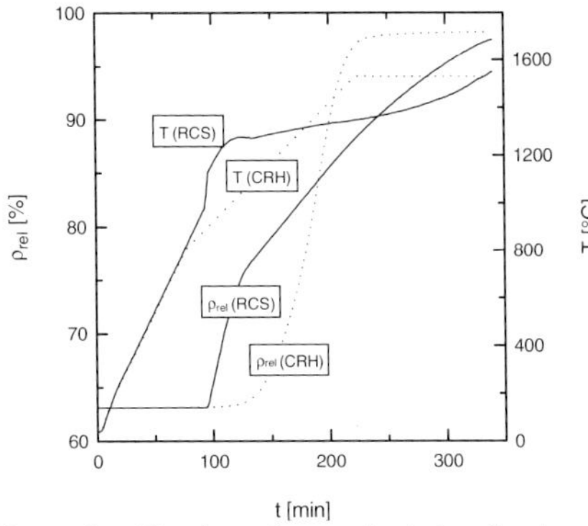

Figure 3. The dependence of relative density of alumina and temperature on time and sintering method

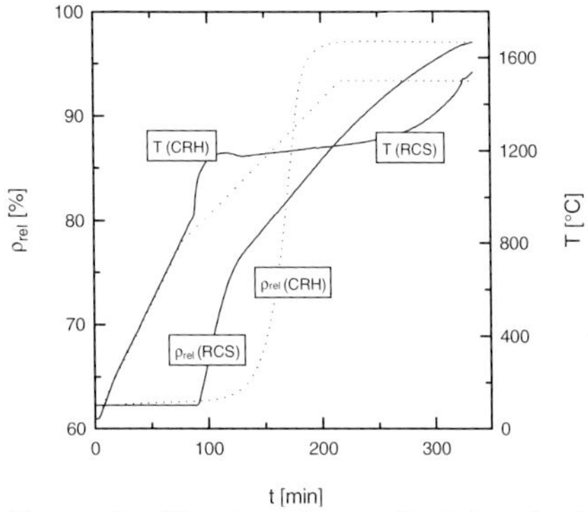

Figure 4. The dependence of relative density of zirconia and temperature on time and sintering method

4 Discussion

Dilatometric measurements of the sintering process of individual types of material revealed that alumina-based materials had sintered at higher temperatures than zirconia-based materials. This phenomenon may have resulted on the one hand from the difference in the structure of the two types of material (zirconia as an ion conductor has higher values of self-diffusion coefficients than is the case with alumina), with the mixtures retaining from this viewpoint the properties of majority phase, on the other hand from the internal microstructure of particles. Unlike alumina, zirconia grains were aggregates of smaller particles, as can be seen in Figures 7 and 8. According to the manufacturer, these primary particles of zirconia were about 0.036 μm in size, which then affected the kinetics of sintering.

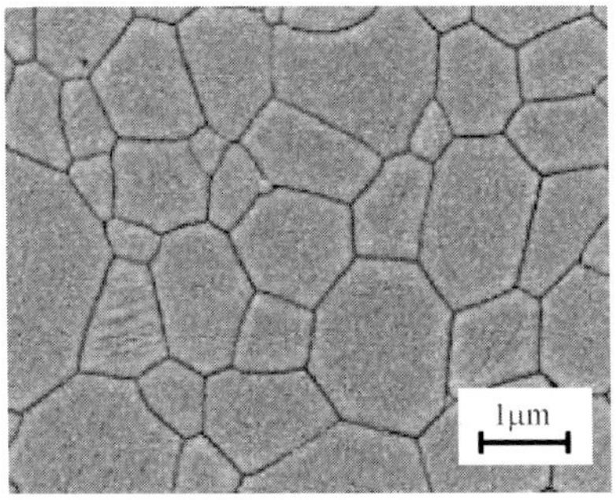

Figure 5. The micrograph of thermally etched alumina after sintering by means of rate-controlled sintering method

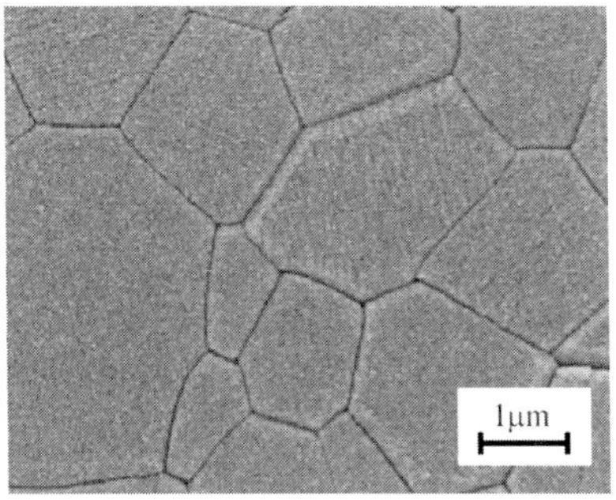

Figure 6. The micrograph of thermally etched alumina after sintering at constant rate of heating (+5°C/min)

Figure 7. The micrograph of thermally etched zirconia after sintering by means of rate-controlled sintering method

Figure 8. The micrograph of thermally etched zirconia after sintering at constant rate of heating (+5°C/min)

As was evident from Figures 1 and 2, zirconia material already reached its maximum relative density before the beginning of the two-hour holding time at 1500°C. Introducing the holding time then led to an unnecessary growth of grains of material and also to a waste of energy. To arrive at this conclusion with the usual sintering of specimens in a furnace and measuring their dimensions before and after firing would require conducting several meticulously planned experiments. At the same time, however, it must be said that a disadvantage of the dilatometric measurements of sintering kinetics lies in the limited specimen dimensions and in the assumption of isotropic specimen shrinkage, which in practice happens with lesser or greater accuracy. From this point of view it seems to be optimal to complement dilatometric measurements of small specimens with sintering experiments with bodies of currently occurring size and shape.

With the CRH method the highest relative densities obtained for all materials were with a temperature gradient of +2°C/min (see, for example, Figure 2). At lower rates of heating the material sinters for a longer period at lower temperatures, i.e. in the open phase porosity, when tubular pores are gradually narrowed down and isolated pores appear. At higher rates of heating, the tubular pores may get narrowed down before they reach their critical radius, the newly appearing isolated pores are larger than in the ideal case and thus need not disappear in the closed porosity phase. At higher rates of heating some pores may break away from grain boundaries. Such pores inside the grains need not disappear due to the lower rate of volume diffusion compared to diffusion along grain boundaries.

Applying the RCS method to the sintering of alumina and zirconia specimens yielded resultant relative densities comparable with those obtained by the CRH method, with the resultant grain size about 1.7 times smaller for the alumina specimen and 1.4 times smaller for the zirconia specimen than when the CRH method was applied. These results are in agreement with results reported on the sintering fine-grain (0.1 – 0.2 µm) alumina with an addition of 0.1% MgO [3]. The different microstructure is evidently the result of the very slow sintering in the area of open porosity. The results obtained for sintering will be complemented with the firing of ceramic disks according to temperature regimes established by the RCS method, the aim being the study of the effect of RCS method on mechanical properties.

The size of particles measured by the linear intercept method cannot be compared with the size of input powder material because this method is only used to establish the mean value of a random intercept of a random plane section through the grain. The grain size value making possible the comparison with the size of powder material can be found approximately by multiplying by a coefficient of 1.62 [1]. The size of alumina particles after RCS sintering was about 1.6 μm, after CRH sintering 2.6 μm; the grains had thus grown 2.3 times and 4.3 times, respectively. With zirconia grains the situation is more complex. Even after recalculation to the spherical shape of grain the post-sintering grain size comes out 0.45 μm (RCS) and 0.65 μm (CRH), which is less than the size measured for input powder (0.7 μm). This can be explained either by the disintegration of agglomerates forming the grains during isostatic pressing or injection molding, or by the shrinkage of grains due to the disappearance of their internal porosity in the process of sintering.

5 Conclusion

Sintering alumina- and zirconia-based materials in a high-temperature dilatometer has yielded the following results:

- alumina-based ceramics sintered at higher temperatures than zirconia-based ceramics,
- the 2°C/min rate of sintering led to highest resultant densities for all the materials considered,
- the rate-controlled sintering method made it possible to obtain similar resultant relative densities as the constant rate-of-heating method but it limited the growth of grains of material.

6 References

1. K. Maca, H. Hadraba, J. Cihlar, Ceramics-Silikaty 1998, 42, 151-158
2. Palmour III et all. US Patent 3 900 542, 1975
3. M.L. Huckabee, H. Palmour III, Am. Ceram. Soc. Bull. 1976, 51, 574-576
4. M. Trunec, J. Cihlar, J. Eur. Ceram. Soc. 1997, 17, 203-209
5. Annual Book of ASTM Standards, Part II, American Society for Testing and Materials, Philadelphia, USA, 1982, E112-82

Rheology of Water-Based Tape Casting Slurries

Bernd Bitterlich[1], Christiane Lutz, Andreas Roosen
University of Erlangen-Nuremberg, Department of Materials Science,Glass and Ceramics, 91058 Erlangen, Germany
[1] Now at: Institute for Nonmetallic Materials, Technical University of Clausthal, Germany

1 Introduction

The tape casting process is an economical method to produce thin, flat ceramic components such as substrates, capacitors, piezoactuators, sensors etc [1, 2]. In comparison with organic solvents water-based systems have the advantage of low toxicity and environmentally safety. The control of the rheological behavior of tape casting slurries is very important to get a homogeneous product which meets the quality criteria of the final product. The rheology of the slurry depends on the type and portions of powder, binder, solvent and other organic additives such as dispersants or wettings agents. The replacement of organic solvents by water has a strong influence on the rheological behavior of the slurry. Water exhibits a higher surface tension than organic solvents. The solubility of binders in water is also limited. In many cases emulsion binders are used, which exhibit a higher viscosity [3].

The contents of water and organic additives in tape casting slurries should be as low as possible to improve the drying and binder-burnout processes. This results in high loaded slurries with high viscosities. For tape casting the viscosity of the slurries must be low enough to carry out the process and the slurry must exhibit pseudoplastic behavior: During passing the blade, the viscosity is decreased due to shear forces. Immediately after the blade the viscosity should rapidly increase again to suppress uncontrolled flow and to prevent sedimentation of the ceramic particles. Thixotropy and any other time dependent behavior are undesired because the rheological behavior becomes dependent on the pre-treatment of the slurry.

The degree of viscosity is determined by small forces, which exist between the suspended particles. In an undisturbed slurry these forces build up an internal structure. With increasing shearing the viscosity is reduced, until a constant level is reached. Dynamical measurements of the viscosity give information about the internal structure and their changes during shearing. For viscoelastic substances the shear module consist of two portions. The storage module G' describes the elastic properties of the substance. The loss module G'' gives the fracture of the applied stress which produces viscous deformation.

In this work aqueous tape casting slurries for the manufacturing of thin electrolytes of the SOFC were examined by rheological measurements. Viscosity, time dependent behavior and strength of the internal structure of the slurries were characterized.

2 Experimental Procedure

Tape casting slurries were prepared from yttria stabilized zirconia (TZ8Y, Tosoh-Zirconia Corp., Japan) as the ceramic filler and an polymer emulsion as the binder (Mowilith DM765 S, Hoechst AG, Germany). The dispersed ceramic particles have an average particle size of 0,26 μm. The binder tangles in the emulsion have a diameter of appr. 0,17 μm. The powder was dispersed with 2 weight-% of a polyelectrolyte (Dispex A40, Allied Colloids Ltd., UK) by ball milling. Slurries with different contents of solid and binder were tested. Table 1 lists the two slurry compositions which were selected for further rheological characterization. The experiments were carried out on a stress-controlled rheometer (UDS 200, Paar Physica, Austria) at 22°C.

Table 1. Composition of examined tape casting slurries (Vol-%).

	Water (total)	Powder	Binder+Additives
Pure Binder	50	-	50
Slurry A	53,4	18,1	28,6
Slurry B	52,7	18,8	28,5

After preshearing the slurry (5 s^{-1}, 60 s), *viscosity curves* were measured by increasing the shear rate continuously in 120 s from 5 to 300 s^{-1}.

The *thixotropy* was characterized by the recovery time of the viscosity after a steep decrease from a high to a low shear rate: After preshearing (300 s^{-1}, 60 s) the shear rate was immediately reduced to 1 s^{-1}. The dependence of the viscosity η with time can be described by an exponential function:

$$\eta(t) = A - B\exp\left(-\frac{t}{THIX}\right) \tag{1}$$

The parameters A and B are given by the viscosities before (=A - B) and after (=A) the steep decrease of the shear rate which occurs at the time t=0. The variable *THIX* (dimension: seconds) characterizes the thixotropic behavior: the greater *THIX* the more thixotropic the slurry behaves.

Dynamical measurements were carried out to characterize the changes of the *internal structure* of the slurry with time and their strength which is given by a critical shear stress [4]. The experimental steps were as follows: preshearing (100 s^{-1}, 60 s), oscillation at a frequency of 1 rad/s with an amplitude of 0.1 Pa. During 25 minutes the value of the storage module was recorded. Directly after this the stress amplitude was increased logarithmically up to 500 Pa at the same frequency.

3 Results and Discussion

Figure 1 shows the viscosity curves of the pure emulsion binder and the slurries. The emulsion binder itself shows the desired strong pseudoplasticity. The addition of ceramic particles increases the viscosity of the slurries, but their general rheological behavior is still like the pure binder.

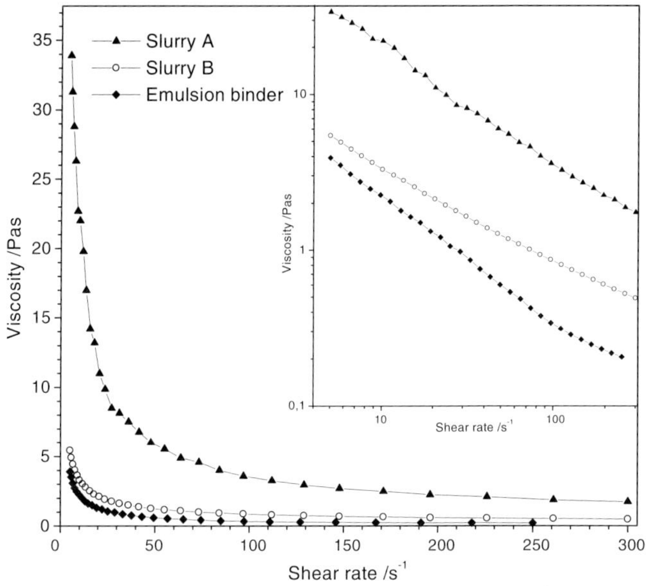

Figure 1. Viscosity curves of the slurries after preshearing at 5 s^{-1} for 60 s.

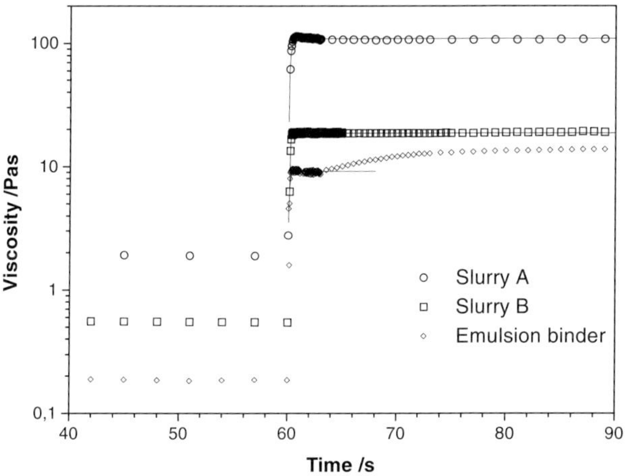

Figure 2. Viscosity increase of the slurries after a steep decrease in shear rate from 300 to 1 s^{-1} at t = 60 s, (line: fitted functions).

The slurries exhibit scarcely no time dependent behavior (Fig. 2). At the steep decrease of the shear rate the viscosity increases almost immediately. The values of the parameter *THIX*, which represents the grade of thixotropy, are very low: 0.12 s for the emulsion binder, 0.16 s for slurry A, 0.06 s slurry B. These results were varified by additional experiments in which viscosity curves are measured with increasing and decreasing shear rates. No hysteresis could be detected which confirmed that the slurries behave non-thixotropic.

During a long time oscillation of 25 minutes almost no changes in the internal structure of the slurries occurred with time. The storage module remains nearly constant during this period. This means that these slurries build up their internal structure very fast and do not change it afterwards. This behavior is preferred in tape casting, because these slurries will not change their properties during the process

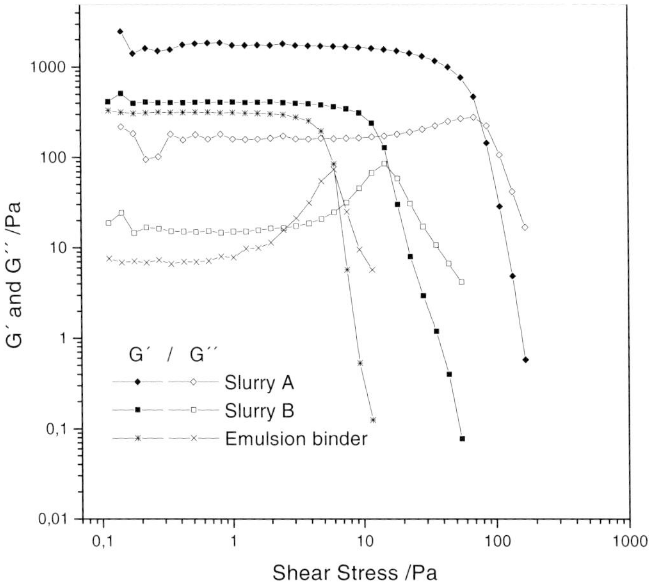

Figure 3. Storage and loss module during oscillation at 1 rad/s with stepwise increased stress amplitude.

Figure 3 shows the modules *G´* and G´´ in dependency on the shear stress amplitude. The storage module *G´* of the slurries is much higher than that of the pure emulsion binder and reaches 2000 Pa. This is caused by the high solid content which lead to greater elastic properties of the slurry. At low shear stresses the modules remain nearly constant. The storage module *G´* exhibit a sudden decrease at a critical shear stress τ_C. This critical stress has its origin in the internal structure of the slurry [5]. Above τ_C the weak forces between the powder and/or binder particles are broken up by the external shear stress. The elastic properties of the slurry become very low and the storage module decreases more than 3 orders of magnitude. The loss module *G´´* reaches a maximum at this point. This is also due to the breaking up of the weak interconnections between the particles in the slurry at these shear rates: energy which is transferred into the system, cannot be stored as elastic deformation like before. The tendency for viscous deformation increases. For tape casting the slurry must viscous enough to carry out the process. That means a shear stress above τ_C must be applied to the slurry.

For tape casting a high critical shear stress τ_C is desired, because then gravitational forces on the powder are too weak to cause viscous deformation in the slurry, which could result in sedimentation of the particles. On the other hand τ_C must not be too high because the slurry must flow under the shear conditions which exist during passing the doctor blades. τ_C of Slurry A has a value of 150 Pa which is quite high. A higher value lead to difficulties in casting the slurry because the slurry does not flow out of its container.

4 Conclusion

The results demonstrate that rheological measurements are an appropriate tool to decide if the rheological properties of the slurries match the demands of the tape casting process. These kind of measurements are of course applicable to other processes like e.g. slip casting. The experimental condition to apply a steep decrease in shear rate to the slurry is a simple and fast test method to examine the grade of thixotropy. Furthermore the results of this method are much more insensitive to the use of different measurement systems.

The examined slurries exhibit almost no thixotropic or other time dependent behavior which is very desired for the tape casting process. The structure, examined by dynamical measurements, showed almost no changes with time which indicates that the properties of the slurries will stay nearly constant during the process. Strong pseudoplasticity was found which is determined by the pseudoplastic behavior of the pure emulsion binder with its very small polymer particles and by the high content of fine grained ceramic powder. The slurries exhibit distinctive elastic properties below a critical shear stress τ_C indicated by a high storage module. Above τ_C the internal structure is broken up. This is necessary to increase the viscous behavior of the slurries to carry out the tape casting process. τ_C should high enough to prevent sedimentation of the ceramic particles after casting the slurry. But it must not be too high so that the slurry is still castable.

5 References

1 R.E. Mistler, D.J. Shanefield, R.B. Runk in Ceramic Processing before Firing (Eds: G. Onoda, L.L. Hench)John Wiley & Sons Ltd., New York, 1978, 411-448

2 A. Roosen, Ceramic Transactions 1988, Vol. 1B, pp. 675-692

3 F. Doreau, G. Tari, C. Pagnoux, T. Chartier, J.M.F. Ferreira, J. Eur. Ceram. Soc. 1998, 18, 311-321

4 J.M. Keller, R.R. Ulbrich, R.A. Haber, Am. Ceram. Soc. Bull. 1997, 76, 89-91

5 J.W. Goodwin, Am. Ceram. Soc. Bull. 1990, 69, 1694-1698

Liquid Phase Sintered SiC with SiO_2 Additive

Haihui Ye , Georg Rixecker and Fritz Aldinger
Max-Planck-Institut für Metallforschung and Institut für Nichtmetallische Anorganische Materialien, Universität Stuttgart, Pulvermetallurgisches Laboratorium, Heisenbergstrasse 5, D-70569 Stuttgart, Germany

1 Abstract

SiC liquid phase sintered with using SiO_2 as the only sintering additive was studied as a model system. The mixture of SiC and SiO_2 was supposed to form a liquid phase at low temperature and thereby promote the densification with the precondition that the decomposition reaction between SiC and SiO_2 ($SiC + 2SiO_2 \rightarrow 3SiO + CO$) could be restricted by using CO gas as sintering atmosphere. The motivation for studying this special system despite its limited thermodynamic stability comes from the theoretical work by D. R. Clarke [1] on the equilibrium thickness of intergranular glass phases in ceramic materials. His calculations showed that only in the pure system SiC/SiO_2 , the siliceous intergranular phase will be unstable because the relatively large van der Waals attractive force between SiC grains outweighs the repulsive structural disjoining pressure of the silica glass phase, causing its equilibrium thickness to become zero. Therefore, the residue of the glass phase, which is necessary in the process of liquid phase sintering, would move to the triple junctions at the end of sintering. The experimental microstructures were observed by scanning electron microscopy (SEM). More detailed investigations will be performed by means of transmission electron microscopy (TEM).

2 Introduction

In this paper, a model system of pressureless liquid phase sintered SiC with SiO_2 as the only sintering additive was studied. The motivation for studying this material comes from the theoretical work by D. R. Clarke[1] who suspected that SiO_2 will be unstable in the grain boundary of SiC particles and finally move to the triple junctions.

Pressureless sintering of SiC was first carried out by Prochazka with small amounts of B and C as sintering aids [2]. Numerous studies were then conducted on the pressureless liquid phase sintering of SiC. Different metals and metal oxides and nitrides such as Al-C[3], Al_2O_3-Y_2O_3 [4], and AlN-Y_2O_3 [5] have been used as sintering additives. All the above studies conclude that an effective sintering additive is extremely important to achieve high density of the SiC compact as a prerequisite for the achievement of high strength in SiC ceramics.

The densities of SiC- Si_3N_4 composites prepared from oxide-covered powder compacts were significantly higher than those of the composites prepared from deoxidized powders [7]. This is because the silica reduces the eutectic temperature and promote the presence of the liquid phase. But when a large quantity of SiO_2 is added to SiC powder, it would cause serious damage on the quality of sintered bodies because of the violent gas-producing reaction between SiC and SiO_2 at high temperatures [8]. This results in the conclusion that the quantity of SiO_2 in the green body should be kept as little as possible. If SiO_2 is to be used as a sintering aid, it can advantageously be applied in the form of an oxide layer coating the SiC particles [6].

In order to perform a detailed microstructure investigation on the characteristics of the grain boundaries by transmission electron microscopy, well-sintered samples with a high enough density are necessary. This paper shows that pretty high sintering densities can be achieved. The mechanism of liquid phase sintering will be discussed.

3 Experiment

As raw material, α-SiC (UF15, H.C.Starck) was used. The amount of silica as sintering additive was zero, 5wt%, 10wt%, or 20wt%. The powder mixtures were prepared by attrition milling in isopropanol with Si_3N_4 milling media for 4h using a polyethylene container and stirrer. The slurry was separated from the milling media via a sieve chain and subsequently dried in a vacuum evaporator. Completely dried mixtures were obtained after 48h at 65°C in a drying chamber. Finally the processed powder was sieved, to obtain granules with a maximum diameter of 160μm, and cold isostatically pressed at a pressure of 800 MPa into cylinders of 15mm∅ × 12mm high. Sintering was performed in a gas-pressure furnace (KCE Kessel GmbH, Rödental) with a graphite heating element. The main sintering conditions are shown in Table 1. To restrict the reaction between SiC and SiO_2, CO was used as sintering atmosphere. CO gas was transferred into the furnace when the temperature had reached 800°C. The CO pressure was released after the end of the isothermal holding period (0.5h).

Table 1. The sintering conditions

No.	Sintering Temperature (°C)	Additions of SiO_2 (wt%)	Sintering Atmosphere and Pressure (MPa)
1	1720	20%	1.5MPa CO
2	1790	20%	1.5MPa CO
3	1860	0, 5%, 10%, 20%	0.1MPa CO & 0.4MPa N_2
4	1950	0, 5%, 10%, 20%	0.1MPa CO & 0.4MPa N_2

The densities of the sintered samples were measured by the Archimedes principle in distilled water at room temperature. The theoretical densities of the composites were calculated using the rule of mixtures, taking the densities of SiC as 3.22g/cm^3 and SiO_2 as 2.20 g/cm^3 . X-ray diffraction patterns (Siemens D5000, Cu K_α radiation) were used to identify the crystalline phases. After sintering, the hexagonal crystal structure of α-SiC was retained, and no crystal SiO_2 could be found. Specimens for scanning electron microscopy (Zeiss DSM982 Gemini, equipped with a Schottky field emission cathode) were polished to 1 μm using routine ceramographic techniques. The microstructure of the sintered samples was observed using SEM with magnifications from ×1000 to ×50000. Preliminary TEM work was done by using a JEOL JEM 4000FX instrument on samples which were prepared with standard techniques, starting from mechanical slicing, polishing, dimpling, and ending with ion-milling until perforation.

4 Results and Discussion

Fig. 1 shows the change of density of the SiC-20wt%SiO_2 specimens with increasing sintering temperature. Quite high density (94.2% of theoretical density) has been achieved at 1860°C, while both higher and lower temperature caused a sharp decrease of the density.

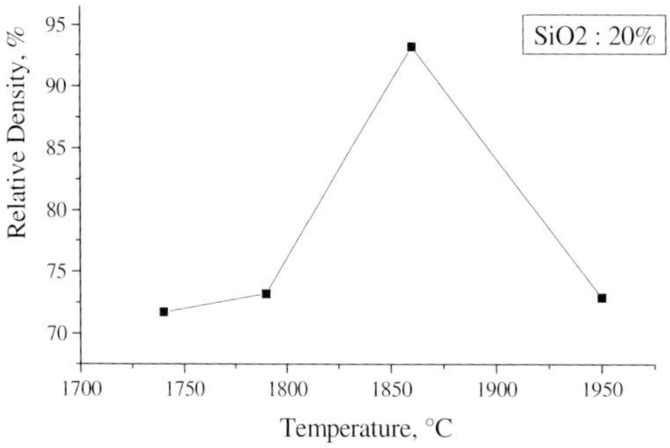

Fig. 1. Change of density of SiC - 20wt%SiO_2 with the sintering temperature.

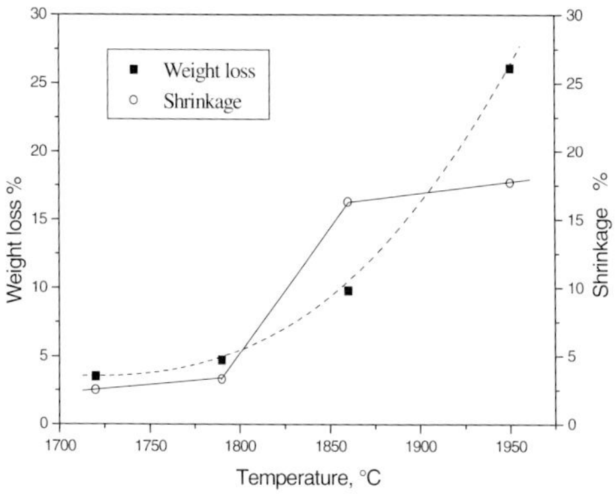

Fig. 2. Variation of weight loss and shrinkage of SiC-20wt%SiO_2 with sintering temperature

The density decrease below 1860°C is considered to be the consequence of an insufficient amount of liquid phase. Although liquid phase sintering may not be the only mass-transfer process, clearly the dominant mechanism which determine the final high density must involve the liquid phase [9]. The lowest melting temperature of SiO_2-SiC is 1723°C, which explains why there is no sufficient amount of liquid phase when sintering at only 1720°C. To explain the low density at the raised sintering temperature of 1790°C, the kinetic aspects of densification should be considered. 30 minutes of sintering time may be not long enough to get the equilibrium amount of liquid phase with a low enough viscosity to complete the rearrangement of the SiC particles. The above consideration is supported by the shrinkage-temperature curve (Fig. 2) which shows that shrinkage increased sharply between 1790°C and 1860°C, indicating that a large quantity of liquid phase formed only in this temperature interval.

Fig. 3. SEM photograph of SiC-20wt%SiO_2 sintered at 1720°C (3a), 1790°C (3b), and 1860°C (3c) respectively.

Through SEM images (Fig. 3) the change of micro-structure caused by the presence of the liquid phase can be monitored. While the microstructure of the sample sintered at 1790°C (Fig. 3b) does not show much difference to that of the sample sintered at 1720°C (Fig. 3a); it merely seems to include some more clusters. In contrast, the microstructure of the sample sintered at 1860°C (Fig.3c) looks strikingly different: the sample is much more dense, the grain size is much larger and the liquid phase can be easily identified between the SiC particles.

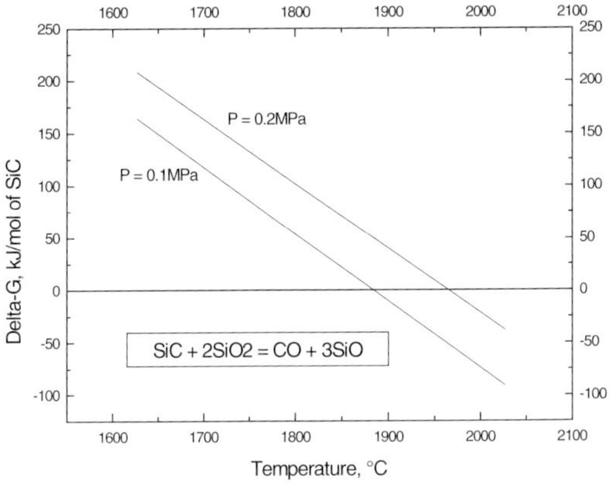

Fig. 4. Free-energy change of reaction (1) with different sintering pressure as a function of temperature.

On the other hand, poor sintering of the samples at 1950°C is thought to be associated with the increase in weight loss due to reactions between SiC and SiO_2 (see Fig. 2), the most possible reaction being [8]:

$$2SiO_2 + SiC \rightarrow 3SiO(g) + CO(g) \quad (1)$$

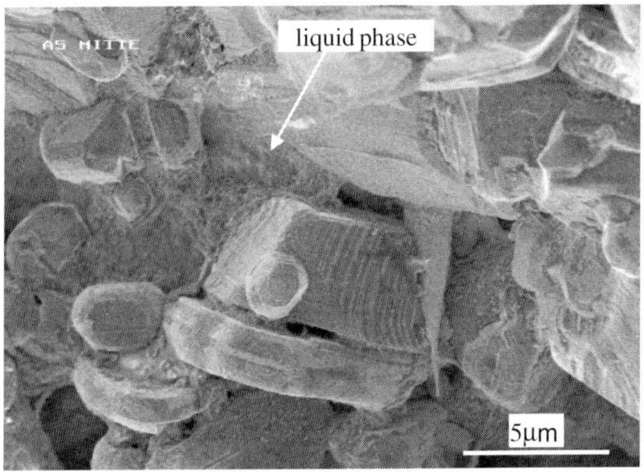

Fig. 5. SEM photograph of SiC-20wt% SiO_2 sintered at 1950°C

At low temperatures (1400K – 1600K) [6], this reaction was found to be very slow to reach the steady state, but at elevated temperatures, it became violent and the volatilization of SiO and CO caused large weight loss and extremely low density. The point where the reaction rate accelerates can be obtained by thermodynamical calculation. Fig.4 shows the free-energy change of reaction (1) at different total pressures as a function of temperature. In a temperature range of 1900°C ~ 1970°C, ΔG changes from positive to negative, i.e. the reaction of SiC and SiO_2 becomes thermodynamically favored. This range of temperature corresponds to the sharp increase of weight loss near 1950°C in Fig. 2.

The scanning electron micrograph of SiC with 20% SiO_2 addition sintered at 1950°C is shown in Fig. 5. Some remnants of solidified liquid phase can be found between the SiC particles (see arrow). More careful observation shows that there are many open gas bubbles on the surface of these solidified liquid phase mattes. These gases can be believed to be produced from the reaction of SiC and SiO_2 according to the above thermochemical analysis.

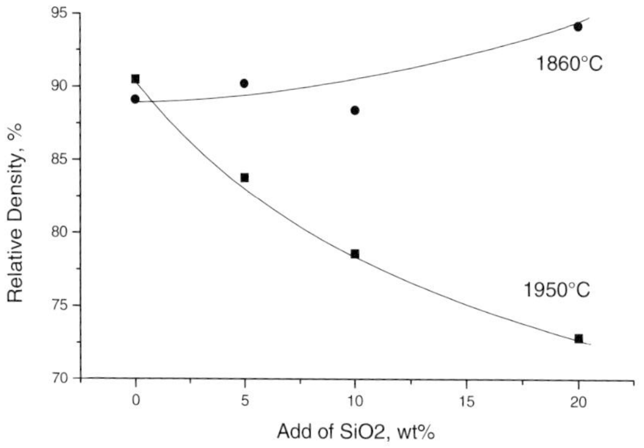

Fig. 6. Variation of density with initial added SiO_2 content in SiC samples sintered at 1860°C and 1950°C respectively.

Although it is reasonable to assume that at a temperature of 1860°C, the reaction of SiC and SiO_2 also occurs to a certain extent, the rate of shrinkage at this temperature obviously exceeds the rate of the gas evolution due to reaction (1), and densification prevails.
The effect of the content of SiO_2 additive on the final density is shown in Fig.6. At sintering temperature of 1960°C, the reaction of SiC and SiO_2 is the main process. Accordingly an increasing SiO_2 content promotes the reaction, and therefore decreased the density of the products. At the reduced temperature of 1860°C, shrinkage caused by the presence of the liquid became dominant, and the increase of SiO_2 is beneficial to sintering.

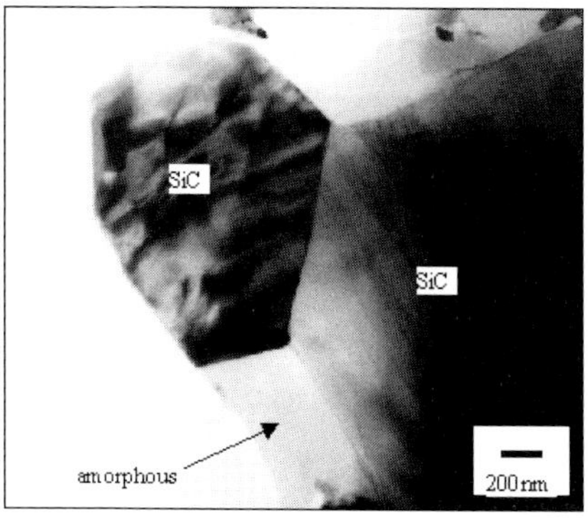

Fig. 7. TEM photograph of SiC-20wt%SiO_2

TEM photograph of SiC-20wt%SiO_2 sintered at 1860°C is shown in Fig.7. The SiC-SiC grain boundary seems to be free of glass phase. Detailed analysis and HREM investigation of the grain boundary regions will be performed in the future.

5 Conclusion

Quite high densities (94.2% of theoretical density) have been achieved when sintering SiC-20wt%SiO_2 at 1860°C. The lower density at sintering temperatures below 1790°C is because of the insufficient amount of liquid phase, while the similar low density at a sintering temperature of 1950°C is due to the violent gas-forming reaction between SiC and SiO_2. SEM and TEM photographs of the sintered samples show that a deeper investigation into the microstructure of this model system is practicable and worthwhile.

6 Acknowledgement

We are sincerely grateful to Mr. Martin Schweizer, Mr. Hartmut Labitzke and Dr. Fritz Phillipp, for help with the sintering, SEM and TEM experiments, respectively.

7 References

1. D. R. Clarke, J. Am. Ceram. Soc. 1987, 70, 15 - 22.
2. S. Prochazka, in Special Ceramics (6, eds. P. Popper and F.Fiee), British Ceramic Research Association, Stoke-on-Trent, 1975, pp. 171 - 181.
3. K. A. Schwetz, A. Lipp, Science of Ceramics, 1979, 10, 149.
4. Young-wook Kim, Wonjoong Kim, J. Mater. Sci. Letters, 1997, 16, 1384 - 1386.
5. M. Keppeler, H. –G. Reichert, J. M. Broadley, G. Thurn, I. Wiedmann and F. Aldinger, J. European Ceramic Society, 1998, 18, 521 - 526.
6. N. S. Jacobson, K. N. Lee, D. S. Fox, J. Am. Ceram. Soc. 1992, 75, 1603 - 1611.
7. S. Turan, K. M. Knowles, J. Microsc., 1995, 177, 287 - 304.
8. M. Nader, Ph. D. thesis, Univ. of Stuttgart, Stuttgart, 1995.
9. R. M. German, Liquid Phase Sintering, 1st ed., Plenum Press, New York, 1985.

Pressureless Sintering of Alumina-Glass Composites

Maria Margarida Lima and Regina Monteiro
Universidade Nova de Lisboa, Cenimat, Departamento de Ciências dos Materiais P-2825 Monte de Caparica, Portugal

1 Introduction

Alumina is a useful ceramic, because of its high mechanical strength, chemical durability, electrical resistivity and melting point. Therefore, alumina is used in a vast number of technological applications, which include electronic devices such as IC packaging [1,2], and composite materials [3].

Sintering of pure alumina requires high temperatures [4]. The most common method for reducing the sintering temperatures of alumina is to induce liquid phase sintering by addition of glass-forming systems. CaO-Al_2O_3-SiO_2, MgO-Al_2O_3-SiO_2 and B_2O_3-SiO_2 are glass-forming system usually added to alumina [3].

Liquid phase sintering and viscous sintering are very important processes for the production of a variety of dense ceramic components from porous powder compacts. During liquid phase sintering of a mixed powder, formed into a green compact, a liquid phase typically 1-20 vol% of the powder compact is formed upon heating, either by melting of one or more of the constituents or by reaction between the constituents [5, 6].

Viscous sintering is the main mechanism of densification in sintered glasses and in most traditional ceramics. It is also important in some technical products in which a large amount of transitory liquid phase is produced during heating [4, 7]. Viscous sintering is driven by the energy reduction brought by the decrease in surface area of the porous body [4, 7]. This phenomenon has been tentatively described by various models, which show the particular importance of three factors: a geometrical factor (the particle size), a kinetic factor (the viscosity) and a thermodynamic factor (the surface tension) [7].

In the present study we investigated the sintering behaviour of ceramic-glass composites, containing alumina and borosilicate glass (25 vol%), in the temperature range 900-1450°C with time ranging from 0 to 360 min. The sintering of the alumina and glass powder compacts was investigated through density measurements, and the microstructure was analysed by scanning electron microscopy (SEM).

2 Materials and Methods

The alumina powder used in this study was SG16 from Alcoa (Alcoa, UK) with a median size of 0.3-0.6 μm and a specific surface area of 12.53 m^2/g.

For comparative purpose, composites prepared with a commercial glass powder, *C*, and with a laboratory-prepared glass powder, *G*, have been used. The composition (wt%) of the commercial glass, as stated by the supplier, (Corning, France), was as follows: 80-80.9SiO_2,

12.5-13.1B_2O_3, 3-4Na_2O, 2Al_2O_3, 1K_2O, 0.5CaO/MgO/Fe_2O_3. Glass powder *G*, of similar chemical composition (81SiO_2, 12.5B_2O_3, 3.5Na_2O, 2.5Al_2O_3, 0.5K_2O) was prepared in the laboratory as explained elsewhere [8]. From each glass, a powder fraction with a particle size smaller than 10 μm was selected. The specific surface areas of powder *C* and powder *G* were 3.82 m^2/g and 6.44 m^2/g respectively as measured by the BET method [8]. The effect of heat on the glasses was determined by differential thermal analysis (DTA) at a heating rate of 10°C/min. X-ray diffraction analysis (XRD) were used to identify the crystalline phases.

For the preparation of the powder mixture, the borosilicate glass powder was added to 75 vol% alumina, mixed with 5 wt% polyethylene glycol, diluted with ethanol and the suspension was deagglomerated in a ultrasonic bath. This was followed by intensive mixing, using a mixer/mill, and the resulting mixture was dried and then ground in an agate mortar. Compacts with 5 mm diameter and about 20mm high were obtained by cold isostatic pressing at 200 MPa.

The heating profile for binder burnout was determined by thermogravimetric analysis (TGA) at a heating rate of 10°C/min in air. As the total weight loss occurred below 400°C, the samples were fired at 1°C/min from room temperature to 400°C and then held for 60 minutes, before the sintering experiments. All the samples were heated in an electric furnace, using a heating rate of 10°C/min up to the required temperature, *i. e.* 900-1450°C.

The densities of the sintered composites were determined by the Archimedes method. Before the density measurements the samples were immersed into water and the open pores were filled with boiling water for 30 minutes. The theoretical densities of the composites were calculated by using the mixing rule, taking into account the borosilicate glass and alumina densities, which were respectively 2.23 and 3.91 g/cm^3.

The microstructure of sintered compacts, either in polished samples or in fractured surfaces, were observed using a scanning electron microscope, Zeiss, model DSM 962.

3 Results

XRD Analysis and SEM observation of glass *C* and glass *G* indicated that both were fully amorphous. The DTA curves showed that the glass transition temperature for these glasses is situated in the range 500-520°C and evidence of devitrification during the heating of the glasses was found at 710-730°C.

Cristobalite was the crystalline phase identified from the XRD results of the plain glass powders when submitted to a crystal growth treatment, *e. g.* 3h at 850°C. However, the number and the intensity of the cristobalite peaks were much more reduced in the X-ray difractogram of an alumina-glass powder mixture, that has been submitted exactly to the same heating schedule, where beside alumina, very few mullite could also be identified, Figure 1. The role of alumina as a devitrification inhibitor in borosilicate glass, and the formation of mullite during sintering of these powders has already been reported by some authors [9].

The effect of temperature on the densification of both series of composites (alumina-glass *G* and alumina-glass *C*) is illustrated in Figure 2. These results show that similar behavior was exhibited by both series of composites. The relative density increased continuously with temperature, although a slower densification rate occurred when the samples were heated up to 1100°C compared to the densification rate achieved in the temperature range 1100-1300°C, which was followed by a decrease in the densification rate in the range 1300-1450°C.

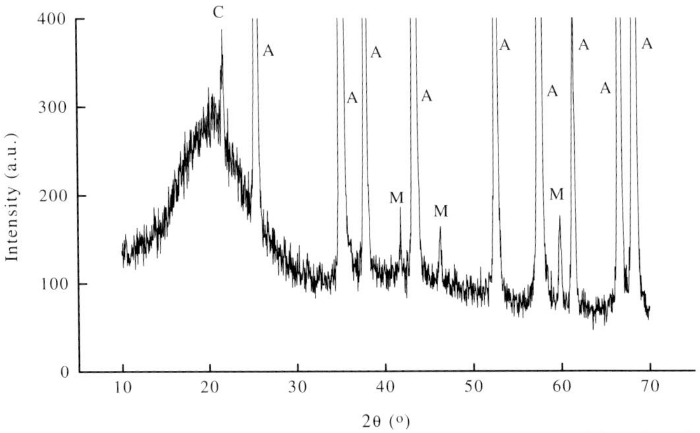

Figure 1. XRD of alumina-borosilicate glass powder mixture fired for 180 min. at 850°C. (C-Cristobalite, M-Mullite, A-Alumina)

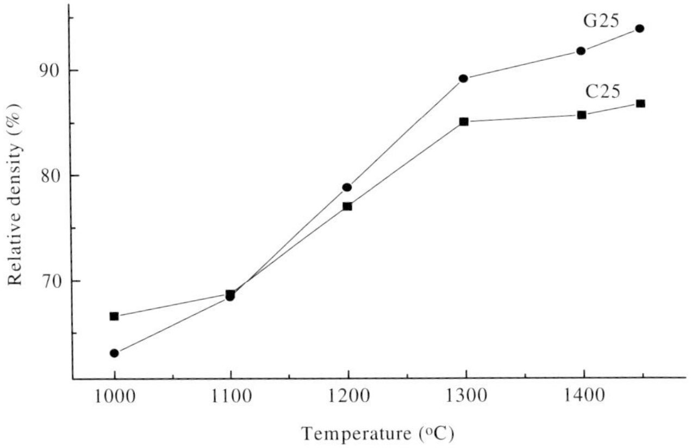

Figure 2. Relative density as function of temperature for the composites with 25 vol% glass *C* (C25) and 25 vol% glass *G* (G25). (Samples fired for 180 min. at each temperature).

From Figure 2 it is also observed that, above 1100°C, composites containing glass *G* achieved slightly higher relative density than those with glass *C*. Due to the higher specific surface area of glass *G*, the composites made from this glass densified better than composites prepared from glass *C*, of larger particle size. This effect is also illustrated in Figure 3, which shows the results for samples that have been heated up to 1200°C and up to 1400°C, at 10 °C/min, and then held at these temperatures for different times. These results indicate that the relative density increased significantly with temperature, but at each sintering temperature a reduced effect of the time on the increment of relative density was observed, and such behavior is particularly evident at the highest temperature.

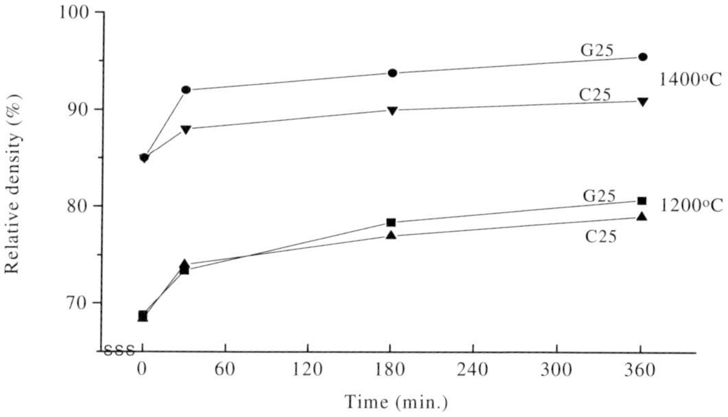

Figure 3. Change of relative density for composites with 25 vol% glass fired at 10°C/min up to 1200°C or 1400°C, and held at each temperature for different times.

From Figure 4 a comparison of the microstructures of samples treated at different temperatures may be done. At 900°C, a large number of very small and continuous pores were present. At 1200°C, the number of small pores decreased, as they were filled by capillary flow of the glass, and large pores appeared, although densification for this composite was higher than for composites sintered at lower temperatures, confirmed by the results shown in Figure 2. At the highest temperature, the microstructure showed very dense areas with large pores, which appeared at the sites of large glass particles in the green compact.

The above results showed three ranges of densification. These separate ranges may be associated to changes in the predominant sintering mechanism. In the first range, 900 to 1100°C, the glass softened and started flowing into the alumina particles. Liquid redistribution into the small capillaries occurred, and densification of the composite may be due to local grain rearrangement. At this temperature range, the devitrification of the glass resulted in the formation of some cristobalite, which reacted at the interface of the alumina particles and caused the appearance of mullite crystallites. The second range began at around 1100°C. During this stage mullite formation was continued, but the remaining glass was fluid enough to penetrate into the small pores between the alumina particles. This may be attributed to a global particle rearrangement and densification of the composite may be due to viscous flow [10]. In the third range, which began at around 1300°C, the densification rate was decreased. This may be attributed to the growth of mullite crystals and pore coalescence.

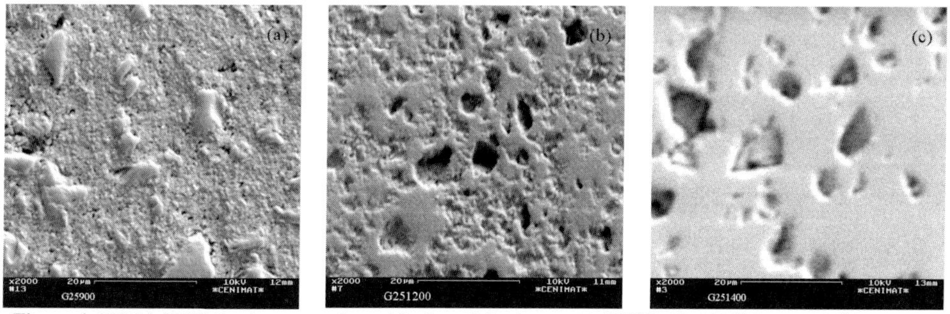

Figure 4. SEM (x2000) for the composites with glass *G* fired up to (a)-900°C, (b)-1200°C, (c)-1400°C.

Figures 5 shows the evolution of the microstructure for samples fired at 1400 °C. All these samples have been heated at 10°C/min up to this temperature and then held for the specific time.

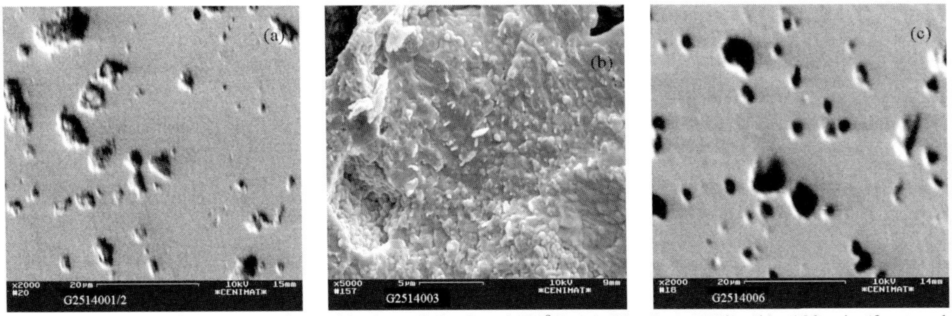

Figure 5. SEM of the composites with glass *G* fired at 1400°C; (a)-30 min.(x2000), (b)- 180 min.(fractured surface, x5000), (c)- 360 min. (x2000)

The densification of the compacts started to occur because at 1400°C the glass was fluid enough to flow between the alumina particles allowing particle rearrangement. Samples exhibited an initial a stage of rapid increase in relative density, which was followed by a stage of slower evolution on the densification process, where pore coalescence and the densification were driven by reduction of surface energy. This was evident from the microstructures which show very dense areas surrounding large pores. Although such dense areas might appear rather uniform when the SEM observations were made in polished samples, the existence of very small (<1µm) needle-like mullite crystals dispersed in the matrix was shown when SEM observations were made in fractured surfaces.

4 Conclusions

The results indicated that liquid phase sintering occurred in the alumina-borosilicate glass composites, with 25 vol% glass, and that densification was likely due to particle rearrangement.

The relative density of the alumina-borosilicate glass composites increased significantly with temperature, but a small effect of the sintering time on the increment of relative density was observed if samples were held at a certain temperature.

Composites sintered at the lowest temperature showed a large number of small pores, which decreased with sintering temperature as pores were filled by capillary flow of the glass. Composites sintered at the highest temperature showed very dense areas with large pores, which appeared at the sites of large glass particles in green compact.

Due to the smaller particle size and to the higher specific surface area of laboratory-prepared glass, composites made from this glass showed higher densification than composites containing the commercial glass.

5 References

[1] A. H. Kumar, R. R. Tummala in Materials in Microelectronic and Optoelectronic Packaging (Ed.: H. C. Ling , K. Niwa, V. N. Shukla), American Ceramic Society, Westerville, Ohio, USA, 1993, Vol.33, p. 83.

[2] R. R. Tummala, Journal of the American Ceramic Society, 1991, 74, 895-908.

[3] J. H. Jean, T. K. Gupta, Journal of Materials Research, 1994, 9, 1990-1996.

[4] M. N. Rahaman, Ceramic processing and sintering, Marcel Dekker, New York, 1995.

[5] O. H. Kwon in Materials Science and Technology, Processing of Ceramics, part II, (Ed.: R. J. Brook) VCH, Weinheim, vol. 17B, 1996, p.99.

[6] W. D. Kingery, H. K. Bowen, D. R. Uhmann, Introduction to ceramics, John Wiley & Sons, New York; 1976.

[7] F. Cambier, A. Leriche, in Materials Science and Technology, Processing of Ceramics, part II, (Ed.: R. J. Brook) VCH, Weinheim, vol. 17B, 1996, p.123.

[8] M. M. R. A. Lima, R. R. C. Monteiro, Materials 99, Portuguese Materials Society, 1999, Vol.3, p. 29.

[9] J. H. Jean, T. K. Gupta, Journal of American Ceramic Society, 1993, 76, 2010-2016.

[10] M. A. Cohn, UMI's Dissertation Abstracts, 1992, 52, 230.

Processing of Mullite Ceramics Using Fly Ash and Alumina

N. Quaranta*, S. Camelli, M. Caligaris, E. Benavidez, R. Caligaris**
Universidad Tecnológica Nacional, San Nicolás, Argentina
A. R. Boccaccini, H. Kern
Technische Universität Ilmenau, Ilmenau, Germany

1 Abstract

The objective of this work was to study the feasibility of using coal power station fly ash as a raw material for the production of ceramic bodies of mullite composition. The fly ash was obtained from a coal power station in Argentina. The as-received ash is composed of water, unburned coal and mineral silicate ash. Mullite-containing ceramic materials from fly ash and alumina mixtures were obtained by cold pressing and reaction sintering. The relative amount of alumina and fly ash and the sintering conditions were optimised in order to obtain high-quality, dense bodies with the possible maximum amount of fly ash. The sintering temperature was in the range 1200 – 1600°C. Fairly dense mullite-containing materials could be produced at a sintering temperature of 1500°C from alumina-fly ash mixtures with 30 wt% alumina content. The materials were characterised in terms of microstructure, density, phase composition and hardness. The present study shows a new way for the utilisation of fly ash in the fabrication of mullite ceramics, which is attractive both from an environmental and from an economical point of view.

2 Introduction

Significant amounts of fly ash are produced permanently as a by-product of the coal combustion in power stations. Currently, only a small part of the fly ash is utilised, mainly in the cement industry [1,2]. The remaining amounts are used for land-filling, which is an unsatisfactory solution from both an ecological and an economical point of view. In a recently published survey [3], the negative environmental impacts of land-filling fly ash have been identified, including the leaching of potentially toxic substances into soils and groundwater, the change in the elemental composition of the vegetation growing in the vicinity of the ash, and the accumulation of toxic elements throughout the food chain. Other more efficient uses of fly ash have been proposed in the last 40 years, including brick and ceramic tile fabrication, lightweight aggregate, road pavements, filler in plastics and paints, mineral wool and for metal recovery [4]. Indeed the attempts to utilise fly ash in new ceramic products are not new. As reported by Butterworth [5], the addition of coal ashes for brick making goes back at least to the eighteenth century in England. The amount of utilised fly ash is, however, still very low

---

* CIPBA Researcher
** CONICET Researcher

and consequently new forms for its utilisation as a raw material resource are continually searched. Recent papers have reported for example on the use of fly ash to synthesise mullite ceramics [4], for the fabrication of ceramic tableware and artware [6], for fabricating mineral polymer composites [7], for reinforcement of aluminum [8] and for the production of cordierite [9] and other glass-ceramics and glass matrix composites [10-14].

Mullite is an alumina and silica compound ($3Al_2O_3.2SiO_2$) with several attractive properties, including chemical and thermal stability, high refractoriness, high strength at high temperatures and good dielectric properties. It is a candidate material for a wide variety of applications, including refractory materials, fibres for composites, infrared transmitting windows, electronic substrates, etc.

The main objective of the present investigation was to demonstrate the feasibility of synthesising mullite from fly ash and alumina powder mixtures by a reaction sintering process. The powder technology route was optimised in order to maximise the sintered density of the products by keeping a high fly ash content and the lowest possible sintering temperature. The sintered bodies were characterised in terms of density, mineralogical composition, microstructure and hardness.

3 Experimental

The starting materials were fly ash from a coal power station (Central Térmica San Nicolás, Argentina) and alumina powder (Aldrich, purity 98.9) having a mean particle size of 10 μm. The as-received fly ash contained 0.14wt% water, 81.98 wt% solid silicate waste and 17.88wt% unburned carbon. The chemical composition of the as-received fly ash is shown in Table 1. As found in other fly ash investigated in the past [14], the as-received fly ash contained small amounts of crystalline phases such as quartz and mullite.

Table 1. Chemical composition of the as-received fly ash (in wt%)

MgO	Al_2O_3	SiO_2	K_2O	CaO	TiO_2	Fe_2O_3
0,80	22,10	59,70	1,42	5,47	2,05	8,50

Based on the compositions of mullite and the fly ash used, different mixtures containing different proportions of fly ash and alumina powder were prepared. The alumina volume fraction of the different mixtures is shown in Table 2. The mixtures were prepared in dry conditions and the fly ash was used in the as-received state.

Table 2. Alumina content of the fly ash – alumina mixtures

Sample	wt% Al_2O_3
CA-10	10
CA-20	20
CA-30	30
CA-40	40

Prismatic samples (40 x 70 x 10 mm^3) were fabricated by uniaxially pressing of fly ash-alumina mixtures with 8% humidity. Pressures of 10 MPa were used. The powder compacts were sintered in air. The heating rate was 5°C/min and the sintering time 120 min. for all samples. The sintering temperature was varied between 1200 and 1600°C. All samples were cooled down in the furnace.

The characterisation of the samples was conducted using standard methods of current use for the testing of ceramic materials. The porosity of the sintered samples was determined by the well-known Archimedes' technique. X-ray diffraction analyses were carried out to reveal the presence of mullite and other crystalline phases in the material. The microstructure was studied by optical microscopy of polished surfaces of selected samples. The same polished surfaces were employed for Vickers indentation tests, which were conducted to determine the hardness (H_v) of the materials. Loads of 300 g applied during 15 seconds were used.

4 Results and Discussion

The main objective of the present study was to develop a cost-effective powder processing route for obtaining dense mullite-containing ceramic materials from mixtures of fly ash and alumina. The sintering behaviour of the compacts was strongly dependent on the alumina content. For alumina contents lower than 20 wt%, the samples showed deformation at temperatures of 1400°C (and at higher temperatures) due to the softening of the silicate phase present in the fly ash. For higher alumina content this effect was not observed, even for sintering temperatures of 1550 °C.

Figure 1 shows the variation of the open porosity with the sintering temperature for compacts with different alumina content. It is seen that open porosity decreases abruptly for sintering temperatures higher than 1400 °C. At this temperature, alumina and silica (this one present in the fly ash) react to yield mullite with a negative volume change.

The X-ray diffraction patterns of samples sintered at temperatures > 1500°C showed peaks belonging to the mullite and alumina phase. The highest content of mullite, as determined by the relative height of the peaks, was observed for the samples containing 30 wt% alumina. For higher alumina content, X-ray diffraction analyses revealed the presence of higher amount of non-reacted α-alumina phase.

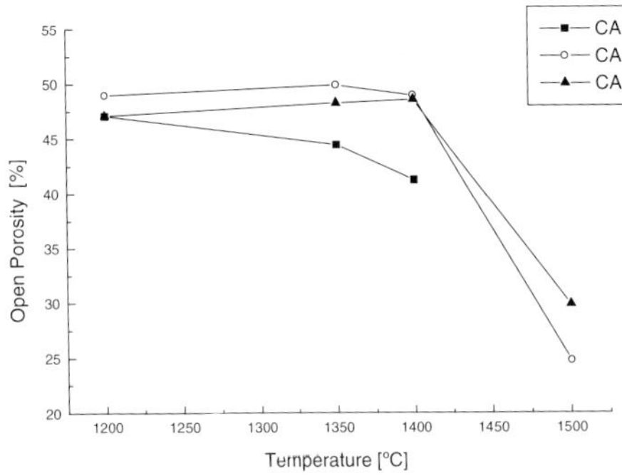

Figure 1. Variation of the open porosity of fly ash-alumina compacts with sintering temperature and alumina content

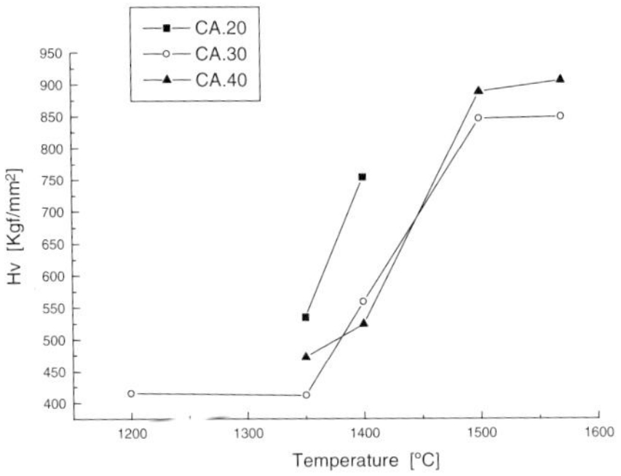

Figure 2. Vickers hardness of fly ash-alumina compacts as function of sintering temperature and alumina content

Achievement of high densities is important because the final properties of the products, especially the mechanical properties, increase with increasing density. Figure 2 shows the results of the measurements of the Vickers hardness for samples with different alumina content sintered at different temperatures. As expected, hardness correlates inversely with the amount of open porosity (Figure 1). According to the chemical composition of the fly ash used, the mixtures containing 30wt% alumina would lead to the formation of stoichiometric mullite. It is expected that an increase of the alumina content will not lead to any further improvement of the mechanical properties. This was confirmed by the hardness measurements (Figure 2). Although open porosity was fairly reduced at sintering temperatures > 1400°C, the sintered samples exhibited a certain amount of closed porosity (about 10%). It is suggested that this was formed by the burn-out of the carbon present in the as-received fly-ash during sintering.

Optical micrographs of polished sections of samples containing 30 and 40 wt% of alumina and sintered at 1570°C are shown in Figures 3 and 4, respectively. In both samples, residual porosity is observed. For the sample containing 30 wt% alumina, the microstructure consists of crystals of mullite (grey colour) embedded in an aluminosilicate glassy matrix. The sample containing 40 wt% alumina shows also the existence of needle-like crystals of non-reacted α-alumina (white colour), as determined by SEM-EDAX analyses [15] (see Figure 4).

5 Conclusions

The present study was conducted to synthesise mullite ceramics from mixtures of coal power station fly ash and alumina by reaction sintering. The lowest porosity and highest hardness were achieved for mixtures containing 30 and 40 wt% alumina and sintered at temperatures of 1500°C. Thus, using a simple, cost-effective powder technology method, fairly dense mullite-containing ceramics can be produced incorporating significant amounts of fly ash. The potential of these ceramic materials for technical applications must be evaluated. In particular, more information concerning the thermomechanical behaviour and the chemical durability of the products is necessary, this being the focus of on-going research.

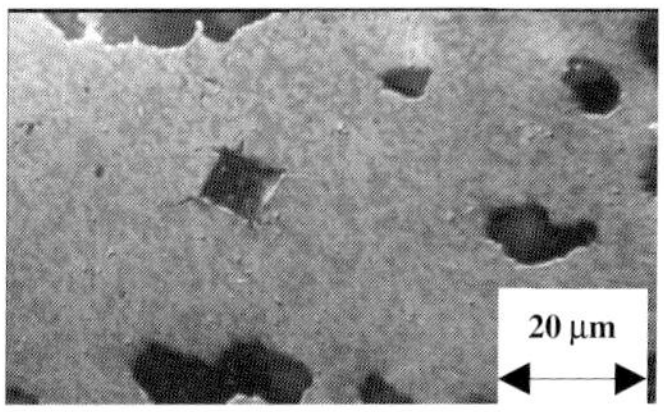

Figure 3: Optical micrograph showing the microstructure of the sample CA-30 sintered at 1570°C. The mullite phase (light grey) is embedded in a silicate glassy matrix. Residual porosity is observed.

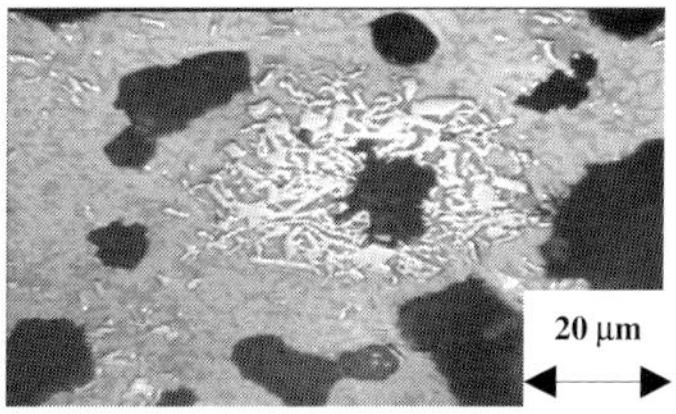

Figure 4. Optical micrograph showing the microstructure of the sample CA-40 sintered at 1570 °C. The white phase is non-reacted α-alumina [15]. Residual porosity is observed.

6 Acknowledgement

The authors would like to thank Central Térmica San Nicolás S. A. for partial financial support.

7 References

1. G. L. Jablonski, and S. S. Tyron, in Proceedings of 5th International Pittsburgh Coal Conference , Pittsburgh, PA, University of Pittsburgh, 1988, pp. 15-21.
2. J. Olek, in Proceedings International Symposium on Brittle Matrix Composites 4, (Eds.: A. M. Brandt, V. C. Li and I. H. Marshall), IKE and Woodhead Publ., Warsaw, 1994, pp. 425-436.
3. C. L. Carlsson, and D. C. Adriano, Journal of Environmental Quality, 1993, 22, 227-247.
4. J.-Y. Hwang, X. Huang, X. and A. M. Hein, JOM , 1994, 47 (5), 36-39.
5. B. Butterworth, Transactions British Ceramic Society, 1954, 53, 293-309.
6. S. K. Mukherji, B. B. Machhoya, British Ceramic Transations , 1993, 92 (1), 6-12.
7. X. Wu, J. Wastiels, S. Faignet, P. Bauweraerts, in Proceedings of International Symposium on Brittle Matrix Composites 4, (Eds.: A. M. Brandt, V. C. Li and I. H. Marshall), IKE and Woodhead Publ., Warsaw, 1994, pp. 437-446.
8. P. K. Rohatgi, R. Q. Guo, H. Iksan, H., E. J. Borchelt, R. Asthana, Mat. Sci. Eng., 1998, A244, 22-30.
9. N. N. Sampathkumar, A. M. Umarji, B. K. Chandrasekhar, Materials Research Bulletin , 1995, 30, 1107-1114.
10. R. Cioffi, P. Pernice, A. Aroni, M. Catauro, and G. Quattroni, Journal of the European Ceramic Society , 1994, 13, 143-148
11. B. Cumston, F. Shadman, and S. Risbud, J. Mat. Sci., 1992, 27, 1781-1784.
12. A. R. Boccaccini, G. Ondracek, J. Janczak, H. Kern, in Proceedings of 4th International Symposium on the Reclamation, Treatment and Utilization of Coal Mining Wastes, Vol. II, (Ed.: K. M. Skarzynska), Krakow, 1993, pp. 719-726.
13. A. R. Boccaccini, M. Bücker, J. Bossert, Tile & Brick Int., 1996, 12, 515-518.
14. A. R. Boccaccini, M. Bücker, et. al., Waste Management, 1997, 17, 39-45.
15. N.E. Quaranta, S.H. Camelli, M.G. Caligaris, E.R. Benavidez, R.E. Caligaris, and A.R. Boccaccini, in Proc. Jornadas SAM, Rafaela, Argentina, 1999, in press.

High Temperature Behaviour of Liquid Phase Sintered Silicon Carbide

Giuseppe Magnani, Gian Loris Minoccari, Luigi Pilotti
ENEA-Innovation Department–Faenza Research Centre, Faenza (RA), Italy

1 Abstract

Liquid phase sintering process of silicon carbide allows to obtain dense sample at lower temperature than solid state process with reduced grain growth and improved mechanical properties. In this paper mechanical properties at RT and at high temperature of liquid phase sintered silicon carbide (LPSSC) was determined. Flexural strength measurements were conducted by means of four-point bend tests in the range RT-1400°C. Flexural strength changed from 532 MPa at RT to 240 MPa at 1400°C. SEM analysis showed that this behaviour is due to superficial oxidation of SiC and formation of yttrium silicate which create bubbles and cavities.

2 Introduction

Silicon carbide (SiC) is widely used for structural applications thank to its good mechanical properties (hardness, fracture strength)[1]. Furthermore it is well known that SiC showed high corrosion resistance and thermo-stability at high temperature[2]. These properties depend strictly on density and grain size[3] and because of the difficulty to obtain SiC parts with high density and small grain size, in these last years methods to sinter silicon carbide with the aim to improve its mechanical properties (fracture strength and toughness) were studied[4-6]. Omori et al.[4] proposed to sinter silicon carbide via liquid phase using alumina and yttria as sintering-aids. Mulla et al.[7] and Do-Hyeong[8] et al. demonstrated that it is possible to obtain SiC parts with high density and high mechanical properties at temperature below 2000°C. At this temperature, crystal structure transformation β-α is prevented. As consequence exaggerated crystal growth due to this crystallographic changing is avoided and thus it is possible to obtain an improvement of mechanical properties.

In this paper mechanical properties of liquid phase sintered silicon carbide (LPSSC) was determined. Flexural strength tests were conducted at different temperature in the range RT-1400°C. To explain flexural strength variation with temperature, corrosion behaviour of LPSSC at high temperature was investigated.

3 Experimental Procedure

β-SiC powder (BF12, H.C. Starck – Germany) was wet mixed in ethanol with 6%wt Y_2O_3 (purity 99.99%, Mandoval Ltd. England) and 4% wt Al_2O_3 (purity 99.99%, Baikalox SM8, Baikowski Chimie France). In Table 1 are reported characteristics of SiC, alumina and yttria powders. Mixing was performed by Turbula mixer using polyethylene bottle and SiC balls as grinding media. After drying and sieving, the powder was compacted by die pressing at 67 MPa and subsequently was pressed at 250 MPa by CIP.

Green body was put in a graphite crucible embedded with powder bed with the same composition of the pellet and covered with a graphite foil. Sintering was performed at 1875°C in a graphite elements furnace in flowing Ar at 1 atm. Thermal cycle was characterised by heating and cooling rate of 22°C/min and by dwell time of 0.5 h at sintering temperature. Density of SiC sintered sample was determined using Archimede's method. Crystalline phases was identified by means of X-ray diffraction (Rigaku – Miniflex CuKα 1.54 Å). Furthermore sintered sample was polished, plasma etched with HF and finally examined by SEM (Stereoscan 250-Cambridge).

Flexural strength was measured by four-point bend tests at different temperature from RT to 1400°C. Samples as bars of 2 x 2.5 x 25 mm (five bars for each temperature) were prepared and tested in according to the standard EN 843-1. Thermomechanical behaviour of LPSSC was investigated through SEM examination of transversal section of samples tested at high temperature. Identification of corrosion products formed on the surface was performed by means of EDS microprobe (QX 2000 Link Analytical).

Table 1. Chemical and physical properties of SiC, alumina and yttria powders

Powder	Purity (wt%)	S.S.A. (m^2/g)	Particle size (μm)
β-SiC	>97.5	11.59	0.72
Y_2O_3	99.99		3.5-4.3
Al_2O_3	99.99	10	<0.3

4 Results and Discussion

4.1 Density and microstructure

In Figure 1 X-ray diffraction pattern of sintered sample is reported. It shows that YAG was formed as liquid phase.

Density of SiC sintered sample resulted 96.3% T.D.. Theoretical density, calculated by means of determination of secondary phase content following method described elsewhere[9], resulted 3.24 g/cm^3. SEM examination of plasma etched sintered sample revealed a fine microstructure with average grain size of 2-3 μm (Figure 2).

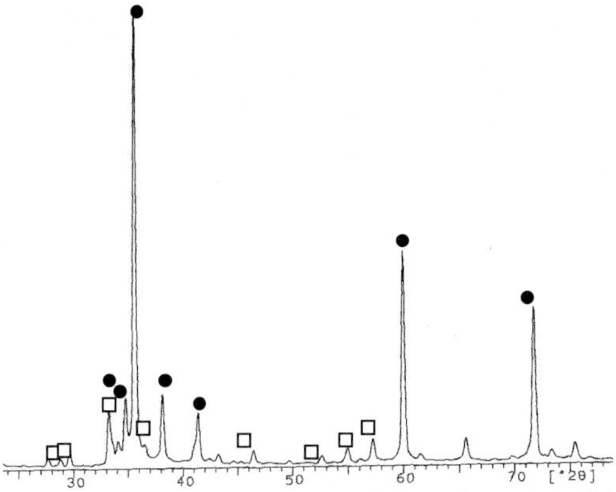

Figure 1. X-ray diffraction pattern of sintered sample (□ YAG, ● SiC).

Figure 2. Microstructure of sintered sample. Black and grey phases are SiC, white phase is YAG.

Large pores are probably due to loss of alumina and/or yttria during sintering process in according to these reactions:

$$SiC(s) + Al_2O_3(s) \Leftrightarrow Al_2O(g) + SiO(g) + CO(g) \quad (1)$$

$$SiC(s) + 2Y_2O_3(s) \Leftrightarrow 4YO(g) + SiO(g) + CO(g) \quad (2)$$

Total weight loss associated to sintering process was 5.4%. In literature are reported several studies concerning methods to reduce weight loss. For example Mulla et al.[10] proposed to sinter in CO atmosphere while Lee et al.[11] used YAG as sintering aid. In our case, presence of powder bed having the same composition of sintered body, allows to control reactions (1) and (2) and thus to reduce weight loss.

4.2 Flexural strength

Flexural strength (σ) variation with temperature is reported in Figure 3.

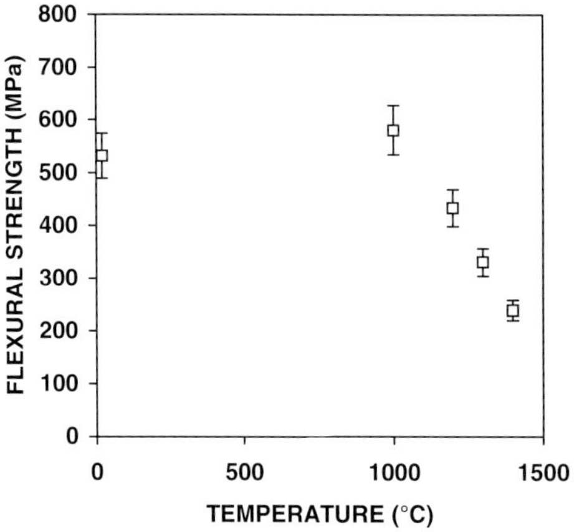

Figure 3. Variation of flexural strength with temperature.

At room temperature σ resulted 532 MPa. At 1200°C it begins to decrease and at 1400°C resulted 240 MPa. This behaviour is due to oxidation of SiC with formation of silica which reacts with yttrium aluminate to form yttrium silicate. Possible reactions are:

$$SiC(s) + \tfrac{3}{2}O_2(g) \Leftrightarrow SiO_2(s) + CO(g) \quad (3)$$

$$6SiO_2(s) + 2Y_3Al_5O_{12}(s) \Leftrightarrow 3Y_2Si_2O_7(s) + 5Al_2O_3(s) \quad (4)$$

This mechanism was confirmed by means of XRD (Figure 4) and SEM-EDS (Figure 5-6) analysis of sintered sample tested at 1400°C. XRD spectrum reveals presence of SiO_2 (α-cristobalite) and γ-$Y_2Si_2O_7$.

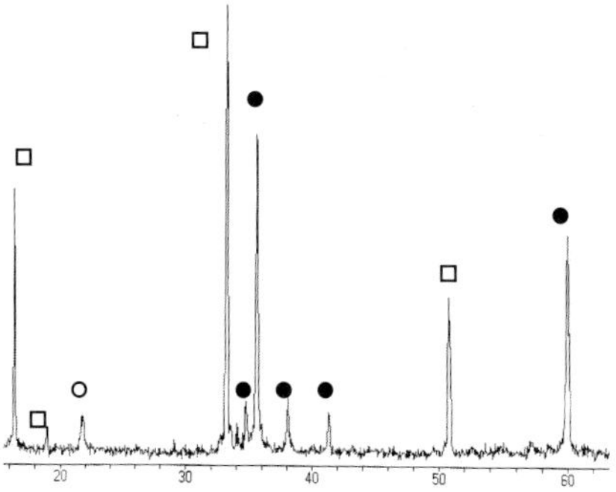

Figure 4. XRD pattern of sintered sample tested at 1400°C (□ γ-$Y_2Si_2O_7$, o SiO_2, ● SiC)

As reported by Drummond et al.[12] at 1400°C polymorph γ of $Y_2Si_2O_7$ is thermodynamically favoured. Free alumina was not identified; probably its content is too low or is present as solid solution. Alumina was detected by SEM-EDS analysis. In fact SEM images of Figure 5a-b show that an oxidation layer of about 4 μm was formed.

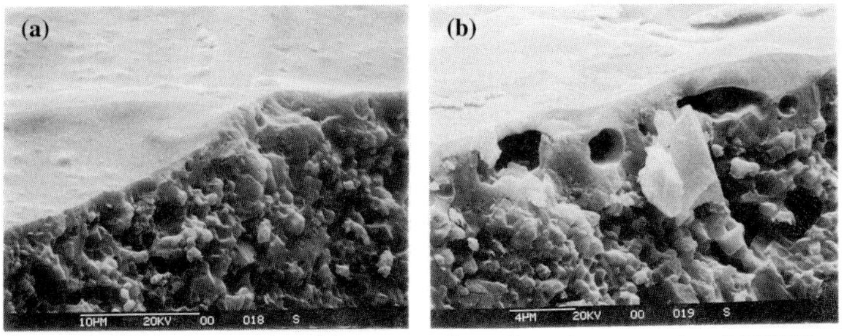

Figure 5. SEM micrographs of fracture surface of specimen tested at 1400°C.In a) are shown bubbles on the surface, while in b) is shown oxidation layer with large holes.

This layer was composed of yttrium silicate (γ-$Y_2Si_2O_7$) and alumina/silica as revealed EDS microprobe investigation (Figure 6a-b). Below the superficial layer, was identified (Figure 6c) silica layer formed by means of reaction (3). The rest of specimen was composed by SiC and YAG (Figure 6d-e).

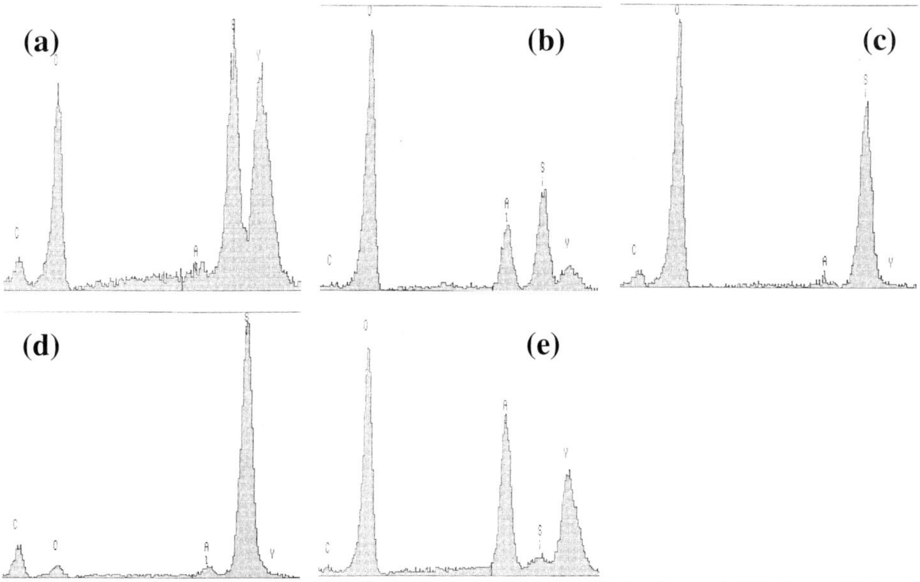

Figure 6. Sintered sample tested at 1400°C: EDS spectra of a)-b) oxidation layer, c) silica layer, d)-e) rest of specimen.

The large holes showed in Figure 5b was due to reactions (3) and (4) and they permit to oxidation to proceed inside specimen. This behaviour is different compared to SiC sintered via solid state. In this last case, oxidation leads to formation of protective silica layer which reduces oxidation rate[13]. In the case of LPSSC, formation of glassy-phase leads to creation of bubbles and holes that allow oxygen permeation through oxidation layer. Thus flexural strength variation with temperature depends on formation rate of glassy phase (γ-$Y_2Si_2O_7$).

5 Conclusion

Mechanical properties of SiC powder sintered via liquid phase were determined.
Flexural strength at room temperature resulted higher than σ of SiC sintered with B and C as sintering aids[13]. At 1400°C flexural strength value is reduced of the 55% compared to the R.T.. This behaviour is due to oxidation of SiC; it proceeds through formation of cristobalite which reacts with YAG to form γ-$Y_2Si_2O_7$. Because of these reactions bubbles and holes were formed and consequently oxidation is not stopped on the surface but proceeds inside sintered specimen.

6 Acknowledgements

This work was funded by Italian Ministry for Technological Scientific Research and University (MURST) under the project PNR-MIA Tema 8. The authors thank Dr. G. Celotti and Dr. A. Bellosi of CNR-IRTEC for providing XRD pattern and sintered specimen plasma etching respectively.

7 References

1. S. Shinrokv (Ed.), Fine Ceramics, Elsevier, England, 1985, pp.188-197.
2. P. Vincenzini (Ed.), High–tech ceramics, Elsevier Science Publishers B.V., Amsterdam, 1987.
3. R.W.Rice, J.Mater.Sci.Lett., 1994, 13, 1408-1412.
4. M. Omori, H.Takei, J.Mater.Sci., 1988, 23, 3744-3749.
5. T. Sakai, H.Watanabe, T. Aikawa, J.Mater.Sci.Lett., 1987, 6, 865-866.
6. S. Dutta, Adv. Ceram. Mater., 1988, 3, 3257-3262.
7. M.A. Mulla, V.D. Krstic, Ceram. Bull., 1991, 70, 439-43.
8. K. Do-Hyeong, K. Chong Hee, J.Amer.Ceram.Soc., 1990, 73, 1431-1434.
9. G. Magnani, L. Pilotti, G. Ronconi, Proceedings of Cimtec '98-Florence 1998, June 14th-19th, in press.
10. M.A. Mulla, V.D. Krstic, J.Mater.Sci., 1994, 29, 934-938.
11. J.K.Lee, H. Tanaka, H. Kim, J.Mater.Sci.Lett., 1996, 15, 409-412.
12. C.H. Drummond, W.E. Lee, W.A. Sanders, J.D. Kiser, Ceram.Eng.Sci.Proc., 1988, 9, 1343-1354.
13. S. Dutta, J.Mater.Sci., 1984, 19, 1307-1313.

Experimental Verification of Mechanical Properties – Porosity Relationships for Porous Silicon Carbide Materials

C. Reynaud, F. Thevenot

Département Céramiques Spéciales, Ecole Nationale Supérieure des Mines de St-Etienne, 42023 ST-ETIENNE, FRANCE.

1 Introduction

Porous ceramic materials are used for several industrial applications and their properties can be affected significantly by porosity. Therefore, their porosity dependence must be predicted accurately. Numerous relationships have been put forward in the literature. Furthermore, most of the models give the property variation with only the porosity content and consequently do not take into account the other porosity features.

The aim of this work was to verify recent models, about porosity dependence of Young's modulus and fracture energy, which take into account several characteristics of the porosity; and to verify a model based on the minimum solid area (MSA) concept.

2 Models

2.1 Young's modulus (E) – porosity (P) dependence

Two models proposed by Boccaccini et al have been considered [1, 2].

"Spheroidal model": This approach is based on a stress analysis around spheroidal pores. The pore structure is described by an array of spheroids which have a mean axial ratio (z/x) and a mean orientation to the stress direction (α_P). For a random orientation of the spheroidal pores, α_P can be taken as 54°.

$$E_P = E_0 \times \left[1 - 1.21 \times S \times P^{2/3}\right] \text{ with } S = \left(z/x\right)^{1/3} \times \left[1 + \left[\left(z/x\right)^{1/3} - 1\right] \times \cos^2\alpha_P\right]^{1/2} \quad (1)$$

A second equation has been obtained by assuming that eq.1 can be seen as the first term of a series expansion of the eq.2 (E_P, E_0 for porous and without pore materials).

$$E_P = E_0 \times \left[1 - P^{2/3}\right]^{1.21 \times S} \quad (2)$$

"Composite model": This second model is derived of a model previously used for two-phase composites. It is based on the two topological parameters of continuity and contiguity. The following equation has been obtained:

$$E_P = E_0 \times f_{\alpha C} = E_0 \times \left[(1-P)^2 \times r\right] \Big/ \left[P + (1-P) \times r\right] \quad (3)$$

where $f_{\alpha c}$ is the volume fraction of continuous solid phase and r is the size ratio between pore and grain.

2.2 Fracture energy (R) – porosity dependence

Two models proposed by Blanks et al have been considered [3].

"MSA approach": This equation is obtained by assuming that the crack propagates through the plane that contains the highest area fraction of pores (i.e. minimum solid area plane). For a cubic array of spherical pores, the highest fraction area of porosity is equal to $\pi \times \left(\frac{}{4\times\pi}\right)$. Furthermore, the fracture energy is directly proportional to the material area broken; then the following equation is obtained:

$$R_P = R_0 \times \left[1 - \pi \times \left(\frac{3 \times P}{4 \times \pi} \right)^{2/3} \right] \quad (4)$$

"H-H model": This model is based on the enhancement of the stress intensity factor according to Hong and Horii, cited in [3]. Assuming that the crack grows from pore to pore, the presence of a pore in front of the crack tip induces an increase of the stress intensity factor, and consequently leads to a decrease of the toughness (and fracture energy) of the material ligament between crack and pore (or pore and pore). Then the fracture energy of the body can be expressed as:

$$R_P = R_0 \times \left(K_0 / K_{MA} \right)^2 \times (1 - P) \quad (5)$$

Where the term $(K_0/K_{MA})^2$ takes into account the decrease of the fracture energy of the ligament. For a cubic array of spherical pores, K_0/K_{MA} can be approximated by the following polynome (after values calculated by Hong and Horii):

$K_0/K_{MA}=1.0002-0.0983P-3.6822P^2+12.263P^3-17.516P^4$ (correlation coef.=0.99993) (6)

2.3 Model developed by Rice based on the Minimum Solid Area (MSA) concept [4]

This model assumes that the different properties have the same dependence with porosity, it predicts a similar evolution of the MSA and the considered property with porosity. The MSA variation with porosity has been calculated for many arrays of pores (Figure 1). A semilogarithmic plot of the log MSA (absolute or relative value) versus the volume fraction of porosity (P) shows a linear decrease of MSA with P for low volume fraction of porosity. The slope of this linear decrease (-b) is a characteristic parameter of the pore stacking. Then an increase of porosity leads to a drop of the MSA. Finally the MSA decreases drastically when the volume fraction of porosity reaches the critical value of porosity P_c, which corresponds to the percolation limit of the solid phase. This second parameter is also characteristic of the porous structure. For the porosity range 0 to 1/3-1/2 P_c, the variation of the MSA (or property) with porosity can be described by the following equation: $A_P / A_0 = e^{-b \times P}$ (7)

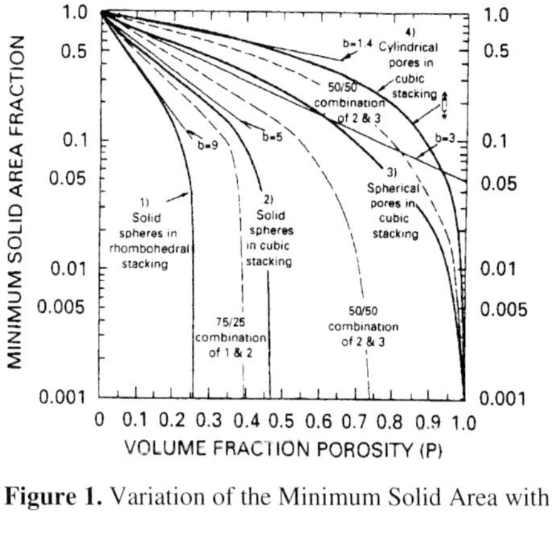

Figure 1. Variation of the Minimum Solid Area with porosity for idealised pore stacking after Rice [4].

3 Experiments

The experiments were performed on liquid phase sintered alpha silicon carbide with 10wt% of YAG-alumina eutectic additive [5]. The porosity content was controlled by varying the relative proportion of a coarse powder (10μm) (Norton AS, Sika Nor III F800) and a fine powder (1μm) (Norton AS, Sika Tech FCP13), and by varying the sintering parameters (time: 1 to 3h, and temperature: 1850 to 1950°C).

The samples were made according to the following procedure. SiC powders were dispersed in aqueous slurry (pH=10.3, dry matter content: 64-75wt%) with 3wt% of dispersant. Before adding 3wt% of Al_2O_3 (Baikowski, CR6) and 2wt% of Y_2O_3 (Rhone-Poulenc) as sintering aids, the slurry was ultrasonically treated during 3min to 8min to deagglomerate SiC powders. Then 3wt% of high molecular polyethylene glycol was added as binder-plasticizer. The slurry was ball-mixed with SiC balls during 20h. After mixing, the slurry was dried during 3h in an oven at 120°C, then sieved to 70-mesh. Five batches have been made (Table 1). Green pieces with disk or rectangular shapes were made by uniaxial compaction of the powder at a pressure of 50MPa. The organic additives burnout was performed under air in two steps (220°C-2h, 420°C-2h). Sintering of the specimens was conducted in a graphite furnace (Vide Appareillage Scientifique) under argon (1 to 3h, 1850°C to 1950°C). The specimens were placed in capped graphite crucible containing a powder bed made of a coarse alpha SiC powder (110μm) (Norton AS, Sika Nor IV F120) and 5wt% of alumina (Alcan Chemical, LS3) to promote a protective atmosphere.

Several characterizations have been performed. Density has been measured by Archimedes method, hence the volume fraction of porosity was obtained ($P = 1 - \rho_{ech}/\rho_{th}$). The theoretical density (ρ_{th}) is taken as 3.29g/cm^3, calculated by using a mixture law with ρ_{SiC}= 3.21g/cm^3 and $\rho_{eutectic}$= 4.36g/cm^3. The porosity has been also characterized by mercury porosimetry (θ=141.3°, γ=480mJ/m^2). Elastic moduli (E, G) and Poisson's ratio (ν) have been measured by using an ultrasonic method based on the propagation velocities measurement of longitudinal and transversal waves performed on disk sample with polished lower and upper faces. Vickers hardness has been measured by a micro-hardness apparatus (load applied: 500g). The toughness of samples has been determined by using the "Indention - Strength of the Beam (ISB)" method developed by Chantikul et al [6]. The bending fracture tests have been realised on smooth surface finished rectangular bars (3.75×4×20mm^3) with a three point cell with a span of 15mm at a transversal speed of 0.2mm/min. Fracture toughness has been computed by using the following relationship [6]:

$$K_{IC} = \alpha \times (E/H)^{1/8} \times \left(\sigma_R \times P_i^{1/3}\right)^{3/4}$$

with α=0.48, where E is Young's modulus, H Vickers hardness, σ_R bending strength, P_i indentation load. The fracture energy of the samples has been calculated using the following equation:

$$R = K_{IC}^2/E \cdot$$

A broad range of porosity was obtained (3vol% to 36vol%) (Table 1.) which permitted significant variations of the different properties.

Table 1. Porosity features.

Fine powder content (wt%)	Range of porosity volume fraction of sintered sample (vol%)	Mean pore radius (μm)
100	3.1-4.9	0.035
75	5.6-11.08	0.1-0.25
50	19.1-23.4	1.2-1.4
25	24.4-31.8	1.2-1.5
0	26.5-35.6	1.4-1.5

4 Results and Discussion

4.1 Young's Modulus – P dependence

"spheroidal model". Parameters which are taken into account in the eq.1 and eq.2 have been adjusted to obtain the best fit with data. Indeed, the porous structure was too complex and it cannot be reliably characterized by image analysis. E_0 has been taken as 470GPa and z/x as 4. Figure 2 shows a good agreement of eq.1 with data. The fitted value of z/x seems in agreement with pore features (elongated pores have been observed) but it cannot be determined experimentally. However a poor agreement with data was obtained for eq.2. Consequently the origin of eq.2 is not valid.

"composite model". The parameters have been adjusted as 450GPa for E_0 and 0.4 for r. This model is in accordance with data for low porosity content (up to P=0.25) (Figure 2). Furthermore, the r experimental values are included between 0.2 and 0.6.

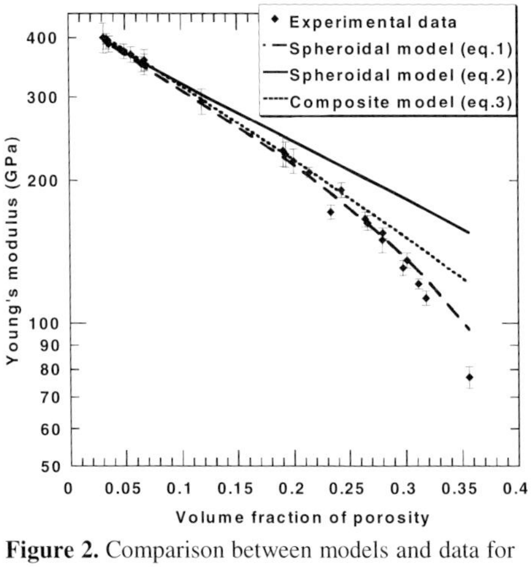

Figure 2. Comparison between models and data for Young's modulus.

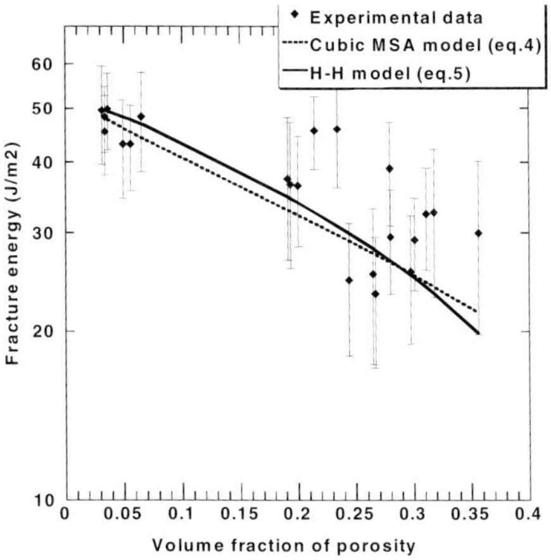

Figure 3. Comparison between models and data for Fracture energy.

4.2 Fracture Energy – P dependence

Figure 3 shows an agreement suitable with data for these two models. However the great fluctuation of data does not permit to conclude about their validity.

4.3 Model based on the MSA concept

Each property has been fitted using an exponential curve fit based on the least square method for the porosity range 0 to 20vol%. Good correlation coefficients were obtained except for the fracture energy and Poisson's ratio fits (Tab. 2). The different properties considered here seem to decrease linearly when the porosity content increases in a semilogarithmic plot (log prop. versus P) for low porosity volume fraction. The values of properties extrapolated at P=0 by the fitted equation are typical values observed for SiC materials. Table 2 shows that the b values of each property are close except for toughness (K_{IC}), fracture energy (R) and Poisson's ratio. Furthermore the b values obtained are in agreement with the average b values reviewed by Rice for pressureless sintered ceramics [7]. Hence the different properties seem to vary similarly with porosity as forecasted by the MSA model.

Table 2. Parameters of an exponential fit performed on each mechanical property for the porosity range 0-20vol%. (a) b values, reviewed by Rice [7], obtained for pressureless sintered samples.

Property	Extrapolated value at P=0	b (preexponential factor or -slope)	Correlation coefficient	Average b value (a)
E	445 GPa	3.48	0.999	3.6±0.8
G	189 GPa	3.45	0.9987	3.2±0.9
ν	0.18	0.19	0.545	0.5±0.5
H	20.2 GPa	3.89	0.959	4.1±1.1
K_{IC}	4.7 $MPa/m^{1/2}$	2.53	0.991	
R	50 J/m^2	1.58	0.915	

A semilogarithmic graph of the log of the relative properties (experimental value divided by the value extrapolated at P=0) versus porosity was plotted to verify this assumption (Fig. 3). A similar evolution of the elastic moduli with porosity was shown. On the opposite, Poisson's ratio is less affected by porosity and it decreases scarcely with P. Hardness appears more sensitive to the porosity volume fraction, whereas toughness seems to be less affected by porosity than elastic moduli. The deviation and the fluctuation shown for toughness and fracture energy properties can be explained by two main reasons. On the one hand, Rice explained this deviation by crack branching and crack bridging mechanisms which appear for large cracks due to pores interactions [4] and induce lower P-dependence. On the other hand, the variation of grain size (GS) with porosity, due to the porosity control performed by varying the proportion of fine and coarse powders, modify the P-dependence of fracture properties. Studies, reviewed by Rice [4], performed on silicon carbide show an increase of R and K_{IC} with GS.

Consequently, the increase of grain size with P leads to a lower P-dependence of fracture properties. Furthermore, a fracture analysis performed on ISB specimen by using SEM showed that fracture mode mainly intergranular became more transgranular for mixes with higher content of fine grains. Consequently, this change of fracture mode contributes to the fluctuations shown for R and K_{IC}.

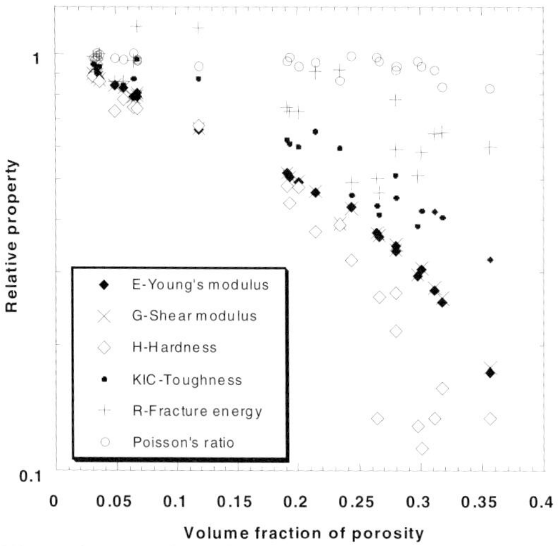

Figure 4. Comparison of porosity dependence of relative mechanical properties.

5 Conclusions

To conclude, the two models proposed by Boccaccini et al about Young's modulus porosity dependence are in accordance with experimental data, except for eq.2. However, their validity cannot be demonstrated. A well description of the pore structure is needed.

The validity of fracture energy porosity relationships from Blanks et al cannot be demonstrated because of the great fluctuation of data. Nevertheless an agreement suitable with data was obtained.

A similar porosity dependence of the mechanical properties was shown. Consequently, the MSA model from Rice, which describes the property - porosity dependence for the entire range of porosity, seems valid. But it was not verified rigorously because the minimum solid area cannot be determined experimentally. The determination of the MSA (i.e. pore array) was more difficult and it was the major disadvantage of this model.

6 References

[1] A.R. BOCCACCINI et al, J. Mech. Behav. Mater. 1993, 119-128.
[2] A.R. BOCCACCINI et al, Ceram. Int. 1997, 239-245.
[3] K.S. BLANKS et al, J. Eur. Cer. Soc. 1998, 1945-1951.
[4] R.W. RICE, J. Mater. Sci. 1996, 102-118.
[5] S. BAUD, F. THEVENOT, Sixth Eur. Ceram. Soc. Conf. Vol. 2, University Press, Cambridge, 1999, p. 17.
[6] P. CHANTIKUL et al, J. Am. Cer. Soc. 1981, 539-543.
[7] R.W. RICE, Porosity of Ceramics, M. Dekker, N.Y., 1998, p. 128.

Processing of SiC-Ceramics with Cellular Microstructure

Christiane Hoffmann, Annette Kaindl, Heino Sieber, Peter Greil, University of Erlangen-Nuremberg, Department of Materials Science (III), Glass and Ceramics

1 Introduction

In contrast to engineering products, materials in biological systems are optimized in a long time evolution process. They are characterized by a complex hierarchical structure, with the structural features ranging from the micro (cell walls) to the macro scale (skeletons). Such structures are interesting for technical applications for example as micro filter in food processing or catalyst in exhaust gas purification. Various biotemplating techniques were developed to convert biological derived materials such as natural fibres and wood into ceramic materials [1-8].

Wood, a natural composite, exhibits a unique porous structure with excellent strength at low density, high stiffness, elasticity and damage tolerance [9-11]. The cell walls of wood are formed by biopolymers such as cellulose, hemicelluloses, pektins and proteins. Several cellulose-macromolecules are bundled to fibrils. Lignin serves as a binder between the cellulose microfibrils and increases the mechanical strength [12]. Different cell types form the heterogeneous tissue wood. The morphology and the arrangement of the different cells vary between deciduous and coniferous wood.

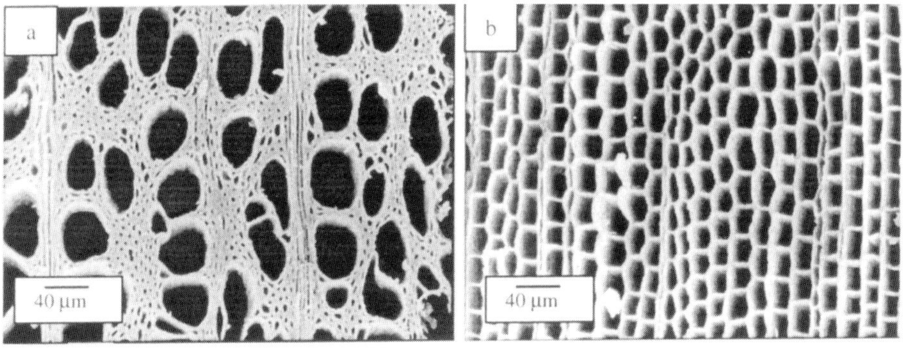

Figure 1: SEM micrographs of pyrolysed wood: (a) beech (deciduous wood), (b) pine (coniferous wood). Section perpendicular to the stem axis (axial direction).

Coniferous wood has a very uniform structure (Figure 1b), it consists of 90-95% tracheids, which are long and slender cells tapered at the ends. The diameter of the tracheids (diameter up to 50 µm) depends on the growth conditions, such as changing temperatures or water supply. Deciduous wood is less homogeneous (Figure 1a).

Tracheary elements (diameter up to 500 μm) form long tubes of a few centimeters. Tracheids and tracheary elements are oriented in the direction of the stem axis. They form the transportation paths for water and minerals within the living tree. Cells arranged in radial direction (rays) and pores in the cell walls (pits, Figure 5) create a three-dimensional network of transport.

There are different technologies for processing of biological structures into ceramic materials. The inherent cellular, open porous morphology of biological structures can be used for liquid and gaseous infiltrations and subsequently for high temperature reaction processes [3]. Biological materials containing silicon such as rice husk or coconut shells can be annealed in inert atmosphere to yield SiO_2 and SiC whiskers or particles [4-5]. Natural fibres like sisal, jute, hump or cotton were infiltrated with $AlCl_3$ or $TiCl_4$ and annealed in air or coated with Si_3N_4 and annealed in Ar-atmosphere to get ceramic fibres [6-8].

It is the aim of the present work to analyse the conversion process from wood to cellular SiC-ceramics by liquid Si infiltration into carbon templates of wood.

2 Experimental Procedure

Different kinds of wood (oak, beech, pine, ebony, maple, balsa) were used to produce biomorphic SiC-ceramics by Si-infiltration (Figure 2). After drying (70°C, 15 h), the samples were pyrolysed in N_2-atmosphere. A slow heating rate of 1°C/min was applied up to 500°C, where the biopolymers (cellulose, hemicellulose, proteins, pectins and lignin) have almost completely decomposed.

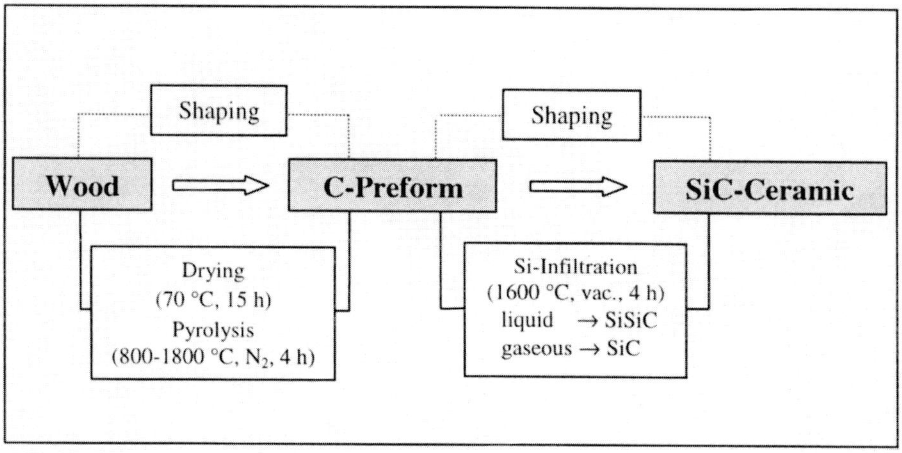

Figure 2: Processing scheme of SiC-ceramics.

Up to the final peak temperature (800-1800°C) an accelerated heating rate of 5°C/min was used. The final temperature was hold over a period of 4 hours. After pyrolysis, the C-preform reproduces the cellullar structure of the initial wood with high precision. Subsequently, the open porosity (tracheids and vessels) was used for liquid and gaseous Si-infiltration processes and the C-preform transformed into SiSiC- and SiC-ceramic.

The conversion of the organic material into the inorganic material was monitored by thermal balance analysis (TGA). The microsructures were characterised by scanning electron microscopy (SEM) and the crystalline phases were determined by X-ray diffraction (XRD).

3 Results and Discussion

3.1 Macroscopical transformation of wood

During heating of wood in nitrogen atmosphere, the biopolymer decomposition is indicated by weight loss starting below 200°C. The decomposition is finished at 500-600°C with a total weight loss of 70-80 wt.%.

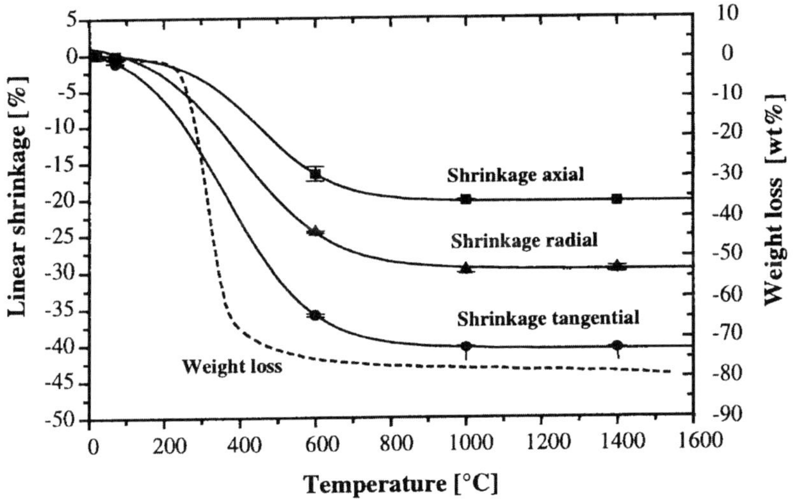

Figure 3: Weight loss and shrinkage of marple during the pyrolysis.

Due to the thermal decomposition of the polyaromatic biopolymers during the pyrolysis, an amorphous carbon structure is formed. Raising the pyrolysis themperature from

600°C up to 1800°C results in a more graphitic like microstructure of the carbon at the highest temperature.

Pyrolysis of wood involves an anisotropic shrinkage of approximately 40% in tangential, 30% in radial and 20% in axial direction (Figure 3), depending on the composition of the different cells in the tissue and the different orientation of the cellulose fibres in the cell walls. Different kinds of wood exhibit a wide range of weight loss depending on the variation of the cellular and molecular composition (e.g. balsa exhibits a weight loss of 73.5 wt%, pine of 73.8 wt% and ebony of 64.6 wt%).

3.2 Transformation of the microstructure

In spite of the high weight loss and the shrinkage of the wood during the pyrolysis, the microstructural details of the initial wood structure were transformed with high precision into the C-preform (Figure 4). However, the porosity of the C-preform is 20-25% higher than of dried wood.

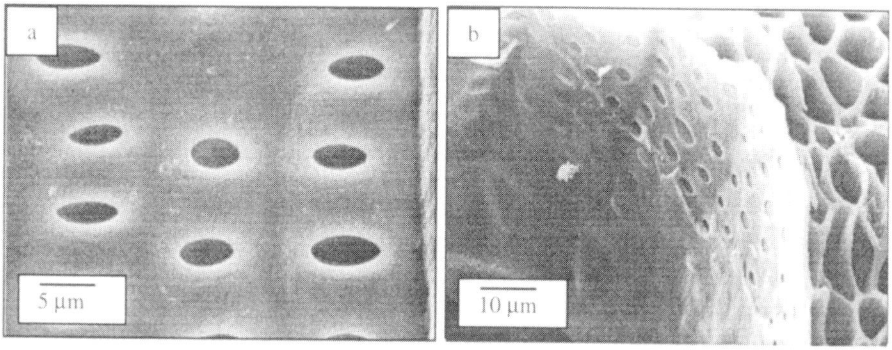

Figure 4: SEM micrographs of pitted walls of vessels (oak): (a) natural wood, (b) pyrolysed wood.

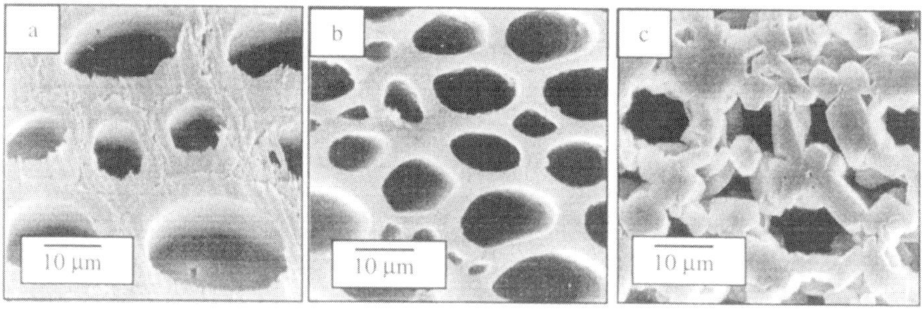

Figure 5: SEM micrographs of cellular structures (marple): (a) natural wood, (b) pyrolysed wood and (c) wood after transformation in SiC-ceramic.

After Si-melt infiltration at 1600°C, tracheids with a small lumina diameter up to 30 µm are completely infiltrated with silicon. β-SiC is formed at the carbon-silicon interface resulting in a mean particle size of 5 µm (Figure 5). Total porosity of the SiC-ceramics depends on the initial cellular structure. Coniferous wood yields SiSiC-ceramics with a high density, high strength and high stiffness in axial direction. The silicon in the tracheary elements can be removed by chemical etching to increase the porosity.

3.3 Mechanical properties

The pyrolysed C-preforms and the SiC ceramics exhibit a distinct anisotropy of the mechanical properties due to the unidirectional cell morphology, Figure 6.

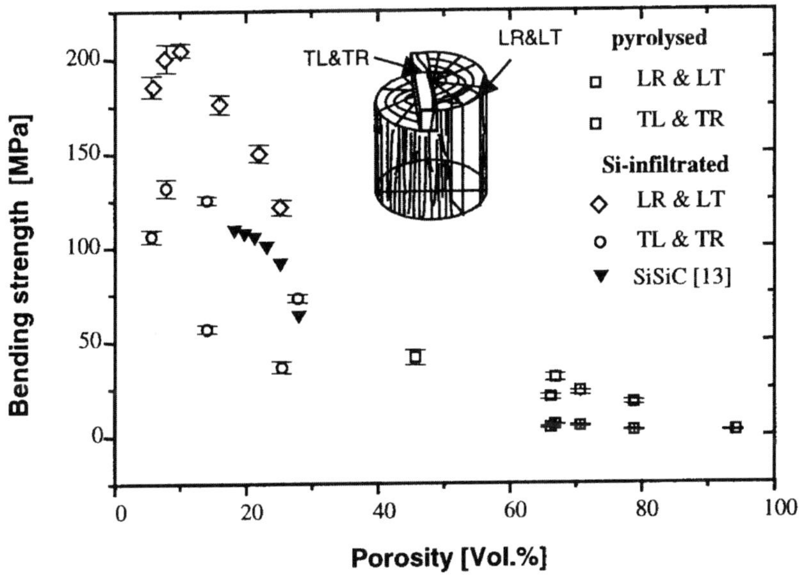

Figure 6: Mechanical properties of a cellular ceramics pseudomorphous to wood in comparison with commercial produced SiSiC materials (▼).

Strength, toughness and Youngs modulus in axial direction are significantly higher than in radial or tangential direction. Compared to conventional (isotropic) SiSiC-ceramics the strength of the wood derived SiSiC-ceramics is superior in axial direction.

4 Conclusions

Carbon preforms derived from wood or other natural structures can be infiltrated with Si-melt to produce novel ceramics and ceramic composites with microstructures pseudomorphous to wood. Different wood structures offer a large variety of ceramics with different porosity and morphology. Anisotropic cellular ceramics could be of interest for technical applications such as heat insulation structures, substrates, filter and catalyst carriers at high temperatures, thermally and mechanically loaded lightweight structures, as well as for medical implant structures (e.g. porous materials for bone-substitution).

Acknowledgement

Financial support from the Volkswagen Foundation under contract I/73 043 is gratefully acknowledged.

References

[1] M. Sarikaya, I.A. Aksay, An Introduction to Biomimetics: A Structural Viewpoint in Microstructure of materials, edited by K.M. Krishnan, San Francisco Press Inc., San Francisco/CA, **1993**.

[2] P. Greil, T. Lifka, A. Kaindl, *J. Europ. Ceram. Soc.* **1998**, 18, 1961 (I), 1974 (II).

[3] T. Ota, M. Takahashi, T. Hibi, M. Ozawa, S. Suzuki, Y. Hikichi, H. Suzuki, *J. Am. Ceram. Soc.* **1995**, 78, 3409.

[4] R.V. Krishnarao, Y.R. Mahajan, T.J. Kumar, *J. Europ. Cer. Soc.* **1998**, 18, 147.

[5] A. Selvan, N.G. Nair, P. Singh, *J. Mat. Sci. Lett.* **1998**, 17, 57.

[6] M. Patel, B.K. Padhi, *J. Mat. Sci.* **1990**, 25, 1335.

[7] M. Patel, B.K. Padhi, *J. Mat. Sci. Lett.* **1993**, 12, 1234.

[8] R.V. Krishnarao, Y.R. Mahajan, *J. Mat. Sci. Lett.* **1996**, 15, 232.

[9] R. Wagenführ, Anatomie des Holzes, 2. Auflage, VEB Fachbuchverlag, Leipzig, **1980**.

[10] E. J. Gibson, *Metals und Materials* **1992**, 8, 333-336.

[11] P.W. Lucas, B.W. Darvell, P.K. Lee, T.D.B. Yuen, M.F. Choong, *Phil. Trans. Roy. Soc. Lond.* **1995**, B 348, 363.

[12] W. Nultsch, *Allgemeine Botanik*, 8. Auflage, Thieme, Stuttgart, **1986**.

[13] H. Osada, A. Kanj, S. Katayama, T. Koga, *Ceram. Trans.* **1993**, 31, 243.

The Conversion Process of Cellulose Fiber into Ceramic Fibers

Evelina Vogli, Holger Friedrich, Heino Sieber and Peter Greil
University of Erlangen - Nuernberg, Department of Materials Science (III), Glass and Ceramics, Martensstrasse 5, D-91058 Erlangen, Germany

1 Introduction

Conversion of biological materials, like woods or plant fibers into ceramic composite materials gained particular interest in the last years [1-4]. Krischnarao et.al [5] and Ohzawa et.al [6] converted cotton fibers into SiC fibers by chemical vapour infiltration of $SiCl_4$ – CH_4 – H_2 and subsequent annealing in Ar-atmosphere at 1200- 1600°C. It was shown that wood can be converted into SiC by liquid or gas infiltration of Si [7]. Cellulose fibers were converted into ceramic fibers by liquid infiltration with different kinds of ceramic powders [8].

The aim of the present paper is to describe the transformational process during the organic ➔ inorganic conversion of cellulose fibers into ceramics. Due to its molecular structure, supermolecular order and fibrillar morphology, cellulose is a typical fiber polymer. It is quite predominantly processed and used in the form of fibers. The average cellulosic fiber dimensions depend on the material, in which cellulose was synthesized, for example: pine has fiber length - 3100μm and fiber width - 31μm; beech - 1200 and 21μm; cotton interns - 9000 and 19μm etc. The cellulose fibers were pyrolysed and the resulting carbon fibers were coated with solid metal particles or metal vapour (Si, Al and a mixture of both). The reaction of the carbide formation and the growth behaviour of carbide particles during the liquid $_{suspension}$ – solid $_{carbon}$ and gaseous $_{metal}$ - solid $_{carbon}$ reaction processes were investigated by XRD, TGA, DTA and SEM measurements.

2 Experimental Procedure

Filter paper, consisting of pure α–cellulose fiber (Whatman International Ltd 541, Springfield Mill, England; fiber diameter 10-40 μm and fiber length 1-4 μm), was used for the investigations. The amount of impurity was negligible (N_2 - 260μg/g, Chlorine 55 μg/g, Calcium 8 μg/g etc.). For preparation of ceramic fibers two different methods were applied: coating of cellulose fiber with metallic powder and metallic vapour infiltration.

Metal powder coated cellulose fibers were prepared by infiltration of the cellulose fiber paper with a 10% slurry of acetone and the respective metal (Al, Si or Al + Si; Al, Si - average grain size 2-4 μm; Si - Fa. Bayer AG, Leverkusen, Germany; Charge CFI 567; purity 98,8%; Al - Alpha, Johnson Matthey GmbH, Karlsruhe, Germany; purity 99,5%). After drying at room temperature, the metallic powder was distributed homogeneously on the surface of the fibers. The fibers were completely coated with metal. Coating layers thickness corresponded to the metal particle size. The specimens were pyrolysed in argon atmosphere at temperatures between 600 and 1600°C for 4h. Up to 500°C a slow heating rate of 2 K/min was applied, where the polyaromatic polycarbon (α-cellulose) has completely decomposed into carbon residue, followed by a heating rate of 5 K/min up to the peak temperature, Fig.1.

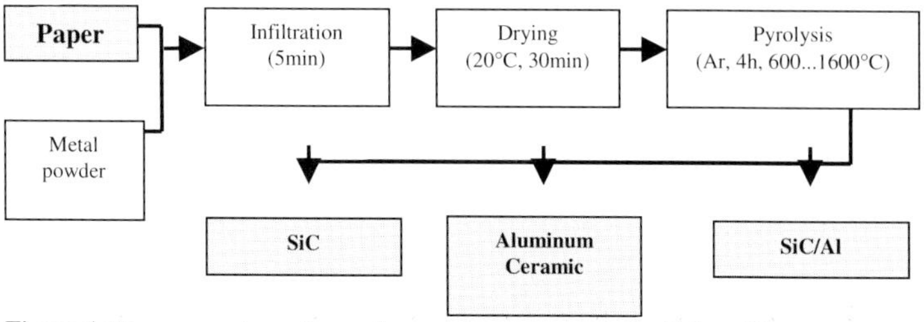

Figure 1: The processing scheme for metal powder coated cellulose fibre.

For conversion of cellulose fibers with metallic vapour, the fibers were first pyrolysed in argon atmosphere at temperatures between 600 and 1400°C for 4h. The heating rate was the same as described before. Subsequently, the specimens were infiltrated with Al vapour at temperature between 600 and 1400°C for 4h in argon atmosphere. In this case, a high heating rate of 5 K/min was applied, Fig.2.

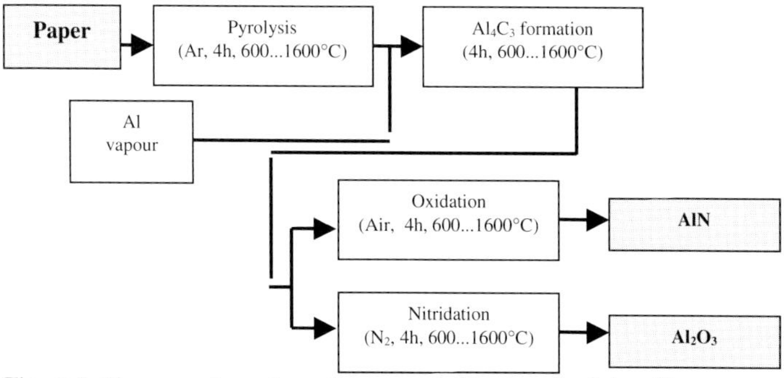

Figure 2: The processing scheme for vapour infiltration of cellulose fiber.

The metal-carbon reactions were studied by thermal balance analysis (TGA, STA 409, Netzsch Gerätebau, Selb, Germany). The microstructure of initial, pyrolysed and ceramic fibers was characterised by scanning electron microscopy (SEM - XL30, Fa. Philips, NL) and by XRD, using monochromatic CuK_{α} radiation (Diffrac 500, Siemens, Mancheim, Germany).

3 Results and Discussion

3.1 Pyrolysis of cellulose fiber

Thermal induced degradation reaction of cellulose fibers to carbon preform can be divided in two steps:

1. Release of water, that results in "dehydroccllulose", followed by the elimination of CO and CO_2 above 260°C.

2. After break down of the -C-C- chains in the biopolymer structures, the formation of carbon structures starts at temperatures above 600°C [9]. Fig. 3 schematically summarises the thermal degradation of cellulose in atmosphere during heating.

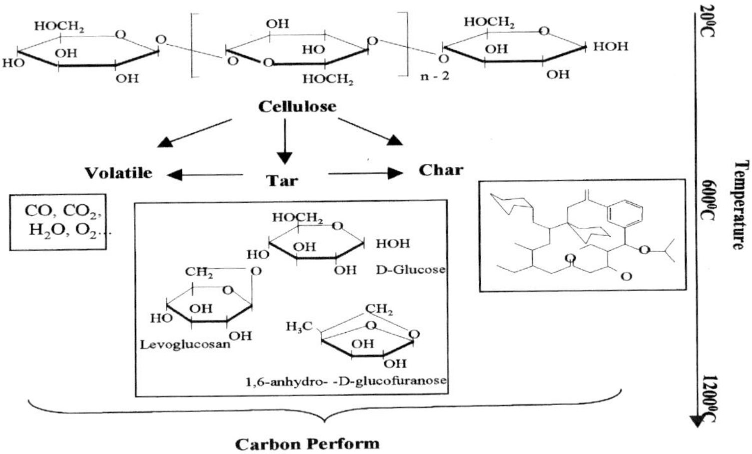

Figure 3: The thermal decomposition of cellulose in inert atmosphere

The aromatic structure of cellulose, which is characterised by oxygen bonded glucose units, shows a significantly high weight loss of about 80wt % during pyrolysis [8].

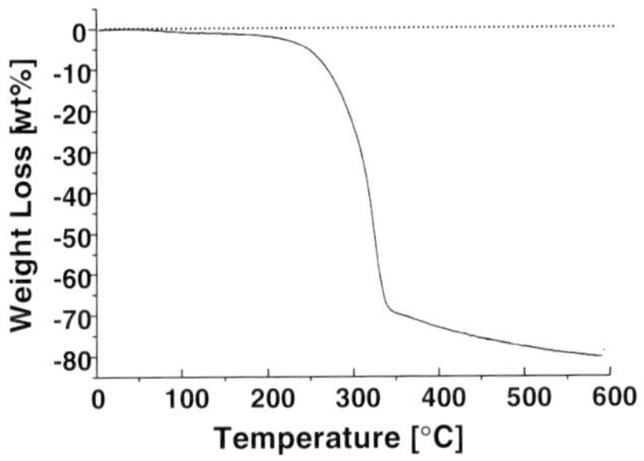

Figure 4: The weight loss of cellulose paper

The thermogravimetric curve of α-cellulose in inert atmosphere shows a rapid weight loss between 260 and 320°C and leaving an amorphous carbon-rich char above 320°C, Fig.4. Rearrangement of the polyaromatic carbon structures at higher temperatures results in an increased crystallinity of the carbon and finally the formation of cellulose fibers with graphitic like crystalline structure [7].
After pyrolysis the cellulose fibers were transformed into pure carbon preforms. The carbon fibers have an average diameter of about 10 μm, Fig. 5.

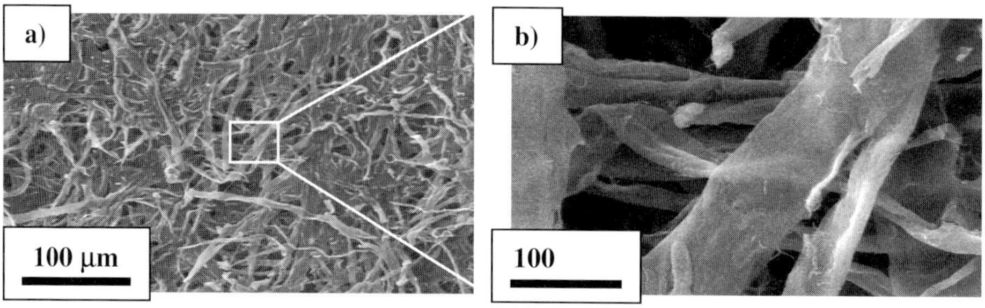

Figure 5: a) SEM micrograph of the pyrolyzed cellulose fiber paper after heating to 1450°C in argon atmosphere, b) microstructure of cellulose derived carbon fiber.

3.2 Pyrolysis of metal coated cellulose fiber

After pyrolysis of Si coated cellulose fiber two different reaction morphologies of SiC were observed: 1- reaction between molten Si and solid carbon; 2- reaction between gaseous Si and solid carbon, Fig. 6b.

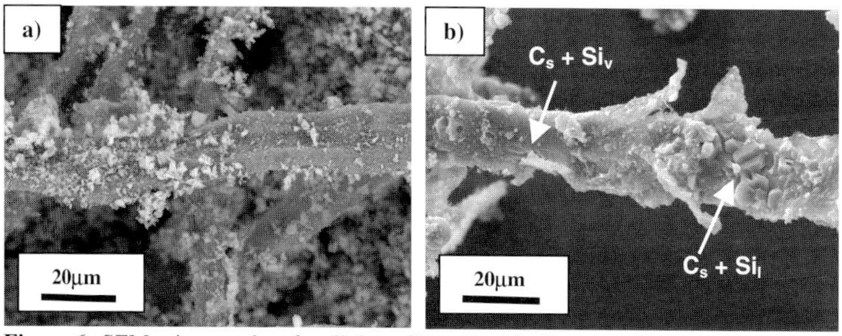

Figure 6: SEM micrographs of cellulose fiber, a) with Si coated fiber, b) in SiC converted cellulose fiber, after heating at 1450°C in Ar atmosphere.

While in direct contact of solid Si to carbon large SiC single grains are formed, fiber areas without direct contact to Si particles reacted with gaseous Si and retained a fiber-like morphology.

The situation is changed for Si/Al coated fibers. Si and Al form an eutectic melt at 577°C at 12.2wt% Si, [8]. XRD measurements reveal the formation of SiC (in addition to Al and Si and very small amounts of Al_2O_3) at about 800°C, but negligible amounts of Al_4C_3. The reaction was finished at 1000°C (approximately the melting temperature of the mixed Al/Si powder).

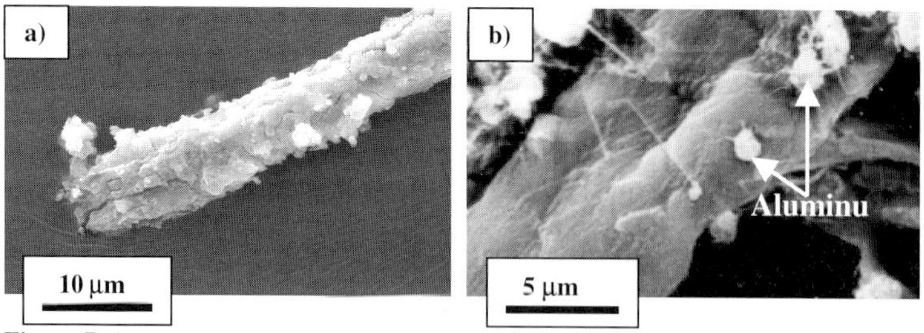

Figure 7: SEM micrographs of pyrolyzed cellulose fiber, a) pyrolyzed fiber coated with Si/Al, b) pyrolyzed paper fiber coated with Al.

In Al coated cellulose fibers, a complete reaction of the carbon fibers into Al_4C_3 at temperature 1200°C was achieved. Some residual molten metal grains were also presented. These grains are visible in the reacted structure, Fig. 7b.

3.4 Vapour infiltrated cellulose fiber

For Al-vapour infiltration of the pyrolyzed cellulose fibers, the starting temperature of the Al_4C_3 formation was observed to be about 1200°C. In contrast to the powder coated fiber, the specimens are unstained by the surplus amount of metal. The cellulose fibers were completely transformed into Al_4C_3 fibers, Fig. 8.

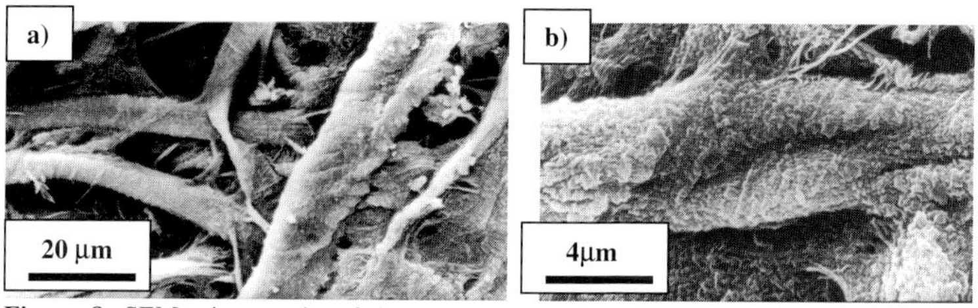

Figure 8: SEM micrographs of the pyrolysed cellulose fiber paper after 1200°C in argon atmosphere in the presence of Al vapour.

However, XRD investigations of the Al_4C_3 fibers did not reveal any crystalline structures and the formation of Al_4C_3 fibers at a temperatures of 1200°C was identified by SEM/EDX-analysis.

4 Conclusions

Cellulose fibers were successfully used as biorganic performs for processing of ceramic fiber morphologies. Coating of cellulose fibers with reacting or gaseous metals like Si and Al yield the formation of carbide/metal composites. The reaction with gaseous infiltrants yield only pure carbide ceramics.

The Al_4C_3 carbide ceramic fibers can be annealed incur to form Al_2O_3. Nitridation of the carbide fibers result in the formation of AlN ceramic fibers, which are interesting for light-weight, high-temperature applications of ceramic composites materials.

5 Acknowledgement

Financial support from the Volkswagen Foundation under contract # I/73 043 is gratefully acknowledged.

6 References

1. M. Sarikaya, Microsc. Res. and Tech., 27, 1994, 360A.
2. M. Patel, B. K. Padhi, J. Mat. Sci., 23, 1990, 1335.
3. M. Patel, B. K. Padhi, J. Mat. Sci., 12, 1993, 1234.
4. K. M. Knowles, M. V. Ravichandran, J. Am. Ceram. Soc. 5, 1997, 1165.
5. R. V. Krischnarao, Y. R. Mahajan, T. J. Kumar, J. Euro. Ceram. Soc. 5, 1997, 1165.
6. Y. Ohzawa, M- Yoshimura, K. Nakane, K. Sugiyama, J. Mat. Sci. Eng. 1998, A-242, 26.
7. P. Greil, T. Lifka, A. Kaindl, J Europ. Ceram. Soc. 1998, 18, 1961.
8. H. Sieber, H. Friedrich, D. Schwarze, A. Kaindl, P. Greil, Proc. of the 101st Annual Meeting of the American Ceramic Society, April 25-28, 1999, Indianapolis, IN, Symposium N: Innovative Processing and Synthesis of Ceramics, Glasses and Composites, in print.
9. D. Klemm, B. Philipp, T. Heinze, U. Heinze, W. Wagenknecht, Comprehensive Cellulose Chemistry, Vol.I, Wiley-VCH, Weinheim, 1998, p. 107.

Preparation and Properties of Novel Nb- and Ti-Based Metal Reinforced Alumina Composites with Interpenetrating Microstructure

R. Günther[1], T. Klassen[1], B. Dickau[3], F. Gärtner[2], A. Bartels[3], R. Bormann[1,3]
[1]Institute for Materials Research, GKSS Research Centre, D-21502 Geesthacht
[2]Institute of Mechanical Engineering, University of the Federal Armed Forces, Germany, D-22043 Hamburg, Germany
[3]Dept. Materials Science and Technology, Technical University of Hamburg-Harburg, D-21073 Hamburg

1 Introduction

Recently, intermetallic/ceramic composites have been developed, with volume fractions of the ceramic phase ranging from 20 % up to more than 90 % [1] [2] [3]. Contrary to conventional particulate- [4] [5] or fiber-reinforced [6] [7] ceramic or intermetallic matrices, this novel composites consist of co-continuous three-dimensional interpenetrating networks of both of the phases. Thus, the favorable properties of ceramics like high-temperature strength and wear resistance can be combined with superior damage tolerance and functional properties of the intermetallic phase, i.e. electrical conductivity. For example, Ni_3Al/Al_2O_3 composites produced by pressure infiltration into a pre-sintered Al_2O_3 body have shown high fracture toughness (11.5 Pa$\sqrt{m}$), even at temperatures ~ 800°C [8]. These novel composites are of particular interest for their high melting points, low overall densities and good corrosion and oxidation resistance, giving these composites a high potential for applications as engine valves or break discs. Furthermore, the Ti-based systems have a high bio-compatibility [9] and are therefore favorable for bone replacements. Based on high-energy milling, a new low-cost production route for these alloys has been developed, which allows for variation of the metallic phase or the ceramic volume fraction in a wide range. In this way, the properties of the composites can be tailored to suit the desired applications.

2 Experimental

Three different systems, Nb-Al/Al_2O_3, Ti-Al-V/Al_2O_3 and Ti-Al-Nb/Al_2O_3, have been prepared. Elemental metal powders were blended together in the ratio 84 at% Nb-16 at% Al and 86 at% Ti-10 at% Al-4 at% Nb, according to the desired overall composition of the (inter)metallic phase and Al_2O_3 powder was added in the desired volume fraction. For preparation of the Ti-Al-V/Al_2O_3 composite pre-alloyed 86 at% Ti-10 at% Al-4 at% V powder was used. Powder blends of 100 g were milled in a planetary ball mill at maximum intensity using 500 g ZrO_2 balls and a 500 ml Al_2O_3 vial. All handling of the powders including milling was carried out in a glove-box under purified argon atmosphere to prevent additional oxidation of the metallic constituents. Green bodies with dimensions of 4 mm × 5 mm × 40 mm for mechanical testing and 5 mm × 40 mm × 50 mm for wear resistance tests were formed by uniaxial pressing at pressures ≤ 50 MPa followed by cold isostatic pressing at 900 MPa. Pressureless sintering was carried out under vacuum (0.1 Pa) at temperatures up to 1580°C. Green densities were measured geometrically, densities of the sintered specimen were determined by Archimedes' method. Mechanical properties were determined by four-point-bending-tests with an inner and outer span of 12 mm and 24 mm, respectively on specimen, that were ground and polished down to 3 μm on the tensile surface. For the fracture-toughness measurements, a Vickers indentation was used for crack initiation and the thoughness was calculated using the ISB method [10]. Hardness (HV10) was obtained with Vickers diamond indenter using a 100 N load and a loading time of 10 s. Wear resistance was investigated using a testing setup according to the Japanese Industry Standard JIS H8615. The electrical resistivities were measured using a four-probe method with seperate voltage and current electrodes. Scanning electron microscopy (SEM) investigations were performed using a scanning electron microscope operated at an accelerating voltage of 15 kV.

3 Results and Discussion

After preparation of the powder blends via high-energy ball milling the powders were pre-compacted. The green compacts have relative densities between 68 % and 70 %. The bending strength of the green compacts ranges between ~ 25 MPa for the Nb-based composites and 41 MPa up to 98 MPa for the Ti-based composites. This allows for easy machining with conventional tools (e.g. milling or drilling) without chipping or breaking. After machining, the specimen were pressurelessly sintered. In this step, the reaction to form the desired phases take place (further details regarding phase formation are published in [11]). Residual porosities of < 2 % have been achieved in this manner. Corresponding to the volume fractions, the desired interpenetrating networks of the ceramic and the metallic phases are formed as can be seen in figure 1. The dark areas correspond to the ceramic phase, the gray areas represent the metallic phases. Due to abrasion of the milling tools during milling of the Ti-based composites, some ZrO_2 is also detected, which appears white (see Fig.1(b)).

The mechanical tests reveal, that the damage tolerance is significantly improved with respect to monolithic Al_2O_3. As summerized in table 1, the bending strength of the Nb-based composites is enhanced by a factor of 4 up to 1393 MPa, in comparison to 350 MPa for monolithic Al_2O_3 [4]. The fracture toughness also shows a significant improvement up to 11.8 MPa√m.

To put the measured values into perspective, figure 2 shows the mechanical data of the Nb-based and the Ti-based composites in comparision with literature data for composites with comparable interpenetrating microstructures. Most of the systems show values in the range of 600 ± 150 MPa and 6 ± 1.5 MPa√m for bending strength and fracture toughness, respectively. While the system Ni_3Al/Al_2O_3 offered the highest fracture toughness so far, i.e. 11.5 MPa√m [8], the bending strength could not be improved over that of monolithic Al_2O_3. The mechanical properties of the Ti-based composites of this study compare well with results of previous investigations of similar systems. However, the Nb-based systems exhibit a drastically enhanced damage tolerance in comparision to all previous results on similar composites.

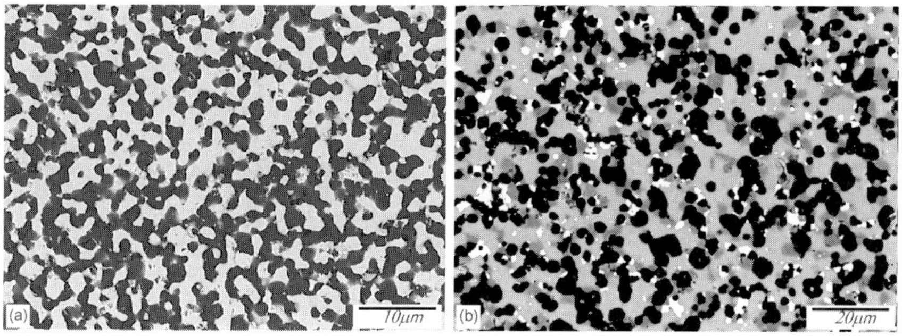

Figure 1. SEM image of a cross section of the systems: (a) $Nb_{84}Al_{54}$/54 vol.% Al_2O_3 and (b) $Ti_{86}Al_{10}Nb_4$/54 vol.% Al_2O_3.

Table 1. Mechanical properties of the Ti-based and Nb-based composites in comparision to monolithic Al_2O_3 [4]

material	σ_B [MPa]	K_{IC} [MPa√m]	hardness [HV10]
$Ti_{86}Al_{10}V_4$ / 50 vol.% Al_2O_3	527 ± 73	5.6 ± 0.4	1329
$Ti_{86}Al_{10}Nb_4$ / 50 vol.% Al_2O_3	524 ± 35	5.1 ± 0.7	1370
$Nb_{84}Al_{16}$ / 54 vol.% Al_2O_3	1393 ± 233	10.5 ± 1.5	805
$Nb_{84}Al_{16}$ / 70 vol.% Al_2O_3	1135 ± 140	11.8 ± 1.5	992
Al_2O_3	350 ±	3.0 ±	1854

For a better understanding of this improved behaviour, microstructural investigations of crack propagation have been performed. Figure 3 shows a crack initialised at the corner of a Vickers indentation in the Nb-based and the Ti-based systems, respectively. The crack is clearly deflected at the phase boundaries of the (inter)metallic and the ceramic phases. In addition, for the Nb-based system crack bridging by (inter)metallic ligaments (center section in Fig. 3(a)) is observed, leading to a significant reduction of the crack length down to 55 % with respect to monolithic Al_2O_3. These observations reveal that the crack energy is efficiently reduced within these interpenetrating microstructures, thus leading to the enhanced damage tolerance.

Figure 4 shows the electrical resistivity of the Ti-based system Ti-Al-Nb/Al_2O_3 in dependence of the ceramic volume fraction as well as on temperature (up to 700°C). As clearly can be seen, the electrical resistivity scales with the ceramic volume fraction. For a given volume fraction of the ceramic phase, the electrical resistivity is nearly independent on temperature up to ~ 400°C. The good temperature stability qualifies these materials suitable for applications in high-power high-temperature resistors.

The wear resistance of the Ti-based and the Nb-based systems is given in table 2 in comparision to monolithic Al_2O_3, conventional steel and selected hard phases. While the value for $Nb_{84}Al_{16}$/54 vol.% Al_2O_3 is comparable to spring steel, $Nb_{84}Al_{16}$/70 vol.% Al_2O_3 reaches values in the range of monolithic Al_2O_3. The best resistance against abrasive wear was measured for the Ti-based systems. The system $Ti_{86}Al_{10}Nb_4$/50 vol.% Al_2O_3 reaches a wear resistance, that is two times better in comparision to monolithic Al_2O_3.

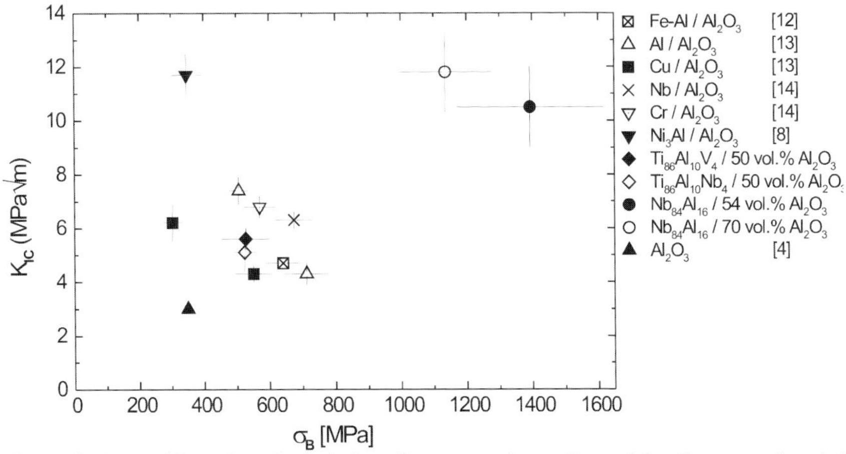

Figure 2. Comparision of mechanical data (fracture toughness K_{IC} and bending strength σ_B) for different composites

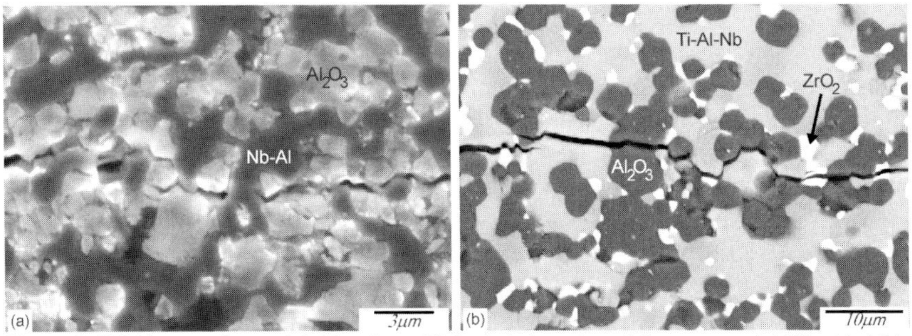

Figure 3. SEM image of a crack initiated at a Vicker's indentation of the systems: (a) $Nb_{84}Al_{54}$/ 54 vol.% Al_2O_3 and (b) $Ti_{86}Al_{10}Nb_4$/54 vol.% Al_2O_3.

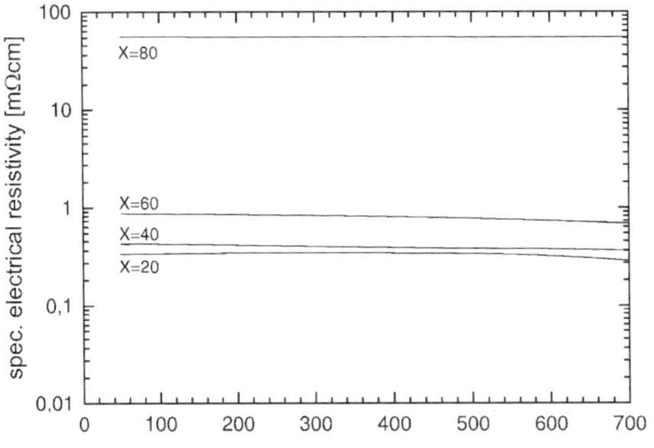

Figure 4. Specific electrical resistivity on dependence of temperature and ceramic volume fraction X

Table 2. Wear resistance of the Ti-based and Nb-based composites in comparision to monolithic Al_2O_3, conventional steel and some hard phases

material	mass abrasion [mg]	volume abrasion [mm^3]
$Ti_{86}Al_{10}V_4$ / 50 vol.% Al_2O_3	7.35 ± 0.21	1.67 ± 0.05
$Ti_{86}Al_{10}Nb_4$ / 50 vol.% Al_2O_3	4.95 ± 0.49	1.11 ± 0.11
$Nb_{84}Al_{16}$ / 54 vol.% Al_2O_3	28.9 ± 3.7	5.24 ± 0.66
$Nb_{84}Al_{16}$ / 70 vol.% Al_2O_3	18.4 ± 3.1	3.54 ± 0.6
Al_2O_3	8.5 ± 2	2.16 ± 0.5
spring steel (DIN: 65 SiCrMoV 5 2)	37.0 ± 2	4.71 ± 0.3
WC-Co (83-14)	~3	~0.21
Galv. Cr hard material	1.0	1.53

4 Conclusions

Novel Nb-based and Ti-based (inter)metallic/ceramic composites have been prepared via a new processing route using high-energy ball milling and pressureless sintering. These composites consist of interpenetrating networks of (inter)metallic and ceramic phases and show improved damage tolerance with respect to monolithic Al_2O_3 in combination with a high wear resistance and electrical conductivity. The maximum bending strength and fracture toughness have been determinded to be 1393 MPa and 11.8 MPa$\sqrt{m}$, respectively, for the Nb-based system. The least abrasive wear is achieved for the Ti-based system Ti-Al-Nb/Al_2O_3, half that of monolithic Al_2O_3. The electrical resistance is nearly constant up to a temperature of 400°C. Due to high wear resistance, these composites are promising for automotive applications, e.g. valves or brake discs. Furthermore, the high bio-compatibility qualifies the Ti-based systems for application in medical implants.

5 Acknowledgments

Helpful discussions with N. Claussen, R. Jansen, S. Schicker, D. García and J. Bruhn are gratefully acknoledged. This work was supported by the German minister for Education, Research and Technology (BMBF).

6 References

[1] N. Claussen, R. Janssen, and D. E. García. *J. Mater. Res.*, 11(11): 2884, 1996

[2] T. Klassen, R. Günther, B. Dickau, F. Gärtner, A. Bartels, R. Bormann, and H. Mecking. *J. Am. Ceram. Soc.*, 81(9): 2504-2506, 1998

[3] R. Günther, T. Klassen., B. Dickau, A. Bartels, F. Gärtner, und R. Bormann. In: J. Heinrich, G. Ziegler, W. Hermel, H. Riedel (Ed.), „Werkstoffwoche 98 – Band VII: Keramik / Simulation Keramik“, 681-686. WILEY-VCH, 1998

[4] J. L. Guichard, O. Tillement, and A. Mocellin. *J. Mater. Sci.*, 32: 4513-4521, 1997
[5] M. R. Ghomashchi. *J. Mat. Sci.*, 30: 2849-2854, 1995
[6] A. Solomah, W. Reichert, V. Rondinella, L. Esposito, and E. Toscano. *J. Am. Ceram. Soc.*, 73(3): 740-743, 1990.
[7] S. Nourbakshs and H. Margolin. *Mater. Sci. Eng. A*, A144: 133-141, 1991
[8] J. Rödel, H. Prielipp, N. Claussen, M. Sternitzke, K. B. Alexander, P. F. Becher, and J. H. Schneibel. *Scripta Met. et Mater.*, 33(5): 843-848, 1995
[9] M. Semlitsch, H. Weber, and R. Steger. *Biomedizinische Technik*, 40(12): 347-355, 1995
[10] G. R. Anstis, P. Chantikul, B.R. Lawn, and D. B. Marshall. *J. Am. Ceram. Soc.*, 64: 539-543, 1981
[11] T. Klassen, R. Günther, B. Dickau, A. Bartels, R. Bormann, and H. Mecking. In *Materials Science Forum Vols. 269-272*, pages 37-46. Trans Tech Publications, Switzerland, 1998
[12] S. Schicker. PHD thesis, Technische Universität Hamburg-Harburgg, 1998
[13] M. Knechtel, H. Prielipp, H. Müllejans, N. Claussen, and J. Rödel. *Scripta Met. et Mater.*, 31(8): 1085-1090, 1994
[14] D. E. García, S. Schicker, R. Janssen, and N. Claussen. *J. Eur. Cerm. Soc.*, 18: 601-605, 1998

Study of Structure and Some Properties of Whiskerizated Silicon Carbide Fibers

L.R.Vishnyakov, V.A. Dubok, L.N.Pereselentseva, K.L.Vishnyakova
Institute for Problems of Materials Science of NAS of Ukraine, Kiev

1 Introduction

High strength high modulus fibrous materials having organized spatial structures still remain topical, in particular, for the design of high strength ceramics based on refractory oxides, nitrides or carbides with high fracture toughness [1-3]. Therefore, despite whisker-like crystals and polycrystalline ceramic fibers as being the most strong fibers, are known for a long time, a search of new technologies of their making still continues [4-6]. The major difficulties which encounter when developing composites reinforced with whisker crystals, are connected with a uniform distribution of fibers within the matrix bulk, and their orientation. This paper studies the method of preparation of a frame structure of SiC fibers by the whiskerization of polycrystalline fibers during their formation of hydrated cellulose precursor. Such structures can be easy used for preparation of composites by liquid phase techniques, e.g. sol-gel technology, impregnation of frames with liquid metals or melted ceramics. A frame-like fibrous structure allows a desired arrangement of the fiber within the bulk of a composite that is undergone mechanical or thermomechanical loading.

2 Methodology

For researches, hydrated cellulose precursor fibers were used of 8-10 micron diameter to form a felt.

The hydrated cellulose fibers were treated with antipyrenes (catalysts) and silicon-containing salts, i.e. alkaline silicate solutions. Then they were heated into a graphite crucible at up to 2000^0C under various media: argon, nitrogen, and in a medium containing silicon compounds in a vapor-gas state. The silicon-containing medium has been formed through initial products (silicon oxides), which were held some centimeters under the felt specimen. The felt was blown through by a carrying gas (Ar or $N_{2)}$) of vapor products of silicate reduction reaction. The phase composition of fibers at different stages of conversion of hydrated cellulose into silicon carbide, was examined using X-ray diffractometer DRON-2 (Cu_{Ka}). The structure of resultant fibers was examined using instruments JEM-100X and Superprobe by methods of electron transmission and scanning microscopy. The mechanical properties of fibers were evaluated indirectly (as results of mechanical testing of composites made thereof) as direct testing was inapplicable because of macrostructural complication of fiber frames obtained. The samples of composites were tested on tensile strength and elasticity modulus by strain measurement techniques.

3 Results and Discussion

The synthesis of silicon carbide is based on carbothermal reduction using carbon fiber of hydrated cellulose precursor as reduction agent. When hydrated cellulose fibers are heated, a range of well-known conversions [7,8] occur resulting in formation of firstly carbonized, then graphitized fiber. The presence of silicate on the hydrated cellulose fibers involves additional physicochemical reactions. At 550 to 600 °C, silicate dehydrates with formation of SiO_2 on the surface. Hydrated cellulose fiber impregnated with silicate solution, after heating at 800 to 900 °C, is a carbonized fiber covered with a SiO_2 film that is distributed rather uniformly upon the fiber surface. X-ray diffraction patterns of hydrated cellulose fiber specimens impregnated with silicate solution and annealed at 750 to 900 °C under nitrogen or argon, show the presence of SiO_2 phase in the form of α-crystobalite. X-ray patterns have faint lines that can be identified as graphite traces. Further temperature increase up to 1450-1500 °C leads to the formation of new phases. The X-ray diffraction patterns of fibers annealed under argon, reveal lines of α-crystobalite and of cubic silicon carbide (β-SiC), halo amorphous carbon, and silicon lines. The X-ray diffraction patterns of fibers annealed under nitrogen also reveal lines that can be identified as hexagonal silicon nitride (β-Si_3N_4). Annealing of fiber specimens at 1750 to 2000 °C results in formation of hexagonal silicon carbide of α-modification with no relation to the annealing medium (nitrogen or argon).

A medium saturated with vapors of carbide-forming compounds (SiO, CO, hydrocarbons) gives a growth of whiskers on the fiber surfaces as will be shown later. Their maximum growing rate is observed within the range of 1750 to 1800 °C. The X-ray patterns of fiber specimens annealed at 1750-2000 °C under such a medium reveal no appearance of new lines that should testify the formation of new phases as compared with specimens annealed at the same temperatures under media containing any silicon compound vapors. It suggests that whiskers belong to the same modification of α-SiC as fibers formed during the process of hydrated cellulose conversion.

Figs. 1 and 2 represent the structures of investigated SiC-fiber specimens produced under optimal temperature/time conditions relative to the rate and completeness of the synthesis. It was found that at 1750-1800 °C, 10-15 minutes are sufficient to completely convert carbon fibers into silicon carbide fibers. Investigation of the microstructure showed that the influence of nitrogen and argon as medium on the formation of the structure of SiC-fibers at these temperatures are equal. The fibers have polycrystalline structures. In the picture by electron microscope in scan mode (Fig. 2) one can see these grains that measure 0.10-0.15 microns. Pore sizes have the same order. The pore value, as determined by the measuring the area of the structural components on the micrographs, is 25 to 30 %, although these data seem to be somewhat exaggerated.

Transmission electron photo and electron diffraction pattern show that the fiber structure is a mixture of single crystals differing in orientations (Fig. 3). Apparent transparent grains have a banded structure and a spherical configuration.

Figs. 4 and 5 present structures of whiskerized SiC fibers as seen in the scan mode. It is seen than whiskers grow from the surface of a polycrystalline fiber. The transmission electron microscope photo (Fig. 5) shows a structure of separated whisker crystals that grow from a polycrystalline fiber The major part of whiskers have a spherical configuration, although separate whiskers may have crystal faces. The structure of whisker crystals like the structure of crystals forming a polycrystalline fiber, has a band-like structure. It is evident that the surface grains of polycrystalline fibers serve as seeding which has whiskers growing on it. The cross-section of whiskers is 0.10-1.15 microns. The length of whiskers can reach high values (several centimeters). The initiation and growth of the whiskers according to the mechanism describing the gas phase - liquid phase - solid phas patterne, were likely contributed by the presence of SiO_2 melt on the fiber surface.

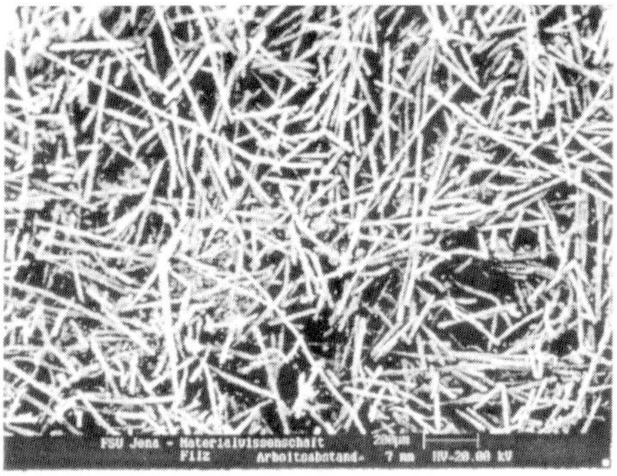

Figure 1. Structure of SiC felt made of polycrystalline fibers

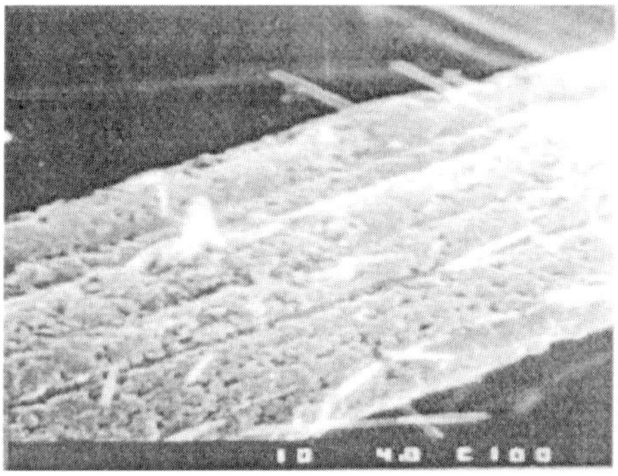

Figure 2. Microstructure of polycrystalline SiC fiber in the scanning mode

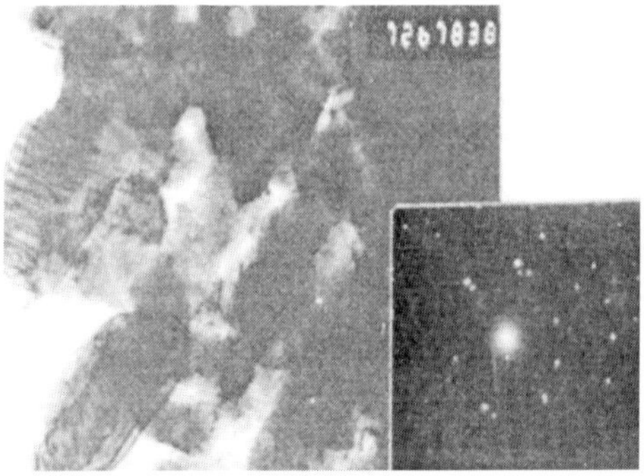

Figure 3. Microstructure and diffraction pattern of the polycrystalline SiC fiber

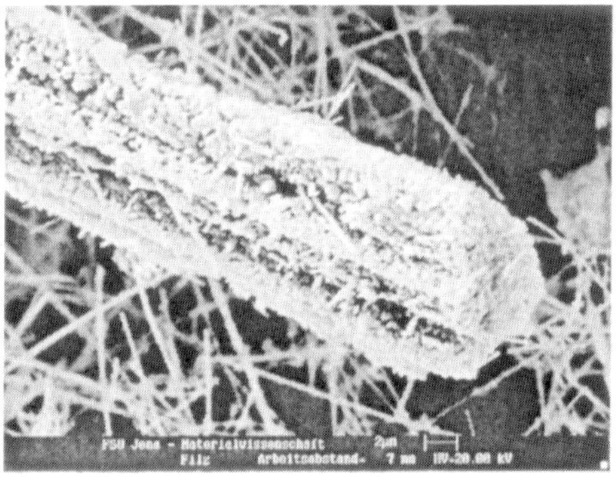

Figure 4. Microstructure of whiskerized SiC fiber in the scanning mode

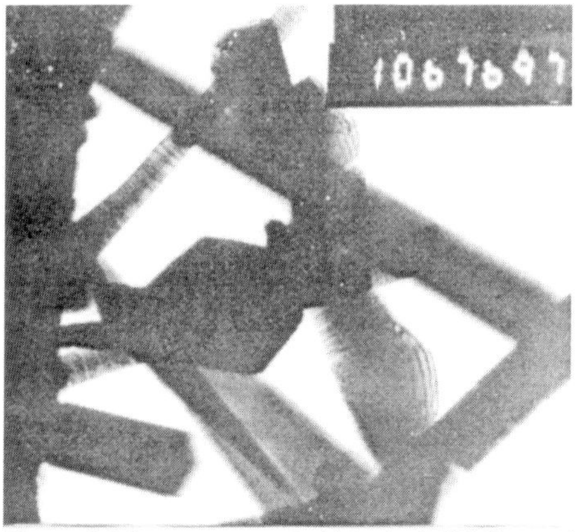

Figure 5. Microstructure of whiskerized SiC fiber and transmission electron microscope

To have estimation of mechanical properties of the fibers studied, composite specimens based on Al-12% Si matrix were prepared by vacuum-compression techniques of felts made of whiskerized (SiC_w) and non-whiskerized (SiC_{pc}) fibers. The composites contained 10 to 12 % of fibers. The whiskerized fibers have had around 50 % of whiskers. The felt has been impregnating for several seconds at 700-800 ^{0}C. Under these conditions, unfavorable reactions have no time to develop on the fiber/matrix boundary. Results needed to measure tensile strength and elasticity modulus are presented in Table.

Table. Strength and elasticity modulus of SiC-fiber-based composites

		Al-12%Si	Al-12% Si+ SiC_{pc}	Al-12% Si+ SiC_w
Elasticity modulus, E	GPa	72	100118	
Tensile modulus, σ_b	Mpa	140	133	145

As can be seen from the Table, the introduction of fibers in question into aluminum matrix considerably improve the composite modulus of elasticity. This effect is especially enhanced when whiskerized fibers are introduced. A rather approximate calculation of elastic properties of reinforced composites by the additivity law allows the estimation of the modulus of elasticity of non-whiskerized fibers as 300-350 GPa, and of whiskerized fibers at least 600-700 GPa. Moderated values of the strength of produced composites can be explained by a poor contact between the matrix and fibers with resultant impossibility to reach strong fibers. This thesis is supported by the investigation of the composite structure, while the development of composite is a subject of other researches.

4 Conclusion

High modulus fibrous frames were prepared by whiskerizing of polycrystalline SiC-fibers during their processing of hydrated cellulose precursor. The fiber can be suggested to the use as reinforcing fillers for metallic or ceramic matrix application.

5 References

1. Braun Hans Joachim, 20th Int. Congr. Hist. Sci., Liege, 1997, 8-10.
2. L.O. Norberg, T. Ekstrom, Xu Fang Fang, J. Mater. Sci. Lett., 1997, 16, 917-920.
3. N. Bamba, Y.-H Choa, T. Sekino, K. Niihara, J. Europ. Ceram. Soc., 1998, 18, 693-699.
4. Super heat resistant silicon carbide fiber produced using electron beam irradiation, process, STA Today, 1995, 12, p.9.
5. Kawai Chihiro, Yamakawa Akira, J. Amer. Ceram. Soc., 1996, 79, 557-558.
6. G.D. Semchenko, A.A. Starolat, N.L. Diakonenko, Glass and Ceramics, 1996, 6, 21-24.
7. A.S. Fialkov, Carbon, interlayer and compounds and composites on its basis, Aspect, Moscow, 1997, p.718.
8. A.A. Konkin, Carbon fibers and other filament materials, Chemistry, Moscow, 1974, p. 375.

In-situ Tensile Testing of SiC_p-Al Metal Matrix Composite Produced by Laser Embedding

Václav Ocelík, Arjan Vreeling, Yutao Pei, Jeff Th.M. De Hosson
Department of Applied Physics, Materials Science Center and Netherlands Institute for Metals Research, University of Groningen, Groningen

1 Abstract

A successful combination of parameters of Nd:YAG laser beam, powder stream and the temperature of Al substrate have been found to form reinforced MMC layer in a single injected laser track, containing 25-30% volume fraction of SiC particles.

Optical microscopy, SEM, TEM, EDS and X-ray diffraction techniques were used to identify morphology and crystallography of phases formed during the process. In-situ tensile observations on V-notched samples in a scanning electron microscope were performed to identify failure initiation and crack propagation mechanisms.

2 Introduction

Several experimental methods can be applied for the formation of protective layers on metal substrates using high power lasers. Layers with very interesting mechanical and tribological properties and also with an excellent bond with the substrate may be prepared [1]. In the present contribution a Metal Matrix Composite (MMC) layer was prepared by the laser assisted injection [1-5] of SiC particles into Al substrate. The layer was mechanically tested using *in-situ* tensile tests with the aim to obtain accurate information about the failure initiation and crack propagation processes by regular inspections in a Scanning Electron Microscope (SEM) during loading.

3 Experimental

Aluminium with 99.6 wt-% purity was used as a substrate material. Single parallel laser tracks with 20 mm length and 10 mm distance were prepared using a 2kW Rofin Sinar Nd:YAG laser by moving a 10 mm thick substrate with respect to the laser beam using a computer controlled X-Y table. SiC particles with the mean size of 60 μm were simultaneously injected into the locally formed melt pool on the substrate surface. A more detailed description of the laser injection process, particular the used parameters as well as more detailed analysis of the whole sample preparation process can be found elsewhere [6].

The flat sheet with a thickness of 1mm was cut from the substrate block by spark erosion. V-type notched specimens for tensile test, 10 mm wide and 40 mm in length, were cut from this sheet by laser cutting in such a way, that an approximately 1 mm wide laser track was situated between two V-notches. The longitudinal axis of the tensile sample corresponded to the moving direction during the laser injection. The specimen surface was mechanically polished.

In-situ tensile tests were performed in a special tensile/compression stage (Kammrath & Co) inside a field emission gun environmental Philips XL30 Scanning Electron Microscope (SEM). Sample elongation data as well as the tensile force were recorded during loading. The displacement speed was 2 μm/s on a sample with 23 mm working distance between the clamps. Optical microscopy, SEM, HRTEM, EDS and X-ray diffraction techniques were used to identify morphology and crystallography of phases formed during the laser injection process. Post-failure SEM observations of fracture surfaces were used to perform a standard fractographic analysis.

4 Results and Discussion

Notable difficulties were superseded during the applied laser injection method for preparation of an Al/SiC MMC layer on an aluminium substrate. Firstly, the laser injection process is complicated by the high laser beam reflectivity of aluminium [7]. Therefore, high energy densities of the laser beam are needed to create a melt pool in the Al substrate. On the other hand, it is necessary to minimize the laser beam energy to prevent thermal decomposition of SiC. Because of the high absorptivity of SiC, powder temperature increases rapidly during the short contact of injected particles with the laser beam in the process. Secondly, the nature of the oxide layer covering the liquid aluminium [8] requires an overheating of the melt above ≈1200 °C to achieve good wetting conditions for the SiC particles and make further penetration into the melt possible. This also requires sufficient amount of laser beam energy, but again it is in a contradiction with the protection of SiC particles against the thermal decomposition.

Pre-heating of the Al substrate over 300 °C solved these contradictory requirements. Nevertheless, an extremely small range of main process parameters, like laser beam energy, powder feeding rate and particle speed was found for successful injection.

4.1 Microstructure

During the laser assisted injection a track containing 25-30% volume fraction of SiC is formed as Fig. 1 demonstrates. The SiC particles are homogeneously distributed between the surface and maximum injected depth of about 0.8 mm.

The microstructure of Al/SiC MMC layer prepared by laser injection differs from the microstructures of MMC prepared by other methods like hot extrusion [9], hot isostatic pressing [10] or squeeze casting press [11]. The grain size of the substrate material inside the laser track is about 0.1-0.2 mm. Grains contain large amount of 5 μm length individual columnar dendrites that are oriented in the radial direction. Silicon was solidified with aluminium as the last part of the melt inside an interdendritic area. An Al/SiC interface reaction layer was identified as short Al_4C_3 hexagonal plates at SiC particle surface. It was confirmed by TEM observations that these Al_4C_3 plates are coherent with the SiC particle. During solidification they prefer to grow from the particle surface into the melt parallel to the base planes of the SiC.

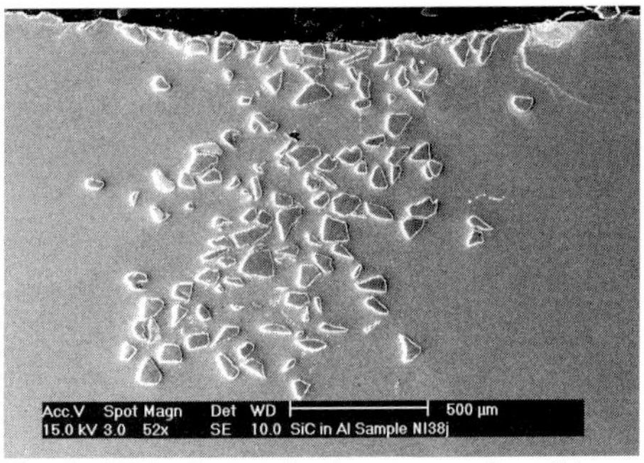

Fig. 1: SEM micrograph of a cross section of laser embedded SiC particles in the Al substrate.

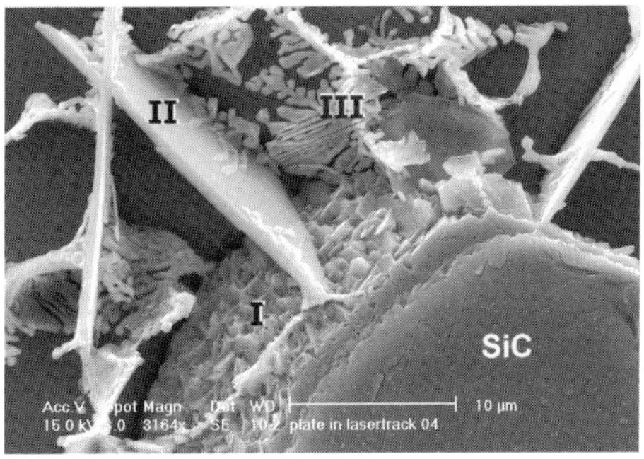

Fig. 2: SEM micrograph of a deep etched Al/SiC MMC prepared by laser injection.

In addition larger Al_4C_3 hexagonal plates (0.5-1.0 µm thick) are randomly distributed in the aluminium matrix, mainly in the sub-surface region. Figure 2 shows: the SiC particle, small Al_4C_3 plates on the Al/SiC interface (I), isolated and large Al_4C_3 hexagonal plates (II) and eutecticly solidified Si (III) present in the observed microstructure after deep etching during which Al matrix is completely etched out.

Free silicon and carbon observed in the microstructure originates from thermal decomposition of SiC and from the reaction of SiC in liquid Al [12]:

$$4Al + 3SiC \longrightarrow Al_4C_3 + 3Si \tag{1}$$

Substrate preheating to 300 °C increases the temperature reached in the melt pool and decreases the cooling rate inside the laser track, which gives more appropriate conditions for the formation of the phase on the SiC/Al interface.

4.2 In-situ tensile test

Figure 3 shows stress-elongation curves observed on four MMC samples compared with the curve observed for substrate material as well as the curve measured on a single laser track without any injected SiC particles.

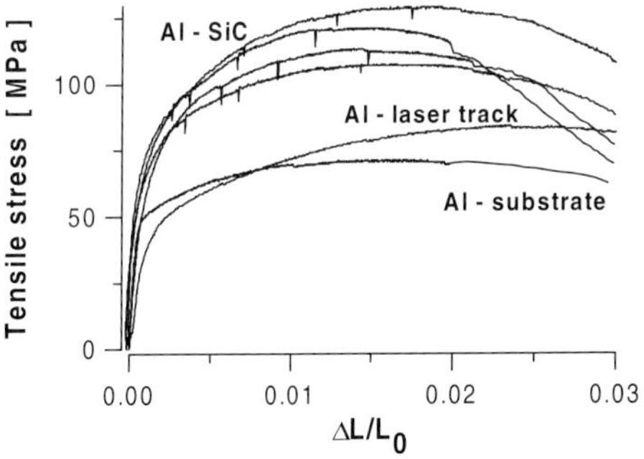

Fig. 3: Stress-elongation curves observed during in-situ tensile tests.

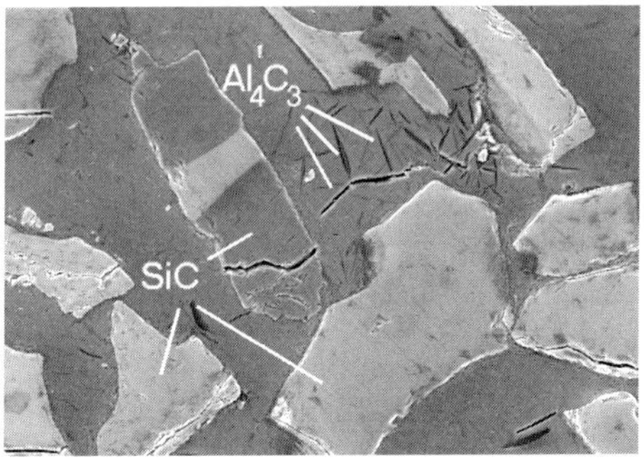

Fig. 4: Failure initiation by the cracking of SiC particles and Al_4C_3 plates decohesion. The tensile stress was applied in vertical direction.

The MMC showed approximately 50% strength improvement compared to the substrate material. From the observed curves it is clear that two main strengthening components are present. The microstructural strengthening originates from the changes of microstructure inside the laser track during the rapid solidification, this involves mainly the structural refining. This strengthening does not change the yield stress considerably but it changes the work-hardening rate at all strains.

The reinforcing particles cause an increase in the flow stress as well as a further increase in the work-hardening rate, which is in accordance with earlier results observed on Al/SiC MMC prepared by a different method [13]. The strength increase is sustained by the reduction of ductility. This increase is comparable with the microhardness increase from HVM 37.1 ± 1.0 for substrate to 51.6 ± 5.1 for inter-particles area. Large increase of variation of this quantity indicates an inhomogeneous distribution of internal matrix stress and deformation, typical for particulate MMC [13].

Small irregularities on MMC stress-elongation curves in Fig. 3 are caused by the stress relaxation during the time periods when the deformation was fixed because of the regular microscopic inspections of sample surface. A quite large variance in stress-elongation behavior of MMC observed in Fig. 3 for different samples may be explained by the differences in local SiC volume fraction just presented between V-notches. Both, tensile strength and work-hardening rate are quite sensitive to this quantity [13].

4.3 Failure initiation and cracks propagation

Immediately after the first manifestation of plastic deformation on stress-elongation curves failure initiates at large Al_4C_3 plates through decohesion between the plates and matrix. Brittle SiC particle cracking is a second type of observed failure initiation. Both these mechanisms appeared simultaneously in the area close to the V-notch, that is working as a stress concentrator. Those initiation places are preferred, which form the cracks approximately perpendicular to the tensile axes, as Fig. 4 shows. A third type of failure initiation, infrequently observed, is decohesion of SiC particles from matrix.

Fig. 5: Propagation of cracks in the advanced stages of fracture.

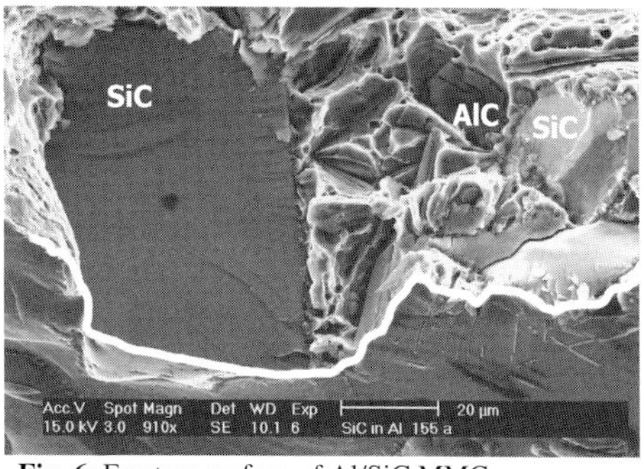

Fig. 6: Fracture surface of Al/SiC MMC.

Figure 4 shows two main in-situ detected failure initiation mechanisms: a) the brittle fracture of SiC particles and b) the simple decohesion between the matrix and hexagonal Al_4C_3 plates oriented more or less perpendicularly to tensile stress. As post-failure fractographic observations confirmed, randomly distributed aluminium-carbide plates are incoherent with Al matrix and therefore simple decohesion from the matrix is observed very frequently. This feature is also important in the crack propagation stage, when cracks propagate through Al matrix along these plates, as Fig. 4 also demonstrates.

The failure initiation and crack propagation stages can not be separated easily, because during the crack propagation many new sub-cracks are initiated. Thus cracks propagate through the already initiated sub-cracks. Figure 5 shows the situation in the advanced stage of the failure. More than one dominant crack propagates in the direction approximately perpendicular to the tensile stress. Finally, one of them prevails and the ultimate fracture surface is formed by the ductile failure of Al matrix.

The fractographic features typical for the two failure initiation processes e.g. cleavage of SiC particles and decohesion of non-coherent Al_4C_3 plates together with the ductile dimples typical for the fracture of Al substrate are observed on the fracture surface, see Fig. 6. The thick white line separates the polished surface observed during in-situ experiment from the fracture surface shown at upper part in Fig. 6.

Lewandowski et al. [14] already had shown, that the brittle fracture of the SiC particles is predominant mechanism in the failure initiation of under-aged aluminium alloy based MMC. Particle cracking has generally been agreed to be a common phenomenon for MMCs with larger (>20 μm) particles [15]. Decohesion of SiC particles is observed infrequently, therefore the bonding between the particles and matrix seems to be good. This is caused by the high surface roughness formed by adjacent Al_4C_3 plates on the SiC/Al interface as Fig. 2 clearly demonstrates.

5 Conclusions

Desirable combination of parameters of Nd:YAG laser beam, SiC_p powder stream and the temperature of Al substrate have been found to form an approximately uniform distribution of SiC particles in a single injected laser track with 1mm width and 0.8 mm depth. The high reflectivity of Al and the wetting property of oxide layer on Al melt make the process parameter window for successful injection very small.

During the laser injection of SiC particles into Al two new phases are formed: Si and Al_4C_3. Silicon is presented in Al-Si eutectic interdendritic regions and hexagonal Al_4C_3 appears in platelet shape both in melt pool matrix and at the SiC/Al interface.

In-situ tensile testing experiments on V-notched samples performed in E-SEM showed approximately 50% strength improvement. Failure initiates mainly by decohesion of Al_4C_3 plates and by cracking of SiC particles. The cracks propagate via sub-cracks initiation, through the matrix along the Al_4C_3 plates and at the final stage of fracture via ductile failure of aluminium matrix.

6 Acknowledgement

The Netherlands Institute for Metals Research and The Foundation for Fundamental Research on Matter (FOM-Utrecht) are acknowledged for the support of research described in this contribution. ESD Delfzijl provided the commercial SiC powder.

7 References:

1. B. Kloosterman, J. Th. M. De Hosson in Surface Treatment (Ed.: M.H. Aliabadi and C.A. Brebbia), Computational Mechanics Publications, Southampton, 1997, 287.
2. J. D. Ayers, T. R. Tucker, Thin Solid Films 1980, 73, 201 - 207.
3. J. H. Abboud, D. R. F. West, J.Mater.Sci.Lett. 1991, 10, 1149 – 1152.
4. T. N. Baker, H. Xin, C. Hu, S. Mridha, Mater.Sci. and Technol. 1994, 10, 536 – 544.
5. B. Kloosterman, B. J. Kooi, J. Th. M. De Hosson, Acta Mater. 1998, 46, 6205 - 6217.
6. J. A. Vreeling, V. Ocelík, Y. T. Pei, J. Th. M. De Hosson, Reinforced SiC/Al composite layer produced by laser particle injection, to appear in: Surface Treatment '99, 20 –22 September 1999, Assisi, Italy.
7. L.F. Mondolfo, Aluminium Alloys, Butter Worths, London-Boston, 1976, p.108
8. G. Kaptay, Materials Science Forum 1996, 215-216, 475-484.
9. C.-S. Kang, K. Maeda, K.-J. Wang, W. Wakashima, Acta Mater. 1998, 46, 1209-1220.
10. P.M. Bronsveld, P. Bruinsma, J.Th. De Hosson, M.A. Sargent, W.H.M. Alsem, Mat. Sci. & Eng. 1991, A135, 77-81.
11. S. Elomari, R. Boukhili, C. San Marchi, A. Mortensen, D.J. Lloyd, J.Mater.Sci. 1997, 32, 2131-2140.
12. J.C. Viala, P. Fortier, J. Bouix, J.Mater.Sci. 1990, 25, 1842-1850.
13. C.Y. Barlow in Metal Matrix Composites – Processing, Microstructure and properties (Ed.: N.Hansen, D. Juul Jensen, T. Leffers, H. Lilholt, T. Lorentzen, A.S. Pedersen, O.B. Pedersen, B. Ralph), RIS∅ National Laboratory, Roskilde, Denmark, 1991, 1-16.
14. J.J. Lewandowski, C. Liu, W.H. Hunt jr., Mater.Sci. & Engng 1989, A107, 241-255.
15. F.J. Humpreys, Mater.Sci.Engng. 1991, A135, 267-273.

Elaboration of Ceramics Composites in the System Alumina-Mullite-Zirconia

F. Belnou, D. Gœuriot, F. Valdivieso, F. Thevenot
Ecole Nationale Supérieure des Mines, Saint-Etienne, France

1 Introduction

Alumina is widely used as electrical insulator. In such applications breakdown strength is a critical parameter. Previous works [1, 2] have shown that, under particular microstructural conditions, zirconia or mullite additions could improve the breakdown resistance of industrial aluminas. This work is an attempt to combine both beneficial effects, using alumina-mullite-zirconia composites obtained from alumina and zircon powders by reaction-sintering.

In the past, such ternary composites have been elaborated by several methods, like mechanical alloying of zircon and alumina before sintering[3], or silica injection in ZTA [4]. But no extended studies have been conducted on the use of reaction-sintering of alumina-rich powder mixtures in order to obtain ternary materials. CARTER and O'CONNOR [5] have sintered alumina and zircon powders in varied proportions (80 to 95 wt% alumina). They obtain ternary microstructures including sometimes elongated alumina grains, equiaxed mullite grains and two different types of zirconia particles: some of them are located at the alumina-mullite grains boundaries, the others, smaller, more rounded and of the tetragonal type, are in intragranular position within mullite. As far as reaction is concerned, they notice zircon dissociation starting at 1400°C and mullite formation around 1450°C: a transient amorphous silica phase exists between these two temperatures.

However the synthesis of mullite-zirconia ceramics through reaction-sintering of alumina and zircon powders has been extensively studied. Authors generally separate the reaction-sintering process into two successive steps: zircon decomposition and slow mullite formation. Zircon decomposes from the surface to the center of the grain, leading to the crystallization of fine zirconia particles and formation of amorphous silica: liberated silica migrates to the outer part of the grain and reacts with alumina. A significant quantity of liquid is produced during the reaction which allows fast densification of the compact [6]. The process is accompanied with a volumic expansion of about 3 to 5% [7]. Both monoclinic and tetragonal zirconias are present in the sintered materials [6, 8].

2 Experimental Procedure

A fine and high purity alumina powder (A6Z, Criceram, d_{50}=1.1μm, around 100ppm impurities) and the finest zircon powder we found (Opasil ZFF Super, Moulin Des Prés, d_{50}=2μm) were used: for a better understanding of the role of zircon additions, alumina has been chosen easily sinterable. Three types of materials were elaborated: A100 was 100 wt% alumina, Z7.5 was 92.5 wt% alumina - 7.5 wt% zircon, and Z15 was 85 wt% alumina - 15 wt% zircon. The chosen proportions of powders were mixed with 33,3 wt% water. Ammonium polyacrylate and ammonium citrate were used as dispersants. If necessary ammonia was added to maintain pH around 10. The obtained mixtures were ball-milled for 15h and attrited at 830rpm for 1h. Binders (PVA 125 and PEG 1500) were then added for better shape forming and the products were ball-milled for 15h again. Every mixtures were spray-dried using a Büchi 190 machine. Samples were preformed by uniaxial pressing (27MPa) and finally isostatically pressed at 250MPa. Binders removal proceeded in three successive soakings of 2h at 180°C and 350°C and 1h at 500°C. Two different types of samples are obtained: cylinders of 12mm height x 8mm diameter for dilatometry experiments, and cylinders of 4mm height x 20mm diameter for sintering and electrical breakdown experiments.

Several dilatometric tests were made: they are detailed in the results part and were performed with a Setaram TMA 92 apparatus. Three different thermal cycles were chosen to sinter the samples with 2h soakings at 1500, 1560 and 1620°C, all carried out at room atmosphere in a Thermolyne 46200 owen. Sintered samples were characterized as follows. Density was mesured by Archimedes method. Phase analysis was made by XRD with local-made Dosophatex apparatus. Microstructures were obtained after polishing and thermal etching with Jeol JSM 840 SEM. Breakdown tests were performed with a Baur DTS machine: samples were maintained between two spherical copper electrodes plunged into Nytro 10GB Nynas dielectric oil to avoid flashover and subjected to 50Hz AC (tension rate was 1kV/s).

3 Results and Discussion

3.1 Phenomenology of reaction-sintering in the system alumina-zircon

Samples of the three types A100, Z7.5 and Z15 were heated with a rate of 2°C/min up to 1700°C. In figure 1, shrinkage curves show a significant difference between pure alumina and alumina-zircon mixes behaviors: alumina shrinkage starts at a lower temperature (100°C). For higher temperatures (1600°C), curves are getting closer to each other but alumina samples shrinkage slows down sooner. The shrinkage rate curves presented on figure 2 exhibit the same tendency.

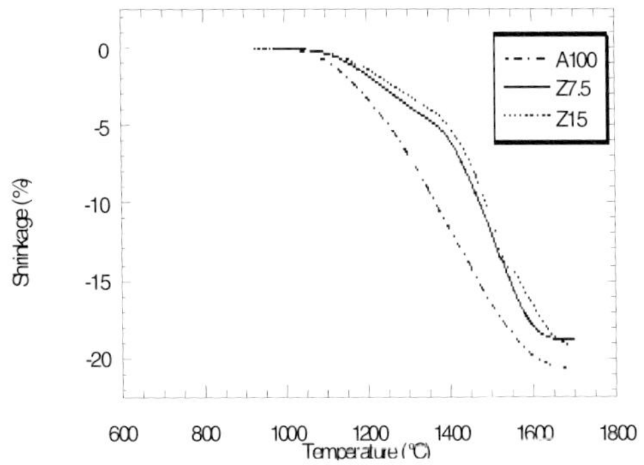

Figure 1. Linear shrinkage for A100, Z7.5 and Z15 samples.

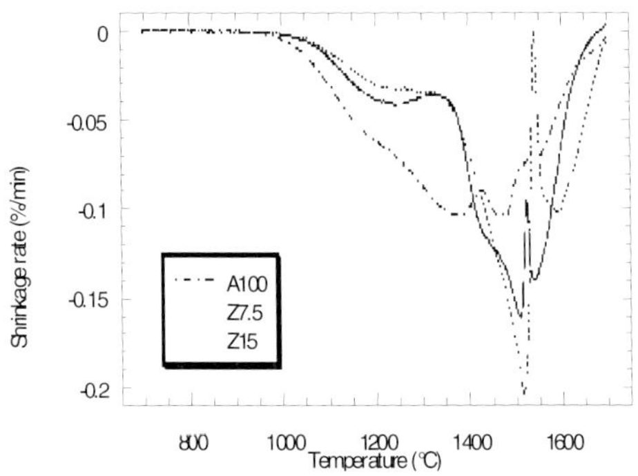

Figure 2. Shrinkage rate for A100, Z7.5 and Z15 samples.

New information comes to light concerning both alumina-zircon mixes: shrinkage accelerates sharply around 1400°C then slows down around 1515°C (for Z15) and reaccelerates at 1545°C (for Z15) to reach a second maximum at 1590°C (for Z15). XRD analysis was performed on Z15 samples heated up to 1515 and 1545°C and fastly cooled. The sample treated at 1515°C contains alumina, zircon and traces of zirconia. After heating at 1545°C, alumina still appears, but zircon has disappeared while mullite and monoclinic and tetragonal zirconias are detected. The sharp shrinkage slowing down that occurs at 1515°C is due to the volumic expansion accompanying the reaction of decomposition of zircon, formation of mullite and crystallisation of zirconia. This reaction is complete at 1545°C. The initial shrinkage acceleration around 1400°C is probably due to viscous phase sintering induced by liberation of amorphous silica during zircon decomposition.

A DTA experiment was performed on Z15 powder, with a heating rate of 15°C/min up to 1650°C: an endothermic peak can be seen at 1590°C. On a dilatometric curve describing the same thermal cycle, this temperature corresponds to the reacceleration of shrinkage after the reaction, that is 1545°C on the curves shown figure 2. Considering the quantity of impurities and the fact that the reaction is locally silica-rich, according to the alumina-silica phase diagram this phenomenon can be characteristic of the melting of a part of mullite.

A Z15 sample was also heated up to 1590°C and then fastly cooled down. XRD indicates a small change in the proportions of monoclinic and tetragonal zirconias at the benefit of tetragonal zirconia (compared to samples treated at 1545°C). If Z15 samples are soaked at least 30min at this temperature before cooling, this becomes significant. After the reaction, tetragonal zirconia crystallises, but only intragranular zirconia, compressed by alumina matrix, remains tetragonal at room temperature. Around 1590°C, the densification rate is high enough to retain the smallest intergranular zirconia particles under tetragonal form after cooling by compression effect.

3.2 Influence of green density

As green density affects the quality and/or quantity of contacts between particles, thus having an effect on sintering and reactional process, Z15 samples pressed uniaxially at 14 and 27MPa and isostatically at 250MPa have been tested. In figure 3, shrinkage rate curves show that reactivity is better for high pressed samples especially after liquid forming at 1545°C: capillary network is finer when green density is increased, which makes liquid spread easier and so enhances sintering kinetics. Final densification is the same for the three samples.

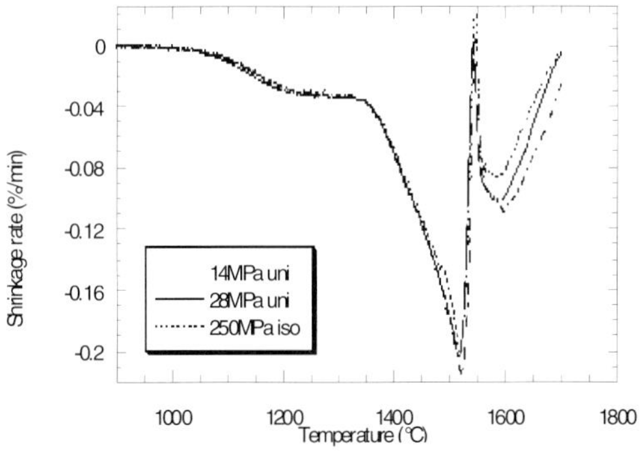

Figure 3. Shrinkage rate for Z15 samples pressed at 14 and 27MPa uniaxial and 250MPa isostatic.

3.3 Influence of sintering temperature

Pure alumina samples are always better densified than alumina-zircon samples undergoing the same heat treatment, and Z15 specimens are very close to open porosity except after 2h at 1620°C. The observed evolution with increasing temperature was expected: densification rates increase as porosity is resorbed, and then exaggerated grain growth limits densification. Figure 4 shows typical microstructures of A100, Z7.5 and Z15 samples for every sintering conditions investigated. Exaggerated grain growth does not occur at lower temperature than 1620°C, which is very high for alumina sintering. Grain shape is different if zircon is present which indicates that liquid phase sintering favors elongated grain growth in this system. Microstructures get finer as zircon quantity increases: this is typical of the growth inhibiting properties of zircon and zirconia. The proportion of intragranular zirconia increases with sintering temperature, because bigger alumina grains trap more zirconia particles. Tetragonal zirconia becomes predominant at high temperature, as indicated by XRD, for two reasons: first, there is more intragranular zirconia, and second, as explained before, better densification allows small intergranular zirconia grains to remain tetragonal after cooling by compressive effect.

3.4 Dielectric breakdown

Results are shown on table 1. Except for Z15 samples that exhibit constant increase of dielectric strength with rising sintering temperature, the general tendency is an improvement of dielectric properties with increasing soaking temperature up to 1560°C, and then a decrease for 1620°C heat treatment. This can be easily correlated with microstructural evolution. Dielectric strength gets better as porosity is resorbed (critical defect size is reduced) and drops as grain growth becomes exaggerated (critical defect size rises). The behavior of Z15 samples is different, simply because they are less densified and exaggerated grain growth does not occur. Dispersion seems important, but it is typical for this type of materials as shown in previous works [1, 2]: it is essential to remember that dielectric breakdown is locally initiated and is consequently very sensitive to microstructural inhomogeneities.

Table 1. Breakdown tests results for every sintered materials (samples are 3mm thick).

Sintering temperature (°C)	Type of material	Breakdown strength (kV/mm)
1500	A100 (0 wt% zircon)	14.8±0.7
1500	Z7.5 (7.5 wt% zircon)	14.8±0.5
1500	Z15 (15 wt% zircon)	14.1±0.5
1560	A100	15.5±0.3
1560	Z7.5	15.3±0.9
1560	Z15	14.3±0.8
1620	A100	15.2±1.3
1620	Z7.5	14.9±0.4
1620	Z15	14.6±1.6

Zircon additions do not improve dielectric strength of this particular alumina, which has good breakdown values due to its purity and fineness. But one must lay emphasis on the fact that here the size and position of zirconia grains are not optimal as this was described in previous works [1, 2].

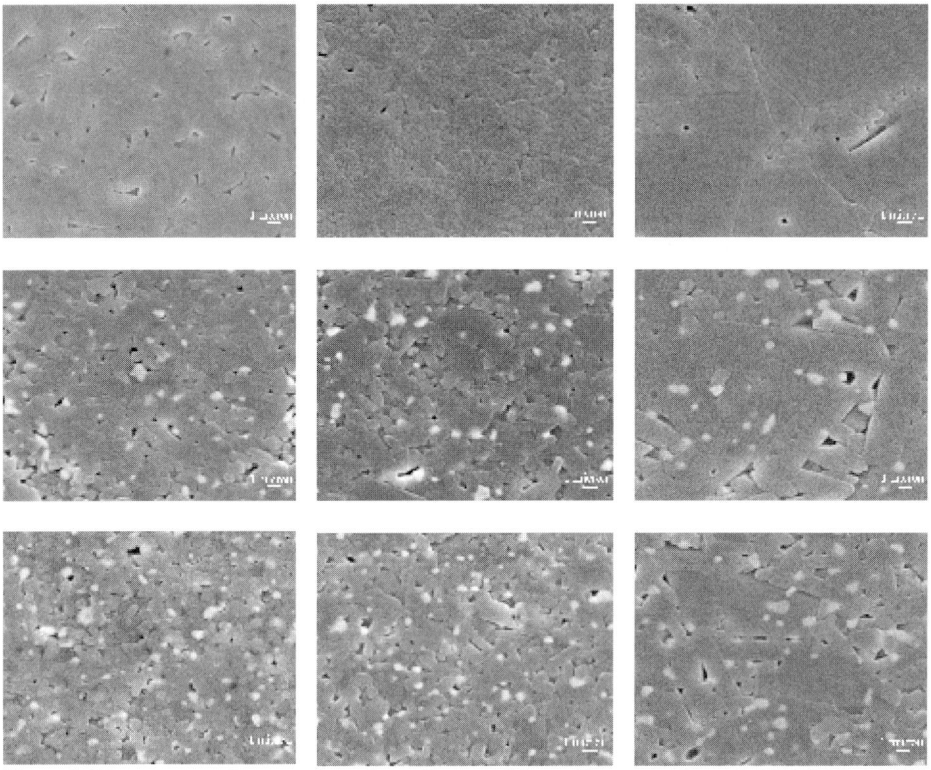

Figure 4. Microstructures for every sintered samples: materials are (from up to bottom) A100, Z7.5 and Z15, sintering temperatures are (from left to right) 1500, 1560 and 1620°C (the small white bar on each photograph is 1μm).

4 Conclusions

In this study, the whole reaction-sintering process of alumina-rich powder mixtures with zircon as additive has been depicted. Considering the sintering inhibiting role of zircon before its decomposition, it is not an efficient sintering aid for the chosen alumina: but this alumina is easily sinterable and zircon should allow better sintering of aluminas of higher grain sizes. Zircon additions also do not improve breakdown strength but microstructures are not optimized. This system is still very interesting because of its ability to produce very varied microstructures. Microstructure control is an interesting perspective. Further investigations will include additions of oxides which should shift the reaction temperatures and change the liquid characteristics: better liquid description should also help us to understand the sintering and grain growth mechanisms.

5 References

[1] J. Liébault, J. Bigarré, D. Tréheux, D. Gœuriot, F. Thévenot, Suppl. of Le Vide: science, technique et applications (Ed.: SFV) Paris 1998, 287, 391-400.
[2] J. Liébault, thesis n° 206TD, Saint-Etienne 1999.
[3] K.A. Khor, Y. Li, Mater. Sci. Eng. 1998, A256, 271-279.
[4] I.M. Low, R. Skala, R. Richards, D.S. Perera, J. Mater. Sci. Lett. 1993, 12, 1585-1587.
[5] J. Carter, B.H. O'Connor, Int. Ceram. Monogr. 1994, 1, 148-153.
[6] A. Leriche, M. Deletter, F. Cambier in Advances in Ceramics (The American Ceramic Society Inc.) 1988, Vol. 24: science and technology of zirconia III, 1083-1089.
[7] A. Leriche, P. Descamps, F. Cambier in Forceram: céramiques composites à particules; cas du frittage-réaction (Ed.: Septima) 1992, 9-38 (in french).
[8] J.M. Rincon, T.R. Dinger, G. Thomas, J.S. Moya, M.I. Osendi, Acta Metall. 1987, 35, 1175-1179.

Microstructural development and high frequency characterization of bulk $YBa_2Cu_3O_{7-x}$ superconductors

A.Brendel, R.Heidinger
Forschungszentrum Karlsruhe, Institut für Materialforschung I
Postfach 3640, D-76021 Karlsruhe

1 Introduction

The physical properties of superconductors are affected by the microstructure. Relevant microstructural features such as grain boundaries and second phases act as regions of weakened superconductivity (weak links). The weak link effects are especially important for the cuprate-based high T_C superconductors. The compound $YBa_2Cu_3O_{7-x}$ (YBCO) is interesting for applications, such as for energy storage or high frequency components. Sintering and texturing of bulk materials as an effective method to develop special microstructural characteristics have been discussed early by Murakami [1] and their influence on the critical current density J_C have been demonstrated. The weak link aspects have been introduced into theoretical models describing the high frequency properties, especially the surface resistance R_S [2]. This parameter is directly related to high frequency losses in the material. The present paper takes up the processing studies on bulk YBCO targeted to identify the relevant microstructural features to R_S. Based on commercial stoiochiometric fine grained $YBa_2Cu_3O_{7-x}$-powders, highly densified materials are produced without sintering additives. By specific choices of the sintering parameters, the grain size (and thus the grain boundary density) and the distribution of secondary phases can be modified. By systematic studies of the achieved microstructure correlations with the observed high frequency properties can be studied.

2 Experimental

The starting powders were stoichiometric grades from the suppliers Hoechst GmbH, Germany (Grade 1) and Solvay Barium Strontium GmbH, Germany (SR 5 grade). The specimens were manufactured by compaction into massive blocks by uniaxial pressing (25 MPa) in a die of 50 mm diameter followed by cold isostatic pressing (300 MPa) and pressureless sintering in air. The temperature cycle of the sintering process consisted of temperature ramps between 920°C and 950°C and a dwell time of 10 h. The melt texturing process comprised a dwell time (10 h/950°C) for sintering, a second dwell time (1 h/1040°C) for melting and different cooling ramps. Following the sintering or texturing process specimens were machined out of the block with a saw in order to achieve a definite specimen geometry and to remove the surface layer which was partially oxygenated during cooling. The specimens then were polished in several steps and exposed to oxygen atmosphere (1bar) at temperatures between 400°C and 500°C and a dwell time of up to 10 days. Before and after oxygenation a microstructural analysis was carried out, the density was measured and the content of oxygen was determined by x-ray diffraction. The value of the oxygen content (typically 6.95) was obtained by calculating the c-axis length using the peak angle of the 006 reflection [3].

The critical temperature T_C was determined by d. c. resistivity measured with the four-probe technique over the temperature interval between 70 K and room temperature. Rod shaped specimens were used with the dimension 20 mm*3 mm*2 mm, which were machined out of sintered material. In order to reduce the contact resistance silver paint was applied at the points where the four wires are clamped to the specimen. The d.c. resistance measurement was performed by measuring the voltage between the two inner probes while a current of 50 mA was induced at the two outer probes.
Experiments with an open resonator at 145 GHz [4] and with a cavity at 10 GHz were carried out in order to determine the surface resistance R_S of the specimens. The specimens were cooled down to 77 K using liquid nitrogen. While warming up to 100 K in the case of the open resonator and up to room temperature in the case of the cavity the resonance curve was registered by transmission measurements. The quality factor of the resonator was determined and from this value the surface resistance was evaluated [5].

3 Results and Discussion

3.1 Powder characterization

The x-ray diffraction pattern for both YBCO powders (Grade 1, SR 5 grade) confirmed the over all stoichiometry. The cation stoichiometry according to the results of the chemical analysis is $Y_{1,000}Ba_{1,991}Cu_{3,009}O_{7-x}$ (Grade 1) and $Y_{1,007}Ba_{2,002}Cu_{2,991}O_{7-x}$ (SR 5 grade). However, copper oxide (CuO) is visible as a white phase in the optical microscope in bright field and as a dark gray phase in SEM. Grain size analysis which was carried out by particle size analyzer revealed a particle size d_{90} of 6 μm for both powders. DTA measurements showed that the temperature of incongruent melting is 1038°C.

3.2 Microstructural analysis

Sintered material
For both powders dense material without residual open porosity in the material was achieved at sintering temperatures exceeding 920°C. The density was typically about 6.05 g/cm³ ≈ 95 % T.D. Sintering in the temperature range of 920°C..950°C causes grain restructuring and grain growth of columnar grains of 10μm to 200μm length. The portion of 10μm-grains amounts to nearly 100% at 920°C, decreases to 50% at 940°C (see Figure 1) and finally reaches 10% at 950°C sintering temperature.

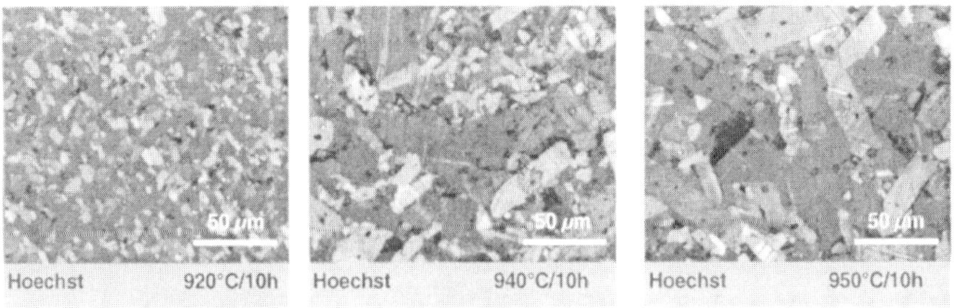

Figure 1. Grain coarsening by increasing the sintering temperature (polarised light)

Sintering in the temperature range of 960°C..970°C again leads to smaller average grain size. The grain size distribution at 960°C is comparable to that at 950°C. The portion of small grains in the microstructure of specimens sintered at 970°C amounts to 50% which is comparable to the result at 940°C sintering temperature. According to the $YBa_2Cu_3O_{7-x}$ - CuO phase diagram, the formation of a copper rich liquid phase together with a solid Y_2BaCuO_5 - phase occurs at 940°C [6]. This accelerates the growth of the YBCO grains by mass transport in the liquid phase. At temperatures above 950°C the simultaneous precipitation of Y_2BaCuO_5-phase impedes further growth despite the enhanced formation of the liquid phase. The liquid phase is efficient in modifying the distribution and shape of the CuO phase. At temperatures up to 920°C, the copper oxide phase is dispersed inside the YBCO grains. With increasing sintering temperatures they appear as thin layers along the grain boundaries as shown by the microstructural analysis in the optical microscope in bright field (see Figure 2).

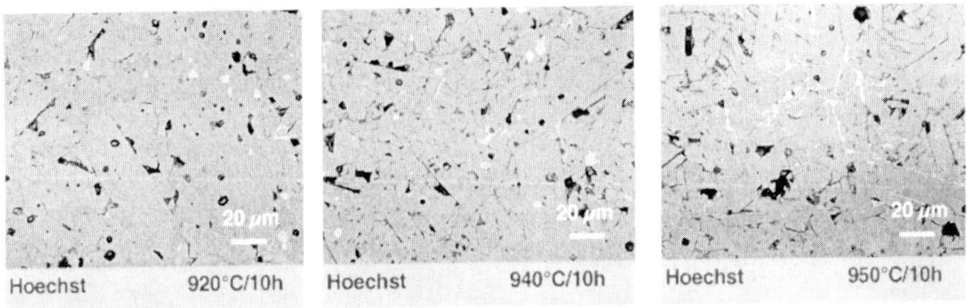

Figure 2. Change of the distribution of the copper oxide phase (white phase) observed in optical microscope (in bright field) as a function of the sintering temperature: 920°C dispersed, 950°C along the grain boundaries.

Textured material

The melt texturing process is based on the incongruent melting of YBCO. For this purpose the specimens were sintered and then heated to 1040°C. The fast cooling to 1020°C produced an undercooled melt where the nucleation of YBCO phase is initiated. In the following transient cooling at very low rates, the orientations with the highest growth rate prevail. This leads to a microstructure with large grains. The average grain size could be varied by the cooling rate between 1020°C and 900°C (see Figure 3). The process induces partially textured specimens with a preferred orientation of the c-axis perpendicular to the specimen surface. The specimen cooled with a ramp of 1°C/h, with grain sizes of typically ≈ 2 mm, showed higher content of c-axis oriented grains than the specimen cooled down with a ramp of 0.5°C/h (grain size ≈ 5 mm). This difference is observed in the light microscope, where the twin boundaries are an indicator of the grain orientation.

A particular problem is the deformation of the massive blocks during the texturing process, as the normal method to increase the viscosity of the melt by addition Y_2BaCuO_5 - phase or platinum cannot be applied without giving up the stoichiometry of the material. This problem has prevented so far the use of the crystal seeds to enhance further the amount of c-oriented grains. Copper oxide segregations do also exist after the melt texturing process. However, in contrast to the sintering process, melt texturing produces a coarse grained microstructure with dispersed copper oxide phase. They are located at preferred locations (such as triple points) of the grain boundaries, but not as intergranular layer (see Figure 4).

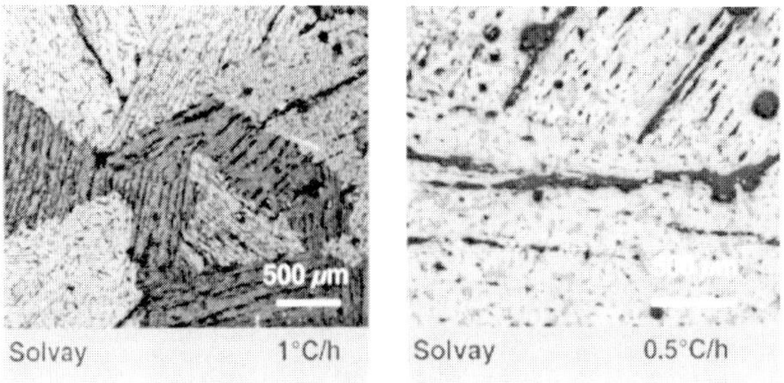

Figure 3. Microstructure in polarised light: a) grain size ≈ 2 mm (left)
b) grain boundary and parts of two grains (grain size ≈ 5 mm)

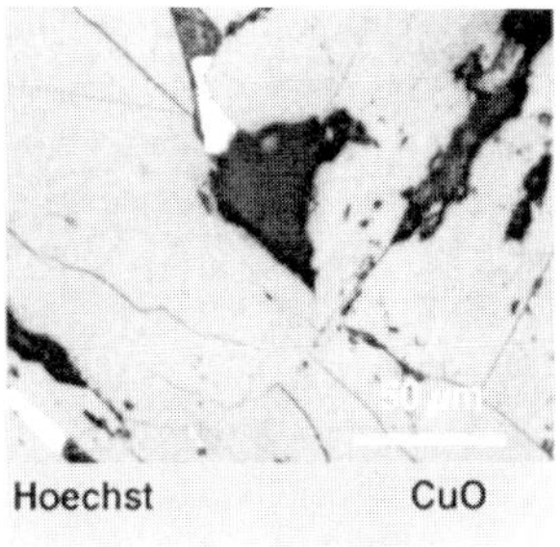

Figure 4. Dispersed copper oxide phase (white phase)

3.3 D. c. resistance measurements

Sintered material

Specimens manufactured with both powder varieties exhibited a dependence of measured d. c. resistivity and transition width on microstructural properties. The lowest d. c. resistivity was measured for the specimens with the smallest grains, whereas the highest resistivity was observed for the specimen with the largest grains. The results exemplified in Figure 5 for the specimens based on the SR 5 grade powder show also that the transition of the superconducting state is strongly broadened for higher sintering temperatures. Apparently the reduction of the concentration of grain boundaries does not play a major role. An explanation for the observed dependence of the d. c. resistivity on the microstructure can be found in the distribution of the copper oxide phase [7]. In the coarse grained specimen sintered at 950°C, the copper oxide forms a layer in the grain boundary and therefore represents a considerable barrier for the current in contrast to the dispersed CuO phase in the fine grained material. This implies a higher d. c. resistivity of the normal conducting phase.

Textured material

Measurements carried out with a textured specimen showed a clearly reduced d. c. resistivity in comparison to the non textured specimens. The microstructure with large, preferentially c-axis-oriented grains leads to low d. c. resistivity in the normal conducting state and to a narrow transition width in the superconducting state. The transition temperature of the textured material is 91 K which corresponds to the sintered (non textured) material (see Figure 5). The copper oxide phase, which is dispersed (see Figure 4), does not affect the d. c. resistivity.

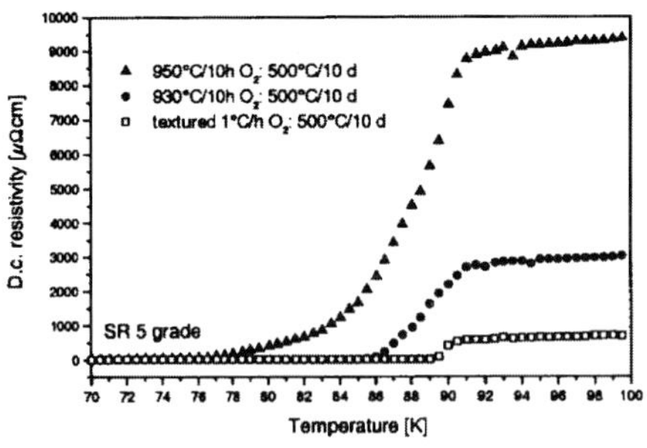

Figure 5. D. c. resistivity near the transition temperature

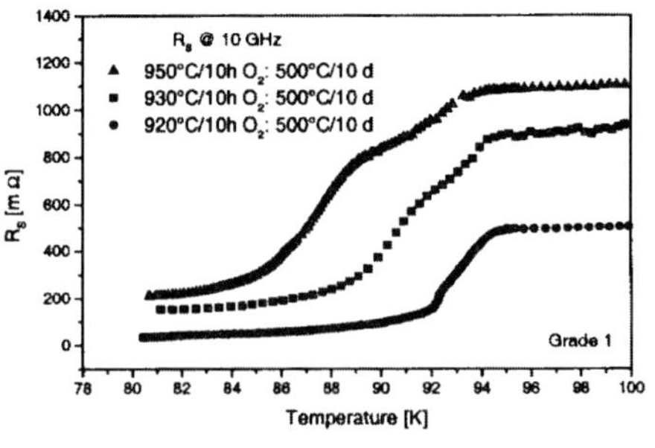

Figure 6. Surface resistance for sintered YBCO

3.4 High frequency characterization

Sintered material

Similar to the case of d. c. resistance, the high frequency surface resistance R_S of the sintered specimens depends also on the location and shape of the copper oxide phase. This relation was confirmed with high frequency experiments carried out with 10 GHz (see Figure 6) and with 145 GHz (see Figure 8). The lowest values of surface resistance were determined in the specimens with the copper oxide dispersed in the microstructure. A higher surface resistance was determined when the copper oxide is located as a thin film at the grain boundaries (see Table 1).

Already in the normal conducting state the dependence of R_S on the sintering temperature can be observed (see Table 1). This trend continues to low temperatures and remains also in the superconducting state (see Figure 6). Evidently the CuO layers form additional weak links, which are effective on the surface resistance. This goes along with the observation of a related reduction in J_C in YBCO sintered above 920°C [1].

Table 1. Rs comparison

Sintering temperature Cooling rate		920°C/10h	930°C/10h	950°C/10h	textured 1°C/h	textured 0.5°C/h
		mΩ	mΩ	mΩ	mΩ	mΩ
R_S (cavity 10 GHz)	300 K	730	1050	1240	1240	3000
	100 K	500	930	1100	810	2200
	77 K	25	150	160	21	340
R_S (open resonator 145GHz)	100 K	2200	2500	2800		
	77 K	820	850	1100	260	

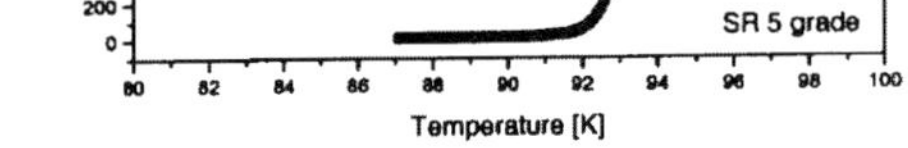

Figure 7. Surface resistance R_S for textured material measured in a cavity at 10 GHz

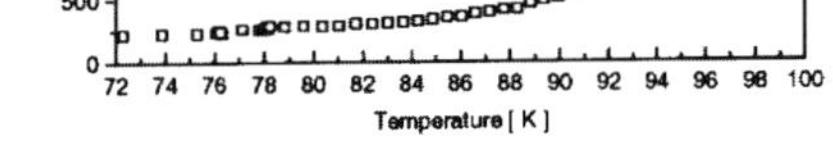

Figure 8. Surface resistance R_S determined in an open resonator at 145 GHz

At 10 GHz differences exist also in the function of the surface resistance versus temperature. Specimens sintered above 920°C/10h revealed a shoulder in the curve (see Figure 6), probably from a contribution of the grain boundaries leading to a second T_C. A higher content of copper oxide at the grain boundaries decreases also the temperature of the second T_C.

The determination of the surface resistance at 145 GHz in the open resonator confirms the results at 10 GHz. The specimen with the copper oxide layers at the grain boundaries showed that the transition width in the normal conducting state is very broad and starts already at 85 K (see Figure 8). The specimen with the copper oxide dispersed in the microstructure shows clearly lower values of the residual resistance. Also the transition width in the normal conducting state is narrower and starts at 92 K.

Textured material

Of the 2 textured specimens measured at 10 GHz the specimen with smaller grains (cooling rate 1°C/h) and a higher content of c-axis oriented grains exhibits comparatively low values of R_S in the normal conducting state as well as in the superconducting state (see Table 1). The R_S of the specimen with the larger grains (cooling rate 0.5°C/h), which has a higher quantity of a- or b- oriented grains, is clearly higher (see Figure 7). In the open resonator at 145 GHz only the specimen textured with a cooling rate of 1°C/h was characterized (see Figure 8). Its surface resistance is clearly lower than the R_S of the sintered material and approaches values, which are not far of the resistance range of thin films. The other specimen could not be measured, due to its to high R_S.
Due to the dispersion of the copper oxide phase in the textured material (see Figure 4) the copper oxide does not affect the resistance as in the sintered (non textured) material. Again the larger grain size did not improve the electrical properties. Apparently the grain orientation is the most effective parameter. The higher content of c-axis oriented grains leads to a lower R_S in agreement with the observations of Wosik et al. [5].

4 Conclusions

Commercial stoichiometric YBCO powder contains a small amount of copper oxide as a second phase in the microstructure. Copper oxide leads to the existence of a liquid phase at temperatures above 930°C . That is characteristic for the sintered material, because the liquid phase supports the sintering process, but on the other hand it causes to the grain wetting with CuO. CuO at the grain boundaries degrades the properties of the superconductor in the d. c. case and as well as in a high frequency field. With the present powders it is not possible to obtain a coarse grained microstructure by sintering without the copper oxide layers. According to the present findings the melt texturing can avoid CuO layers for coarse grained materials.

References

[1] Murakami, Supercond. Sci. Technol. 1992,5,185-203
[2] J.Halbritter, Journal of Superconductivity 1995,8, 691-703
[3] M.S.Osowsky, J.L.Cohn, E.F.Skelton et al., Physical Review B 1992, 45
[4] R. Schwab, R. Heidinger, Conference Proceedings, MIOP 1997
[5] J.Wosik, R.A. Kranenburg, J.C. Wolfe, J. Appl. Phys. 1991, 69, 874-879
[6] G.Krabbes, W.Bieger, U.Wiesner et al., Physica C 1991, 503-504
[7] P. Majewsky, N. Rüffler, G. Petzow, Z. Metallkd. 1992, 83, 866-870

Structural, Electrical and Magnetic Properties of $(Y_{1-y}Ca_y)Ba_2Cu_{3-x}Ni_xO_{7-\delta}$ Ceramics

Iosif Gr. Deac, Emil Burzo, Viorel Pop and Romulus Tetean,
Faculty of Physics, "Babes-Bolyai" University, 3400 Cluj-Napoca, Romania

1 Abstract

Polycrystalline samples of nominal composition $(Y_{1-y}Ca_y)Ba_2(Cu_{3-y}Ni_y)O_{7-\delta}$ were investigated by using electrical resistivity and ac susceptibility measurements. X-ray powder diffraction has been used to find the lattice parameters of the samples. Electrical resistance and ac susceptibility measurements showed that the T_c values decrease from 91 K to 68 K, for the samples with $y = 0$, as the level of doping, x, increases from 0.0 to 0.25. As the Ca content increases, the T_c increases, from $y = 0.05$ (72 K) to $y = 0.2$ having T_c of 76 K. Analyzing the resistivity and ac susceptibility data, the following conclusions are drawn about the role of Ca and Ni doped simultaneously in the YBCO sample. (1) The conduction mechanism of electrons appears to follow a crossover from the purely metallic regime to the localization regime. The results are explained on the basis of hole localization and magnetic pair-breaking mechanism.(2) The co-substitution of calcium tends to counteract the some of the effect induced by Ni, but it cannot recover the decrease of the critical temperature.

2 Introduction

The study of impurity substitutions and their effect on the transport, magnetic and superconducting properties of high-T_c superconductors (HTS) may provide clues to understanding unusual normal-state properties and offer some insight into the mechanism of high-T_c superconductivity.[1-9] By far the most extensive research performed to date, has been for the 3*d* transition-metal doped at the copper sites.[1,2,7,8] As reported in the literature, these substitutions rapidly depress T_c. Divalent Zn substitution for Cu results in unusually decrease of T_c, with typical values over 10 K/at.%. The behavior of $YBa_2Cu_{3-x}Ni_xO_{7-\delta}$ is, somehow, surprising. In spite of the fact that Ni is divalent and paramagnetic, the superconductivity is suppressed less than in the case of nonmagnetic Zn. The carrier concentration and the defects in CuO_2 planes affect the superconducting and the normal properties of HTS systems.[1,2,11,12]The carrier density can be enhanced when Y^{3+} is partially replaced by lower-valence-state cations. Some earlier works[3] have considered that the substitution of Ca^{2+} for Y^{3+} is expected to generate holes. The results reported in the literature are however varied and some of them reveal that such a substitution creates oxygen vacancies, in CuO_2 planes thus reducing the holes generation effect.[3,4]

If the suppresion of T_c in $YBa_2Cu_{3-x}Ni_xO_{7-\delta}$ system were entirely due to the reduction of mobile holes, then co-substitution of Ca^{2+} for Y^{3+} might compensate for the loss of holes and restore superconductivity. Therefore it is of interest to study the effect on the superconducting and normal state properties of Ni and Ca doped simultaneously at the Cu and respectively Y sites, in the $(Y_{1-y}Ca_y)Ba_2Cu_{3-x}Ni_xO_{7-\delta}$ compounds.

In this paper we report the results of X-ray diffraction, electrical resistivity and ac susceptibility measurements on the series of compounds having the stoichiometric composition $YBa_2Cu_{3-x}Ni_xO_{7-\delta}$ for $x = 0.0 - 0.25$ and $(Y_{1-y}Ca_y)Ba_2Cu_{2.85}Ni_{0.15}O_{7-\delta}$ for $y = 0.05 - 0.3$.

3 Experimental Details

Samples with nominal compositions $YBa_2Cu_{3-x}Ni_xO_{7-\delta}$ ($x = 0.0; 0.05; 0.1; 0.15; 0.2; 0.25$) and $(Y_{1-y}Ca_y)Ba_2Cu_{2.85}Ni_{0.15}O_{7-\delta}$ ($y = 0.05; 0.10; 0.15; 0.2; 0.25; 0.3$) were prepared by a solid state reaction from a stoichiometric amount of high-purity (better than 99.99%) dry powder of Y_2O_3, $BaCO_3$, CuO, CaO and NiO.

The phase and lattice parameters of the compounds were determined by X-ray powder diffraction using CuKα radiation. The X-ray analysis revealed that all the samples were mainly single phase with an impurity level of less than 3%.

Electrical resistivity measurements were carried out, in the interval from 77 K to 300 K. The resistivity of the samples was measured by a standard four-probe method, with a multipurpose Oxford Instruments MagLab System 2000 equipment. High quality electrical contacts were made by applying silver loaded paint and using silver leads.

The ac susceptibility measurements were performed by using the same Oxford Instruments MagLab System 2000 equipment. The complex susceptibility was measured at 1 kHz as a function of temperature at a field amplitude H_{AC}=0.1 Oe. Bar specimens were cut from the sintered samples and used for ac susceptibility measurements.

4 Results and Discussion

The X-ray diffraction patterns of $YBa_2Cu_{3-x}Ni_xO_{7-\delta}$, and $(Y_{1-y}Ca_y)Ba_2Cu_{2.85}Ni_{0.15}O_{7-\delta}$ suggest that these compounds have the YBCO structure. The orthorhombicity, $(b-a)/(b+a)$, slightly decreases with increasing Ni content but, for the Ni-Ca doped samples, it does not change significantly. Quantitative data for the investigated samples are given in Table 1.

The resistivity results from the Ni and Ni-Ca doped samples are displayed in Figs. 1 and 2, respectively. The resistivity of $YBa_2Cu_{3-x}Ni_xO_{7-\delta}$ samples with $x \leq 0.10$ gradually decreased with decreasing temperature and then abruptly dropped when they transformed from the normal to the superconducting state. $T_c(R)$ was estimated from the maximum of $d\rho/dT$. The values of $T_c(R)$ obtained in this way are presented in Table 1. Above 120 K the resistivity shows linear temperature dependence for all the samples and therefore we assume that the normal state resistivity is given by $\rho = \rho_0 + \alpha.T$. Here $\rho_0 = \rho_n(T = 0)$ is the residual resistivity and α is the resistivity slope in the normal state. For Ni doped samples with x = 0.15, 0.2, 0.25 the most noticeable feature of the resistivity data is the change to a semiconductor-like behavior, as shown in Fig. 1.

Table 1. Quantitative data for the $(Y_{1-y}Ca_y)Ba_2(Cu_{3-y}Ni_y)O_{7-\delta}$ system

Sample (x, y)	a (Å)	b (Å)	c (Å)	(b-a)/(a+b) %	T(R) (K)	T(χ) (K)
(0.00, 0.00)	3.831	3.881	11.672	0.681	91.4	91.0
(0.05, 0.00)	3.831	3.879	11.671	0.623	82.8	83.7
(0.10, 0.00)	3.832	3.878	11.673	0.597	80.1	80.8
(0.15, 0.00)	3.833	3.878	11.673	0.584	78.2	77.0
(0.20, 0.00)	3.834	3.877	11.674	0.558	75.3	74.0
(0.25, 0.00)	3.834	3.877	11.674	0.558	67.5	68.0
(0.15, 0.05)	3.833	3.878	11.675	0.584	70.1	72.0
(0.15, 0.10)	3.836	3.878	11.675	0.544	74.1	73.2
(0.15, 0.15)	3.836	3.879	11.676	0.557	74.9	74.4
(0.15, 0.20)	3.837	3.88	11.677	0.557	76.3	75.8
(0.15, 0.25)	3.837	3.881	11.678	0.570	75.1	75.5
(0.15, 0.30)	3.838	3.881	11.679	0,597	71.3	73.5

It resembles a decrease in the mobility of the mobility of the current carriers that occurs in oxygen-deficient YBCO samples.[12] Such a behavior suggests that the magnetic moments of the Ni ions, by pair breaking and localisation effects could depress the superconducctivity and could change the electrical conduction in the normal state from metallic to semiconductor.

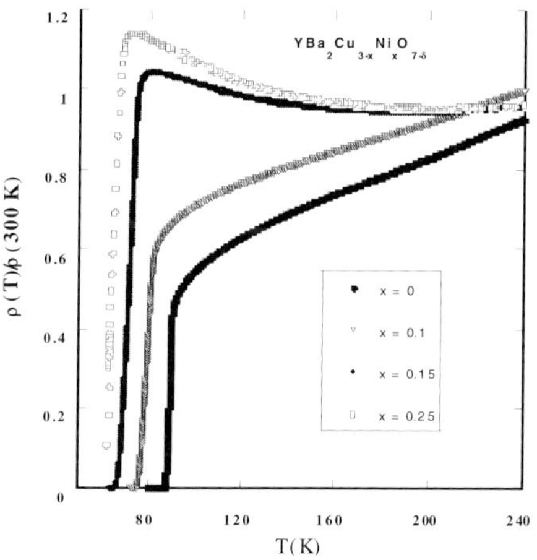

Figure. 1. Normalized resistivity $\rho(T)/\rho(300K)$ versus temperature of some $YBa_2Cu_{3-x}Ni_xO_{7-\delta}$ compounds

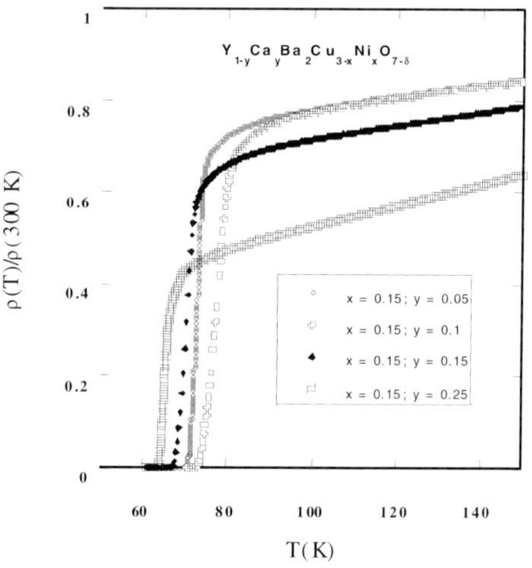

Figure. 2. Normalized resistivity $\rho(T)/\rho(300K)$ versus temperature of some $(Y_{1-y}Ca_y)Ba_2Cu_{2.85}Ni_{0.15}O_{7-\delta}$ compounds

For the $YBa_2Cu_{3-x}Ni_xO_{7-\delta}$ samples the T_c decreases with increasing Ni content from 91.3 K, when $x = 0.0$ to ~58 K, when $x = 0.25$. The resistivity of the Ni-Ca doped samples behaves very differently. Its temperature dependence is characterised again, by a wide linear regime, while the T_c's continue to by lower than the T_c. The Ca incorporation in $YBa_2Cu_{3-x}Ni_xO_{7-\delta}$ compounds, leads to the recovery of the linear conduction only, and not to the increase of T_c, contrary to some previous reports.[5,6] But, as seen in Table 1, $T_c(y)$ reaches a maximum value, 76.1 K for $y = 0.2$. It seems, that the holes generation effect induced by Ca could restore the linear behavior of $\rho(T)$ but could not counteract the decrease of T_c induced by Ni doping. The superconductivity can not be completely recovered, maybe, due to the oxygen vacancies created mainly in the CuO_2 planes by Ca doping.

The dependence of the real part $\chi'(T)$ and of the imaginary part $\chi''(T)$ of the complex susceptibility, $\chi(T)=\chi'(T)+i\chi''(T)$, for the $(Y_{1-y}Ca_y)Ba_2Cu_{3-x}Ni_xO_{7-\delta}$ compounds at $H_{AC} = 1$ Oe and f = 1 kHz is shown in Fig. 3 and 4.

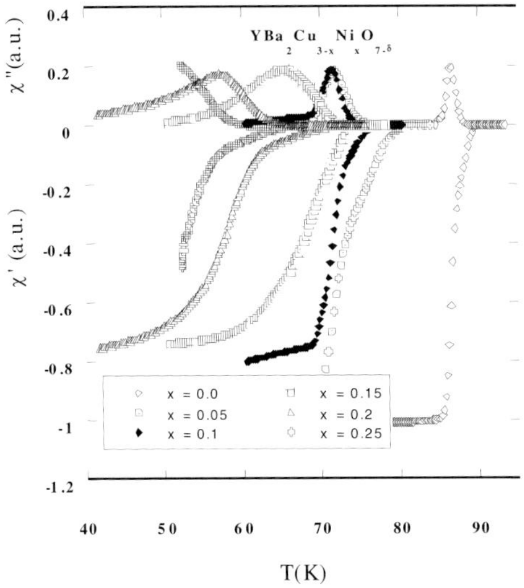

Figure 3. The temperature dependence of the real $\chi'(T)$ and the imaginary $\chi''(T)$ parts of the complex susceptibility for $YBa_2Cu_{3-x}Ni_xO_{7-\delta}$ ceramics for H_{AC}=0.1 Oe and f = 1kHz

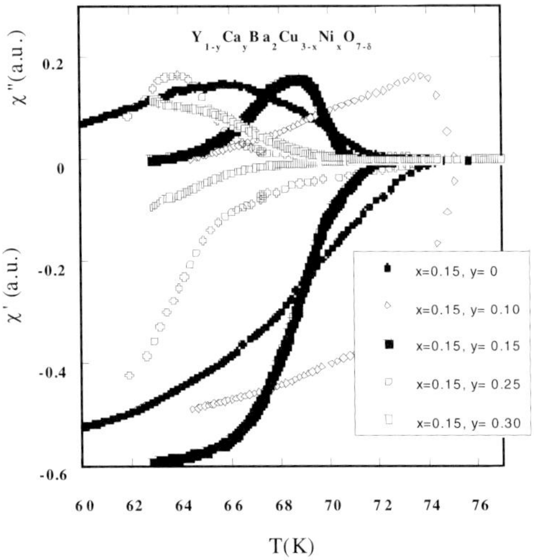

Figure 4. The temperature dependence of the real $\chi'(T)$ and the imaginary $\chi''(T)$ parts of the complex susceptibility for $(Y_{1-y}Ca_y)Ba_2Cu_{2.85}Ni_{0.15}O_{7-\delta}$ ceramics for H_{AC}=0.1 Oe and f = 1kHz

For all the investigated samples χ' starts to decrease below T_c and at lower temperatures it begins to saturate. The first step corresponds to exclusion of magnetic field from the bulk superconducting grains and the second comes from magnetic shielding by the supercurents flowing between these bulk grains. The imaginary component of the ac susceptibility, χ'', measures the absorption of the flux inside the weak-links formed in the intergranular regions. As seen in Figs. 3 and 4, χ'' exhibits only a very smooth single peak at a temperature T_p ($T_p < T_c$). The onset of bulk superconductivity is indicated by the drop in electrical resistivity to zero and the onset of the peak in χ'' associated with hysteresis resulting from the trapping of flux between grains when sample can support a supercurrent. It is evident from Table I that $T_c(R)$ and $T_c(\chi')$, agree very well with each other confirming the consistency of the measurements. The concentration-dependent $T_c(\chi')$, values given in Table I correspond to the temperatures where the ac susceptibility becomes diamagnetic. The transition width is increased in the Ni doped compounds and the peak in $\chi''(T)$ is broadened as can be seen from Fig.3. In Fig. 4 are depicted $\chi'(T)$ and $\chi''(T)$ for the $(Y_{1-y}Ca_y)Ba_2Cu_{2.85}Ni_{0.15}O_{7-\delta}$ compounds. For comparison, in Fig. 4 there are also shown $\chi'(T)$ and $\chi''(T)$ for the $YBa_2Cu_{2.85}Ni_{0.15}O_{7-\delta}$ sample. It is interesting to remark that Ca doping reduces the width of the transition, and consequently the width of the peak in $\chi''(T)$. This is another adverse effect of Ca and Ni when they are incorporated in YBCO ceramics.

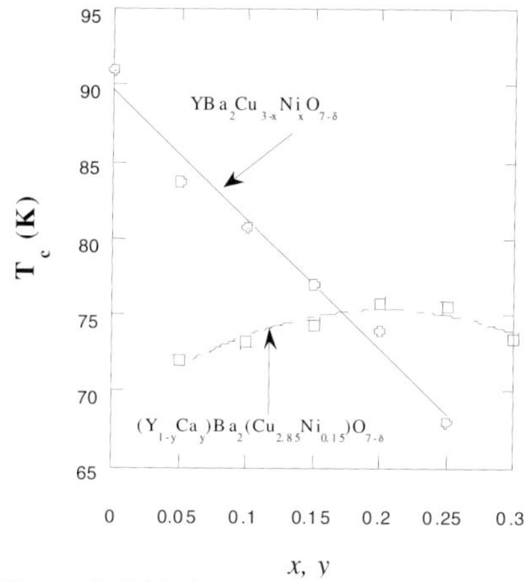

Figure. 5. Critical temperature dependence vs. dopant content for $YBa_2Cu_{3-x}Ni_xO_{7-\delta}$ and $(Y_{1-y}Ca_y)Ba_2Cu_{2.85}\,Ni_{0.15}O_{7-\delta}$ compounds

The T_c variation versus dopant content for the $YBa_2Cu_{3-x}Ni_xO_{7-\delta}$ samples was found to be linear, as plotted in Fig.5, and agree with some other reports.[1,5,6,8,10] Such a behavior was explained in terms of pair-breaking theory such as Abrikosov-Gor'kov theory, in which the depairing of superconducting electrons occurs via exchange scattering of mobile holes in CuO_2 plane by Ni magnetic moment. The behavior of the normal state electrical resistivity also suggests the presence of some localisation effects. The T_c of the Ca-Ni doped samples passes through a broaden maximum T_c = 75.8 K, at y = 0.2, when the Ca content increases (also shown in Fig.5) from 0.05 to 0.3 that could arise as a result of the competition between the holes doping and oxygen vacancies doping in the CuO_2 in the presence of the Ni ions. This assumption could explain, at least in a qualitative way, the electrical and magnetic behavior of these compounds in the superconducting and in the normal state.

6 Conclusions

The superconducting compounds $YBa_2Cu_{3-x}Ni_xO_{7-\delta}$, and $(Y_{1-y}Ca_y)Ba_2Cu_{2.85}\,Ni_{0.15}O_{7-\delta}$ with the oxygen-defect perovskite structure were obtained for x = 0.0; 0.05; 0.15; 0.2; 0.25 and y = 0.05; 0.10; 0.15; 0.20; 0.25; 0.3. The orthorhombicity of the Ni-doped samples is slightly decreased with increasing the Ni content. The incorporation of Ca does not change the orthorhombicity, substantially. Superconducting transition was established in these compounds by both resistivity and ac susceptibility techniques.

There is a significant change in T_c for the Ni-doped samples $YBa_2Cu_{3-x}Ni_xO_{7-\delta}$, and for $x \geq 0.15$ the electrical conduction in the normal state change from metallic-like to semiconductor-like behavior. Ca doping restores the linear dependence of $\rho(T)$ in the normal state and improves the coupling between the superconducting grains. The T_c of the Ca-Ni doped samples passes through a broaden maximum that is lower than the T_c of the

$YBa_2Cu_{2.85}Ni_{0.15}O_{7-\delta}$ sample. The co-substitution of calcium tends to counteract some of the effects induced by Ni, but it cannot recover the decrease of the critical temperature.

7 References

1. Y. Kitaoka, K. Ishida, and K. Asayama, J. Phys. Soc. Jpn. 1994, 63, 2052-2054.
2. S. K. Tolpygo et al. Phys Rev. B 1996, 53, 12462-12474, and references therein.
3. B. Fisher, J. Genossar, C.G.Kuper, L. Patlagan, G. M. Reisner and A. Knizhnik, Phys. Rev. B 1993, 47, 6054-6060.
4. V. P. S. Awana and A.V.Narlikar, *Phys. Rev.* B 1994, 49, 6353-6357
5. J. J. Neumaier, T. Bjørnholm, M.B. Maple and I. K. Schuller, Phys. Rev.Lett. 1989, 63, 2516-2523.
6. J. L. Tallon, C. Bernhard, G. V. M. Williams, and J. W. Loram, Phys. Rev.Lett. 1997, 79, 5294-5299.
7. S. Zagoulaev, P. Monod and J. Jégoudez, Phys. Rev. B 1995, 52, 10447-10458.
8. N. Peng, W.Y. Liang, Physica C 1994, 233, 61-66.
9. T. R. Chien, Z. Z. Wang and N. P. Ong, Phys. Rev.Lett. 1991, 67, 2088-2093.
10. R.A. Gunasekaran, R. Ganguly, J.V. Yakhmi, Physica C 1995,243, 160-166.
11. T. Ito, K. Takenaka, and S. Uchida, Phys. Rev. Lett. 1993, 70, 3995-4001.
12. K. Tolpygo et al. Phys Rev. B 1996,53,12454-12466

Formation and Properties of Poly(siloxane) Derived Ceramic Foams

T. Gambaryan-Roisman, M. Scheffler, T. Takahashi*, P. Buhler, P. Greil
University of Erlangen-Nuernberg, Department of Materials Science (Glass and Ceramics, *Polymer Materials), Martensstrasse 5, D-91058 Erlangen, Germany

1 Introduction

Ceramic foams possess unique set of useful properties, such as low density, low thermal conductivity, high surface area, high permeability, excellent thermal stability, high thermal shock resistance, high specific strength, high resistance to chemical corrosion. These materials may find various technological applications, e.g. as thermal insulations, filters, lightweight structures, catalyst supports [1], in porous medium burner technology [2], etc. Conventional methods for ceramic foam production include the immersion of an open-cell polymeric foam into ceramic slurry with subsequent burning out of the organic skeleton [3]; the sol-gel technique [4]; sintering of hollow spheres [5]; the replication technique [6]. Recently the production of silicon oxycarbide foam by co-blowing a solution containing a preceramic polymer and polyurethane precursors and subsequent pyrolysis has been reported [1].

In our recent work [7] we have reported about a novel method of production of ceramic foams from phenylmethyl poly(siloxane). This method utilizes intrinsic components of the starting polymer for precursor foam formation. The gases released as a result of the crosslinking of the polymer serve as an intrinsic blowing agent for the polymer foam formation. This polymer foam was pyrolyzed in the temperature range 1000°C-1500°C.

In the last several years a significant attention have been paid to investigation of influence of various fillers on processing and properties of polymer-derived ceramics [8]. Filling the polymer matrix with inert powders, like SiC, or with reactive particles, like Ti, Cr, Si, B, significantly expands the possibility of this processing technique by allowing to tailor the microstructure of the final product. The metal filler particles, most frequently used as the active fillers, react during the pyrolysis with the polymer decomposition products and form the carbide phases. Since the formation of the new phases is accompanied by the change of specific volume, the microstructure of the final ceramic products is also significantly affected by active fillers. The active-filler-controlled pyrolysis of preceramic polymers has been used for production of bulk components, the matrix in fiber-reinforced composites, for coatings, etc. [8].

Filler powders were added to the poly(siloxane) powder before foaming, which lead to modification of the foams morphology and phase composition [7].

The foam morphology is very sensitive to the kind of the starting poly(siloxane), to the thermal regime of foaming, and to the type and the concentration of the fillers. While the formation of the polymer foam by injection molding is governed by the polymer melt surface tension and viscosity, the solubility and diffusivity of the blowing agent within the melt [9], the foam formation during the poly(siloxane) crosslinking is strongly influenced by the crosslinking reaction rate. The curing rate depends on temperature and on the concentration of the terminating groups in the polymer molecular structure. Investigation of the foaming mechanism with and without fillers is very important in order to fabricate foams with tailored microstructure and physical properties.

2 Experimental

A phenylmethyl poly(siloxane) silicon resin (H44, Wacker GmbH Nuenchritz) and methyl poly(siloxane) silicon resin (MK, Wacker GmbH, Nuenchritz) were used as starting polymers for the foam formation. The H44 resin contains 2 wt % of OH groups and 1 wt % of etoxy groups, as reported by the manufacturer. The ratio between the phenyl and methyl groups in the H44 resin was determined by measurement of the weight loss of the resin during oxidation of the powder and its complete conversion to SiO_2 and was found to be 2. Elemental powders (B, Mo, Si), $MoSi_2$ intermetallics and oxide powders (Al_2O_3, MgO, Cu_2O) were added as fillers and thoroughly mixed with the H44 powder. The mixture was placed in a preheated furnace to allow foaming. The foaming temperature of the polymer was 300°C. Oxide fillers normally increased the rate of the foam formation and shifted the foaming towards lower temperatures. To control the foam morphology, the polymer was in several cases pretreated at 200°C for a period up to two hours, which lead to partial crosslinking of the polymer and to significant change of the melt properties at the moment of beginning of the foaming.

Foaming of pure poly(siloxane) begins from heterogeneous nucleation of gas bubbles at the bottom of the vessel and their rising to the surface of the melt. At the beginning the bubbles break up at the surface, and their content releases to the atmosphere. After several minutes the bubbles begin to remain at the surface. From this moment the thickness of the foamed layer begins to increase due to the arrival of the new bubbles from the bottom and due to the growth of the already existing bubbles. In contrast to pure poly(siloxanes), the nucleation and growth of bubbles in suspensions of oxide micron and submicron particles (Al_2O_3, MgO) in poly(siloxane) melts takes place “homogeneously” in the suspension volume.

The most probable crosslinking reaction mechanisms in the H44 silicone resin are the condensation of Si – OH and Si – OC_2H_5 terminating groups with release of water and ethanol, respectively:

$$-\overset{|}{\underset{|}{Si}}-OH + HO-\overset{|}{\underset{|}{Si}}- \rightarrow -\overset{|}{\underset{|}{Si}}-O-\overset{|}{\underset{|}{Si}}- + H_2O\uparrow , \quad (1)$$

$$-\overset{|}{\underset{|}{Si}}-OH + C_2H_5O-\overset{|}{\underset{|}{Si}}- \rightarrow -\overset{|}{\underset{|}{Si}}-O-\overset{|}{\underset{|}{Si}}- + C_2H_5OH\uparrow . \quad (2)$$

The process of foaming was studied by the thermogravimetrical analysis (TGA 2950 TA Instruments). The gases released during the crosslinking reaction and foam formation were analyzed by FT-IR spectroscopy using TGA interface (Nicolet Magna-IRTM Spectrometer 750). The kinetics of the water release was studied on the Karl Fischer apparatus (Mettler KF-Coulometer DL 37), which can detect a very small amount of generated water by an electro-chemical method.

The pure poly(siloxanc) foams and foams with metallic fillers were pyrolyzed in argon atmosphere for 3-4 hours in the temperature range 1000°C-1500°C or in nitrogen atmosphere at 1000°C with a heating rate 60 K/h. The poly(siloxane) foams with Al_2O_3 filler were pyrolyzed in air in the temperature range 1000°C-1500°C. The average linear shrinkage of the polymer precursor without fillers after the pyrolysis in argon at 1000°C was about 15%. The pyrolyzed pure poly(siloxane) foams are composed of SiO_2, SiC, and free carbon. The ratio between these three components has been determined by means of a redox analysis [10].

The morphology of the foams was studied by Scanning Electron Microscopy (SEM), and the phase analysis was performed by the X-ray diffraction analysis (XRD). To evaluate the stability of the foams to oxidation, the TGA analysis of the foams was performed in air with heating rates of 6 and 12 K/h up to 900°C-1000°C. The compressive strength of the specimens has been measured using the Instron apparatus (Model 4204, Darmstadt) with crosshead speeds 0.2 mm/min and 0.5 mm/min.

3 Results and Discussion

The filler-free H44-derived ceramic foams are characterized by a bulk density lying in the range 0.1-0.2 g/cm^3, a porosity of 70-80%, and tensile strength of 0.3-0.4 MPa. Figure 1 presents micrographs of the filler free foams pyrolyzed at 1200°C. Both foams is characterized by an open-cell structure. The cell size of the foams is determined by the thermal regime during foaming. Filler-free H44 specimens foamed at 300°C after pretreatment at 200°C for two hours are characterized by the cell sizes in the range 350-800 μm (Fig. 1a), while specimens foamed at 300°C without pretreatment are characterized by cell sizes in the range of 800-1500 μm (Fig. 1b). Two factors may be responsible for that: increasing viscosity during partial crosslinking of the polymer melt at 200°C, and decreasing concentration of the blowing agents due to their evaporation at 200°C. The foams containing 20 vol % Al_2O_3 filler are characterized by wider cell size in the range 200-2000 μm [7].

Figure 2 presents the results of the FT-IR analysis of the gases released by the H44 silicon resin performed during heating at 10 K/min in flowing nitrogen. Although the crosslinking of the phenylmethyl poly(siloxane) starts already at about 110°C, the gas release from the polymer at 200°C is still not significant. The gas release intensifies at 250°C, which is also confirmed by the TGA analysis. Besides the peaks corresponding to the water and ethanol, which could be expected from the most probable reaction schemes (Eqs. 1, 2), one can identify the clear benzene peaks, probably a product of the polymer decomposition.

The TGA analysis revealed a total weight loss of the sample in the range 4-8% during crosslinking at 300°C. The quantitative analysis of released water has shown that the weight loss due to the water release is 1.4 wt %. The other components are ethanol, benzene, detected by the FT-IR spectroscopy, and also the low-molecular fractions of polymer, as, for example, $[(CH_3)_2SiO]_3$ or $(CH_3)_6Si_2O$ [11].

Table 1 summarizes the data about the phase composition of the H44 and MK–derived ceramic foams, measured by the redox analysis and their total weight loss after heating at 12 K/h to 900°C in air. The increased SiC content of H44-derived ceramic foam pyrolyzed at 1500°C may be attributed to the carbothermal reduction of SiO_2 [8]. One can see that the foams derived from the MK silicon resin (containing only methyl functional groups) contain much less carbon and are much more stable to oxidation than the H44-derived foams (phenyl and methyl groups).

Table 1. Phase composition of H44 and MK – derived ceramic foams and their weight loss after heating at 12 K/h to 900°C in air.

Starting Polymer	Pyrolysis Temperature, °C	SiO_2/SiC/C wt %	Weight loss after heating in air, %
H44	1000	53.4/10.6/36.0	32
H44	1500	12.6/73.5/13.9	15
MK	1000	76.0/16.0/8.0	0.06

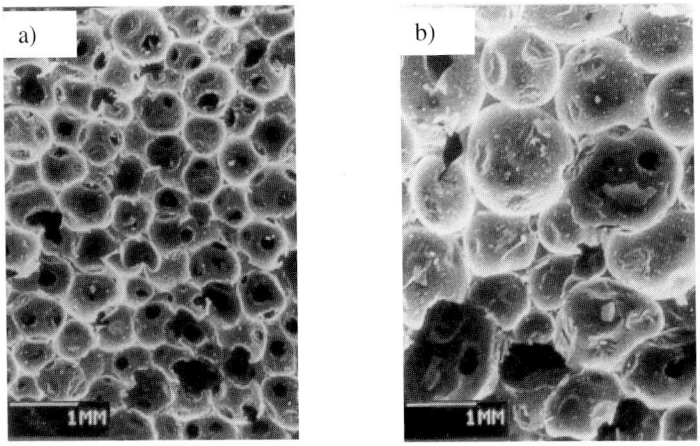

Figure 1. SEM micrographs of H44 foams pyrolyzed at 1200°C in argon. Mean cell size: a) 600 μm; b) 1200 μm.

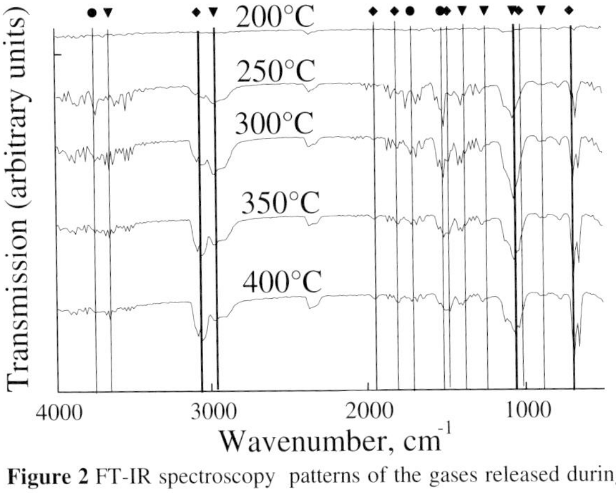

Figure 2 FT-IR spectroscopy patterns of the gases released during heating of the H44 powder at the rate of 10 K/min in flowing nitrogen. •- water, ▼ ethanol, □ benzene (strong peaks are shown as thicker corresponding lines).

The XRD analysis of the pure H44 foams pyrolyzed in argon at 1000°C and 1200°C revealed no crystalline phase, indicating the amorphous or nanocristalline nature of the pyrolysis residue. However, clear β-SiC peaks can be observed for the foam pyrolyzed at 1500°C, which can be attributed to the growth of the SiC or to the carbothermal reduction of SiO_2. The XRD analysis of the 20 and 30 vol % Al_2O_3 – filled H44 foams pyrolyzed in air at

1500°C revealed the clear mullite peaks. Using Mo filler lead to formation of the carbide phase (Mo_2C). The $MoSi_2$ filler (20 wt %) did not react with the products of the polymer decomposition. This kind of filler increased the electrical conductivity of the foams by four orders of magnitude at room temperature (ρ= 60 Ω cm), in comparison with filler-free polymer ($\rho = 1.9 \cdot 10^5$ Ω cm).

4 Conclusions

Ceramic foams are produced using intrinsic components of poly(siloxanes) for foaming. Foam formation is accompanied by release of water, ethanol, and benzene. Addition of fillers allows to vary physical properties of foams over a wide range.

5 Acknowledgements

Financial support from Minerva Foundation for T. Gambaryan-Roisman is greatfully acknowledged.

6 References

1. P. Colombo and M. Modesti, J. Am. Cer. Soc. 1999, 82, 573-78.
2. O. Pickenäcker, K. Pickenäcker, K. Wawrzinek, D. Trimis, W.E.C. Pritzkow, C. Müller, P. Goedtke, U. Papenburg, J. Adler, G. Standke, H. Heymer, W. Tauscher, F. Jansen, Keramische Zeitschrift 1999, 51, 190-199.
3. J. Saggio-Woyansky, C.E. Scott and W.P. Minnear, Bull. Am. Ceram. Soc. 1992, 71, 1674.
4. T. Fujiu, G.L. Messing and W. Huebner, J. Am. Cer. Soc.1990, 73, 85-90.
5. D.J. Green, J. Am Cer. Soc. 1985, 68, 403-409.
6. T.J. Fitzgerald, V.J. Michaud, A. Mortensen, J. Mater. Sci. 1995, 30, 1037-1045.
7. T. Gambaryan-Roisman, M. Scheffler, P. Buhler and P. Greil, in Proceedings of the 101st Annual Meeting of ACerS 1999, Indianapolis, USA (in print).
8. P. Greil, J. Am. Cer. Soc. 1995, 78, 835-48.
9. Arefmanesh, S.G. Advani and E.E. Michaelides, Pol. Eng. Sci. 1990, 30, 1330-1337.
10. M.J. Wild and P. Buhler, J. Mat. Sci. 1998, 33, 5441-5444.
11. S. Schicker, "Polymer abgeleitete Dentalkeramik", Dipl. Thesis, Institute of Materials Science (Glass and Ceramics), University of Erlangen-Nuernberg, 1995.

Development of Advanced High-Temperature Vacuum Furnaces

Gerhard Rothlauf*, Vitaliy Petrovich Grictenko**, Vladimir Dmitrievich Karpenko**, Sergey Aleksandrjvich Uvarov**, Vadim Arkadievich Borkovskih**, Natalia Nikolaevna Borkovskih**.

*BAYTEX Handels-GmbH. Unterschleißheim.

**State Plant "Uglecomposite". Zaporozhye.

1 Furnace

Successful combination of such properties of various carbon carbon composite materials, as thermal resistance, little variation of coefficients of thermal expansion, low vapor pressure and the rate of evaporation of up to temperatures 2500 - 2800°C makes it possible to create on their basis electrical vacuum furnaces with a working zone of up to 12 m^3 and inner temperature of up to 2800°C, with the maximum temperature to be maintained for 10 hours without deterioration of operational characteristics.

The design of this furnace uses three types of carbon carbon composite materials. The first one is low-density carbon carbon TIB material which has thermal insulating properties. The second material, UKPM-K, ensures surface protection for TIB material from radiation and ensures stability of the lining. The third one, UKPM-E, is used as electrical heaters instead of traditionally used graphite materials.

TIB material is manufactured on the basis of viscose and phenol-formaldehyde resins. Molded from shortly cut viscose fiber as well as finely ground resin the material undergoes carbonization and heat treatment up to a temperature dependent on the operating conditions. The basic purpose of TIB material is to create the greatest possible thermal resistance between the working zone of the furnace and the metal body which ensures the required vacuum. So, the main TIB parameter, is thermal conductivity. The thermal conductivity of TIB material in vacuum is shown in Fig.1. It is clear from the drawing, that with the increase of density thermal-insulating properties somewhat decrease. On the other hand the low density of this material causes low strength.

The following types of protective coating are often applied on the surface of the materials: based on resins, graphite foil or made by carbon deposition from gas phase [1]. But the most successful protection of the surface of low-density carbon materials against erosion and radiation ensures carbon composite material with rather high mechanical properties.

UKPM-K material is based on several types of low-module viscose graphite fabrics and phenol formaldehyde resins. The carbonized blanks are subjected to densification by pyrolytic carbon up to bulk density of 1.2 - 1.3 g/cc. The mechanical properties of this material are given in Fig. 2 and Fig. 3. From Fig. 2 and Fig. 3 it is clear that bending strength, compression strength, and shear strength grow with the increase in the material density. It is necessary to note, that the increase in density of the samples by 30 % favors growth of the tested properties by almost two-fold. Practically, their level does not depend on the filler type. With the identical density of the material the specified properties reach equal values for different fillers.

The conditions for obtaining samples for determining dependence of strength properties on density, were selected so as to eliminate influence of all possible technological parameters, except for various contents of pyrolytic carbon. Therefore we can assume, that the increase in their mechanical properties is associated just with the indicated factor. Such an assumption agrees with the data obtained by the authors [2], who subjected MPG-8 graphite to pyrolytic densification. The higher relative enhancement of mechanical properties obtained in the present paper in comparison with the data [2] can be explained by large contents of pyrolytic carbon in UKPM-K samples, which is connected with greater open porosity of UKPM-K carbonized blank in comparison with MPG-8.

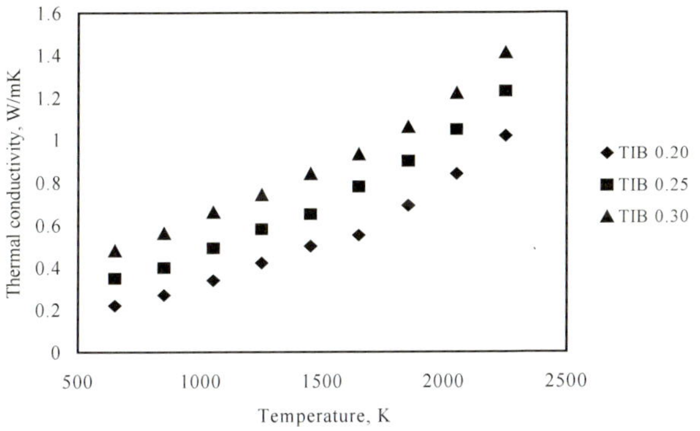

Figure 1. Thermal conductivity of TIB material as function of temperature. Bulk density of TIB 0.20 -0.20 g/cc, 0.25 -0.25 g/cc, 0.30 -0.30 g/cc

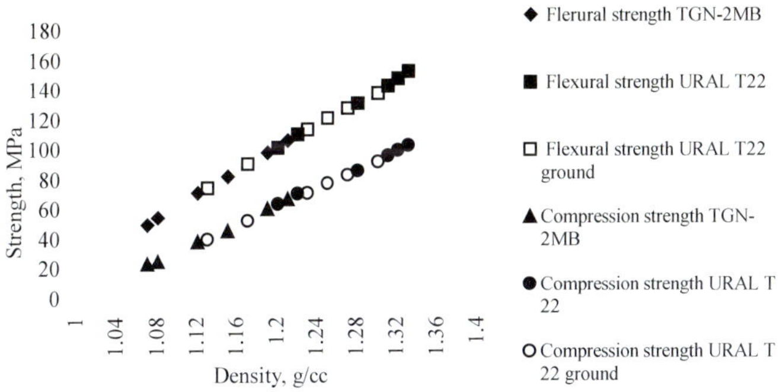

Figure 2. Flexural strength and compression strength as function of UKPM material density.

From the data represented in Fig. 3 it is clear that tensile strength depends not only on density of the material, but to a greater degree is determined by the properties of filler. For samples from UKPM-K with similar density tensile strength increases almost by 1.5 times when passing from TGN-2MB filler to "Ural" T-22. It correlates rather well with the change in properties of initial fillers (breaking load for TGN-2MB three times less than that for " Ural " T-22, see Table 1). The difference in tensile strength in composite material samples with different fillers and breaking load of fillers can be explained by greater inhomogeneity of TGN-2MB fabric, which disappears when it is used as layer-by-layer reinforcement of the material samples.

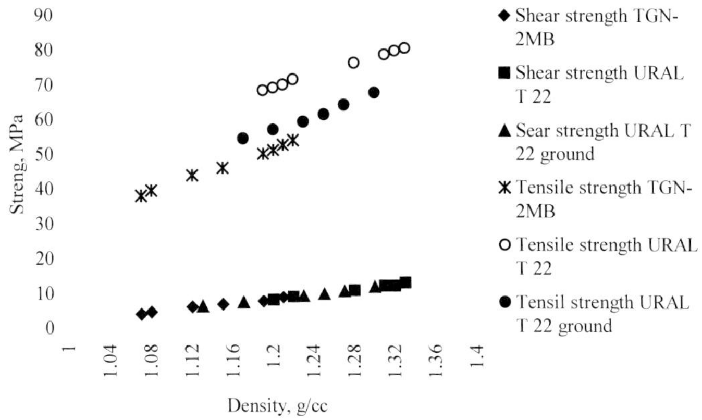

Figure 3. Interlayer shear strength and tensile strength as function of UKPM material density.

Table 1. Strength characteristics of carbon fabric samples.

Type of filler	Breaking load on a strip with width of 50 mm, warp / weft, N	Surface density g/m^2
"Ural" T-22	980/343	290
TGN-2MB	294/49	350

This inhomogeneity can be determined by a considerable amount of microdefects in TGN-2MB fabric. For the confirmation of the above-mentioned the research of physical and mechanical properties in the composite material whose filler were square pieces of prepreg with a side from 30 up to 40 mm, on the basis of "Ural" T-22 fabric. The research results of this material samples are also given in Fig. 2 and Fig 3. It is notable that bending strength, compression strength and shear strength dependent on density practically precisely coincides with the appropriate curves obtained for samples based on integral filler. The tensile strength decreases by half in comparison with integral filler.

UKPM-E material is manufactured similarly to UKPM-K. But during manufacturing it is subjected to treatment at higher temperature. The material is characterized by apparent density of 1.35 to 1.45 g/cc and lower strength in comparison with UKPM-K. But its most important parameter is resistance. The dependence of its specific electrical resistance on temperature is shown in Fig. 4. As a comparison the characteristics of some graphite materials [3] are also shown there. It is clear, that the character of change of UKPM-E material resistance does not differ greatly from that of graphites.

Each of lining components is of layered structure: the layer of TIB material is lined from both sides with UKPM-K, and the side oriented to greater temperature is lined up with thicker layer of UKPM-K material. The stressed parts are the layers of UKPM-K material, but not those of thermal insulating material. This creates conditions for making the lining repair easier. In the case under consideration, any element of the lining can be removed without removing the whole lining or its larger part. This is achieved at the expense of appropriate designing solutions. Besides, the suggested designing solutions make it possible to use soft carbon felts as thermal insulating materials.

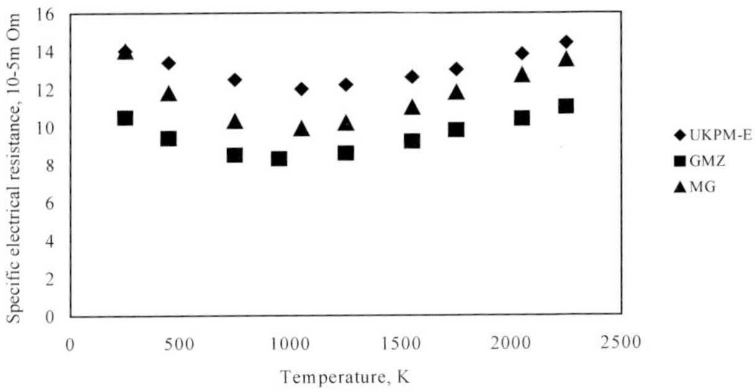

Figure 4. Specific electrical resistance of UKPM-E and some graphites as function of temperature. The data regarding graphites are taken from [3].

Long maintenance of temperature in the working zone exceeding 2500°C results in evaporating carbon from lining materials and especially from heaters. The greater part of the weight loss falls at protective layer from UKPM-K and UKPM-E material for heaters. The properties of these materials are restored by introducing pyrolytic carbon from gas phase into opened or enlarged pores. Carbon from the matrix which is distinguished by the presence of defects in crystallographic structure is subjected to evaporation to a greater degree. It consists of phenol formaldehyde resin coke and pyrolytic carbon. Graphitized fiber has more perfect structure, for this reason it does not considerably loses its weight: so it can be neglected. Therefore after pyrolytic process has finished the properties of furnace lining materials and heaters are practically completely restored. All this can be described roughly in the following way. The operation at temperatures exceeding 2500°C results in increase of the heater resistance at the expense of the weight loss. If operation of the heaters stops before reaching resistance, which makes operation with the existing power supply impossible, the subsequent introduction of pyrolytic carbon will again result in the resistance decrease. The perfection of structure of the introduced pyrolytic carbon during graphitization under the influence of operating temperatures has also impact on the resistance decrease.

The lining material has also lower mechanical characteristics, partially at the expense of the weight loss, partially at the expense of the partial graphitization of the component phases of the material. Basically it occurs in the surface layer facing the hot zone. The operation without repair stops prior to disintegration of graphite fiber and for this reason the filler structure is conserved. Introducing pyrolytic carbon into this area, which is coherently bonded with this fiber and has excellent mechanical properties, results in the creation of the composite material, which has the same characteristics as the initial material.

2 References

1. Calcarb Ltd brochure. 1996. Schottland.
2. J. C. Wirilew, L. L. Lyschow, W. K. Petschik. Chimia twordoro topliwa. 6, 1989. p.126-130.
3. O. S. Gurewitsch, J. P. Lachin, S. I. Sobolew. Wyssokotemperaturnye elektropetschi s grafitowymi elementami. Energija. Moskwa. 1974. p.103

New Aluminium/Carbon-Based Ceramic Compounds

Falko D. Meyer and Harald Hillebrecht
Laboratorium für Anorganische Chemie, Universität Bayreuth, Germany

1 Introduction

Ternary ceramic compounds Al-C-X with X = B, N, O, Si play an important role in different fields of materials sciences.

Ternary compounds can be formed during processing of cermets with Al-matrix like B_4C/Al or SiC/Al. Composite materials made of an Al based matrix reinforced with B_4C, SiC, AlN, Al_2O_3 and carbon fibres are promising candidates for aerospace, ligthweigth armor and electronical applications. One important factor influencing the properties of such composites is the emerging of ternary phases during processing in the contact zone between matrix and ceramic component.

Some ternary phases are the active part during the sintering of SiC, B_4C, Al_2O_3 or AlN with a sintering aid like Al, B, C, AlB_2, B_4C or SiC. For optimization of the sintering process, a better understanding of the reactivity and the stability range of the compounds is necessary.

The ternary phases themselves are promising ligthweigth ceramic materials because of their high chemical and thermal stability and their semiconducting behaviour. Some of them are candidates for high-temperature semiconductors and for thermoelectric applications.

In view of optimizing the production of MMC's by melt infiltration, a general study of the reactivity of liquid aluminium towards boron carbide, silicon carbide, aluminium nitride/carbon and aluminium oxide/carbon has been undertaken [1]. The aim of our investigation was the synthesis of pure phases in the systems Al/B/C, Al/Si/C, Al/C/N and Al/C/O, growth of single crystals, characterization of the pure compounds and a better understanding of the complex reactivity in the ternary systems.

2 Synthesis

The compounds were synthesized from powders of the elements which were pressed to rods and placed in Al_2O_3 or graphite crucibles. Often a flux of aluminium was used, other methods were solid state reaction or high-temperature synthesis through the gas phase. The synthesis took place in Al_2O_3-tube furnaces, in a furnace with graphite heater and by arc melting. The samples were heated fast up to a temperature range of 1400°-2200°C and cooled slowly, depending on the sample. Excess of Al was removed with diluted HCl or NaOH.

3 Characterization

All samples were characterized by X-ray powder diffraction. Pure phases were characterized by the means of FT-IR- and FT-Raman-spectroscopy, UV/Vis and DTA measurements. Quantitive analyses were done by WDX measurements. X-ray crystal structure analyses were carried out on single crystals with good quality.

4 Results

4.1 System Al/B/C

We managed to synthesize the pure ternary compounds $Al_4B_{0.66}C_{2.34}$, Al_3BC_3, Al_3BC, $B_{48}Al_3C_2$ and $B_{51}Al_2C_8$. On selected single crystals of each phase X-ray crystal structure analyses could be carried out. The results are shown in Fig. 1 and Fig. 2.

$Al_4B_{0.66}C_{2.34}$ is a new phase and results from Al_4C_3 by partial substitution of the octahedral coordinated carbon atom with boron. By increasing the boron content, the stability against hydrolysis increases.

Al_3BC_3 is a carbide carboborate with the isolated anions $[CBC]^{5-}$ [2]. These anions are the 16-electron system with the highest negative charge known up to now.

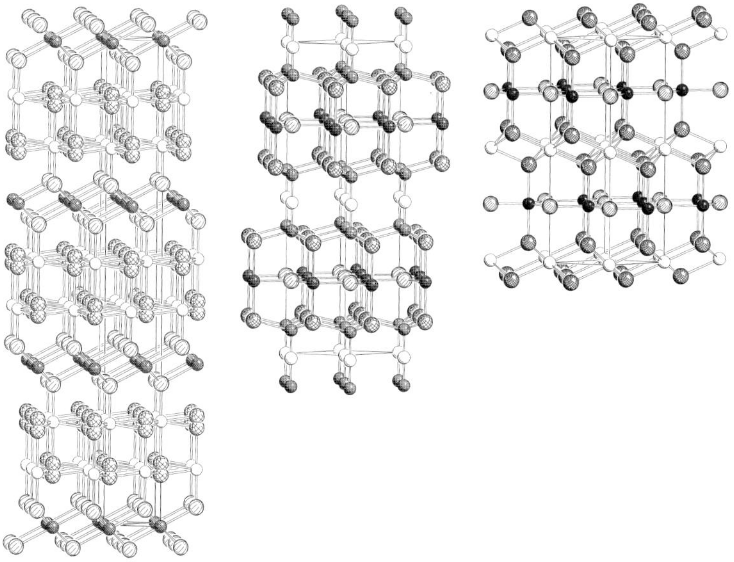

Figure 1. Crystal structures of $Al_4B_{0.66}C_{2.34}$, Al_3BC_3 and Al_3BC (from left to right).

Al_3BC is a boride carbide with isolated boron and carbon atoms [3]. By replacing the anions CBC^{5-} in Al_3BC_3 with boron atoms, one can see the relation between these two compounds.

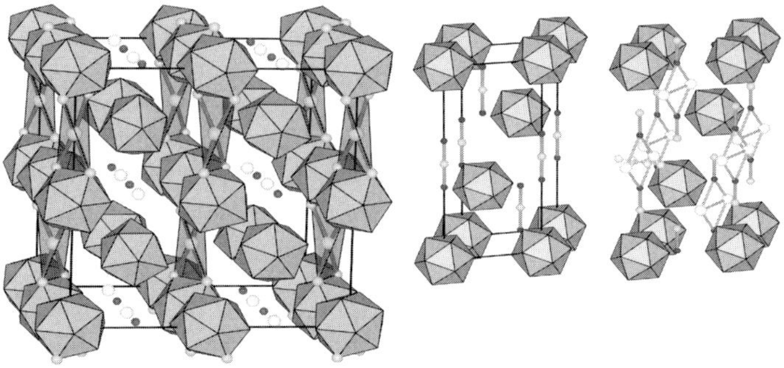

Figure 2. Crystal structures of $B_{48}Al_3C_2$, boron carbide $B_{13}C_2$ and $B_{51}Al_2C_8$ (from left to right).

The crystal structure of $B_{48}Al_3C_2$ is related to that of „tetragonal boron I". Al- and C-atoms are located in octahedral and tetrahedra voids of the tetragonal B_{12}-icosahedra packing. $B_{48}Al_3C_2$ shows a phase transition at ca. 650°C from a tetragonal HT-phase to a orthorhombic LT-phase. This is the reason for the described complex twinning patterns [4].

$B_{51}Al_2C_8$ results by the inclusion of aluminium in boron carbide, $B_{13}C_2$. Contrary to the model published by Neidhard et al. [5], we found a partial replacement of the boron position in the CBC-chain by disordered Al_2-pairs and a partial occupation of a new boron position [10].

4.2 System Al/Si/C

Pure phases of the compounds Al_4SiC_4 and Al_8SiC_7 could be prepared. The result of a crystal structure analysis on Al_4SiC_4 is shown in Fig. 3, a crystal structure analysis on Al_8SiC_7 is under current investigation.

The crystal structure of Al_4SiC_4 is isotypic to that of Al_5C_3N. The compound $Al_4Si_2C_5$ was yielded as a by-product, the optimization of synthesis and crystal growth is aim of our current work.

Phases with a higher silicon content [6] could not be obtained. Like in the system Al/C/N, very fast cooling rates seem to be necessary, which are not possible to archive with our experimental setup.

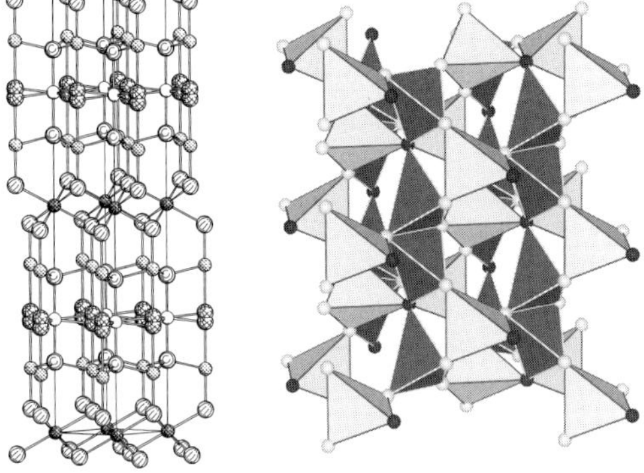

Figure 3. Crystal structures of Al_4SiC_4 (left side) and Al_4CO_4 (right side).

4.3 System Al/C/O

Single crystals of the phase Al_4CO_4 could be obtained. The result of a crystal structure analysis on Al_4CO_4 (Fig. 3) is in good agreement with that of Jeffrey and Slaughter [7].

From Al_2CO only powder samples could be obtained, single crystal growth is under current investigation.

4.4 System Al/C/N

Single crystals of the pure ternary compounds $Al_{3.89}C_{2.66}N_{0.34}$ and Al_5C_3N could be grown. Of selected single crystals of each phase X-ray crystal structure analysis could be carried out. The results are shown in Fig. 4.

$Al_{3.89}C_{2.66}N_{0.34}$ is a new phase and results from Al_4C_3 by the partial substitution of the [4+1] coordinated carbon atom with nitrogen, accompanied with Al vacancies.

The crystal structure of Al_5C_3N shows small differences compared to that published by Jeffrey and Wu [8,9].

$Al_6C_3N_2$ could be identified as powder, single crystals could not be obtained. Other phases with higher nitrogen content, mentioned in literature [8,9], could not be obtained. One problem in the synthesis of these compounds seems to be the need of a very fast cooling rate, which is not possible to archive with our experimental setup.

During our preparative work in the system Al/C/N we found blue, transparent single crystals of AlN. The result of a crystal structure analysis shows a good agreement with that of colourless AlN, blue AlN shows slightly enlarged cell parameters. Despite thorough investigations, the reason for the blue colour is yet not understood.

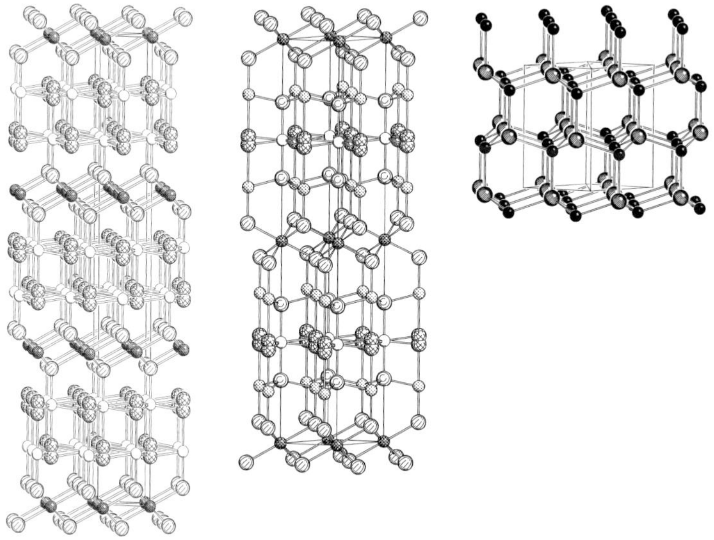

Figure 4. Crystal structures of $Al_{3.9}C_{2.66}N_{0.34}$, Al_5C_3N and blue AlN (from the left).

5 Summary

Most of the crystal structures of the ternary phases in the systems Al/C/X with X = B, N, O, Si show a strong relation to the structure of aluminium carbide (Al_4C_3) and to the wurtzite structure type (AlN and SiC). Aluminium is always tetrahedrally coordinated, the AlC_4-tetrahedra are linked through shared corners and edges.

6 Literature

1. F. D. Meyer, PhD thesis, Albert-Ludwigs-Universität, Freiburg, Germany 1998.
2. H. Hillebrecht, F. D. Meyer, Angew. Chem. 1996, 108, 2655-2657; Angew. Chem. Int. Ed. 1996, 35, 2499-2500.
3. F. D. Meyer, H. Hillebrecht, J. Alloys Comp. 1997, 252, 98-102.
4. V. I. Matkovich, J. Economy, R. F. Giese Jr., Z. Kristallogr. 1965, 122, 108-115.
5. H. Neidhard, R. Mattes, H. J. Becher, Acta Cryst. 1970, 26, 315-317.
6. R. J. Oscroft, P. Korgul, D. P. Thompson in Br. Ceram. Proc. (Ed. R. Stevens, D. Taylor), Vol. 42, Institute of Ceramics, Stoke-on Trent, U.K. 1989, 33-47.
7. G. A. Jeffrey, M. Slaughter, Acta Cryst. 1963, 16, 177-184.
8. G. A. Jeffrey, V. Y. Wu, Acta Cryst. 1963, 16, 559-566.
9. G. A. Jeffrey, V. Y. Wu, Acta Cryst. 1966, 20, 538-547.
10. H. Hillebrecht, F. D. Meyer Z. Kristallogr. 1998, Suppl 15, 33.

Large-Scale Ceramic Composite Structures via Near-Net-Shape Processing of Carbon Preforms

C. Müller, S. Walter, U. Papenburg and W. Pfrang
Industrieanlagen-Betriebsgesellschaft mbH (IABG) Einsteinstr. 20, D-85521 Ottobrunn, Germany

1 Abstract

Processing of ceramic matrix composites (CMC) presented in this article is based on the conversion of porous carbon preforms into C/SiC-ceramics by reaction infiltration of liquid silicon. For the processing itself as well as for the materials properties of the derived composites, microstructure and architecture of the used preforms play a key role.

Composites are fabricated by starting from chopped, carbon short fiber reinforced carbons (C/C) which are machined by conventional CNC milling. Accordingly, very complex component geometries can be shaped. After subsequent reaction infiltration by a silicon melt, silicon carbide (SiC) is formed. In the as-processed composites, remnants of unreacted carbon as well as excess silicon are also present. The infiltration process is accompanied by no significant volume change. This near-net-shape feature allows to fabricate comparatively large components. At the same time also very thin-walled, light-weight structures can be processed. Furthermore, the near-net-shape character of the process allows a joining technique to be applied for the fabrication of multi-segmented components.

This contribution presents C/SiC composites for specific applications that are derived from different types of carbon/carbon starting materials. Examples include large-dimensioned optical light-weight structures, large retorts for processing of metals and tube components to be used in combustion technology.

2 Near-Net-Shape Processing

Many conventional ceramic processing routes are based on powder consolidation and sintering. They have to deal with linear shrinkages up to 10% or even above [1]. Hence, conventional fabrication of large-dimensioned ceramic parts is often limited by the risc of crack formation upon processing as well as by expensive machining after heat treatment due to inhomogeneous shrinkage.

However, Industrieanlagen-Betriebsgesellschaft (IABG), Ottobrunn, Germany, follows another processing route, fig. 1. Instead of powders, IABG uses porous carbon/carbon composites as starting material. These composites consist of a carbon matrix reinforced by carbon short fibers (C/C).

C/C-raw material blocks are machined into the final component shape by conventional CNC-milling. Standard steel cutting methods are applied. Accordingly, preforms with very complex geometries or thin-walled structures can be shaped in the greenbody state. By the following thermal vacuum process the C/C-preform is heated up to above 1400°C and brought into contact with molten silicon (Si). Silicon penetrates into the porous preform and reacts with carbon to form silicon carbide (SiC). After cooling to room temperature, a C/SiC composite ceramic is derived, due to the presence of unreacted carbon fibers [2].

Taking into account that silicon is provided in excess, the theoretical overall volume change of the reaction infiltration Si + C → SiC is positive. Accordingly, no shrinkage should occur. Experimental analysis revealed that the linear shrinkage of the components during processing is in the range of 0 to 0.5 %. The exact amount of shrinkage depends on starting material and process parameters (see sections below). The very low to zero shrinkage is characteristic for all of IABG's innovative components (see fig. 3a).

Striking advantages of the described net-shape processing approach are:

- post-treatment (e.g. grinding) after ceramization can be minimized
- comparatively large components can be fabricated
- very thin-walled, light-weight structures can be processed
- a joining technique can be applied to process complex, multi-segmented components by one heat treatment.

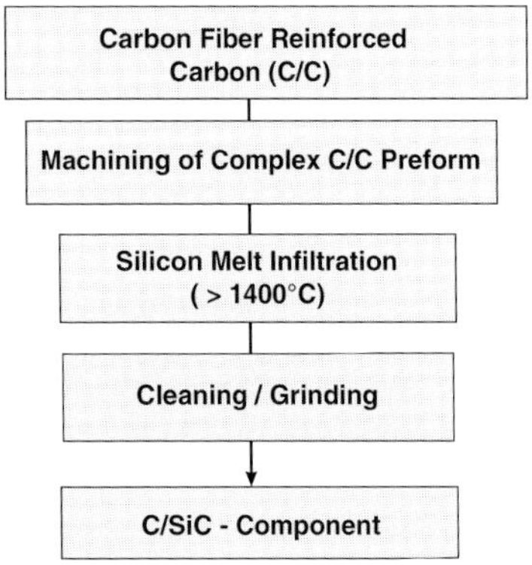

Fig. 1: Near-net-shape processing of C/SiC composite ceramics

2 Microstructure

For specific applications different material properties are required, e.g. low thermal expansion and high isotropy for optical applications [3, 4, 5] or high thermo-mechanical strength and damage tolerance for high temperature use [6].

Architecture and microstructure of used C/C starting material strongly influence processability as well as properties of the derived C/SiC composites. Key parameters are, for example, carbon fiber volume content, fiber length („short"/chopped versus „endless"/continuous), pore structure, thickness of fiber bundle and fiber orientation. Fig. 2 shows examples of C/SiC microstructures after ceramization.

The components presented in this article consist of SiC as the phase with the highest volume content. Unreacted carbon as well as silicon are also present. For composites derived by providing excess silicon, a dense microstructure can be achieved with molten silicon having filled up residual pores.

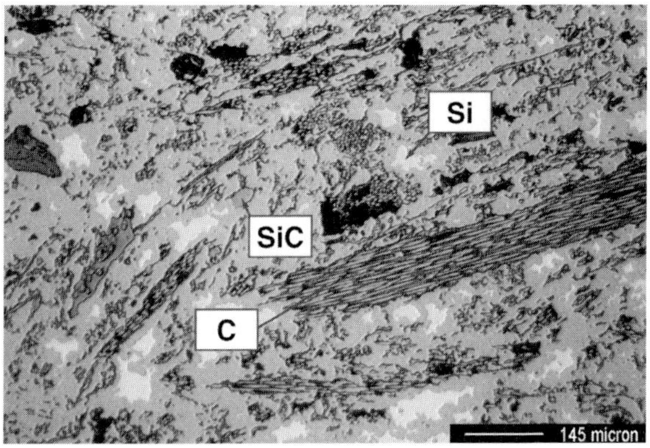

(a)

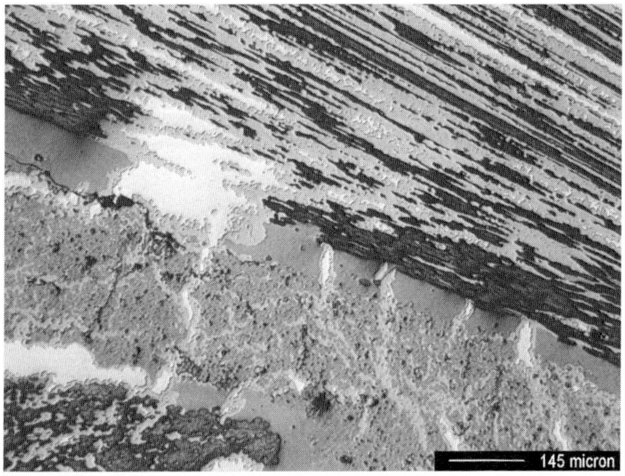

(b)

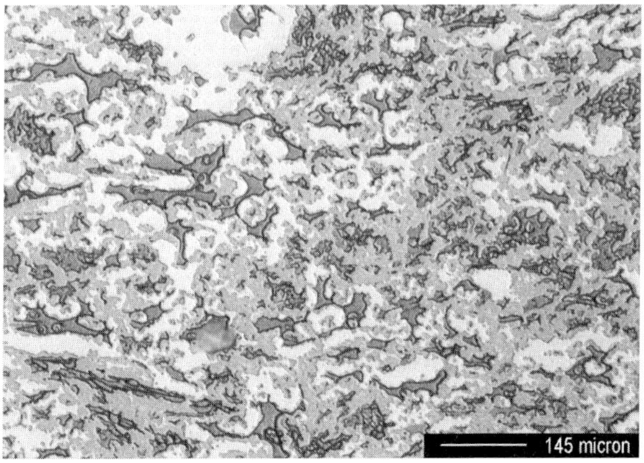

(c)

Fig. 2 a, b, c: Various C/SiC microstructures derived from different types of C/C raw materials: carbon phase (C) is shown in dark, SiC in medium, Si in bright color.

3 Fabrication of Large-Scale C/SiC Structures

The near-zero-linear shrinkage during processing (0 - 0.5 %, depending on C/C starting material and process parameters, see above) allows the fabrication of comparatively large and complex C/SiC parts. At the same time very thin-walled, light-weight structures can be processed. Examples of components for specific applications are shown in fig. 3.

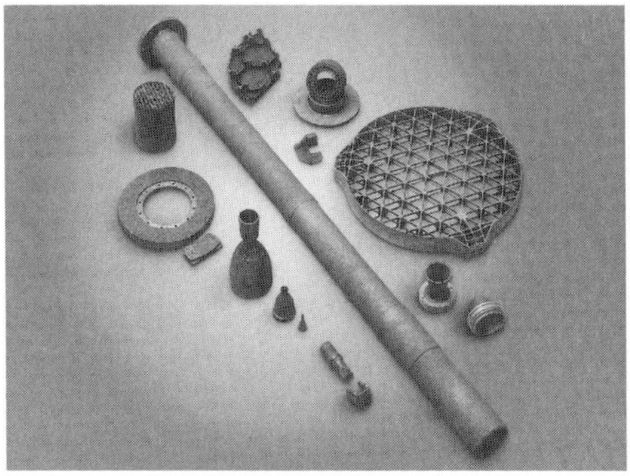

(a) Overview of C/SiC products; center: high temperature probe tube length 210 cm; right: lightweight mirror substrate diameter 72 cm

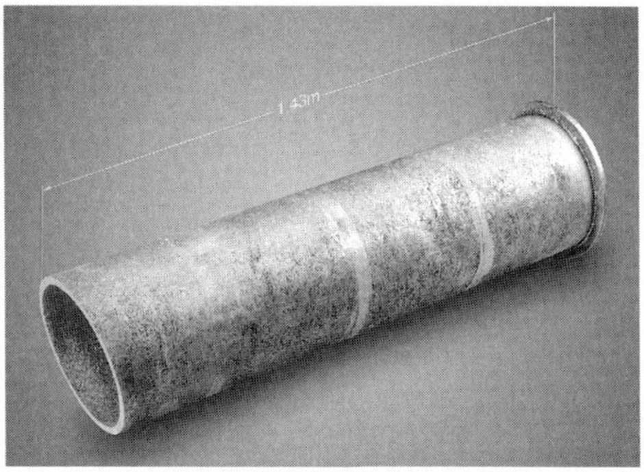

(b) Burner protection tube, ∅ 38 cm, length 143 cm; fabricated for ECM, Munich

(c) Retort for high temperature processing, length 133 cm

(d) Optical lightweight mirror (cladded, polished and reflective-coated), 80,5 x 52 cm; fabricated for DaimlerChrysler Aerospace, Munich

Fig. 3 a,b,c,d: Fabricated large-scale C/SiC components for specific applications

4 References

[1] D. W. Richerson, „Modern Ceramic Engineering" – Properties, Processing and Use in Design, 2nd Edition by Marcel Dekker Inc., New York (1992)

[2] U. Papenburg, E. Blenninger, B. Kunkel und M. Deyerler, „Optomechanische Leichtgewichtsstrukturen aus kurzfaserverstärkter Keramik", *Fortschrittsberichte DKG*, Seminar 'Festigkeit Keramischer (Verbund-)Werkstoffe', Wien, 29.-30.09.1994, 1-43

[3] U. Papenburg, W. Pfrang, G.S. Kutter, C. Müller, B. Kunkel, M. Deyerler and S. Bauereisen, „Optical and Optomechanical Ultra-Leightweight C/SiC Components", *Proc. SPIE Int. Symp. Optical Science, Eng. and Instrumentation*, Vol. 3782-16, (1999) pp

[4] C. Müller, S. Walter, W. Pfrang und U. Papenburg, "Potential von Leichtgewichts-spiegeln aus C/SiC-Verbundkeramik für die Anwendung in der Weltraumoptik", in *'Verbundwerkstoffe und Werkstoffverbunde'*, Herausgeber K. Schulte und K. U. Kainer, Wiley-VCH, Weinheim (1999) 352-357

[5] R. A. Paquin, „Materials for Mirror Systems: an Overview", *Proc. SPIE Int. Symp. Optical Science, Eng. and Instrumentation*, Vol. 2543 (1995), 2-11

[6] O. Pickenäcker, K. Pickenäcker, W. Wawrzinek, D. Trimis, W.E.C. Pritzkow, C. Müller, P. Goedtke, U. Papenburg, J. Adler, G. Standke, H. Heymer, W. Tauscher und F. Jansen, "Innovative Ceramic Materials for Porous Medium Burners", *Interceram,* Vol. 48 [5] (1999) 326-330 (Part I) and [6] (1999) 424-434 (Part II)

II Ceramics Processing at Low Temperatures

Colloidal Processing of Nanocrystalline Cerium Oxide

Andreas Tschöpe, Jürgen Markmann, Rainer Birringer
Universität des Saarlandes, Saarbrücken, Germany.

1 Abstract

Nanocrystalline cerium oxide was synthesized by homogeneous precipitation. A colloidal aqueous sol of the precipitate was prepared by ultrasonification and was stabilized electrostatically. Transparent green bodies with >45% of theoretical density were generated by centrifugal casting. These green bodies could be sintered to >95% of theoretical density at temperatures as low as 900°C. This temperature is remarkably lower than the corresponding sintering temperatures of 1600°C for microcrystalline ceria. Studies on the effect of yttrium doping on cerium oxide grain growth showed that a doping concentration of 7 at-% was effectively retarding grain growth such that an average grain size of <50 nm could be preserved during sintering for 10 hours at a temperature of 1000°C. The stabilizing effect of Y-doping on the grain size and the high sinter activity of green bodies which were generated by centrifugal casting of a colloidal suspension could be combined to produce high density nanocrystalline cerium oxide ceramic through pressure-less sintering.

2 Introduction

Because of its high ionic conductivity, cerium oxide is a potential candidate for applications of solid electrolytes such as fuel cells or catalytic membrane reactors. However, cerium oxide also exhibits a distinct redox chemistry which results in a considerable electronic conductivity at high temperature and low partial pressure. This propensity for reduction imposes some limitations on the application of ceria as solid electrolyte, but it also opens the way for other applications, e.g. in heterogeneous catalysis, conductivity gas sensors or as electrode material in electrochemical devices. Some of these applications require nanostructured morphologies and may only be realized if improved processing with a high degree of microstructural homogeneity is available for such materials. A critical example is the synthesis of nearly dense, nanocrystalline cerium oxide. Various studies demonstrate the high sinter activity of nanocrystalline cerium oxide as compared to microcrystalline materials [1-3]. However, densification by pressure-less sintering of powder compacts require temperatures of at least 1100°C to achieve >90% of theoretical density. At these temperatures, the nanocrystals have grown to micrometer size. As an alternative, uniaxial hot pressing has often been successfully employed for densification of nanocrystalline ceramic powders. However, the resulting materials often suffer from an inhomogeneous density distribution. The objective of the present study was to investigate the potential of colloidal processing with regard to the preparation of nanocrystalline ceramic bodies of high density. The efficacy of this processing route has been shown in various studies on other metal oxides, e.g. YSZ [4,5].

A central issue in colloidal processing is the control over the repulsive forces (e.g. steric or electrostatic forces) between the particles in a colloidal solution. If steric stabilizers are applied, the volume fraction of organic matter in the green body increases rapidly as the grain size becomes smaller. In order to reduce the risk of excessive residual impurities in the grain boundaries, electrostatic stabilization may be preferred. However, DLVO theory predicts, that electostatic stabilization becomes less effective with smaller grain sizes [6] and requires the use of rather dilute ceramic suspensions. This limits the choice of the consolidation method. Centrifugal casting is compatible with this restriction and in contrast to filter pressing or electrophoretic deposition, it can be utilized to produce massive ceramic objects.

3 Experimental

Nanocrystalline cerium oxide was synthesized by homogeneous precipitation [1]. Aqueous solutions of 0.0375M Ce(III)nitrate and 0.5M hexamethylenetetramine (HMT) were combined and stirred for 24 h at a temperature of 30°C. For doping with yttrium, the appropriate amounts of yttrium nitrate was added to the solution. The precipitate was centrifuged, washed twice with 2-propanol and dispersed in water. Nitric acid was added to the solution until pH 2 was reached and the particles were dispersed by moderate ultrasonification over night. Residual larger agglomerates were separated by centrifugation of the colloidal suspension. The supernatant of this process was centrifuged at accelerations of 1.500 to 15.000 g. After removing the residual liquid, the deposit was slowly dried at room temperature for a period of 2 weeks.

X-ray diffraction studies were performed using a Siemens D-500 with Cu K radiation. The average grain size was determined from the peak broadening with respect to a coarse-grained reference sample, using the FORMFIT software package [7]. Dynamic light scattering (ALV-4000) was employed for characterization of the hydrodynamic radius of the particles in solution.

Nitrogen-adsorption isotherms at 77K were obtained using a Micromeritics ASAP 2010 and the pore size distribtuion in the micro- and mesopore size range were calculated with the help of the DFT-software package.

Sintering of centrifuged green bodies in air was performed by heating the samples at a rate of 2°/min to the desired temperature followed by a 4 h hold time and a cooling ramp at a rate of 2°/min. The density of the sintered samples was measured by the Archimedes method.

4 Results

For structural analysis of the washed product, the precipitate was dried in air at room temperature. The X-ray diffraction pattern confirms the crystallization of the precipitate in the fluorite structure of CeO_2, Fig. 1. The analysis of the diffraction pattern yielded a lattice constant of 542.28 pm and a volume-weighted average grain size of 8.6 nm. The hydrodynamic diameter of the particles in the electrostatically stabilized suspension as determined by dynamic light scattering was 40 nm.

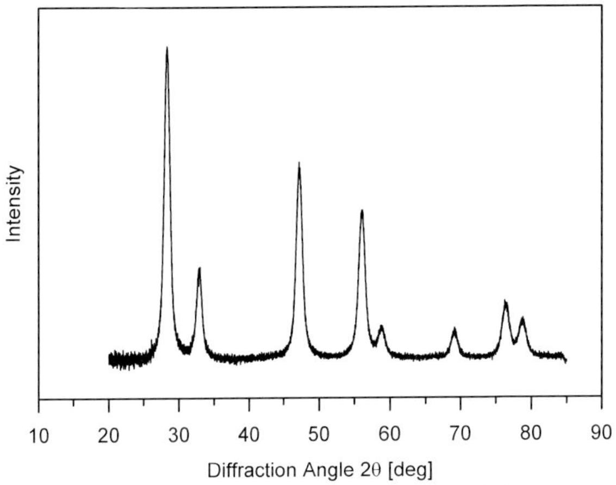

Figure 1. X-ray diffraction pattern of the washed precipitate.

Nitrogen adsorption isotherms at 77K in the partial pressure range from $p/p_0 = 10^{-4}$ - 1 were measured after sample degasing at 300°C. As shown in Fig. 2, the green body generated by centrifugal casting exhibited a bimodal pore size distribution with maxima at 15Å and 45Å, i.e. micro- and mesopores. Apparently, the centrifuged green bodies did not contain larger volume fractions of macropores as evident from the transparency of the sample. From the culmulative pore volume, a green density of 45% of theoretical density was obtained for this specific sample after degasing at 300°C.

A number of samples were sintered in air, and the density of the ceramic specimen as function of maximum hold temperature during sintering is shown in Fig.3. Relative densities of >95% were achieved by sintering at temperatures above 900°C.

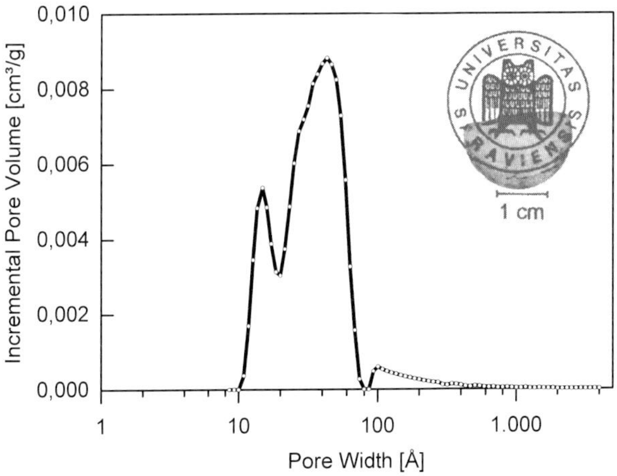

Figure 2. Pore size distribution and photograph of a centrifuged green body.

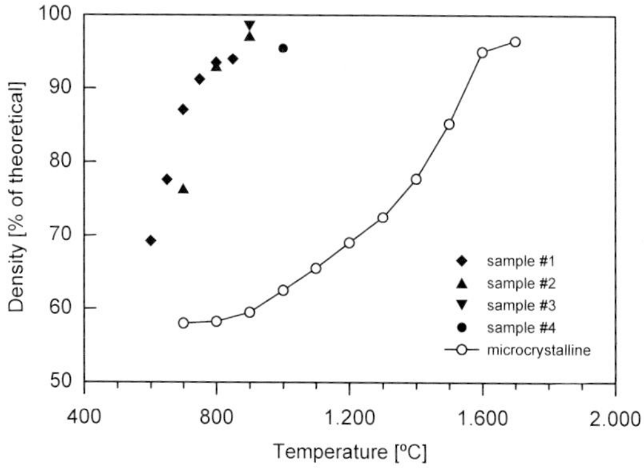

Figure 3. Density of various centrifuged green bodies and of microcrystalline ceria [2] after sintering at indicated temperatures. Sample 4 was doped with 10 at% yttrium.

The effect of high temperature annealing on the grain size was investigated for undoped and 7 at%-Yttrium doped cerium oxide, Fig. 4. Whereas the grain size of the undoped sample increased to >50 nm after annealing at 1000°C, the grain size of the yttrium-doped sample was only 30 nm. However, yttrium doping did not affect the sinter activity of the cerium oxide significantly, as shown by sample #4 in Fig. 3.

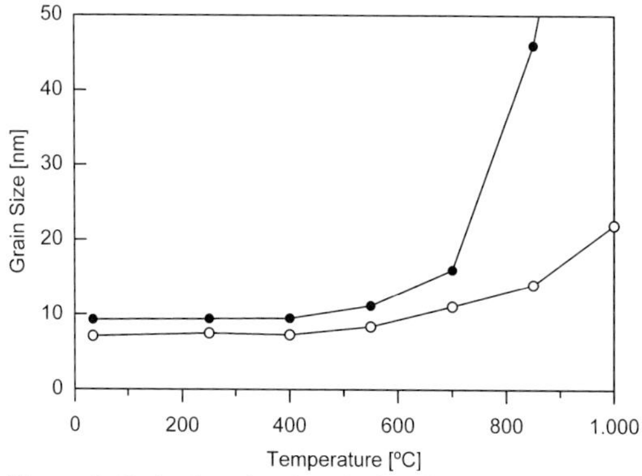

Figure 4. Grain size of undoped (closed circles) and 7 at% Y-doped (open circles) cerium oxide as function of annealing temperature.

5 Discussion

Complete crystallization of the washed precipitate was confirmed by X-ray diffraction in agreement with [1]. The grain size of 8.6 nm was significantly smaller than the hydrodynamic diameter of 40 nm for the particles in solution as determined by dynamic light scattering. This discrepancy points to a moderate degree of agglomeration. This conclusion is further supported by the bimodal pore size distribution. One may associate the mesopores of 4.5 nm size with void spaces between 40 nm agglomerates and the micropores of 1.5 nm size with the voids between 8.6 nm primary particles within the agglomerates. As a first conclusion, the cerium oxide particles could not be dispersed into the primary particles in aqueous solution.

However, the agglomerate size of 40 nm is still rather small and translates into very high sinter activity. The density of various sintered green bodies, as shown in Fig. 3, increased rapidly at sintering temperatures of 600°-800°C and values >95% of theoretical density were obtained after sintering at temperatures ≥ 900°C. These very low sintering temperatures indicate that in spite of a moderate degree of agglomeration, centrifugal casting of colloidal cerium oxide provides ceramic green bodies with extraordinarily high sinter activity. In principle, the sinter activity could be further increased, if a dispersion of primary particles could be realized, for instance by the application of surfactants during synthesis or colloidal stabilization.

Doping cerium oxide with yttrium was found to have an important effect on grain growth during high temperature sintering, Fig.4. Whereas the grain size of undoped cerium oxide increased to more than 50 nm after annealing at 1000°C, the corresponding values of a 7 at% Y-doped sample remained below 30 nm. It can be assumed that yttrium segregation to the grain boundaries is retarding grain growth for two reasons. First, the segregation of yttrium inevitably causes a decrease in grain boundary free energy according to the Gibbs adsorption isotherm [8] and hence reduces the driving force for grain growth. Second, the yttrium impurities segregated at the grain boundary may reduce the mobility of the grain boundary, known as solute drag effect [9,10].

In conclusion, colloidal processing and centrifugal casting of nanocrystalline cerium oxide can be used to generate green bodies which densify at temperatures below 1000°C. Taking advantage of the retardation effect of yttrium on grain growth, it further becomes possible to densify the centrifuged green bodies while preserving the nanocrystalline structure.

6 Acknowledgment

The authors wish to thank Dr. W. Härtl for his help with the dynamic light scattering measurements. This project was supported by the Deutsche Forschungsgemeinschaft (SFB 277).

7 References

1. P.-L. Chen, I-W. Chen, J. Am. Ceram. Soc. 1993, 76, 1577-1583.
2. Y. C. Zhou, M. N. Rahaman, J. Mater. Res. 1993, 8, 1680-1686.
3. Y. Zhou, R. J. Phillips, J. A. Switzer, J. Am. Ceram. Soc. 1995, 78, 981-985.
4. W. H. Rhodes, J. Am. Ceram. Soc. 1981, 64, 19-22.
5. M. J. Mayo, D.-J. Chen, D. C. Hague in Nanomaterials: Synthesis, Properties and Applications (Eds.: A.S. Edelstein, R.C. Cammarata), The Institute of Physics, Bristol, 1996, Chapter 8.
6. G. D. Parfitt in Dispersions of Powders in Liquids, 3rd ed., Applied Science, London, 1981, p. 1.
7. FORMFIT, R. Haberkorn, Universität des Saarlandes, FB 11 Analytische Chemie.
8. E. D. Hondros, M. P. Seah, S. Hofmann, P. Lejcek in Physical Metallurgy, 4th ed., (Eds.: R. W. Cahn, P. Haasen) North Holland, Amsterdam, 1996, p. 1201.
9. J. W. Cahn, Acta Metall. 1962, 10, 789- .
10. P.-L. Chen, I-W. Chen, J. Am. Ceram. Soc. 1996, 79, 1793-1800.

Direct Coagulation Casting and Gel-Casting: Two Innovative Ceramic Forming Technologies[1]

Wolfgang Burger*
CeramTec AG, Plochingen
Andreas Krell
FHG-IKTS, Dresden
Dietrich Stock
Kliniken HEH, Braunschweig
Lutz Claes
Unfallchir. Forsch. u. Biomech., Ulm

1 Introduction

The material properties of high purity alumina have been improved significantly during the past 20 years. This effect has to be related mainly to the development of new, chemically derived, high purity alumina powders and also to improved sintering technologies. Alumina powders for high strength alumina ceramics are no more based on Bayer-process derived powders, because the chemical impurity level of such powders is relatively high. Even purification processes of the hydrates do not lead to a purity required i. e. for biomedical applications. In a very first approach to high purity powders single crystal corundum particles have been milled off to a grain-size of about 0,6 μm and purified by leaching in hydrofluoric acid [1]. High purity alumina ceramics based on these powders have been qualified in 1974 for biomedical applications as ball-heads in THR-prosthesis, which became well known under the trade mark Biolox®. By the synthesis of salts like alaun $NH_4Al(SO_4)_2{*}12H_2O$, aluminiumchloride $[Al(H_2O)_6]Cl_3$ or alcoxides $Al(OR)_3$ high purity chemical precusor salts for the direct production of alumina powders became available. These precursor salts are decomposed chemically to very well defined hydroxides followed by calcination in order to achieve the very stable and reproducible α-phase of alumina as well as a controlled morphology.

Besides the significant improvements in the raw materials powders, body preparation methods and pressing technologies have been improved, too. However new sintering technologies, especially the hot isostatic pressing (HIP) technique applied to alumina has been very effective for further improvement of the mechanical properties. During the HIP treatment microstructural defects of a certain size are eliminated. However big microstructural defects cannot be eliminated by this method. The average mechanical strength is increased, but the Weibull modulus may be lowered compared to sintered parts in an oxidizing atmosphere. As a consequence of this, the safety of ceramic parts manufactured by this technology should have been proof-tested in order to reject those ceramic parts with large defects and holes within the microstructure.

[1] This work has been funded by the German research ministry under project No. 03N4010

Although the material properties have been improved significantly, there are still limitation for the general application of high strength ceramic materials due to the limited and expensive forming technologies. In addition, a high Weibull modulus is required for many applications, where proof-testing is too expensive. Up to date classical powder technologies are applied for the manufacturing of ceramic parts: preparation of ready-to-press powders, pressing, sintering.

Injection molding has been qualified about 40 years ago for small and complex parts. However up to date this processing technology is limited for thick-walled parts due to de-bindering problems. Recently a method related to injection molding, Gel-Casting has been proposed for the manufacturing of complex ceramic parts [2]. Additional near-net-shape forming technologies have been published during the past years, which are based on coagulation of a slurry by changing the surface properties of the particles by chemical reactions: Direct Coagulation Casting (DCC, 1992), Hydrolysis Assisted Solidifaction (HAS, 1995), Temperature Induced Forming (TIF, 1997) [3-5].

With respect to biomedical applications, i. e. ceramic knee prosthesis, near-net-shape forming technologies are required and furthermore chemically derived high purity alumina powders have to be qualified. In order to achieve further improvements of the materials properties, the grain-size of the sintered parts have to be reduced further. Since pressing of extremely fine grained powders is difficult, all processes starting from a suspension are attractive in order to achieve defect-free microstructures with sub-micron grain-size.

2 Forming Methods

2.1 Injection Molding (IM) and Gel-Casting (GC)

Injection molding is well known in the ceramic industry for many years. Today it's main application is related to small and complex ceramic parts. A typical flow-chart of this forming technology is shown in figure 1. The composition consisting of a high volume load of powder and organic polymers are homogenized by typical mixing processes at higher temperatures. After cooling down, granules are formed. For injection the mixture is re-heated to a temperature of about 130 °C and cast into a mould. After cooling down again, the plastic part is easy to remove from the mold. De-bindering takes place under controlled atmosphere and temperature. In this processing step all organic components are removed from the part. Finally sintering takes place.

As it is shown in figure 2, Gel-Casting is related to injection molding. However this process starts usually from a water-based suspension, which contains the inorganic powder particles and organic monomers. After homogenization an initiator is added and cast into a mold. The polymerization reaction is started by either adding a catalyst or by higher temperatures of about 60 °C. After finalization of the reaction the part is de-molded, de-bindered and sintered.

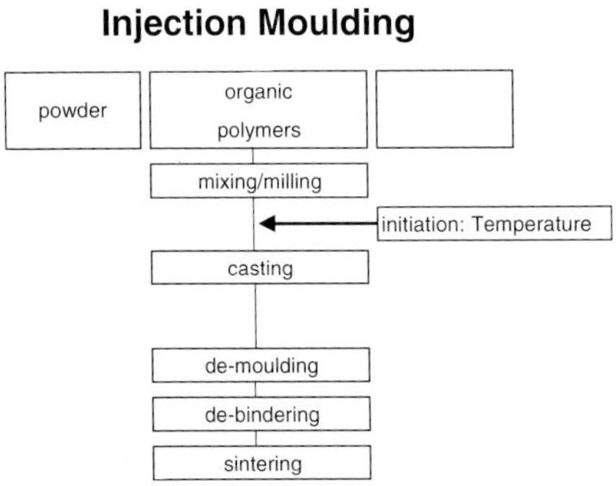

Figure 1. Flow-chart of injection molding process

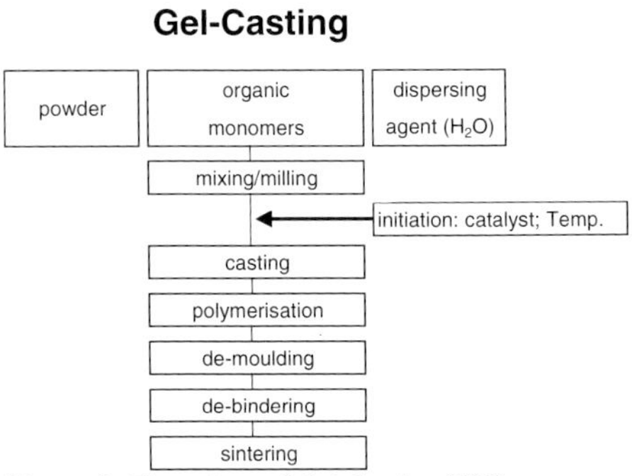

Figure 2. Flow-chart of Gel-Casting (GC) process

2.2 Direct Coagulation Casting

The coagulation of slurries is based on the DLVO theory [6]. According to this theory in a slurry stabilized at pH4 the powder particles are positively charged. Next to the positive charged ions the corresponding negative ions are placed in order to reach neutrality. Besides this so-called "Stern-layer" there is a diffuse double layer in between two particles as long as the slurry is dispersed. Schematically this is demonstrated in the figure 3 [6]. As long as the surface charge of the particles is strong enough the electrostatic repulsive forces are strong and the slurry remains dispersed. Repulsive forces between the particles are lowered by reduction of the surface charges. At the isoelectric point there are no surface charges and in this case the van-der-Waals attractive forces can act. The strength of these forces are directly depending on the distance between the particles and in addition on the ionic strength and the Stern-potential as well as depending from the particle size. Figure 4 describes the situation for an ionic strength of 0,06 mol/l, a Stern-potential of 0,12 V and a particle size of 0,5μm [6].

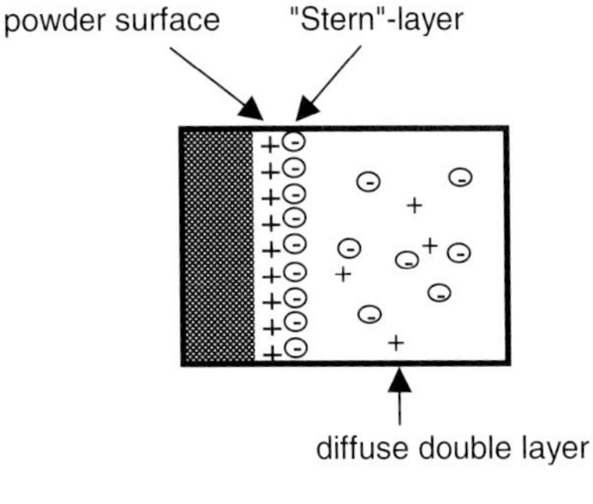

Figure 3. Surface charge of particles

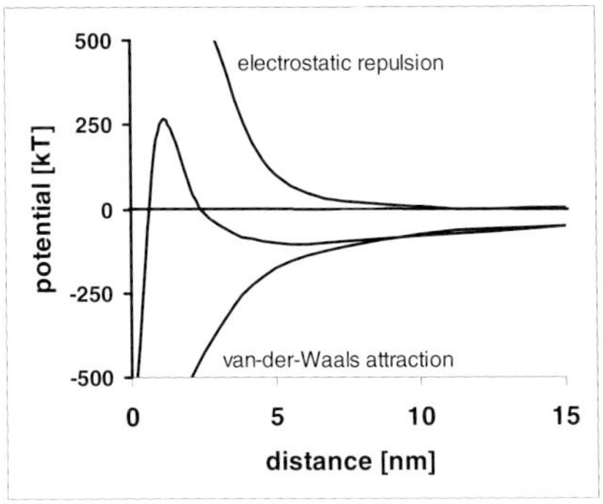

Figure 4. Attractive and repulsive forces in a water-based Al_2O_3 powder dispersion

Based on these principles and the overlapping curve in figure 4, it is demonstrated that in inter-particle distances between the secondary minimum and 15 nm the overall potential is negative and this means the slurry is coagulated. In the case of the primary maximum the suspension is stable against coagulation. The stability of a suspension is influenced significantly by the ionic strength. Small ionic strength values guarantee a high stability against coagulation, while high ionic strength values at a critical concentration lead to coagulation. This means that coagulation may be achieved by changing the ionic strength.

Another possibility to reach coagulation is in the reduction of the surface charge of the suspension by shifting the pH-value. Both possibilities for coagulation are demonstrated in figures 5 [6].

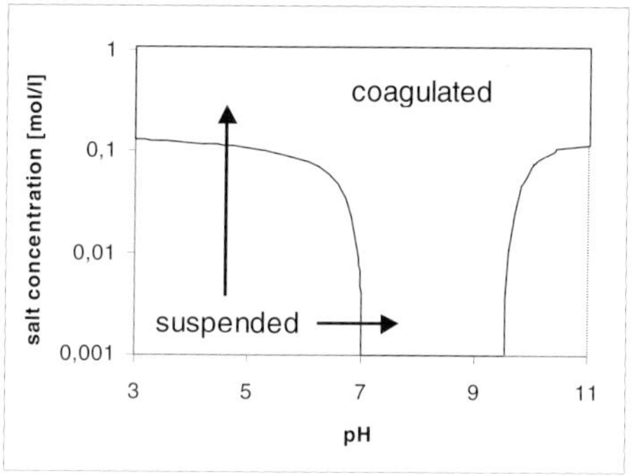

Figure 5. Coagulation of supspensions

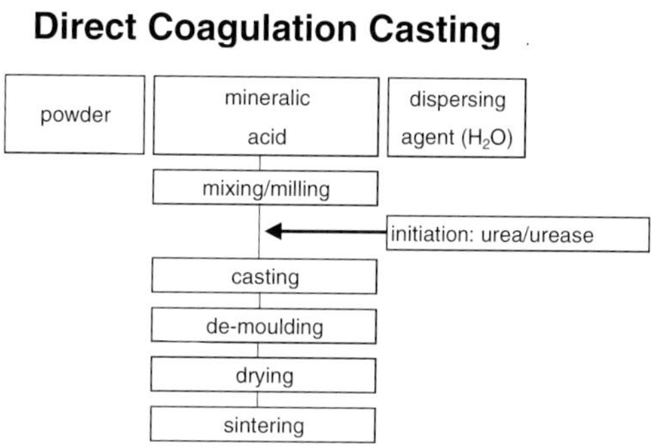

Figure 6. Flow-chart of DCC process

In principle all coagulation technologies are based on these principles. In the HAS-process ionic strength is changed by the temperature induced oxidation of AlN and the forming of hydroxides and ammonia, shifting the pH of the suspension. TIF forming technology is based on the temperature dependent solubility of hydroxides and the forming of new hydroxides especially in the alkaline pH. Furthermore by application of the temperature the surface of the particles is influenced by substitution of small citric ions by polymeric arcrylic chains. Even in a water-based slurry stabilized at pH 4, which contains urea a thermal decomposition of urea, which occurs at about 100 °C leads to a coagulation reaction by shifting the pH due to ammonia forming. From chemical synthesis routes it is well known that temperature may cause secondary exothermal reactions and temperature control and homogeneity overall the part to be coagulated is difficult to realize in practice.

DCC, which has been published first in 1992, is an enzyme catalyzed forming technology, which may be operated at room temperature or even lower. For highly loaded alumina slurries, stabilized at pH 4 the coagulation is induced by adding urea and the enzyme urease. A homogeneous distribution of urea and urease guarantees the homogeneous decomposition of urea starting from a lot of different centers. This finally leads to a very homogeneous defect free green body after coagulation. A flow chart of the principle process is shown in figure 6.

3 Materials and Methods

The experiments have been based on a selected high purity alumina powder derived by thermal-dissolving method from $NH_4AlCO_3(OH)_2$. The raw material powder has been characterized by established methods: chemical analysis by X-ray spectroscopy (Siemens SRS300) and mass spectroscopy (GCE-AET/Kratos MS7, GB); crystallographic analysis by

X-ray diffraction (Siemens D500); specific surface area by BET method (Micromeritics ASAP2000) and grain-size distribution by laser scattering (Malvern Instruments, GB).

Preparation of the slurry for Gel-Casting was made according the following procedure: adjust pH 4 in water by adding nitric acid, addition of monomers, addition of alumina powder up to 43 Vol-% loading, desagglomeration and homogenization of the suspension, addition of initiator, casting into a mold, degassing, raising up the temperature to 60 °C, deforming, debindering and sintering. For the evaluation of the mechanical strength values 4-point test-bars according DIN EN 841-1 from plates by sawing and grinding the parts to the required dimensions of 4x3x40 mm³.

The slurry preparation for DCC was according to the following method: the pH of distilled water was adjusted to pH 4 using HCl, alumina was added slightly under stirring until 55 Vol-% were reached. This suspension was filled into a ball mill. After milling for 24 hours desagglomeration was completed. The remaining slurry was sieved. Then urea and urease were added. After de-gassing the slurry was cast into a mold. After coagulation took place, the wet green body was removed from the mold, dried under defined atmosphere and finally sintered in an electrically heated furnace. Test-bars have been cast directly into a mold, deformed after coagulation and ground after sintering to the required dimensions for 4-point-testing. 4-point-testing has been performed for both materials in the same equipment under identical conditions.

4 Results

4.1 Powder Characterization

The chemical impurity level of the selected powder is summarized in table 1. As it can be taken from this table, the overall impurity level is much lower than 100 ppm, the highest single value obtained in the chemical analysis is Si with 12 ppm. A loss of ignition of 0,5% has been detected. The diffraction pattern of the X-ray diffraction pattern is related to α-alumina. The intensity of the pattern observed was quite high. In the particle size analysis a very narrow distribution of the particles has been observed with d_{90} = 0,44 μm, d_{50} = 0,22 μm and d_{10} = 0,11 μm. The particle size of about 0,2 μm has been confirmed by SEM images, too. The specific surface area of the powder was measured with 13,5 m²/g.

Table 1. Chemical impurities in α-Al_2O_3 powder

element	Na [ppm]	K [ppm]	Mg [ppm]	Ca [ppm]	Fe [ppm]
impurity level	3	3	< 15	4	6
element	Si [ppm]	Ti [ppm]	Cr [ppm]	Ni [ppm]	Mn [ppm]
impurity level	12	2	2	3	< 1
element	V [ppm]	Cu [ppm]	Zn [ppm]	Zr [ppm]	Ga [ppm]
impurity level	8	3	3	3	2

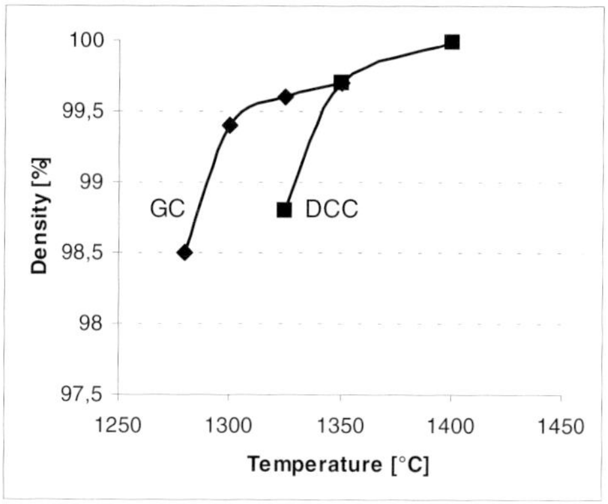

Figure 7. Sintering behavior of samples

4.2 Sintering behavior

Sintering took place already at relatively low temperatures. Densities of more than 99,5 % have been achieved already at a temperature of 1300 °C for GC-samples. Surprisingly, samples prepared by Gel-Casting showed a higher sintering activity than the DCC samples at lower temperatures. At a sintering temperature of 1350 °C DCC and GC samples have shown similar density values. Figure 7 summarizes the sintering results.

4.3 Microstructural and mechanical results

Gel-Casting test-bars prepared for microstructural and mechanical investigations have shown a density of 3,94 g/cm³ - 3,96 g/cm³. The typical microstructure of the sintered body is shown in figure 8. From this figure it is clearly seen that the mean grain-size is in the sub-micrometer region. The quantification of the microstructure has shown a mean grain-size of 0,66 μm with minimum grains of a size of 0,27 μm and maximum grains with 1,95 μm. The results of the 4-point bending test have shown a mean strength of 520 MPa with a corresponding Weibull modulus of 9. A minimum strength level of 385 MPa and a maximum value of 637 have been measured.

The DCC-samples prepared for mechanical and microstructural investigations have shown densities between 3,965 g/cm³ and 3,975 g/cm³. In the samples a mean grain-size of 0,9 μm has been measured. Maximum grains up to 2,63 μm have been detected during the grain-size analysis, while the minimum grains remained at 0,28 μm. A mean strength of 620 MPa at a Weibull modulus of 5 has been detected. A minimum strength 315 MPa and a maximum value of 825 MPa have been obtained.

5 Discussion

The high purity commercially available alumina powder is characterized in it's extremely fine grain-size, high crystallinity and α-content. Based on these properties sintering takes place at relatively low temperatures. However the sintering interval is very narrow. Similar densities have been measured for samples sintered at 1350 °C. From the results summarized in figure 7 it has to be concluded that at lower sintering temperatures GC-processes samples show a higher sintering activity than DCC-samples. This effect is to be related to remaining organic components within the raw material, characterized in the relatively high loss of ignition for such a highly crystallized powder and it's smelling of organic components, probably used as milling aid during powder production.

Figure 8. Microstructure of GC-sample

Figure 9. Microstructure of DCC-sample

The higher sintering activity in the low temperature region of GC-samples finally leads to a smaller grain-size of about 0,7 μm compared to 0,9 μm derived by DCC. On the other hand, GC-samples used for mechanical investigations have shown a higher variation of density compared to DCC-samples. Taking into account the density variation in GC-samples from 98,8 % - 99,3 % it has to be concluded further that the asymptotic region for sintering has not been achieved. Opposite to these results, all DCC test-bars have shown a density of more than 99,4 %. Big defects within the microstructure of both materials have not been observed by SEM-analysis and the microstructure is uniform.

The mechanical strength level of the test-bars obtained by GC is on a reasonable height. Although this material has been sintered only under atmospheric pressure, the strength level achieved is comparable to conventionally processed parts after a hot isostatic densification (HIP). DCC-test-bars have shown a significantly high mechanical strength of 620 MPa. Single values of more than 800 MPa have been measured, too. However some test-bars have shown very weak lower levels in between 300 MPa – 400 MPa. Fracture analysis of these samples have shown cracks within the surface. These cracks may be induced either during the hard-machining process, where these very high density sub-micron materials are difficult to be machined or during de-forming from the mold. Meanwhile it has been shown that de-molding and further handling of the wet samples is critical with respect to induce cracks on the surface. Such an effect is avoided, when test-bars are prepared from higher volume plates, as done for the GC-samples. Further work will have to be spent to de-molding and handling of wet samples in order to avoid such defects, which finally will result in a higher Weibull modulus and also to higher mean strength values.

6 Literature

[1] E. Dörre, personal communication
[2] A. C. Young, O.O. Omatete, M. A. Janney, J. Am. Ceram. Soc., 1991, 74, 612 – 618
[3] T. J. Graule, F. H. Baader, L. J. Gauckler, Ber. CFI, 1994, 71, 317 – 323
[4] T. Kosmač, S. Novak, M. Sajko, J. Europ. Ceram. Soc., 1997, 17, 427 – 432
[5] N. S. Bell, L. Wang, W. M. Sigmund, F. Aldinger, to be published
[6] F. H. Baader, Dissertation ETH Zürich, 1995

Molecular Routes to Advanced Materials: Synthesis of $NdAlO_3$ Ceramic and $NdAlO_3/Al_2O_3$ Composite From Single Source Precursors

Sanjay Mathur,[a] Michael Veith,[a] Nicolas Lecerf,[a] Nicole Donia,[a] Volker Huch,[a] Hao Shen [b] and Stefan Hüfner [b]

[a] Institute of Inorganic Chemistry,

[b] Institute of Experimental Physics

University of Saarland, D-66041 Saarbrücken, Germany

1 Introduction

The success of chemical synthesis routes is largely attributed to the availability of molecular derivatives which can be transformed via solution (Sol-Gel) or gas phase (CVD) reactions into high-purity ceramics or composites at temperatures lower than those required for the conventional (mixing, grinding and calcining) methods.[1,2] Among the solution based methods (sol-gel, micro-emulsion, coprecipitation, etc.), the sol-gel process based on the hydrolytic decomposition of metal alkoxides has been successfully used for the synthesis of a large number of monometal oxide phases (Al_2O_3, TiO_2, ZrO_2, CeO_2, Nb_2O_5, etc).[3] However, the extension of this molecular strategy to multicomponent oxides is limited to a few systems(e.g., $MgAl_2O_4$, $BaTiO_3$, $LiNbO_3$), mainly due to the (mis)belief that the heterometal molecule would break-down into the homometal constituents during the hydrolysis. Moreover, the extension of molecular routes to multi-metallic ceramics demands well-defined molecules with metal ratios compatible to the targeted material.[4]

To this end, we have shown that precursors with a predefined reaction (decomposition) chemistry enforce a molecular level homogeneity in the obtained materials.[5] We report here, on the *single source* sol-gel processing of $NdAlO_3$ ceramic and $NdAlO_3/Al_2O_3$ ceramic-ceramic composite which have important applications as solid-state laser materials, phosphors for fluorescent lighting, optical amplifiers and non-reactive substrate for high-temperature superconductors.[6,7]

2 Experimental

2.1 General and Instrumentation

The synthesis of alkoxides was performed using standard vacuum line techniques and taking stringent precautions against atmospheric moisture. For the sol-gel synthesis of the ceramics, the appropriate precursor, dissolved in isopropyl alcohol, was activated by a controlled hydrolysis (Alkoxide/Water < 3). The sols were stirred (4-6 h) at 50 °C to promote the condensation reactions and then allowed to stand overnight which resulted in the formation of monolithic gels, for both the precursor systems. The green powders obtained after the removal of solvent were calcined, in air, at different temperature to obtain the crystalline ceramics.

The TGA/DTA measurements were performed, on a STA 490 analyser (Netzsch Germany), in quartz crucibles which also served as the reference. Powder X-ray diffraction measurements were performed at room temperature on a Siemens D-500 diffractometer operating with a Cu Kα radiation. The SEM and EDX analysis were performed on CAM SCAN S4 microscope. The XPS data were recorded on a Surface Science Instrument SSI-M-Probe using Al Kα radiation. The data were referenced to O 1s at a binding energy of 530.5 eV. Optical absorption measurements were carried out using a BENTHAM double-monochromator (1.35nm/mm dispersion) and a 250 W halogen lamp as the light source.

3 Results and Discussion

3.1 Single Source Precursor Concept

The synthesis of mixed-metal ceramics is commonly performed from simple and readily available raw materials (inorganic salts, oxides, metal-organic compounds, etc) with the consequence that purity and homogeneity is difficult to control. The atomic mixing of the components is generally not achieved due to different chemical behavior of the components. For example, inhomogeneity can occur due to the differential hydrolysis tendency (an intrinsic property governed by the electropositivity of the metal) of individual components in a mixture of alkoxide precursors. These inherent difficulties lead to phase separation, imprecise control of metal stoichiometry and poor reproducibility. A major development, in this context, is the use of ***single-source precursors*** where the requirements of the material are incorporated in a ***single molecule***.[5] Some of the salient features of this approach are: (i) the possibility of tuning the metal and organic content in the molecule according to the requirements of the material, (ii) low decomposition temperatures since the metals are already mixed on a molecular scale and no diffusion of ions is necessary (iii) nearly carbon-free ceramics or composites because of the neat decomposition processes. In the context of $NdAlO_3$ and $NdAlO_3/Al_2O_3$ systems, ideal precursors would be heterometal assemblies based on Nd-Al ratios 1:1 and 1:3, respectively.

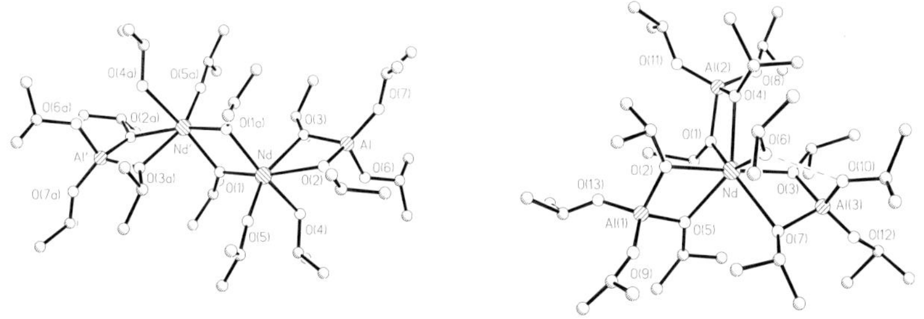

Figure 1. Molecular structures of the mixed-metal Nd-Al alkoxides **1** and **2**

The molecular structure of the two precursors, confirming the heterometal nature and the metal stoichiometry, are shown in Figure 1. The synthesis and structural details of the heterometal precursors [$\{NdAl(OPr^i)_6(Pr^iOH)\}_2$] (**1**) and [$NdAl_3(OPr^i)_{12}(Pr^iOH)$] (**2**) will be published elsewhere.

3.2 Sol-Gel Process

The sol-gel technique using alkoxides is based on inorganic polymerisation reactions in which partially hydrolysed alkoxide species condense to form multi-molecular aggregates assembled via oxo- or hydroxo groups. Further cross-linking of such building blocks favors the gelation process. Since alcohol is the only organic source in the gels, carbon-free ceramics are obtained at quite low temperatures (700 °C). The overall sol-gel reaction path starting from the source molecule $[NdAl(OPr^i)_6(Pr^iOH)]_2$ to the $NdAlO_3$ phase is represented in equation 1.

$$\tfrac{1}{2}\,\mathbf{[NdAl(OPr^i)_6(Pr^iOH)]_2} \xrightarrow{\text{Hydrolysis}} \text{'}NdAl(OH)_x(OPr^i)_{6-x}\text{'} \xrightarrow{\text{Alcoholation/ Oxolation}}$$

$$\text{'}NdAl(OH)_x(O)_y(OPr^i)_{6-x-y/2}\text{'} \;(z \ll x \text{ and } y) \xrightarrow{\text{Polycondensation}} NdAl(OH)_x(O)_y(OPr^i)_z\text{'} \xrightarrow{\text{Calcination}} \mathbf{NdAlO_3} \qquad \text{..... (1)}$$

Owing to the presence of Nd–O(R)–Al bridges in **1**, the hydrolytic decomposition and the formation of $NdAlO_3$ ceramic occurs at a molecular level and crystalline ceramic is obtained at significantly low temperature (800 °C). However, the sol-gel synthesis of $NdAlO_3$ phase from a mixture of neodymium and aluminum isopropoxides[8] produced crystalline ceramics at much higher temperature (1150 °C). The solution of alkoxide species are known to be composed of several species present in complex equilibria. We believe that the isolation of heterometal precursor is a key step for obtaining pure ceramics at low temperatures.[5b] In contrast to **1,** the hydrolysis of **2** gave, after calcination, a mixed-metal ceramic of formulation $NdAlO_3/Al_2O_3$ (eq. 2) which is a novel example of a biphasic nano/nano composite formation from a single source precursor.

$$\mathbf{[NdAl_3(OPr^i)_{12}(Pr^iOH)]} \xrightarrow{\text{Hydrolysis}} \mathbf{NdAlO_3/Al_2O_3} \qquad (2)$$

3.3 Characterisation of the Powders

EDX analysis of the gels and powders gave neodymium:aluminium ratios corresponding to those present in the molecular precursors giving the compositions $NdAlO_3$ and $NdAl_3O_6$. This observation was supported by the determination of Nd and Al contents by chemical methods which confirmed the intended stoichiometry. The thermal behaviours (Fig. 2) of the dried gels obtained from **1** and **2** was investigated by thermogravimetry (TG) and differential thermal analysis (DTA). A small weight loss below 150 °C corresponds possibly to the removal of adsorbed and chemically bonded water and alcohol molecules in the gels. The TG curves in both the cases demonstrate that the major weight loss corresponding to thermal decomposition of precursor, as indicated by prominent exotherms in this region, occurs in a single step (up to 400 °C). The observed decomposition process supports our contention that the single source properties of the molecular precursors are carried forward to the gels resulting in the formation of a green body where Nd and Al centers are chemically linked. The exothermic peaks observed around 810 and 840 °C indicates the crystallisation of $NdAlO_3$ phase in the gels obtained from **1** and **2**, respectively. The slightly higher crystallisation temperature of $NdAlO_3$ ceramic in the later system may be an influence of the alumina matrix. In the case of **2**, a second exothermic peak is observed around 1300 °C (Fig. 2) and, since no corresponding weight loss is observed, this can be assigned to the crystallization of amorphous alumina in $NdAlO_3/Al_2O_3$ composite. This observation is supported by the XRD measurements (see later) on the $NdAl_3O_6$ sample heat treated at 1400 °C.

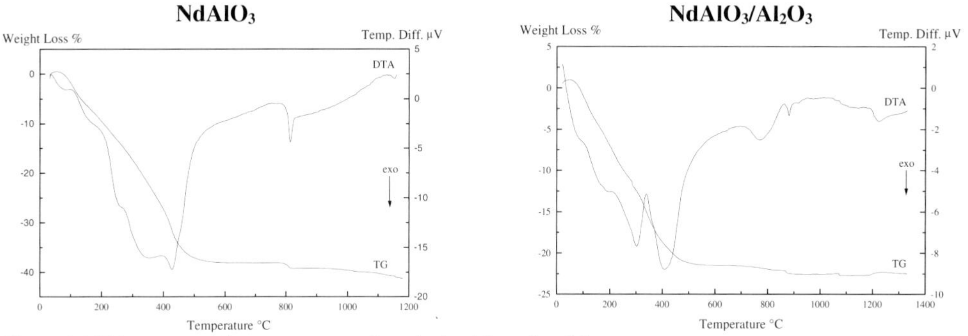

Figure 2. DTA curves of the precursor powders obtained from **1** and **2**

The XRD studies were performed on precursor and heat treated (600–1400 °C) powders of both the systems. Representative XRD traces are shown in Fig. 3. The precursors are X-ray amorphous up to 600 °C however, broad peaks indicative of the initiation of crystallisation and some local ordering are observed at this temperature. The neodymium aluminate precursor powder calcined at 800 °C produced monophasic $NdAlO_3$ (~ 20 nm, Scherer's Formula). The XRD patterns at higher temperatures reflected the crystallite growth and no formation of any undesired phase was observed till 1400 °C (Fig. 3). The precursor obtained from **2** showed, after calcination at 800 °C, $NdAlO_3$ to be the only crystalline phase. Since the formal composition of the oxide material was determined (EDX and chemical analysis data) to be $NdAl_3O_6$, the residual amorphous content, was expected from stoichiometric balance, to be Al_2O_3. In fact, the crystallisation of transition aluminas (mixture of γ, κ and δ phases) was observed around 1300 °C (Fig. 3). The two oxides are present as discrete phases even at 1500 °C and no sign of the formation (by solid state reactions) of phases based on other Nd:Al ratios was observed. This shows a new facet of molecular precursor approach to achieve *nanoheterogeneity* in ceramic composites.

The transmission electron micrographs (not shown) for both the systems revealed nano-crystalline materials with a narrow size distribution. In the case of $NdAlO_3/Al_2O_3$ composite, well-faceted grains of $NdAlO_3$ (ca. 150 nm) were found to be regularly dispersed in an alumina matrix made up of tiny crystallites (ca. 20 nm). The average grain sizes were close to the values determined by Scherer's Formula.

The X-ray photoelectron spectroscopy corrobrated the elemental composition andexistence of two chemically different Al species observed in XRD measurements, in the case of **2**. Figure 4 shows the Al 2p peaks of $NdAlO_3$, $NdAlO_3/Al_2O_3$ and Al_2O_3 (reference sample) ceramics. In comparison to $NdAlO_3$, the Al 2p peak in $NdAlO_3/Al_2O_3$ composite is asymmetric in nature. A deconvolution of the peak formrevealed the presence of a second component (Fig. 4). The presence of two components in 1:2 ratio can be attributed to the aluminum ions located in different chemical surrounding, namely $NdAlO_3$ and Al_2O_3.

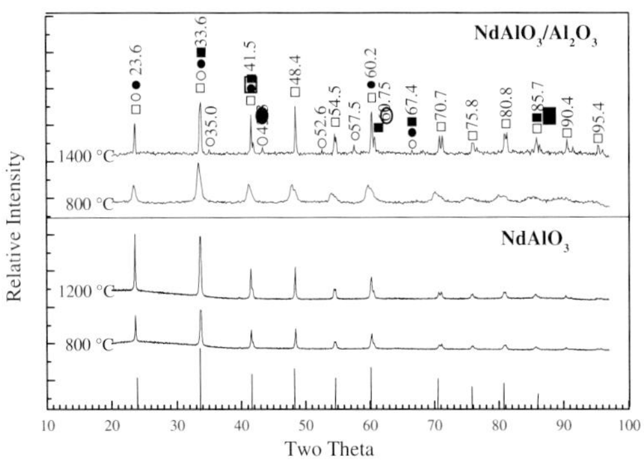

Figure 3. Room temperature XRD patterns of $NdAlO_3$ and $NdAlO_3/Al_2O_3$ ceramics at various calcination temperatures. The vertical lines correspond to the JCPDS indexing for $NdAlO_3$ ([29-56]); the alumina phases detected at 1400 °C are γ, κ and δ (JCPDS card nos.[10-425], [04-878] and [04-877]).

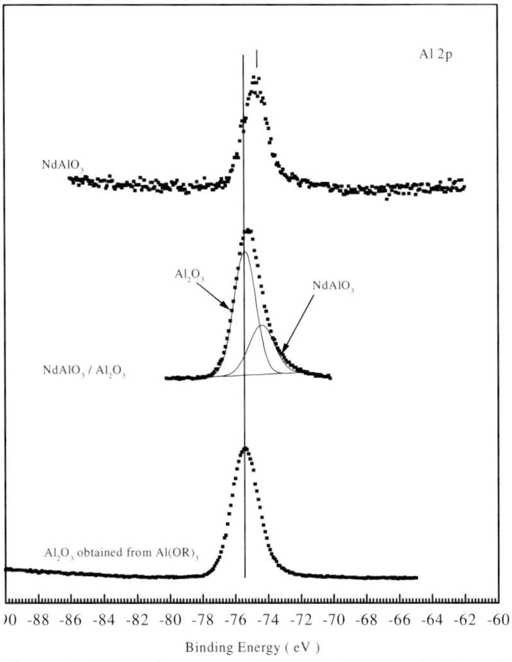

Figure 4. XPS Al 2p spectra of $NdAlO_3$, $NdAlO_3$:Al_2O_3 and Al_2O_3 ceramics

The absorption spectra of $NdAlO_3$, $NdAlO_3/Al_2O_3$ and Nd doped YAG ceramics, recorded (350–850 nm) at room temperature, are shown in Figure 5. The absorption intensity of the ceramics containing $NdAlO_3$ phases is much higher than that of Nd:YAG possibly because of the lower neodymium ion concentration in YAG system. The difference between the lattice parameters of YAG and $NdAlO_3$ is not large. Furthermore, the nearest neighbors of the Nd^{3+} ions in YAG are the same as in $NdAlO_3$ and it might be considered that the crystal field acting upon the neodymium ions in these two crystals are nearly the same. Experimental studies[9] show that the crystal field upon the rare earth ions in garnets depends strongly on the nature of the next-nearest neighbours which explains a similar absorption behavior for the two different systems.

A comparison of the absorption spectra of $NdAlO_3$ and $NdAlO_3/Al_2O_3$ ceramics reveals that the absorption coefficient for the transition from $^4I_{9/2}$ multiplet ground state to higher energy states ($^4F_{9/2}$ or $^2H_{11/2}$ or $^4G_{7/2}$ + $^4G_{5/2}$ or ^{2} $K_{13/2}$+ $^4G_{7/2}$ + $^4G_{9/2}$)is much larger for the $NdAlO_3$ crystallites incorporated in an Al_2O_3 matrix than that observed for pure $NdAlO_3$. However, the absorption coefficients of $^4F_{7/2}$+$^4S_{3/2}$ and $^2H_{9/2}$+$^4F_{5/2}$ bands in the two ceramics are nearly the same. The enhanced absorption bands, located at higher energy level of conduction band, can be attributed to the influence of the Al_2O_3 matrix which probably affects the electronic band structure of the embedded $NdAlO_3$ by electron screening and transferring processes.

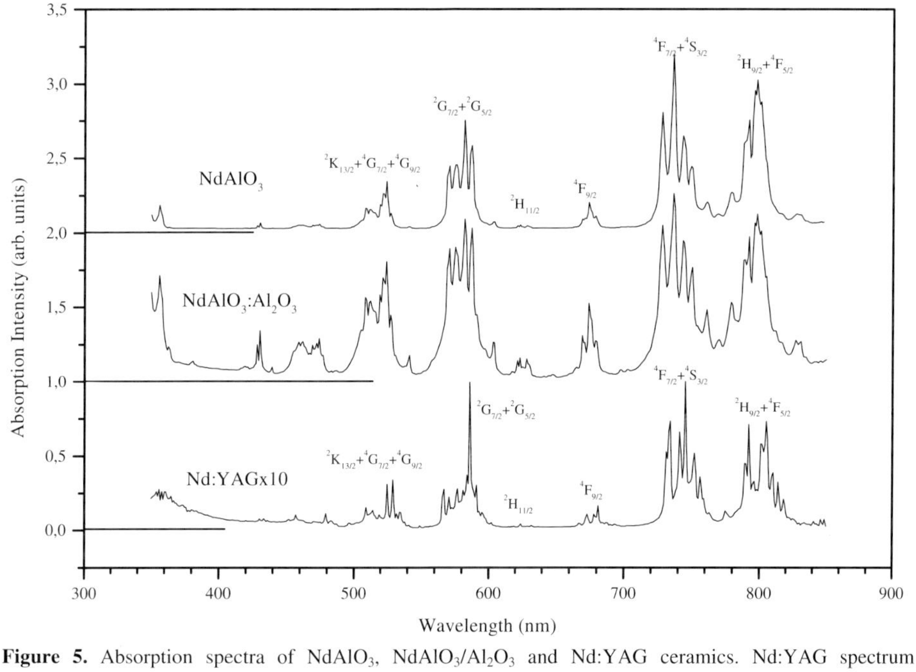

Figure 5. Absorption spectra of $NdAlO_3$, $NdAlO_3/Al_2O_3$ and Nd:YAG ceramics. Nd:YAG spectrum is multiplied (X 10) because of the weaker intensity.

4 Conclusion

Heterometal Nd-Al alkoxides have been used as *single source* for the sol-gel synthesis of nano-scaled $NdAlO_3$ and $NdAlO_3/Al_2O_3$ ceramics. The principal conclusion to be drawn is that the well-defined heterometal alkoxides form gels by a controlled polymerisation without breaking down into separate constituents. Apparently, the retention of Nd – (O) – Al bridges in the gels is responsible for the easy and low-temepratur conversion of green precursors to crystalline ceramics.

5 References

1. L. L. Hench, J. K. West Chem. Rev., 1990, 90, 33.
2. C. Jones Chem. Vap. Deposition 1998, 4, 169.
3. J. Brinker, G. W. Scherer Sol-Gel Science: The Physics and Chemistry of Sol-Gel Processing, Academic Press, San Diego, CA, 1990.
4. M. Veith, S. Mathur, C. Mathur, Polyhedron 1998, 17, 1005.
5. M. Veith; A. Altherr; N. Lecerf; S. Mathur; K. Valtchev, E. Fritscher Nanostructured Materials 1999, 12, 191. (b) F. Meyer; R. Hempelmann; S. Mathur; M. Veith J. Mater. Chem. 1999, 9, 1755.
6. Blasse, G. Chem. Mater. 1989, 1, 294.
7. T. Morelli, J. Mater. Res. 1992, 7, 2492.
8. S. S. Shoup, M. Paranthaman, A. Goyal, E. D. Specht, D. F. Lee, D. M. Kroeger, D. B. Beach, J. Am. Ceram. Soc. 1998, 81, 3019.
9. V. Nekvasil; M. Guillot; A. Marchand, F.Tcheon, J. Phys. C., 1985, 18, 3551.

Deposition of Titania Thin Films on Different Functionalized Organic Self-Assembled Monolayers

Thomas P. Niesen, Joachim Bill, and Fritz Aldinger
Max-Planck-Institut für Metallforschung and Institut für Nichtmetallische Anorganische Materialien, Universität Stuttgart, Stuttgart, GERMANY

1 Abstract

Self-assembled monolayers (SAMs) on single-crystal Si wafers have been used as substrates for the deposition of oxide thin films. The organic surface has been shown to be effective for promoting the growth of films from aqueous solutions at temperatures below 100°C. Titania films have been previously synthesized by hydrolysis of $TiCl_4$ in 6 M HCl solutions at 80 °C on sulfonated SAMs. In the present study, the formation of a titanium chelate complex in presence of H_2O_2 is used to stabilize an otherwise spontaneously precipitating aqueous titanium solution. Uniform titania films have been formed at 80°C on sulfonated SAMs, while hydroxyl and amine functionalities led to inhomogeneous coatings. The films were characterized by a variety of techniques, including ellipsometry, RBS, AFM, and SEM, to determine thickness, topography, microstructure and chemical composition of the films. Based on these and previous results, the current understanding of the film formation is discussed.

2 Introduction

From the viewpoint of energy saving, environmental protection, and unsufficient thermal stability of the underlying substrates, there has been a growing interest of low-temperature materials synthesis during the last decades. Organic self-assembled monolayers (SAMs) have been shown to be effective surfaces for promoting the formation of oxide thin films from aqueous solutions at temperatures below 100°C.[1] SAMs are ordered molecular assemblies formed by the chemical adsorption of an active surfactant on a solid surface. The oxide films – including nano-crystalline TiO_2, ZrO_2, V_2O_5, ZnO, Fe_3O_4, and SnO_2 – are adherent and microstructurally uniform.

Titania is well known for its high refractive index, permittivity and transmittance of visible light. These properties make it suitable for many applications such as dielectric layers in microelectronic devices,[2] as oxygen sensor[3] and as high-efficient catalyst[4]. In previous depositions of titania thin films from aqueous solutions,[5] highly acidic solutions – 0.5 M $TiCl_4$/6 N HCl – were necessary to prevent the solution from spontaneous bulk precipitation. Films with a typical thickness < 100 nm could be formed in 2-4 h on sulfonated SAMs. The films consisted of densely packed, randomly oriented TiO_2 (anatase) crystals, 2–4 nm in diameter. AFM studies revealed uniform coatings with a typical rms roughness value less than 10 nm.[6] Recently, Baskaran *et al.*[7] successfully applied an aqueous titanium lactate solution to deposit titania films on sulfonated SAMs and plastic surfaces with high deposition rates.

We used the formation of an alternative chelate to stabilize an aqueous titanium solution: In presence of hydrogen peroxide, $Ti(O_2)^{2+}$ is formed through η^2 coordination (side-on) of O_2^{2-} to titanium.[8] First results have recently been reported.[9] In the present study, the influence of different functionalized SAMs on the nucleation and growth of titania thin films was investigated.

2 Experimental

2.1 Substrate preparation and SAM depositions

The substrates were p-type {001} "single crystal" Si wafers, polished on one side, approximately 10 mm x 10 mm in size. These wafers were cleaned and oxidized in "piranha" solution (70 vol% of H_2SO_4, 30 vol% of 30 wt% H_2O_2 aqueous solution). The wafers were then dipped into dicyclohexyl containing 1 vol% of the surfactant (trichlorosilylhexadecane thioacetate or -nitrile, and ω-heptadecenyltrichlorosilane), at room temperature under inert atmosphere for several hours, during which the SAM formed spontaneously. The wafers were finally thoroughly washed in chloroform to remove all traces of excess surfactant.

To convert the thioacetate to a sulfonate ($-SO_3H$) functionality, the wafers were immersed in oxone, an aqueous persulfate oxidant, for a minimum of 4 h at room temperature. The nitrile functionality was transformed to an amino ($-NH_2$) functionality by reduction with a saturated solution of lithium aluminium hydride in diethylether for 15 h at room temperature and subsequent quenching with 5 vol-% hydrochloric acid. The vinyl group was hydroborated with 1 M boran-tetrahydrofuran complex solution for 2 h and subsequently quenched to an –OH functionality with 30% H_2O_2/0.1 N NaOH. (Further details of the synthesis of the SAMs on Si can be found elsewhere.[10-12])

2.2 Synthesis of TiO_2 thin films

Freshly distilled $TiCl_4$ (0.1 mol) was added dropwise to an ice-cooled aqueous solution containing H_2O_2 (0.2 mol). Stock solutions of 10 mM $[Ti(O_2)^{2+}]$ and 0.14 N HCl (excluding acid from $TiCl_4$ hydrolysis) were prepared from this solution and hydrochloric acid. The SAM-coated substrates were immersed in 10-ml aliquots of this solution, covered and then placed in an oil bath at 80 °C. Samples were immersed between 4 and 6 h, in vertical orientation (no visible bulk precipitation occured during such depositions). Finally the samples were washed with distilled water, ultrasonicated and dried with a stream of dry argon gas. Control experiments were done with piranha-oxidized Si wafers (*i.e.*, without SAMs).

3 Results

Figure 1 shows the surface topography of a ~100 nm thick titania film deposited in 6 h from 10 mM $[Ti(O_2)^{2+}]$ and 0.14 N HCl at 80 °C on sulfonated SAM. XPS and RBS analysis (data not shown) indicated a fully covered surface, and all spectroscopic data identified the film as titanium (IV) oxide. The AFM image revealed a dense packing of particles of 50–100 nm in width and height variations of 50–60 nm. The statistical roughness parameter remained low (rms = 15–18 nm, depending on the investigated area), indicating the uniformness of the coating's topography.

Figure 2 shows SEM micrographs of titania precipitates deposited under similar conditions on different functionalized SAMs. On sulfonated SAMs, a uniform film of densely packed particles was formed in 4 h. The morphology of the films did not change during further film growth, consistent with the previously described AFM investigation.

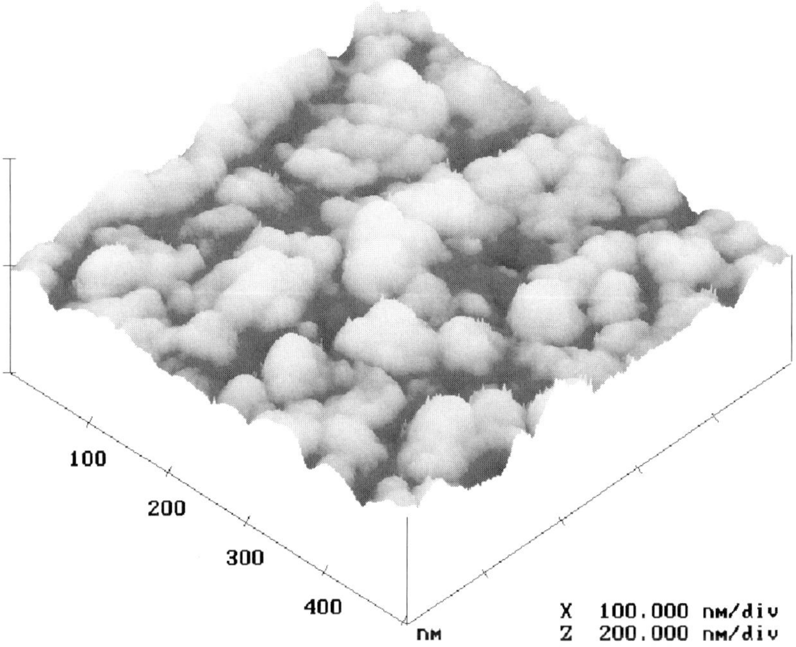

Figure 1. AFM image of a titania thin film on sulfonated SAM, deposited in 6 h at 80 °C, indicating roughness on a sub-micron scale.

In contrast, hydroxyl and amino functionalities led to island-like growth. The islands were 70–200 nm in size after 4 h and seem to grow together if the deposition was continued. The final films consisted of domains and were partially tarnished, while films on sulfonated SAMs were always specular reflecting. No attempts to measure the adherence of the films have been made yet, however, all samples were ultrasonicated before drying to remove loosely attached particles. In contrast, no films were formed on bare silicon.

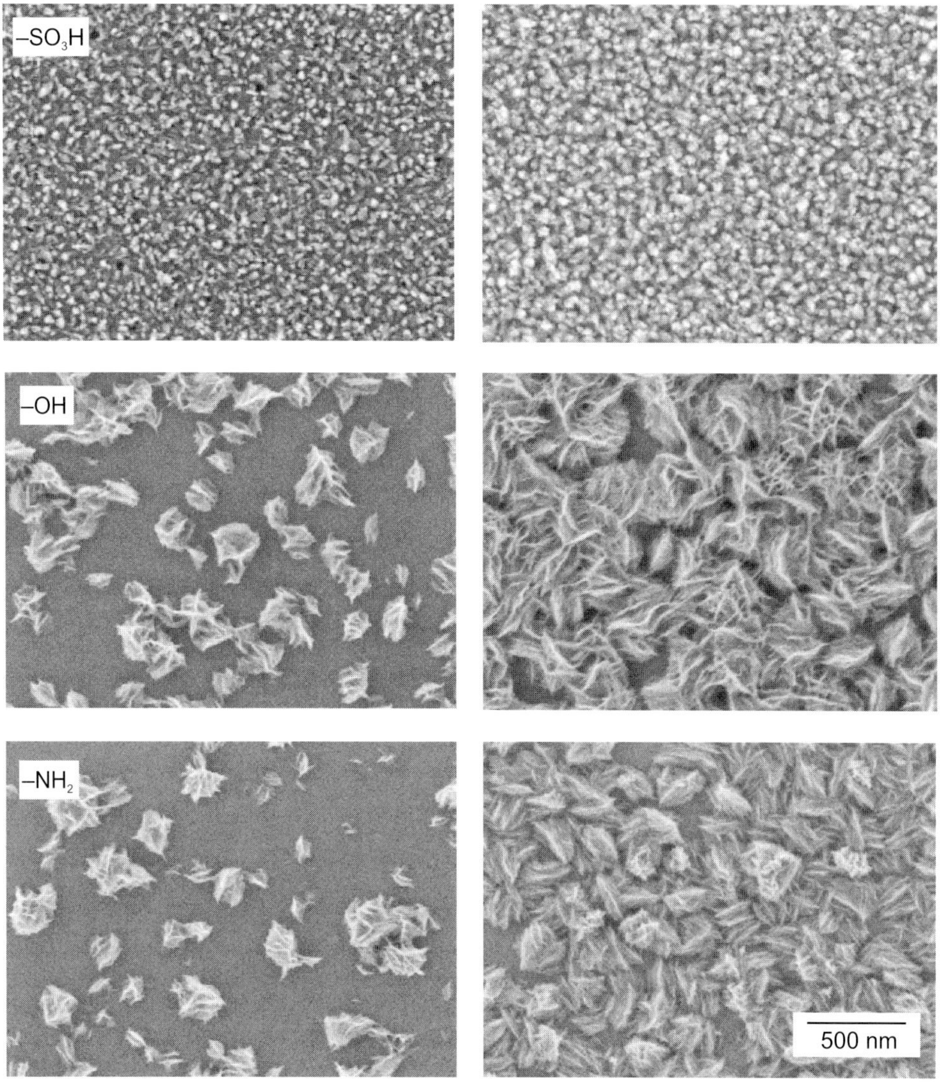

Figure 2. SEM micrographs of TiO_2 thin films on different functionalized SAMs after four (left) and six hours (right) deposition at 80°C.

4 Discussion

Defect- and crack-free titania films with uniform surface roughness on a submicron scale were formed on sulfonated SAMs. In contrast, film formation on hydroxyl- or amine-functionalities started from individual islands and the final films consist of interconnected domains. As is discussed in more detail elsewhere,[13] we believe that films on sulfonated SAMs are formed via attachment to the substrate of nano-sized particles formed in the bulk solution: The net attraction will result from the summation of electrostatic, van-der-Waals and hydration interaction. While van-der-Waals forces are always attractive and hydration forces repulsive, the electrostatic term depends on the functionalization of the surface: For the positively charged titania colloidal particle, the sulfonate group – deprotonated even in acidic solution – leds to an attractive term, while the protonated amine group leds to a repulsive term. (For negatively charged colloids, as recently demonstrated for vanadia,[9, 14] film formation was not possible on $-SO_3H$, but on $-NH_2$.) In case of hydroxyl functionalization, no electrostatic term has to be considered. In comparison, the net attraction is highest for sulfonated SAMs, in agreement with the experimental results: While the film formation starts early on sulfonated SAMs from small particles, on both other SAMs induction time is large and film formation seems to start either from preformed larger agglomerates or, likewise plausible, from nucleation *on* the surface. It has to be pointed out that films were not formed directly on "piranha"-oxidized silicon (*i.e.* also hydroxyl-terminated), indicating a higher reactivity of the SAM's headgroup.

In contast to the present titania films on sulfonated SAMs, films deposited from strong acidic aqueous solution (6 M HCl) showed a background roughness on a finer scale (< 10 nm).[6] Such surface roughness is in agreement with the size of the nanocrystallites that form the films. While these nanoparticles agglomerate in solution during the deposition, the SAM seems to be able to selectively attract individual nanoparticles. Probably larger particles are attracted in the present work. TEM investigations of the film's microstructure are currently in progress.

5 Conclusions

Thin titania films were successfully deposited from an aqueous titanium peroxide solution at 80°C. The predominant effect of the sulfonic acid surface group for promoting the film growth has been shown, which is in agreement with the role of acidic groups in protein structures in biomineralization. The presence of an uncharged or positively charged surface does not generally exclude the film formation because of strong van-der-Waals attraction forces. However, changes in deposition mechanism occur that led to ununiform, partially tarnished films.

6 Acknowledgments

The ongoing cooperation with Profs. A. H. Heuer and M. R. De Guire, Case Western Reserve University, Cleveland/OH, USA, is gratefully acknowledged. The authors thank Mr. H. Labitzke for SEM and Mr. D. Cantarutti for AFM characterization. This work was financially supported by the Deutsche Forschungsgemeinschaft.

7 References

[1] M. R. De Guire, T. P. Niesen, S. Supothina, J. Wolff, J. Bill, C. N. Sukenik, F. Aldinger, A. H. Heuer, M. Rühle, Z. Metallk. 1998, 89, 758-766, and references therein.
[2] G. P. Burns, J. Appl. Phys. 1989, 65, 2095-2097.
[3] M. Nabavi, S. Doeuff, C. Sanchez, J. Livage, Mater. Sci. Eng. 1989, B3, 203-207.
[4] T. Carlson, G. L. Griffin, J. Phys. Chem. 1980, 90, 5896-5900.
[5] H. Shin, R. J. Collins, M. R. De Guire, A. H. Heuer, C. N. Sukenik, J. Mater. Res. 1995, 10, 692-698.
[6] T. P. Niesen, M. R. De Guire, J. Bill, F. Aldinger, M. Rühle, A. Fischer, F. C. Jentoft, R. Schlögl, J. Mater. Res. 1999, 14, 2464-2475.
[7] S. Baskaran, L. Song, J. Liu, Y. L. Chen, G. L. Graff, J. Am. Ceram. Soc. 1998, 81, 401-408.
[8] F. P. Rotzinger, M. Grätzel, Inorg. Chem. 1987, 26, 3704-3708.
[9] T. P. Niesen, J. Wolff, J. Bill, M. R. De Guire, F. Aldinger in Organic-Inorganic Hybrid Materials II (Eds. L. C. Klein, L. F. Francis, M. R. DeGuire, J. Mark), Vol. 576, Materials Research Society, Warrendale, PA, 1999, pp. 197-202.
[10] N. Balachander, C. N. Sukenik, Langmuir 1990, 6, 1621-1627.
[11] R. J. Collins, C. N. Sukenik, Langmuir 1995, 11, 2322-2324.
[12] S. R. Wasserman, Y. T. Tao, G. M. Whitesides, Langmuir 1989, 5, 1074-1087.
[13] H. Shin, M. Agarwal, M. R. De Guire, A. H. Heuer, Acta Mater. 1998, 46, 801-815.
[14] T. P. Niesen, J. Wolff, J. Bill, T. Wagner, F. Aldinger in Proc. 9th CIMTEC-World Forum on New Materials, Symposium III - Surface Engineering (Ed. P. Vincen-cini), Advances in Science and Technology 20, Techna Publishers, Faenza, Italy, 1999, pp. 27-34.

Effect of Li, Ca and Sn Doping on Anatase Stability

Xim Bokhimi, Antonio Morales, Manuel Aguilar
Institute of Physics, The National University of Mexico (UNAM), A. P. 20-364, 01000 Mxico D. F., Mexico
Amado García-Ruiz
UPIICSA, COFAA, The National Polytechnic Institute (IPN), Té No. 950 Esq. Resina, 08400 México D. F., Mexico.

1 Introduction

Titanium oxide has three polymorphs: anatase, brookite and rutile. They all exist in a wide range of temperatures determined by their synthesis conditions; for instance, anatase can exist from 70 to 900 °C [1], bookite only at temperatures lower than 500 °C, but rutile can exist between 70 and 1850 °C [2].

In the literature related to the formation and stability of anatase the data about its transformation are controversial [3-7]. For example, when anatase is prepared by using the sol-gel technique with different hydrolysis catalysts and the samples are annealed at 600 °C [1], those prepared with hydrochloric acid contain only impurities of anatase, but it is the only phase present in those samples prepared with acetic or oxalic acid; the synthesis with ammonium hydroxide produces a mixture of anatase and rutile. Systematic studies in our laboratory have shown that the hydrolysis catalyst not only catalyze hydrolysis and condensation reactions, but they also provide anions that combine with titanium and oxygen to stabilize anatase [8].

Anatase stabilization can also be changed when in the synthesis other cations different to titanium are present. For example, platinum causes the transformation of anatase into rutile even at 70 °C [2], at the same temperature copper promotes the formation of brookite instead of anatase [9]. In both cases the dissolution of the cation in anatase was negligible. Because of anatase importance in electronics and catalysis, it is of interest to study its stability when during its synthesis are present cations soluble in its crystalline structure; for instance, lithium [10], aluminum [11], calcium [12], vanadium [13], niobium[13], ruthenium [14], tungsten [15], and tin [16].

To analyze some of the aspects about anatase stability, we have doped it with three different cations: lithium, which occupies interstitial sites in anatase unit cell, calcium, which substitutes for titanium atoms in anatase but has a smaller valence than titanium, and tin, which also substitutes for titanium atoms, has the same valence, but its stable oxide has a rutile-type crystalline structure. The analysis was performed by using X-ray powder diffraction.

2 Experimental

2.1 Sample preparation

Titanium *n*-buthoxide (Acros 99 %) was dissolved at room temperature in ethanol (Baker 99.9 %). After homogenizing the solution, while stirring, HCl was added to it until obtaining pH 3. Thereafter, the appropriated amount of an aqueous solution of $LiNO_3$ (Baker 99.9%), for doping with lithium, or $SnCl_2.2H_2O$ (Aldrich 98.9 %), for doping with tin, or $CaCO_3$ (Alfa, 99.99 %), for doping with calcium, was slowly added to the above solution drop by drop under continuous stirring until gelling (it typically occurred after 2 h). The water to alkoxide ratio was 2.4, which would produce an oxo reaction of 98 % [17]. Gels were dried at room temperature in a dynamic vacuum of 10 Torr, giving rise to a white powder.

2.2 X-ray diffraction analysis

Crystalline structures were analyzed with X-ray powder diffraction, and refined via the Rietveld method by using DBWS-9411 code [18]. X-ray diffraction patterns were measured at room temperature with a BRUKER D8-ADVANCE diffractometer using CuK_α radiation on specimens prepared by packing sample powder into a glass holder. Intensity was measured by step scanning in the 2θ range between 10° and 110°, with a step of 0.02° and a measuring time of 2 s per point. Diffraction peak profiles were modeled with a pseudo-Voigt function that uses the average crystallite size as a fitting parameter [19]. The standard deviations, given in parenthesis, correspond to the variation of the last figures of the corresponding number. Since these deviations are produced from the refinement, they are not estimates of the analysis as a whole, but only of the minimum possible probable errors based on the normal distribution of the parameters [20]. Before the X-ray analysis, Lithium and calcium doped samples were heated in flowing argon (1.8 L/h) at 200, 400, 600, or 800 °C for 12 h; those doped with tin were annealed in flowing air (1.8 L/h).

3 Results and Discussion

The fresh samples of all three systems were amorphous. This indicates that a good condensation occurred during the synthesis, avoiding the formation of oxide crystals. The presence of the residues associated to the titanium and doping precursors hindered the contact between metal oxide clusters to form crystals, and stabilized the amorphous structure even after annealing the samples at 200 °C.

When the samples were annealed at 300 °C, they crystallized in the characteristic structures of titania polymorphs: anatase, brookite and rutile. It is important to note, however, that the observed phases were not the polymorphs of titania, but, titania solid solutions with the same structure. The concentration of these phases and their crystallography was obtained by refining their crystalline structures, starting from the crystallography reported for titania polymorphs [1].

Anatase solid solutions were modeled with a tetragonal unit cell with the symmetry described by space group $I4_1/amd$. Brookite solid solutions were modeled with an orthorhombic unit cell described by space group *Pbca*; the solid solutions with the rutile structure were also modeled with a tetragonal unit cell, but described by space group

$P4_2/mnm$. For each particular system, the structure of the observed crystalline phases different to the above mentioned ones were also refined. In the following paragraphs we describe in detail each system with emphasis on anatase stability.

3.1 Li_2O/TiO_2 system

For the present study two lithium oxide concentrations were used: 10 and 20 mole %. When the samples were annealed at 400 °C, they contained anatase and brookite (Fig. 1); in the samples with 20 mole % besides anatase, they also had rutile and brookite in large concentrations, indicating that lithium doping in large amounts promotes the formation of these phases. The lithium effect on rutile formation was more evident when the samples were annealed at 500 °C (Fig. 2); in this case the samples with 20 mole % lithium oxide were composed of 94(2) wt % rutile, 3.2(2) wt % $Li_2Ti_2O_4$, and 2.4(2) Li_2TiO_3. Because the concentration of these last two compounds was very small, the tick marks identifying them were not indicated in the corresponding figure.

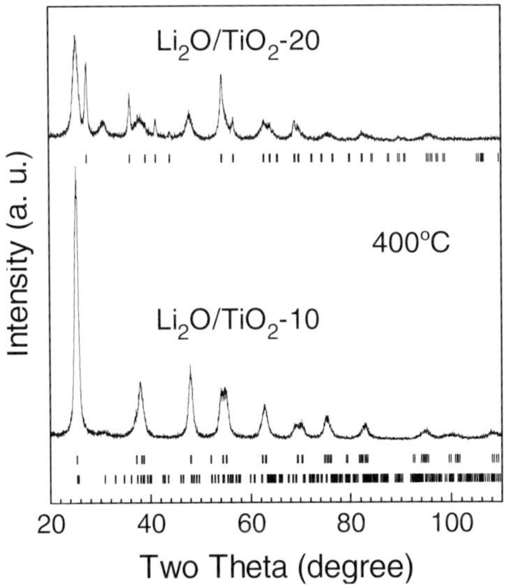

Figure 1 X-ray diffraction patterns of the lithium doped samples after they were annealed at 400 °C in argon. The tick marks under the pattern of the sample with 20 mole % L_2O correspond to rutile. The upper tick marks under the diffraction pattern of the sample with 10 mole % L_2O correspond to anatase; the lower ones, to brookite.

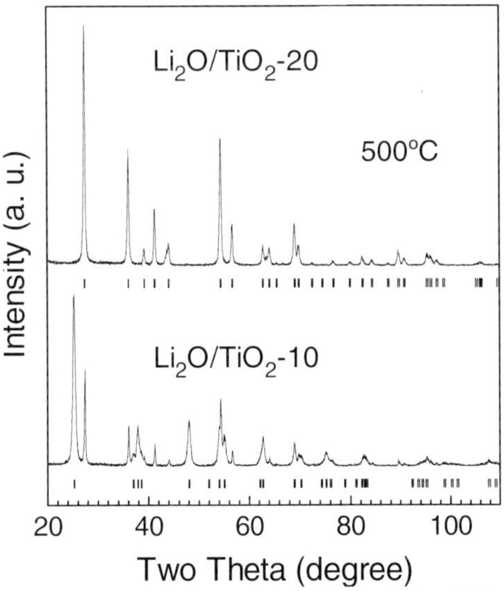

Figure 2 X-ray diffraction patterns of the lithium doped samples after they were annealed at 500 °C in argon. The tick marks under the pattern of the sample with 20 mole % L_2O correspond to rutile, and those under the diffraction pattern of the sample with 10 mole % L_2O correspond to anatase.

The above results indicate that lithium atoms destabilized the crystalline structure of anatase. With X-ray diffraction it is impossible to determine the positions that lithium atoms occupy in the unit cell of anatase. R. J. Cava et *al.* [21], however, report about the positions, determined by neutron diffraction, of lithium atoms doping anatase. Their lithium concentration was higher than the one used in the present work. They find that the tetragonal structure of anatase is transformed into an orthorhombic one with space group *Ima2*, and the lithium atoms occupying at random the vacant octahedral interstitial *4e* sites with atomic fractional coordinates $x = 0.0$, $y = 0.25$ and $z = 0.343$.

In the present work, we did not observe a significant distortion of the tetragonal structure of anatase, probably because the lithium amount dissolved in its structure was smaller than in Cava's samples. The transformation of anatase, which is the aim analysis of the present work, occurred between 400 and 500 °C, which is lower than the transformation temperature of non-doped titania [1]. This can be a consequence of the proximity of lithium, titanium and oxygen atoms, which form clusters that are precursors for building phases more stable than anatase, like $Li_2Ti_2O_4$ and Li_2TiO_3 (in the present case) and the spinel $LiTi_2O_4$ (in Cava's work where titanium was reacting with valences 4+ and 3+).

Finally, we want to notice that the solid solution of lithium in anatase is a good precursor to synthesize lithium titanates at relatively low temperatures; the final phases, however, will depend on lithium concentration and titanium valence.

3.2 CaO/TiO₂ system

Titania was doped with 10 and 20 mole % CaO, which are representative of two doping concentration ranges. The first concentration represents the maximum solubility of calcium in anatase (Fig. 3); and the second, the range where calcium titanates were formed, in addition to the anatase solid solution (Fig. 4).

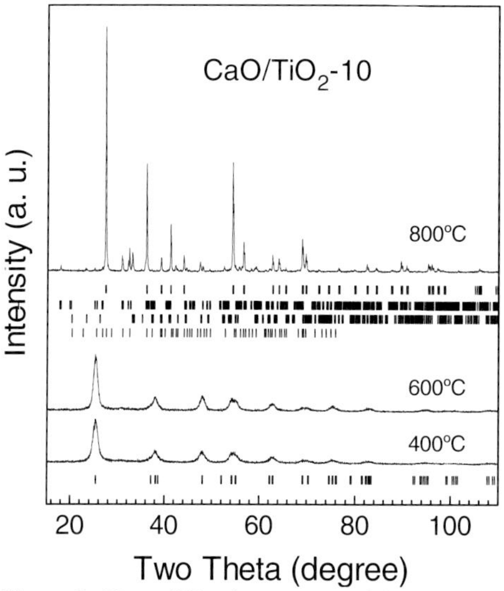

Figure 3 X-ray diffraction patterns of the samples doped with 10 mole % CaO and annealed in argon at different temperatures. From top to botton, the tick marks under the pattern of the sample annealed at 800 °C correspond to rutile, Ti_3O_5, $CaTiO_3$, and $CaTi_4O_9$ respectively. The tick under the pattern of the sample annealed at 400 °C correspond to anatase.

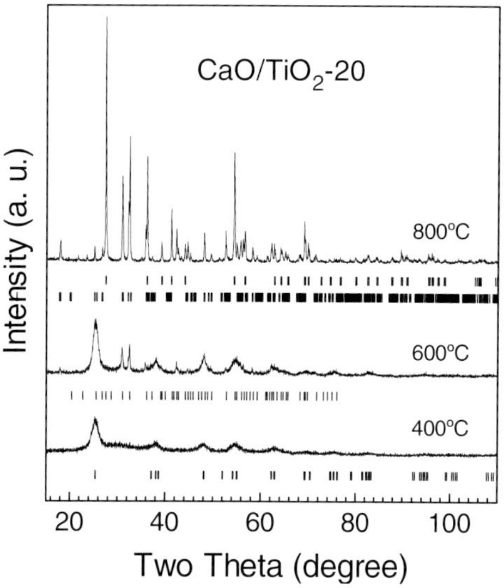

Figure 4 X-ray diffraction patterns of the samples doped with 20 mole % CaO and annealed in argon at different temperatures. The upper tick marks under the pattern of the samples annealed at 800 °C correspond to rutile, and the lower ones to Ti_3O_5. The tick marks under the pattern of the sample annealed at 600 °C correspond to $CaTi_4O_9$; those below the pattern of the samples annealed at 400 °C correspond to anatase.

Calcium solubility in anatase depended on the sample annealing temperature. When it was 400 °C, even 20 mole % calcium oxide was soluble (Fig. 4). At this concentration the anatase solution was partially transformed into $CaTi_4O_9$ and the reduced Ti_3O_5 oxide, when the samples were annealed at 600 °C. The position of the calcium atoms in unit cell were determined from the X-ray diffraction analysis: they substituted for the titanium atoms in anatase. In contrast to lithium doping, in the present case calcium, titanium and oxygen atoms did not form stable new clusters; therefore, anatase stability was similar to the one observed without doping.

The solubility concentration of calcium was high although it has a lower valence than titanium. As mentioned above, this concentration was as large as 10 mole %, which contrasts with the low solubility found when titania is doped with magnesium [22] or copper [9], two other elements that also have valence 2+, but a smaller ion size. Because of the high solubility of calcium, the crystalline structure of the solid solution must be highly oxygen deficient to compensate the valence difference between titanium and calcium ions.

For 10 mole % CaO (Fig. 3), the transformation of anatase not only gave rise to rutile but also to $CaTi_4O_9$, Ti_3O_5 and perovskite ($CaTiO_3$). This last phase was not produced in the samples with 20 mole % CaO, because its formation was probably inhibited by the $CaTi_4O_9$ and Ti_3O_5 crystals formed at 600 °C (Fig. 4).

3.3 SnO_2/TiO_2 system

Although tin oxide has a of rutile-type crystalline structure, when it was dissolved in titania in concentrations below 10 mole % and the samples were annealed at 400 °C, the obtained solid solution had anatase crystalline structure (Fig. 5), with tin atoms substituting for the titanium atoms in unit cell. For higher tin concentrations (Fig. 6), however, at this temperature also a solid solution with the rutile-type crystalline structure was formed, together with a low concentration of brookite with a small crystallite size and which was probably also a solid solution.

When the samples were annealed at 600 °C, the solid solution with anatase structure partially transformed into a more stable solid solution with rutile structure (Figs. 5-6); this transformation was complete at 800 °C. For SnO_2 concentrations higher than 10 mole %, besides the anatase solid solution, the samples also contained a solid solution with the structure of cassiterite; it appeared since crystallization beginning, and was insoluble in the other phases at temperatures below 800 °C (Fig. 6). This is in accordance with the miscibility gap observed in this system by other authors [23].

It is remarkable that for the present solid solution, as for lithium doping, the local environment formed by tin, titanium and oxygen atoms was adequate to produce an stable phase: the solid solution of tin in rutile. This local order promoted the transformation of anatase at relatively low temperatures.

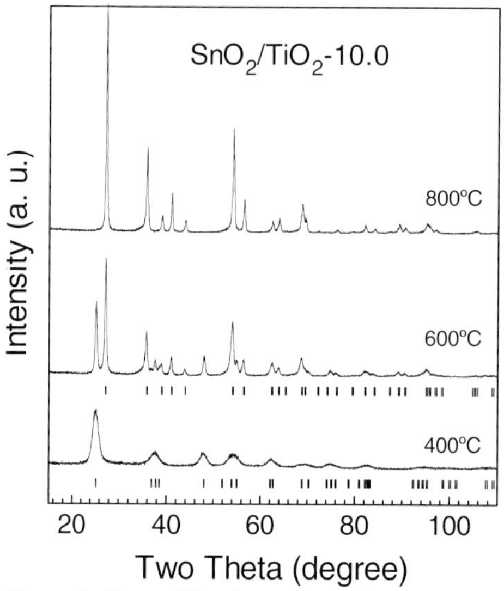

Figure 5 X-ray diffraction patterns of the samples doped with 10 mole % SnO_2 and annealed in air at different temperatures. The tick marks under the pattern of the sample annealed at 600 °C correspond to rutile; those under the pattern of the sample annealed at 400 °C correspond to anatase.

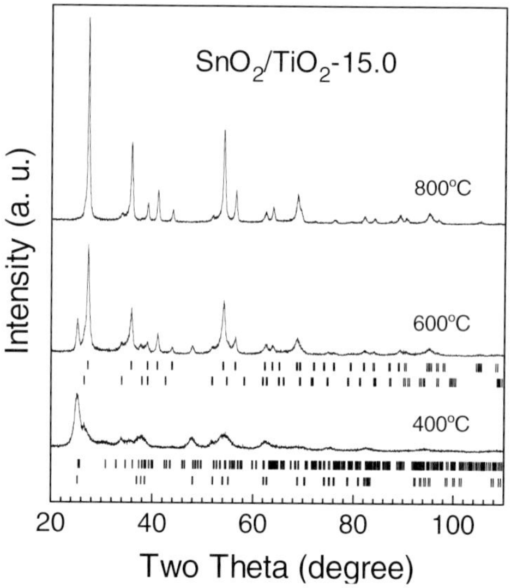

Figure 6 X-ray diffraction patterns of the samples doped with 20 mole % SnO_2 and annealed in air at different temperatures. The upper tick marks under the diffraction pattern of the sample annealed at 600 °C, correspond to rutile; the lower ones, to cassiterite. The upper tick marks under the pattern of the sample annealed at 400 °C correspond to brookite, and the lower ones to anatase.

4 Conclusions

Lithium, calcium and tin, were soluble in anatase for concentrations of the order of 10 mole % of their respective oxides. In these solid solutions, lithium atoms occupied interstitial sites in the unit cell, but calcium, and tin atoms substituted for titanium atoms. When lithium and tin atoms were the doping cations, the transformation of anatase into rutile and into other titanates occurred at lower temperatures than in the non-doped anatase. This could be caused by the fact that in these solid solutions the doping, titanium, and oxygen atoms formed local stable clusters that were precursors of the phases in which anatase was transformed. This is a proposition that could be of interest to those who calculate the stability of atom clusters. The results of the present work show the usefulness of anatase for the synthesis of titanates at relatively low temperatures.

5 References

1. Bokhimi, A. Morales, O. Novaro, T. López, E. Sánchez, R. Gómez, J. Mater Res. 1995,10, 2788-2796.
2. E. Sánchez, T. López, R. Gómez, Bokhimi, A. Morales, O. Novaro, J. Solid State Chem. 1996, 122, 309-314.
3. P. Murugavel, M. Kalaiselvam, A. R. Raju, C. N. Rao, J. Mater. Chem. 1997, 7, 1433-1438.
4. J. Livage, Mat. Res. Soc. Symp. Procc. 1986, 73, 717-724.
5. O. I. Micic, Y. Zhang, K. R. Cromack, A. D. Trifunac, M. C. Thurnauer, J. Phys. Chem. 1993, 97, 7277-7283.
6. S. Yin, Y. Inoue, S. Uchida, Y. Fujishiro, T. Sato, J. Mater. Res. 1998, 13, 844-847.
7. K. Yanagisawa, M. Sasaki, M. Nishioka, K. Ioku, N. Yamasaki, J. Mater. Sci. Lett. 1994, 13, 765-766.
8. X. Bokhimi, E. Muñoz, J. L. Boldú, E. Adem, O. Novaro, T. López, F. Rojas, R. Gómez, A. García-Ruiz, J. Sol-Gel Sci. and Tech. 1999, submitted.
9. X. Bokhimi, A. Morales, O. Novaro, T. López, O. Chimal, M. Asomoza, R. Gómez, Chem. Mater. 1997, 9, 2616-2620.
10. D. W. Murphy, M. Greenblatt, S. M. Zahurak, R. J. Cava, J. V. Waszczak, G. W. Hull, Jr., R. S. Hutton, Revue de Chim. Miner. 1982, 19, 441-449.
11. X. Ding, L. Liu, X. Ma, Z. Qi, Y. He, J. Mater. Sci. Lett., 1994, 13, 462-464.
12. X. Bokhimi, A. Morales, in preparation, to be submitted to the J. Solid State Chem. 1999.
13. M. Schneider, U. Scharf, A. Wokaun, A. Baiker, J. Catal. 1994, 150, 284-300.
14. K. Kameyama, Sh. Shohji, S. Onoue, K. Nishimura, K. Yahikozawa, Y. Takasu, J. Electrochem. Soc. 1993, 140, 1034-1037.
15. C. Cristiani, M. Bellotto, P. Forzatti, F. Bregani, J. Mater. Res. 1993, 8, 2019-2025.
16. T. J. Bastow, L. Mursgaski, M. E. Smith, H. J. Whitfield, Mater. Lett. 1995, 23, 117-121.
17. B. E. Yoldas, J. Mater. Science 1986, 21, 1087-1092.
18. R. A. Young, A. Sakthivel, T. S. Moss, C. O. Paiva-Santos, J. Appl. Crystallogr. 1995, 28, 366-367.
19. P. Thompson, D. E. Cox, J. B. Hastings, J. Appl. Crystallogr. 1987, 20, 79-83.
20. E. Prince, J. Appl. Crystallogr. 1981, 14, 157-159.
21. R. J. Cava, D. W. Murphy, S. Zahurak, A. Santoro, R. Roth, J. Solid State Chem. 1984, 53, 64-75.
22. X. Bokhimi, J. L. Boldú, E. Muñoz, O. Novaro, T. López, J. Hernández, R. Gómez, A. García-Ruiz, Chemistry of Materials 1999, in press.
23. T. C. Yuan, and A. V. Virkar, J. Am. Ceram. Soc. 1988, 71, 12-21.

Contribution to the Determination of Residual Stresses in Carbon-long-fibre-reinforced Ceramics

Bernhard Wielage, Uta Zesch
Institut of Composite Materials, TU Chemnitz

1 Introduction

During production of carbon-long-fibre-reinforced ceramics through infiltration of fibre rovings with Si-organic precursors residual stresses in matrix and fibre develop. The reason for this is, on the one hand, the shrinkage that occurs during the pyrolytic transformation of the Si-organic polymer into SiC, which is dependent on the composition of the precursors of the matrix material, and, on the other hand, the different thermal expansion coefficients of the composite components. Furthermore, the level of the residual stresses is considerably influenced by the condition of the interface between fibre and matrix, in particular the adhesive properties existing there.

It is possible to determine the existing residual stresses of the individual components of the composite with the help of stress measuring through x-ray and neutron diffraction. The aim of this investigation is to obtain information for process optimisation in order to improve the mechanical properties of the composite.

2 Residual Stresses in Composite Materials

Due to their heterogeneous structure, residual stresses in composite materials are particularly critical. Because of the different thermal expansion coefficients of matrix and reinforcing component it is impossible to avoid the formation of residual stresses during the manufacturing of the composite materials. Other influence factors are the fibre content and the matrix strength.

The residual stresses in the composite materials determine their static and dynamic strength behaviour. Residual stresses either reduce or increase the loadability of the individual components. Tensile residual stresses in the matrix material lead, for example, to corresponding compressive stresses in the fibres and limit their compression loadability, but increase their tensile strength at the same time. In this way, the condition of the residual stress of one composite partner influences the loadability of the other composite partner and the areas of tolerable compression load and tensile stress can be shifted.

In order to properly dimension the component parts, taking into account and if need be, specificly use the residual stresses, it is necessary to have exact knowledge about the existing residual stress conditions. Compression loads, for example, have a positive effect on the life of the component parts under fatigue stress.

The most important factor influencing the development of residual stresses in the matrix of the investigated composite materials can be said to be the content of pyrolyticly converted

SiC, which is dependent on the pyrolysis end temperature, the temperature regime and the content of further matrix components in the precursor (inert and/or reactive fillers). As far as the fibres are concerned, differences are expected regarding the development of residual stresses through modification of the fibre surfaces.

3 Experiments

The production of carbon-long-fibre-reinforced composite materials takes place through the infiltration of aligned fibre rovings and the following lamination to specimen size. The procedure is based on thermal decomposition in inert gas above 1000 °C. This technology offers several advantages. Because of the use of the conventional forming techniques for the manufacture of fibre-reinforced polymers (winding, lamination), short processing times can be realised. The interface between fibre and matrix, the development of which is decisive for the mechanical properties of the composites, can be influenced specifically by means of various methods.

Inert and reactive fillers are added to the dissolved siliceous precursor PSS. SiC-powder is used as inert filler. Additionally, the effects of Ti, Zr and V as reactive fillers are investigated. High-strength carbon long fibre T 800 – sized, desized and also deposited with a pyroC-SiC layer system - are used as reinforcing components.

The results of the investigations of the specimen give information about the correlation between the structure of the composite and the mechanical properties. The qualitative as well as the quantitative structure analysis by means of light microscopy, image analysis and SEM is used for the general assessment of the structure of the composites.

Thermographimetric investigations are carried out in order to determine the ceramic yield of the applied matrix precursors, the determination of the phase composition is carried out with the help of x-ray diffraction spectra.

Additionally, mechanical testing methods, the 3-point-bending test in particular, are used for the determination of strength and fracture behaviour. In the course of this, the specimens are loaded vertically to the fibre orientation. The load normally to the fibre layers and also the specimen size are chosen in such a way that failure occurs if tension or pressure is applied.

For the determination of the internal stresses in the composite materials x-ray stress measuring (RSM) according to the $\sin^2\Psi$ method was used /1/. The x-ray stress determination has the advantage that it is possible to measure the strain phase-selectively and it is therefore suitable for multi-phase materials such as composite materials.

The measurements were carried out in a Ψ-diffractometer with Co-Kα-radiation ($\lambda_{K\alpha}$ = 0,178892 nm), vertically as well as parallel to the fibre orientation. The neutron diffractions were carried out at the E3-spectrometer of the BER II reactor at HMI Berlin.

4 Results and Discussion

The specimens show average fibre volume contents of about 40 to 50 % and high porosities of about 20 to 40 Vol-%. The porosities are reduced through the application of inert and especially reactive fillers. The thermogravimetric investigations of various matrix mixtures show that the addition of reactive fillers considerably increases the ceramic yield of the mixture. The specimens with Zr or Ti in the precursor mixture achieve the best results. Adding relatively coarse-grained V-powder that is less suitable for the reaction pyrolysis, does, however, not reveal a fundamental improvement of the ceramic yield compared to the matrix mixtures without reactive fillers. /2/

Results regarding the influence of the matrix/fibre variations on the strength and the fracture behaviour of the composites are provided by the bending test /2/. When comparing composites with the same matrix composition (precursor with 20 mass-% SiC-powder as inert filler) but with different fibre conditions it can be said that there is

- highest resistance to fracture when using sized fibres, pseudo-plastic fracture propagation under fibre-pull-out and delamination;
- lowest bending strength when using desized fibres, but no fibre fraction at highest bend, no transmission of stresses to the fibres, because there exists no sufficient bond;
- maximum bending strength when using deposited fibres (pyroC/SiC) already at relatively low bend and early break down, strong fibre-matrix-bond, simultaneous fracture of fibre and matrix, brittle fracture with smooth and even fracture surface.

The different reactive fillers cause different matrix formations and consequently influence the fracture behaviour as well:

- When adding reactive Ti-powders, a strong bond between matrix and fibres develops which causes a relatively brittle fracture.
- Zr-powder in the precursor mixture leads to a small-grained matrix with many small pores. The specimens show fibre pull-out and quasi-ductile fracture behaviour.
- The V-powder used is relatively coarse-grained, which impedes the infiltration and only a few large pores develop in which area the fracture occurs at first.

The high roughness of the specimen surfaces from the fabrication process causes problems with regard to the measurement of the residual stresses. Measurements at grind surfaces showed no or only very low residual stresses, which were in the sphere of measurement uncertainty. In order not to influence the state of the residual stresses it was therefore abstained from preparation of the surface.

The x-ray stress measuring as well as the neutron diffraction experiments with the composite specimens basically reveal high tensile residual stresses in the matrix parallel to the fibres, but low compressive stresses vertically to them.

The reason for this type of residual stress formation can be said to be the different thermal expansion coefficient of the composite components. For the ceramic matrix the expansion coefficient is the same for all directions. However, for carbon fibres in longitudinal direction it is slightly negative, whereas it is positive in vertical direction, but considerably higher than that of the matrix material.

During the production process, the fibre is therefore extending in longitudinal direction when being cooled down to room temperature, whereas the matrix is contracting, the more so, since, depending on the composition of the precursor, it is also subject to a heavy conversion shrinkage that releases fluid split-off products. Thus, high tensile residual stresses develop in the matrix, which at high fibre-matrix-bonds can possibly lead to damages such as crack formation and delamination (Figures 1 and 2) already during the production process.

Vertical to the fibre axis, a stronger contraction of the fibres is caused because of the different thermal expansion coefficient. However, the matrix is coming closer to it by additional shrinkage through precursor pyrolysis. Because of this the remaining compressive stresses are low.

Increasing contents of inert fillers drop the tensile residual stresses in the matrix, because they reduce the share of residual stresses that is caused by the shrinkage of the polymer during the pyrolysis. The modification of the precursor through the addition of reactive fillers has the same effect.

Comparing specimens with different fibre surface modifications, residual stresses in sized fibres are the highest, and the lowest in coated fibres. Because of the high fibre-matrix-interface-bond of these specimen the developing residual stresses already exceed the tolerable load of the matrix during the manufacturing, so that it leads to its stress relaxation through crack formation and delamination (Figure 1 and 2).
The resulting residual stresses are only marginally influenced by the pre-damaging of the composites through the application of low mechanical loads during 3-point-bending. The first matrix cracks, which are visible in the curvature, obviously lead only to local stress relaxation. Only when using desized fibres a stress reduction occurred. Since there is no bond between fibre and matrix, developing cracks in the matrix can spread unhindered and lead to failure and therefore to stress relaxation already at lowest loads.

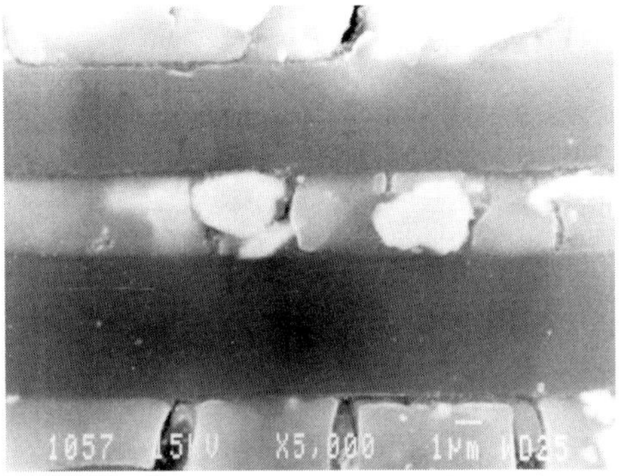

Figure 1. C/SiC-composite material with crack network, longitudinal section

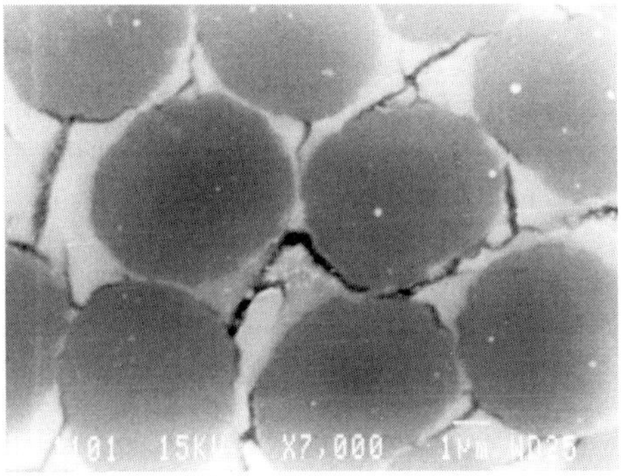

Figure 2. C/SiC-composite material with crack network, cross section

5 Conclusion

Carbon-fibre-reinforced SiC ceramics with fibre contents of about 40 to 50 Vol-% were produced through infiltration and pyrolysis of Si-organic precursors. Sized, desized as well as carbon-longfibres coated with pyroC/SiC were used as fibre material. Stress measurements at the composite specimens revealed high tensile residual stresses in the matrix parallel to the fibres. Vertical to it, there are lower compressive residual stresses in the matrix. The precursor material, the filler inclusions and also the condition of the fibre surface are of particular importance for the formation of the interface between fibre and matrix and thus for the mechanical behaviour and the state of the residual stress.

6 Literatur

1. E. Macherauch, P. Müller, *Zeitschrift angewandte Physik* 13 (1961) 7, S. 305-312.
2. B. Wielage, U. Zesch in *Werkstoffwoche 98, Bd. 7* (Ed.: Heinrich, J. u.a.), WILEY-VCH Verlag Weinheim, 1999, S. 631-635

7 Acknowledgement

We would like to thank the Deutsche Forschungsgemeinschaft for the financial support of the research work and also BENSC for the available beam time and the support with the diffraction experiments.

Influence of Al Reactive Filler on the Carbon Redistribution During Pyrolysis of Poly (Siloxanes)

M. Scheffler, P. Buhler and P. Greil
University of Erlangen-Nuernberg, Department of Materials Science (Glass and Ceramics),Martensstrasse 5, D-91058 Erlangen, Germany

1 Introduction

Polymer derived ceramics offer the opportunity to establish the plastic shaping technology for manufacturing of near net shape ceramics [1] or can be used for alternative processing routes [2, 3]. Applying the Active Filler Controlled Pyrolysis, reactive powders e.g. elements, oxides, carbides or nitrides undergo a reaction with the amorphous ceramic residue formed during pyrolysis of the preceramic polymers in order to reduce shrinkage and to tailor resulting material properties [4, 5]. For the preparation of polymer ceramics the knowledge of the composition of the starting polymers, their reaction pathes, their ceramic yield and the distribution of their pyrolysis products is essential. However, there is a lack of investigations concerning the elemental distribution in the solid and in the gas phase, respectively.

After pyrolysis of polysiloxanes $[RSiO_{1.5}]_n$ with R = CH_3, C_3H_7, C_6H_5 or C_2H_3SiC, SiO_2 and free carbon can be found [6, 7]. The element ratio Si/O/C in the ceramic residue depends on the starting composition of the polysiloxanes. This starting ratio influences the formation of the ceramic phases at higher temperatures and can be controlled by the organic substituents, commercially available as methyl-, phenyl-, phenylmethyl-, phenylpropyl- or methylvinyl poly(siloxanes). The aim of this work is to show the influence of elemental fillers on the total carbon content in the solid phase after pyrolysis in the temperature range up to 1 000°C.

2 Experimental Procedure

A methyl poly(siloxane) (PMS, NH2100, Wacker GmbH Nuenchritz) or a phenyl methyl poly(siloxane) (PMPS, H44, Wacker GmbH Nuenchritz) were thoroughly mixed with 10, 30, 50 and 70 wt % of reactive filler powder (Al) in a ball mill. After mixing the specimens were isostatically pressed and cross linked at 300°C for 3 hours. Pyrolysis was carried out with a heating rate of 10 K/min from room temperature to 1000°C and kept for 1 hour. After pyrolysis the carbon content of the solid samples was determined by a method described in [8]. About 50 mg of the pyrolyzed samples were mixed with lead oxide (PbO) at a weight ratio 1:100, placed in a silica tube, evacuated and heated up to 800°C. At this temperature redox reaction takes place and CO_2 and elemental lead is formed. The amount of CO_2 was measured by the pressure increase and used to calculate the total carbon content in the samples by applying the ideal gas equation.

Simultaneous thermal analysis (STA 409, Netzsch GmbH, Selb) was carried out in Ar to investigate the influence of the fillers on the decomposition reaction and the weight loss of the poly(siloxanes). About 100 mg of ground samples were heated up to 300°C, kept for 3 hours and heated up to 1050°C with a heating rate of 5 K/min. Phase formation was examined by XRD (Siemens Diffrac 5000).

3 Results and Discussion

Figure 1 shows the powder diffraction patterns of the phenyl methyl poly(siloxane)-aluminum samples (PMPS-Al) after pyrolysis at 1000°C.

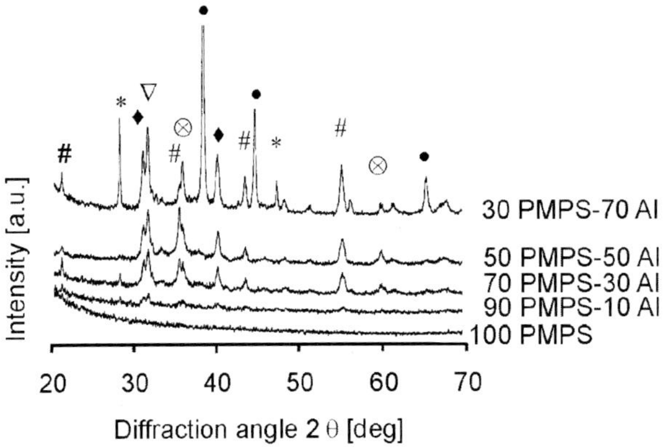

Figure 1. Powder diffraction patterns of PMPS-Al samples after pyrolysis at 1000°C ♦ Al_4C_3 * Si • Al ⊗ β-SiC ∇ Al_4SiC_4 # α-Al_2O_3

Crystalline phases occur at aluminum concentrations above 10 wt % (70PMPS-30Al). PMS leads to similar phase formation as can be seen in figure 2. While in PMS-Al samples more SiC, Al_2O_3 and Si was found, Al_4C_3 and Al_4SiC_4 phases show higher concentrations in the PMPS-Al samples.

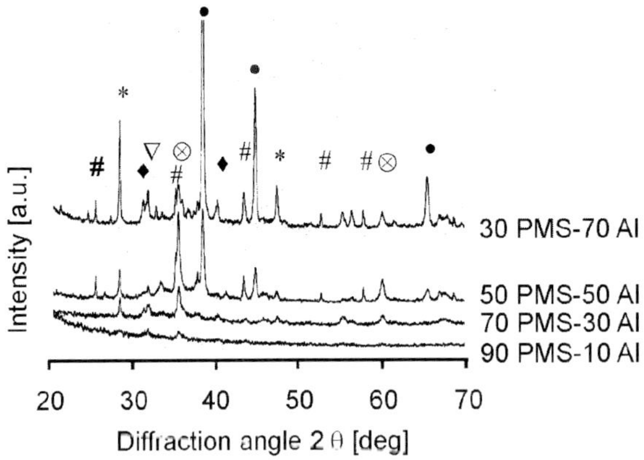

Figure 2. Powder diffraction patterns of the PMS-Al samples after pyrolysis at 1000°C ♦ Al_4C_3 * Si • Al ⊗ β-SiC ∇ Al_4SiC_4 # α-$A_{l2}O_3$

These differences are attributed to the Si/O/C ratio of the starting resins: PMS consists of a Si/O/C molar ratio of about 1/1.5/1, PMPS has a starting ratio of 1/1.5/3.5.

Figure 3 shows the DTA/TG investigations of three aluminum containing PMS-Al samples. The DTA curves show an endothermic peak at 661°C, which is the melting point of aluminum. This peak is subsequently followed by a strong exothermic peak, which is related to an accelerated weight loss. At a temperature of about 800°C a second exothermic peak occurs. In samples with higher aluminum content this peak is shifted to lower temperatures and, finally, disappears with highest aluminum loading.

Figure 4 shows XRD patterns of PMS-Al samples pyrolyzed at 750°C and 1000°C. In the sample pyrolyzed at 750°C Al, Si, Al_4C_3 and Al_4SiC_4 are found. At 1000°C the peak intensities of Si, Al_4C_3 and Al_4SiC_4 decreased while Al_2O_3 and SiC are formed.

Correlation of DTA/TG and XRD analyses suggests the following tentative reactions to take place during heating:

Increasing Temperature

$$[CH_3SiO_{1.5}]_n \xrightarrow{>400°C} (n-3)SiO_2 + (n-1)SiC + (n-2)\,C + (n-1)\,CH_4\uparrow + (n-4)\,H_2\uparrow \quad (1)$$

$$Al_{(s)} \xrightarrow{661°C} Al_{(l)} \quad (2)$$

$$8Al + 3SiO_2 + SiC + 2C + CH_4 \longrightarrow 2Al_2O_3 + 3Si + 0.5SiC + 0.5Al_4C_3 + 0.5Al_4SiC_4 + H_2\uparrow \quad (3)$$

$$Al_4C_3 + Al_4SiC_4 + 6SiO_2 \longrightarrow 4Al_2O_3 + 7SiC \quad (4)$$

$$nAl_{(l)} + [CH_3SiO_{1.5}]_n \longrightarrow 0.5n\,Al_2O_3 + nSiC + 1.5nH_2\uparrow \quad (5)$$

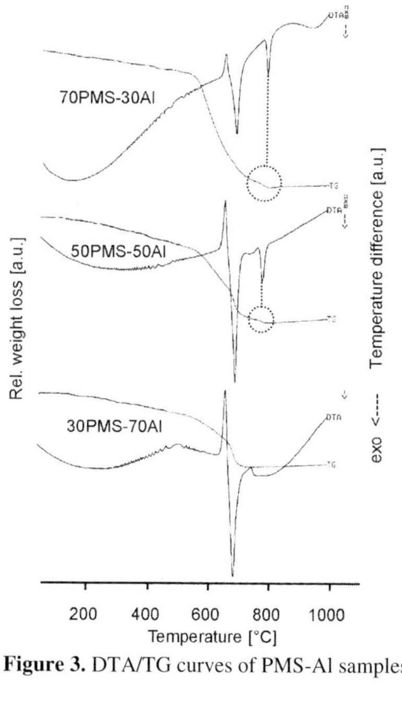

Figure 3. DTA/TG curves of PMS-Al samples with different compositions

Analysis of the total carbon content in the solid ceramic residue showed, that the carbon content increased in the polymer derived Si/O/C matrix phase with increasing aluminum content, figure 5.

While in filler free PMS 67 wt % of the initial carbon is retained after pyrolysis, 76 wt % were observed in the specimens containing 70 wt % Al filler.

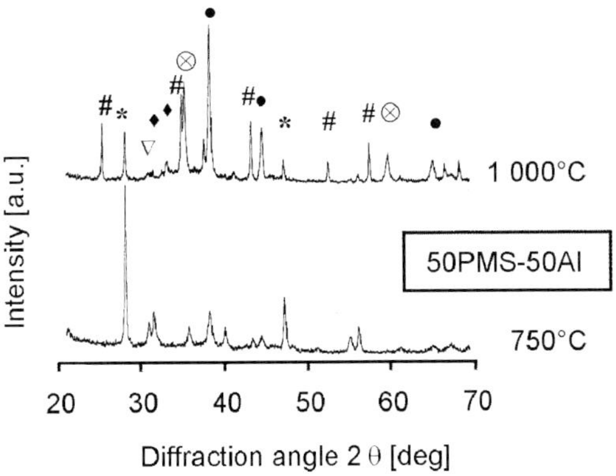

Figure 4. Powder diffraction patterns of the 50PMS-50Al sample after pyrolysis at 750 and 1000°C ♦ Al_4C_3 * Si • Al ⊗ β-SiC ∇ Al_4SiC_4 # α-$A_{l2}O_3$

In PMPS-Al specimens the carbon content ranges from 65 % at low Al concentrations (90PMPS-10Al) to 92 % (30PMPS-70Al) at highest Al loading, compared to the Al free sample.

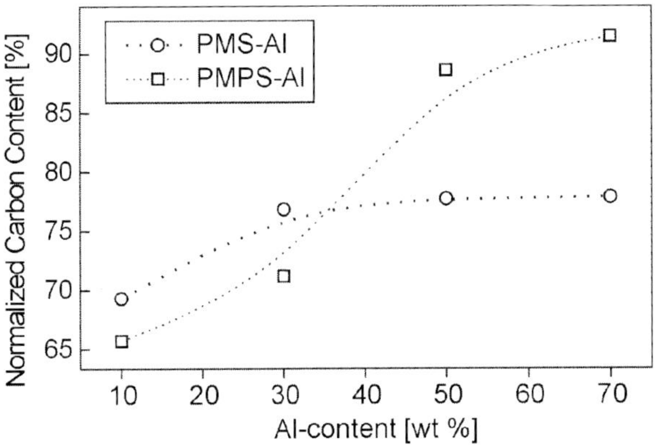

Figure 5. Normalized carbon content formed in the polymer derived Si/O/C matrix after pyrolysis with Al fillers at 1000°C

While in filler free samples the release of methane (CH_4) is observed up to > 700°C aluminum carbide formation takes place at the melting point of aluminum (661°C). $Al_{(l)} + 3CH_4 \longrightarrow Al_4C_3 + 6H_2\uparrow$ results in an increase of the carbon content in the final composite material.

4 Conclusions

Al as active filler has a pronounced infuence on the carbon content in pyrolyzed poly(siloxane)/filler composites. With increasing aluminum content an increasing fraction of the initial carbon content is retained. It could be shown, that the element ratio of the starting polymer plays an important role for the phase composition in the ceramic residue. While the amount of carbon can be controlled in a small range by specific active fillers, the control by mixing poly(siloxanes) with different organic substituents seems to be more effective.

5 Acknowledgement

We greatfully acknowledge the Deutsche Forschungsgemeinschaft for financial support.

6 References

1. P. Greil, Key Eng. Mat. 132-136 (1997) 1981-1984
2. P. Colombo, M.O. Abdirashid, A. Maddalena, A. Marchetti in Fourth Euro Ceramics, Vol. 4, pp.117-124
3. D. Suttor, H.-J. Kleebe and G. Ziegler, J. Amer. Ceram. Soc., 80 (10) 2541-48 (1997)
4. P. Greil, M. Seibold, Ceram. Trans., 19 (1991) 43
5. P. Greil, Key Eng. Mat. 159-160 (1999) 339-346
6. L. Bois, J. Maguet and F. Babonneau, Chem. Mater. 6, (1994) 796
7. H. Schmidt, P. Buhler and P. Greil in Fourth Euro Ceramics, (Ed.: C. Galassi) Vol. 1, 1995, p. 229
8. M. J. Wild, P. Buhler, J. Mater. Sci. 33 (1998) 5441-5444

Elaboration of Low Temperature and Shrinkage Mullite Derived from Filled Siloxane

Thibaut Michalet[1], Michel Parlier[1], Jacques Crampon[2]
[1]Office National d'Etudes et de Recherches Aérospatiales, DMSC, Châtillon, France
[2]Université des Sciences et Technologies de Lille, Villeneuve d'Ascq, France

1 Abstract

Among the numerous techniques for the processing of ceramic matrix, synthesis via filled polymers have been used for the synthesis of mullite monoliths. This paper describes the preparation of mullite, with low shrinkage and low processing temperature, based on the reaction-bonding of Al and Al_2O_3 powders with the pyrolysis of a silica polymer precursor. The behaviour of the polymer/Al/Al_2O_3 mixture is compared with those of polymer/Al_2O_3 mixture, by means of thermoanalytical and microstructural techniques.

2 Introduction

Oxide-oxide ceramic composites are promising candidates for aircraft applications because of their inherent stability in air at high temperature. The elaboration of these ceramic composites via sintering of powders requires two conditions : first, the matrix shrinkage during the sintering should remain limited, and second, relatively low sintering temperature should be used in order to preserve the fiber integrity.

Recently, a new elaboration method of monolithic and composite structural ceramics has been developed from filled polymers [1-3]. In this process, the filler identical to a 'skeleton', may react with the product of the polymer pyrolysis, and thus, limit the shrinkage. The advantages of this method when applied to mullite are : possibility of complex shapes elaboration, homogeneity improvement, moderate processing temperature due to the high reactivity of amorphous silica (SiO_2) yielded by the polymer precursor [4].

In this work, mullite monoliths are processed, associating the reaction-bonding of Al and/or Al_2O_3 powders with the pyrolysis of the silica precursor.

3 Experimental procedure

A solid silicon resin (polydimethysiloxane NH2000, Hüls, Germany) was used as a silica precursor. Aluminium (A1183, Cerac, USA) and α-Al_2O_3 (CR1-AS1, 500wt-ppm MgO-doped alumina, Baikowski Chimie, France) with a median particle size of 0.8 and 1μm, were used as filler components. By variation of content of Al and α-Al_2O_3, suitable polymer/filler ratios, with respect to stoechiometric 3/2 mullite, were determined for two compositions :

polymer/α-Al_2O_3 (PAO) and polymer/Al/Al_2O_3 (PAAO) (Table1). Powder mixtures were processed in organic medium (propanol-2/acetone, 75/25 vol%) with magnetic and ultrasonic homogenization. After evaporating the solvent, the residue was granulated and then uniaxialy pressed (130°C, 15 MPa) to get solid samples with low porosity (3 to 7%). The heat-treatment of the sample was performed at temperature up to 1550°C (3h) with heating rates in the range 1-5°C/min. Differential thermal analysis/thermogravimetric analysis (DGA/TGA 92, Setaram) and dilatometry (DI24, Adamel-Lhomargy) were used to monitor the associated phase, mass and dimensional changes during the reaction-bonding process. Reaction products were identified, between 700°C and 1550°C, by powder X-ray diffraction (XRD, Philips) patterns of crushed samples. Microstructure of the sintered samples was observed by scaning electron microscopy (SEM, Gemini Zeiss) and the porosity was characterized by Hg porosimetry (Pore sizer 9310, Micromeretics).

Table 1. Polymer/filler ratios for the two compositions

Specimen	wt%	vol%
NH2000/α-Al_2O_3 (PAO)	32.5/67.5	59/41
NH2000/Al/Al_2O_3 (PAAO)	39.5/24.5/36	62/19/19

4 Results and Discussion

4.1 Thermoanalytical characterization

The oxidation sequences of the PAO system are fully similar to those of a polydimethylsiloxane one. The specific mass change in air is of about -5.5% and is only due to pyrolysis of the polymer. Degradation and thermal oxidation of the polymer start at temperature above 200°C and is completed at 800°C

In the case of siloxane/Al/Al_2O_3 mixture, the reaction steps are significantly different (figure 1). The specific mass change (TGA) in air shows that the weight loss due to the thermal degradation of the polymer is progressively compensated, between 350 and 700°C, by a mass increase due to the oxidation of Al, in solide state, to alumina. After a hold at 700°C during 2h, the PAAO sample shows another mass gain in the temperature range 700-900°C, due to the oxidation of Al in liquide state. The mixture is fully oxdized at 1400°C. The two exothermic peaks of the differential thermal analysis (DTA), characteristic of the degradation/thermal oxidation of the polymer, appear respectively at 290 and 550°C. The oxidation process of Al is evidenced at first by a sharp exothermic peak at 610°C, where oxidation of solid Al particles takes place. After a soaking time at 700°C (2h), the process accelerates with an oxidation exothermic peak at 830°C. The rupture of the oxide scale under the thermal expansion of the melt results in a rapid oxidation of Al in liquid state [5].The weak endothermic peak at 660°C corresponds effectively to the melting of Al metal. An other endothermic peak appears at 1410°C and is attributed to the melting of free Si metal. This peak is absent in the case of the PAO system and thus, the presence of free Si in the PAAO system at high temperature is correlated to the consumption of the oxygen, in air and from the polymer, by the oxidation of Al.

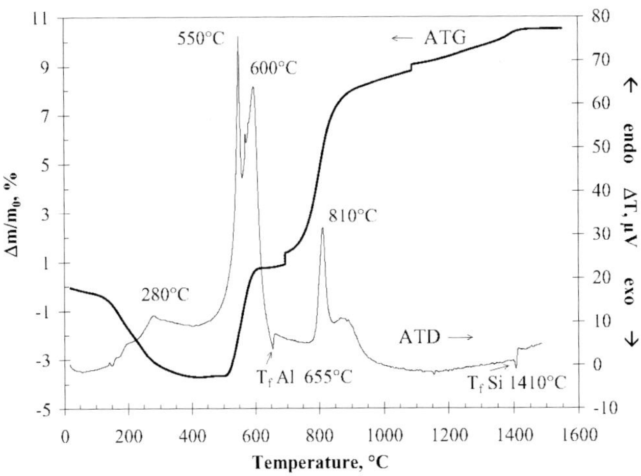

Figure 1. TGA and DTA curves of polydimethysiloxane/Al/Al_2O_3 mixture

4.2 Phase development

The phases evolution for the siloxane/Al_2O_3 system were determined by XRD as shown in figure 3 (left). Up to 1200°C, only the α-Al_2O_3 phase is identified. The small hump, in the $2\theta = 20°$ range, indicates the presence of amorphous SiO_2 at this temperature. Amorphous silica crystallizes into cristoballite above 1200°C. After maximum crystallization of the cristobalite at 1350°C, the mullite formation starts between 1350°C and 1400°C. At this temperature three phases coexist :α-Al_2O_3, cristobalite and mullite. Between 1400°C and 1550°C, α-Al_2O_3 and cristobalite disappear progressively. Finally at 1550°C sample consists mainly of 3/2 mullite and only traces of α-Al_2O_3 is left.

In the case of siloxane/Al/Al_2O_3 compound (figure 1., right), the evolution is slightly different. The oxidation of Al occurs at temperature up to 1200°C and results in the formation of transition alumina (γ-alumina), as it appears on the XRD pattern at 700°C. Mullitization begins at 1200°C along with the crystallization of cristobalite. This coexisting temperature of mullite, cristobalite and alumina is somewhat lower than that of sample PAO, which presents the same pattern at about 1400°C. The subsequent reaction behaviour, between 1200°C and 1400°C is comparable to the PAO system, and also yields a 3/2 mullite with apparently no excess alumina after three hours at 1550°C.

Temperature and mullite formation rate during reaction sintering of alumina-silica mixtures depend on various parameters. The most important of them are nature (amorphous or crystalline), particle size [6] and scale of mixing precursors [7]. In the case of PAO the mullite precursors mixture consists of fine and stable α-Al_2O_3 and amorphous silica yielded by the polymer but the temperature of mullitization should be higher than 1350°C, which is characteristic of inhomogeneous mixture. On the other hand, in the PAAO mixture the beginning of 3/2 mullite formation was achieved at lower temperature (~1200°C), which indicates that the scale of mixing of the precursors was finer than in the PAO one. The presence of transition alumina and the lower reaction temperature may be indicative of a dual mechanism of mullitization :on one part, formation of mullite directly from the reaction between transition Al_2O_3 and the amorphous silica phase, on the other part, mullitization through the individual crystallization of α-Al_2O_3 and cristobalite, as in the case of PAO.

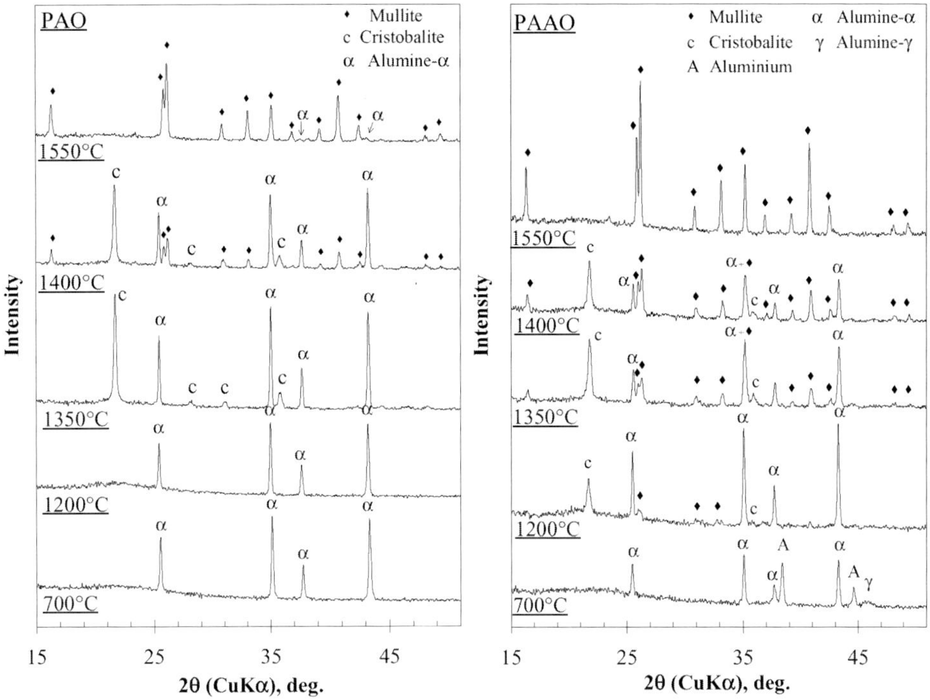

Figure 2. Powder X-Ray diffraction patterns of the PAO and PAAO samples, phase evolution during heat treatment

4.3 Sintering of green compacts

The dimensional changes associated to the reaction process are shown in figure 3. Up to 400°C, both mixtures expand, followed up to 700°C by a shrinkage of about 3-3.5% that is caused primarily by the thermo-oxidation of the silicon resin.

In the case of siloxane/Al_2O_3 system, notable densification is observed for temperatures above about 1000°C, although shrinkage starts at temperature higher than 800°C, and extends to the temperature at wich mullitization starts (1350-1400°C). Sintering probably takes place by a similar mechanism as was proposed for the sintering of heterogeneous mixture of amorphous SiO_2 with α-Al_2O_3 [8, 9] (i.e. by viscous flow sintering of the amorphous silica phase). Therefore, densification occurs prior to mullitization. A germination period for mullitization can be identified at 1350°C, with a corresponding shrinkage compensation up to 1550°C which is caused by the volume expansion that is due to mullitization (+4.6 %vol). A final linear shrinkage of 7.2% was measured for the PAO system after 3 hours at 1550°C, for final pore volume of about 18%.

For the PAAO mixture, the shrinkage associated to pyrolysis of the polymer is partly compensated by the volume expansion due to oxidation of Al (+28 %vol), in the temperature range 700-1200°C. Here again densification probably proceeded by the transient viscous flow sintering mechanism which occurs prior to mullitization. The final pore volume is only 20% for a linear shrinkage of about 3.6%.

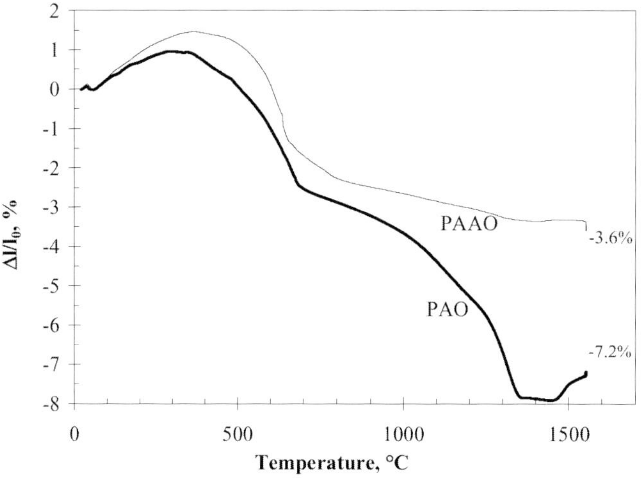

Figure 3. Relative linear shrinkage of PAO and PAAO compositions during oxidation and sintering under air

4.4 Microstructure

The results of Hg porosimetry are shown in figure 4.

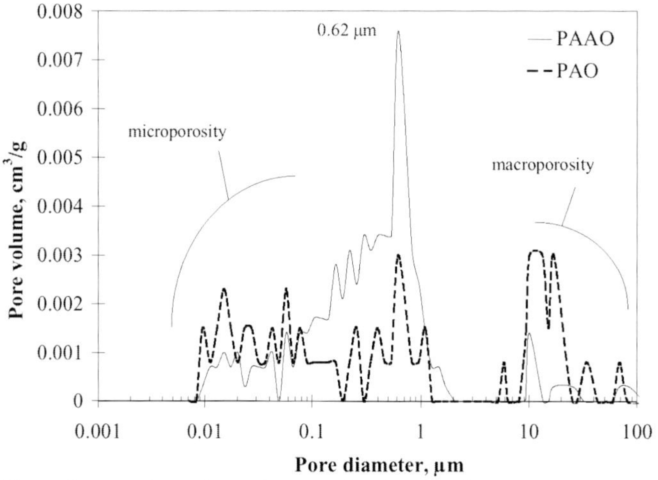

Figure 4. Comparison of pore diameter of the 2 compositions

Two families of pores are evidenced for the two systems after heat treatement at 1550°C (3h) : the first class is in the range 10-100μm and for the second in the range 0.01-1μm. It seems that the amount of the second class of pore increases with the presence of aluminium in the initial mixture and moreover there is tendency for a pore size distribution centred on 0.6μm. The observed morphological pore structure difference between the two samples (figure 5) confirms the result of Hg porosimetry.

The improvment obtained with the PAAO system is shown on SEM micrographs, were large pores have been replaced by a fine-distributed pore network. From the defect point of view, the mixture PAAO seems to be a good system as it presents minimum macroporosity. The mullite matrix is constituted of granular-shaped grains of about 1μm in size.

Figure 5. SEM microstructures of PAO and PAAO samples

5 Conclusion

In comparison with PAO system, addition of Al in the PAAO mixture enables both to limit sintering shrinkage of about 50%, for the same porosity, and to decrease mullitization temperature of 200°C. Moreover, the microstructure is finer and more homogeneous. PAAO mixture seems to be a good compromise as a precursor for mullite matrix oxide/oxide composites.

6 References

[1] P. Greil, M. Seibold, T. Erny, Mat. Res. Soc. Symp. Proc., 1992, 274, 155-156
[2] T. Erny, M. Seibold, O. Jarchow, P. Greil, J. Am. Ceram. Soc. 1993, 76, 207-213
[3] P. Greil, J. Am. Ceram. Soc., 1995, 78, 835-848
[4] D. Suttor, H. J. Kleebe, G. Ziegler, J. Am. Ceram. Soc., 1997, 80, 2541-2548
[5] S. Wu, D. Holz, N. Claussen, J. Am. Ceram. Soc., 1993, 76, 970-980
[6] F. Kara, J. A. Little, J. Eur. Ceram. Soc., 1996, 16, 627-635
[7] J. A. Pask, A. P. Tomsia, J. Am. Ceram. Soc., 1975, 58, 507-512
[8] H. Schneider, B. Sahuran, D. Voll, L. Merwin, A. Sebald, J. Eur. Ceram. Soc., 1993, 11, 87-94
[9] M. D. Sacks, N. Bozkurt, G. W. Scheiffele, J. Am. Ceram. Soc., 1991, 74, 2428-2437

Direct Electroless Plating of ZrO_2 Powder

G. Wen[1,2], Z. X. Guo[1,*], and C. K. L. Davies[1]

[1]Department of Materials, Queen Mary and Welstfield College, University of London, Mile End Road, London E1 4NS,UK

[2]School of Materials, Harbin Institute of Technology, Harbin 150001,China

1 Abstract

A study was carried out on the self-reaction of a pure Ni electroless plating bath with N_2H_4 as a reductant and on the direct plating of Ni on ZrO_2 powder particles. The Ni-coated composite powders were characterised by means of optical microscopy, SEM and XRD. The effects of solution composition and processing parameters on the plating rate, solution stability and coating structure were evaluated. Methods for further improving the coating are discussed.

2 Introduction

Composite materials and coatings have been developed to overcome the shortcomings of each constituent. Many of such materials are produced via a powder metallurgy route. The common method to prepare metal/ceramic composite powders for composite materials and coatings is mechanical mixing, which is effective and simple, but has its own weakness. For instance, it is very difficult to obtain a homogeneous particle distribution, especially when there is a large difference in grain sizes or in volume fractions between different constituents.

An electroless plating technique can be used to solve this problem by the formation of a metal coating on the surface of ceramic powder particles. Electroless platings of Cu and Ag have been achieved for this purpose on Al_2O_3, SnO_2, CdO, SiC, SiO_2 and $BaTiO_3$ powders, SiC whiskers, and C fibres [1-6]. Electroless Ni-P plating has been used to coat ceramic particles such as Al_2O_3, SiC and SiO_2, and graphite fibres for preparation of aluminium matrix composites [7-9]. By means of Ni coating, the wettability of aluminium on graphite fibres can be improved, the interfacial reactions between fibre and aluminium matrix are inhibited, and so a C_f / Al composite may be obtained with uniform microstructure and a high tensile strength [7].

The aim of the present study is to develop a pure nickel coating route for ZrO_2 powder particles by means of electroless plating. The coated powder is very desirable for the fabrication of high quality ZrO_2/Ni composites and thermal barrier coatings. For this purpose, N_2H_4 was selected as a reductant for the plating bath. The self-reaction of the bath and the direct plating of ZrO_2 powder were studied The effects of solution composition and processing parameters on the plating rate, solution stability and coating structure were evaluated.

---

* Correspondence: Tel: ++44-20-7882-5569; Fax: ++44-20-9804;
e-mail: x.guo@qmw.ac.uk

3 Experimental

Commercially available ZrO_2 (with 6wt%Y_2O_3) powder was used as the starting material. The powder particles possess an angular morphology with a particle size of 50-150μm. The chemical N_2H_4 was elected as the reductant because it does not involve any contaminants, such as P and B, which can be introduced in the coating, leading to coating embrittlement. Here the bath was composed of the water solution of $Ni(AC)_2$, N_2H_4, Lactic Acid (LA), EDTA(Ethylenediaminetetraacetic acid), and NaOH. Self-reaction of the bath was firstly studied without ZrO_2 addition, and the plating of ZrO_2 powder was then directly conducted without any pre-treatment or activation. The composition of the plating bath and the operating parameters are listed in Table I.

Table I Composition of Plating Bath and the Operating Parameters

Composition of the bath		Operating parameters	
$Ni(AC)_2$	50-60 g/l	Temperature	80-96 °C
EDTA	20-30 g/l	PH value	8.0-11.5
Lactic Acid	40-60 ml/l	Time	5-60 min
N_2H_4	80-100ml/l	Powder load	10-20 g/l
NaOH	35-50 g/l		

The resulting composite powders were characterised by means of weighing, XRD, SEM, and optical microscopy. The effects of solution composition, PH value and temperature on the plating rate and the structure of the coatings were investigated. The plating rate was measured in terms of the weight gain (g) of Ni coating on 1g of ZrO_2 powder per unit time (hour) of plating.

4 Results and Discussion

4.1 Self reaction and stability of the bath

In order to understand and control the plating process, the reaction of the plating bath itself without the addition of ceramic powder particles was firstly evaluated. The phenomena of gas release, formation of brown sediment and deposition of nickel have been seen to depend on the temperature and the composition of the bath. In general, the following main reactions were involved:

$$2Ni^{2+} + N_2H_4 + 4OH^- \rightarrow 2Ni + N_2 + 4H_2O \quad (1)$$

$$Ni^{2+} + 2OH^- \rightarrow Ni(OH)_2 \quad (2)$$

The formation of $Ni(OH)_2$ was mainly dependent on the NaOH content or PH value of the bath. $Ni(OH)_2$ could not be formed when the PH value was lower than 10. $Ni(OH)_2$ could be formed at a lower temperature and a higher PH value, and it could be desolved by stirring and heating to a temperature above 55°C. The formation of $Ni(OH)_2$ should be avoided because it not only consumes Ni^{2+} but also results in the adhesion of powder to the sticky $Ni(OH)_2$.

Reaction (1) was evidently affected by both the PH value and the temperature of the solution. It seems that there was a critical temperature below which the reaction did not take place, but above which the reaction took place and progressed rapidly. This means that there is a large activation energy for this reaction when the catalyst is absent, and the reaction is self-catalysed once triggered. The critical temperature is in the range of 86-97°C, which decreases with increasing PH value. This reaction was so rapid that the solution "boiled" with the release of a large quantity of the gaseous products, particularly when conducted at temperatures higher than the critical point. Such produced Ni mainly deposited on the wall of the container or in the form of large porous particles or agglomerates. Such Ni agglomerates was then released by the agitation of the solution, which has the typical silvery metal colour.

4.2 Direct plating of ZrO_2 powder

The plating of ZrO_2 powder without activation was attempted because of the simplicity of the process. The result showed that Ni can be coated on the surface of ZrO_2 powder by electroless plating without an activation step, indicating that the surface of ZrO_2 particles can reduce the activation energy of the reaction. But the plating was not ideal both in terms of the coating structure and the plating operation. For example, the plating rate was very slow at a lower temperature, typically lower than 0.02g/gh. On the other hand, a large quantity of porous pure Ni particles and Ni pieces were produced at a higher temperature, as shown in Figure 1a. This means that the reaction is still difficult to control. Moreover, the coating is not continuously and uniformly distributed on the particle surface, which implies that the coating had a strong selectivity to the surface morphology, only the coarse surface and the concave area can be effectively coated, as further shown in Figure 1b. It was also noted that the coating was porous, Figure 2, and the porosity increased with increasing temperature. The porosity was believed to be resulted from the rapid release of the nitrogen gas during the reaction.

The composite powder particles consisted of Ni and ZrO_2 only, no other impurities, such as $Ni(OH)_2$, were involved, as confirmed by XRD, shown in Fig.3. It was noted that such deposited nickel possesses a crystalline structure, instead of being amorphous. The latter was the case of Ni-P and Ni-B electroless platings[10-12].

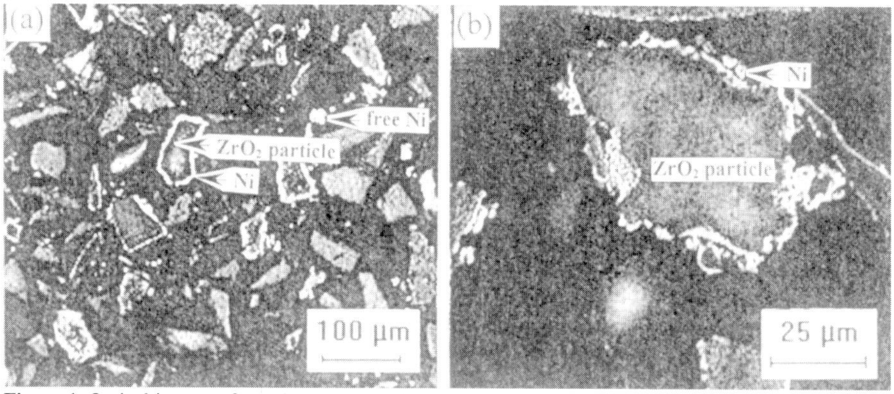

Figure 1. Optical images of powder cross –sections showing: (a) plated ZrO_2 powder and free Ni particles; and (b) selective deposition on the surface of the particles

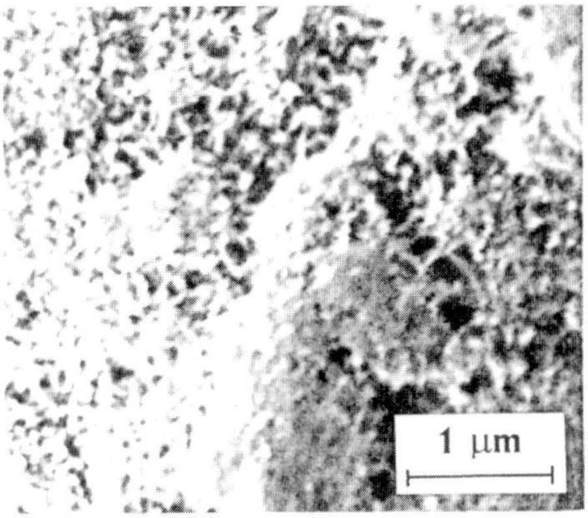

Figure 2. An SEM image showing the porous structure of the coating surface

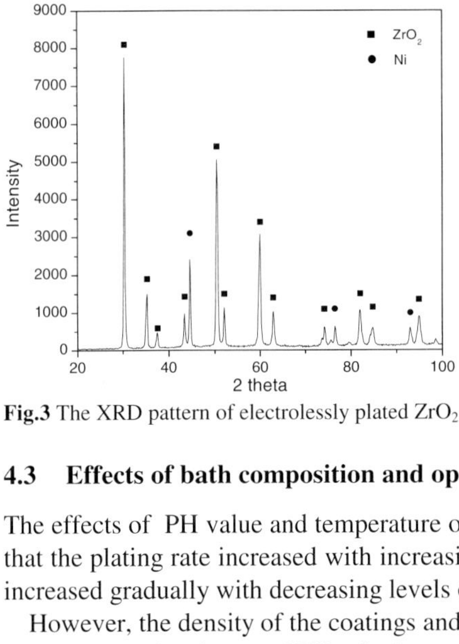

Fig.3 The XRD pattern of electrolessly plated ZrO_2 powder

4.3 Effects of bath composition and operating temperature on the plating rate

The effects of PH value and temperature on the plating rate are shown in Fig.4. It can be seen that the plating rate increased with increasing PH value and temperature. The plating rate also increased gradually with decreasing levels of EDTA and Lactic Acid.

However, the density of the coatings and the stability of the bath (including the formation of $Ni(OH)_2$ and the change of PH value) were reduced with increasing PH value and temperature and with decreasing levels of EDTA and LA. Therefore the electroless plating should be conducted at an optimum condition so that a rapid plating rate and a desirable microstructure can be attained at the same time. In the present study, the PH value of 10-11 and the temperature of 87-92°C have been found to be the optimum. The optimum contents of EDTA and Lactic Acid are about 27g/L and 60ml/L, respectively. Further improvement of the plating technique is under investigation. Particular attention is paid to surface treatment and activation of the ZrO_2 powder prior to plating.

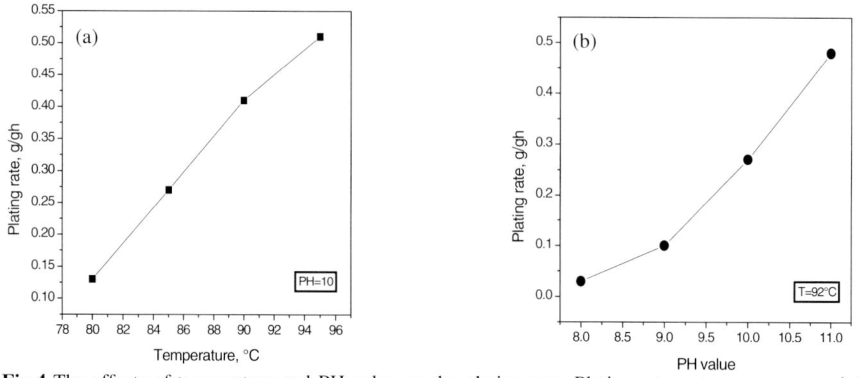

Fig.4 The effects of temperature and PH value on the plating rate: Plating rate vs temperature; and (b) plating rate vs PH value

5 Conclusions

A self-reaction of electroless plating route has been developed for the synthesis of a pure Ni coating on ZrO_2 powder particles. It was noted that the bath can spontaneously react in an uncontrollable manner at temperatures higher than the critical value, resulting in the formation of porous Ni particles and Ni film on the wall of the container. The ZrO_2 particles can be directly plated without any pre-treatment, but the plating operation was difficult to control and the coating structure was highly porous and non-uniform. In the plated ZrO_2 powder there also are free nickel particles. PH value and temperature largely influence the plating rate and the coating microstructure. With increasing PH value and temperature, the plating rate can be improved but the coating becomes more porous. The optimum composition and operating temperature have been determined as the PH value of 10-11, the temperature of 87-92°C and EDTA of 27g/L and Lactic Acid of about 60ml/L, respectively.

6 Acknowledgement

The project was kindly supported by The Royal Society under a FCO Chevening Fellowship, awarded to G.Wen.

7 References

[1] H. Chang, C. H. Pitt, G. B. Alexander, Journal of Materials Science, 1993, 28, 5207-5210.
[2] S. Y. Chang, S. J. Lin, Scripta Materialia, 1996, 35, 225-231.
[3] P. Yih, D. D. L. Chung, Journal of Materials Science, 1996, 31, 399-406.
[4] S. Y. Chang, J. H. Lin, S. J. Lin, and T. Z. Kattamis, Metallurgical and Materials Transactions A, 1999,30A, 1119-1136.

[5] P. K. Lee, IEEE Trans. CHMT. 1980, 3, 4.
[6] X. Liu, W. Y. Shih, and W. H. Shih, J. Am. Ceram. Soc. 1997, 80, 2781-88.
[7] Z. Y. Shi, X. Z. Wang, and Z. M. Ding, Applied Surface Science, 1999, 140, 106-110.
[8] Y. J. Lin, B. F. Jiang, J. Am. Ceram. Soc. 1998, 81, 2481-84.
[9] R. L. Deuis, C. Subramanian, J. M. Yellup, K. N. Strafford and P. Arora, Scripta Metallurgia et Materialia, 1995, 33, 1217-1224
[10] C. D. Young, J. G. Duh, IEEE Transactions on Components, Packaging and Manufacturing Technology-Part A, 1993, 17, 644-653.
[11] K. H. Yoon, H. S. Park, S. O. Yoon, H. I. Song, Journal of Materials Science Letters, 1989, 8, 1442-1444.
[12] Y. J. Lin, B. F. Jiang, J. Am. Ceram. Soc., 1998, 81, 2481-84.

Structural Evolution of a Stoichiometric Silicon Oxycarbide from Gel Through Glass to Decomposition

H. Brequel[1] / L. Schiffini[1] / S. Enzo[1,2]
[1]Dipartimento di Chimica, Sassari (Italy)
[2]Unità INFM Dipartimento di Chimica Fisica, Venezia (Italy)

1 Abstract

The appropriate mixture of two polymeric precursors methyldiethoxysilane $HSi(OEt)_2CH_3$ (MDES) and triethoxysilane $HSi(OEt)_3$ (TREOS) leads, by sol-gel method to the true stoichiometric silicon oxycarbide compounds $SiC_{0.3}O_{1.5}$. Pyrolysis of the specimen at 1000°C makes it possible to obtain the corresponding glass, and further heat treatment leads to decomposition. Wide Angle X-ray Diffraction is presented for each step of pyrolysis. The short-range order is extracted by the Radial Distribution Function approach until 1000°C, when a single amorphous phase is observed. The relation between short range order and phase separation is discussed with reference to results obtained at temperatures above 1000 °C.

2 Introduction

In the synthesis of ceramic materials exhibiting new or improved properties, silicon oxycarbide glasses $Si_xC_yO_z$ are becoming the subject of intense studies. Unfortunately, they are difficult to prepare without formation of a quite large amount of carbon dispersed in the network, even by sol-gel method [1-3]. A way to obtain a true stoichiometric compound with the general formula $SiC_xO_{2(1-x)}$ is to use a mixture of suitable precursors in the following proportion : two parts of methyldiethoxysilane $HSi(OEt)_2CH_3$ (MDES) for one part of triethoxysilane $HSi(OEt)_3$ (TREOS) [4-6]. The key step of the process is the pyrolysis of the crosslinked polymeric precursors to form a glass (at 1000°C). The understanding of the synthetic route is made difficult by the non-crystalline character of the compounds and various reactions and mechanisms involved, and the non-crystalline character of the compounds. After extended heat treatment of the glass up to 1500°C, the composition is slightly changing and a phase separation leading to crystalline phases can be observed [7].

In this work, we will use X-ray Diffraction (XRD) at wide angle to study five different steps of evolution of the materials : the dried gel, the glass obtained after pyrolysis of the gel at 1000°C, the glass after further pyrolysis at 1200°C, 1400°C and 1500°C. When the XRD patterns show characteristic features of an amorphous phase the interpretation can be done conveniently by the Radial Distribution Function (RDF) approach, which gives information on the short-range order in the sample. On the other hand, when different crystalline phases start to appear, a Rietveld analysis of the semicrystalline patterns will be conducted, with emphasis on quantitative analysis of SiC.

3 Experimental Procedure

A mixture of TREOS and MDES in the molar ratio 2 for 1 respectively was prepared. Ethanol and distilled water were added to perform the hydrolysis/condensation process [6]. The gel is obtained after drying in air. Then it is pyrolyzed for one hour under Ar flux at 1000°C. The resulting glass was studied by RDF and subsequently heat treated at 1200, 1400 and 1500°C under inert atmosphere for the study of crystallization phenomena. For each case, the powder obtained is ground down to a grain size between 100 and 350 μm. The powder was pressed to form a pellet of ca. 14 mm in diameter with thickness between 2 and 3 mm and subjected to XRD. The general composition of each sample was determined by chemical analysis and is presented in Table 1.

3.1 XRD patterns recording

X-ray diffraction patterns were collected with a Bragg-Brentano powder diffractometer using Cu Kα radiation ((λ= 1.54178 Å) and Mo Kα radiation (λ= 0.7103 Å) for the classic XRD analysis and the RDF calculation respectively. In the later case, the molybdenum source was used to cover a wide interval of scattering vector S. A graphite monochromator was used in the diffracted beam, and the data collection was carried out from 2θ= 6 to 145° with a step of 0.02° (with Cu) or 0.2° (with Mo).

3.2 RDF procedure and calculation

The calculation procedure was computed in an home-made program including the following steps : corrections of air scattering, polarization and absorption, normalization process by the method of Krogh-Moe-Norman [8, 9] ; calculation of the reduced scattering intensity i(S) and the interference function Si(S) ; data extrapolation to the origin and interpolation at equispaced values in S, application of a damping factor to limit the error in the following Fourier Transform (FT) procedure. This gives the reduced RDF Q(r), and the RDF itself D(r) is easily finally calculated.

Table 1. Chemical analysis of the sample after each step of treatment.

Samples	Empirical formula	Composition (Wt. %)			
		Si	C	O[(a)]	H[(b)]
Dried gel	$SiC_{0.354}O_{1.283}H_{1.946}$	51.24	7.75	37.46	3.55
Glass	$SiC_{0.300}O_{1.491}$	50.57	6.49	42.94	0.00
Glass pyr. at T=1200°C	$SiC_{0.267}O_{1.493}$	50.90	5.80	43.30	0.00
Glass pyr. at T=1400°C	$SiC_{0.258}O_{1.406}$	52.33	5.76	41.91	0.00
Glass pyr. at T=1500°C	$SiC_{0.260}O_{1.419}$	52.11	5.78	42.11	0.00

a) the amount of oxygen is calculated by difference.
b) under the detection limit of 0.3 Wt. %, the hydrogen content is considered to be zero.

4 Results

4.1 RDF calculation on data from dried gel and glass

The reduced RDF obtained for the gel and the glass are summarized in Figure 1. Even if significant features are found up to 6 Å or more, we will consider here only the information attributed to the two first shells of neighbors i.e. for the radial distance up to 3.6 Å. They are quite similar to what is found in the literature for various forms of amorphous silica [10-12], so we assumed that the description of the network by connected tetrahedral units $Si(C_wO_{4-w})$ is valid.

In the gel, the average positions of the first and second shells are found around 1.63 and 3.02 Å respectively, while in the glass the positions for the first and second shell are 1.66 Å and 3.05 Å respectively. From these values it is possible to calculate the average Si-O-Si bond angle : 135.8° for the gel and 133.5° for the glass. It is important to keep in mind that the differences in local order observed between the gel and the glass can be due to the presence of H atoms, which are invisible with XRD, but the effect of which can be appreciated on the local arrangement of Si, O and C atoms.

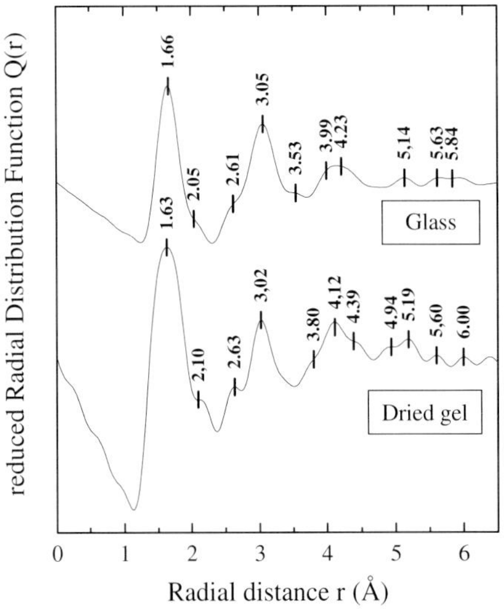

Figure 1. Comparison between reduced RDF obtained for gel and glass, in the range 0 – 6 Å

4.2 X-ray study of the phase separation during heat treatment

The phase separation occurring in silicon oxycarbide glasses was already studied as a function of the carbon content in a previous work [7]. Here we are focusing our attention on the true stoichiometric compound, , in order to correlate the local order found by RDF with the mechanism of decomposition. In the partially crystallized glasses we have carried out a Rietveld refinement, which was able to account quantitatively for also the amorphous matrix component.

Table 2. Quantitative analysis, crystallite size and microstrain of SiC in samples pyrolyzed calculated from Rietveld analysis of XRD.

Samples	Amount (Wt. %)	Crystallite size (nm)	Microstrain ($\times 10^3$)
Glass pyr. at T=1200°C	5 ± 5	1.0 ± 0.3	8.3 ± 3.5
Glass pyr. at T=1400°C	23 ± 5	0.9 ± 0.3	2.8 ± 1.4
Glass pyr. at T=1500°C	22 ± 5	1.5 ± 0.3	0.4 ± 0.2

In relation to the SiC nanocrystalline component, average crystallite size and microstrain were also obtained. The experimental data and the calculated profiles are shown on Figure 2 as a function of the scattering vector for the comparison among patterns taken with different wavelength. The values obtained for quantitative analysis, crystallite size and microstrain are summarized in Table 2. At 1200°C the amount of SiC is hardly detected, and is actually within the range of experimental significance. At higher temperature, the presence of SiC becomes evident, top-level of 22 Wt. %, but seems to remain constant between 1400 and 1500 °C. Conversely, the average crystallite size remains constant (about 1 nm) at 1200 and 1400°C and increases at higher pyrolysis temperatures.

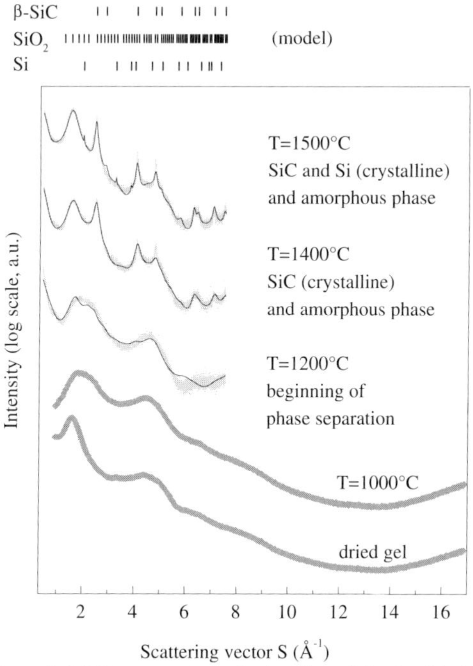

Figure 2. XRD paterns (gray dots) collected for the dried gel and the gel pyrolyzed at T, with the corresponding Rietveld fit (black line) when T>1000°C.

5 Discussion

The RDF gives only an average of local neighboring of each atom, which is in any case a starting point for the description of the amorphous network. We know from the chemical formula of the precursors that the typical units forming the polymer are SiO_3H and $HSiO_2CH_3$ respectively called T^H and D^H. The ideal composition of the polymer corresponding to the mixture in a ratio 2:1 is $SiC_{0.33}O_{1.33}H_2$ which is close to the chemical analysis of the gel. Thus, it seems reasonable to consider the gel made of uniformly distributed units T^H and D^H, in the ratio 2:1. Then the RDF inversion will give the correlation function of such units. Using 1.46 Å, 1.62 and 1.88 Å for the bond lengths of Si-H, Si-O and Si-C, the two latter being taken from distances in SiO_2 and SiC respectively, we obtain a typical distance of first neighbors to be around 1.60 Å. The experimental result of 1.63 Å suggests that this approach may be rather crude. This value obtained for the dried gel is close to the first neighbor distance 1.62 Å reported for amorphous silica. The value of the average bond angle Si-O-Si (135.8°) is low with respect to the range 144 – 151° found in the literature for various forms of amorphous silica [12-14]. This suggests a tendency of the gel to form smaller chains of tetrahedra, which is induced by the presence of H directly linked to Si. However, these chains are longer than those formed in the glass.

It is admitted that the main role of the pyrolysis process at 1000°C is to remove the hydrogen atoms from the gel, even if the mechanism is complex and involves recombination reactions with radicals[15-17]. This is easily understood since the weakest bonds in the gel are Si-H, C-H, and Si-C with dissociation energies from less than 300 to 450 $kJ.mol^{-1}$. The chemical analysis of the glass gives not exactly the stoichiometric composition expected : this corresponds to the presence of a free carbon phase representing 13 % of the total amount of carbon in the glass. This means that simultaneously to the release of H, C atoms manifest a tendency to arrange in small domains within the silicon oxycarbide network, for example in polyaromatic graphene rings as described by Pantano and al. [3]. The effect of these domains is hardly detectable in the RDF, and their presence will not be taken into account for the interpretation of the RDF at 1000 °C.

Assuming that the units describing the network are only SiO_4 and $SiCO_3$ with an average bond distance of 1.685 Å for the latter, we determined 46.2% of SiO_4 units in the glass. Again this example is perhaps too simple but it confirms in a first approximation that the atoms of carbon which are not involved in the free phase, are well dispersed in the glass (more than 50% of the tetrahedral units are including one C atom). During further pyrolysis, these units could act as starting nuclei for the phase separation, (as it is suggested by the results of the Rietveld analysis), the growth of the crystalline phase being limited, even at high temperature, by the surrounding glassy network.

6 Conclusions

It was shown that the bonding topology in the gel is made of tetrahedral units typically SiO_3H and $HSiO_2CH_3$ (and certainly minor quantities of other species) with an average first neighbor distance of 1.63 Å, although it is not clear how they are connected one to each other. An estimation of the bond angle between the first and the second shell gives 135.8°, supporting the idea that rings of tetrahedra are formed, wider than in amorphous silica. After the first pyrolysis at 1000°C, the tetrahedral units have an increased average bond length and a smaller Si-O-Si bond angle, 1.66 Å and 133.5° respectively. Moreover, the C atoms are uniformly dispersed in the glassy network. This configuration with numerous Si-C bonds already formed makes the initial stage of the phase separation easier, but the surrounding Si-O links are limiting the growth of SiC up to 1500°C.

It will be useful to confirm these results by neutron diffraction, especially in the case of the gel which contains a large amount of hydrogen. Another possibility is to extract the RDF from the XRD pattern obtained at 1200°C, which represents the very beginning of the phase separation. Finally, a comparison with results obtained from ^{29}Si NMR and MET would help to confirm the mechanism of separation described here.

7 Acknowledgements

This work is supported by the European Community under the TMR contract n.° ERB FMRX-CT980161. Thanks are due to the Laboratoire Central d'analyse - Vernaison (France) for chemical analysis, and to Dr. L. Lutterotti for providing a copy of Rietquan. The authors are grateful to J. Parmentier and G.D. Sorarù for preparation of the samples, and to S. Walter and J. Latournerie for density measurements.

8 References

[1] J. Lipowitz, Am. Ceram. Soc. Bull. 1991, 70, 1888 - 1894
[2] R. Campostrini, G. D'Andrea, G. Carturan, R. Ceccato, G.D. Sorarù, J. Mater. Chem. 1996, 6, 585 – 594
[3] C.G. Pantano, A.K. Singh, H. Zhang, Journal of Sol-Gel Science and Technology, 1999, 14, 7 - 25
[4] F. Babonneau, K. Thorne, J.D. Mackenzie, Chem. Mater. 1989, 1, 554 - 558
[5] G.M. Renlund, S. Prochazka, R.H. Doremus, J. Mater. Res. 1991, 6, 2716 - 2722
[6] G.D. Sorarù, G. D'Andrea, R. Campostrini, F. Babonneau, G. Mariotto, J. Am. Ceram. Soc. 1995, 78, 379 - 387
[7] H. Brequel, J. Parmentier, G.D. Sorarù, L. Schiffini, S. Enzo, Nanostructured Materials, 1999, to be published
[8] J. Krogh-Moe, Acta Crystallogr. 1956, 9, 951 - 953
[9] N. Norman, Acta Crystallogr. 1957, 10, 370 – 373
[10] H.E. Robson, L. Broussard, Advances in X-ray Analysis 1967, 8, 108 - 116
[11] J.H. Konnert, J. Karle, Acta. Cryst. 1973, A29 702 - 710
[12] G.S. Henderson, M.E. Fleet, G.M. Bancroft, J. Non-Cryst. Solids 1984, 68, 333 – 349
[13] R.L. Mozzi, B.E. Warren, J. Appl. Cryst., 1969, 2, 164 -
[14] J. Zarzycki, Verres Réfractaires, 1957, 1, 3 -
[15] L. Bois, J. Maquet, F. Babonneau, H. Mutin, D. Bahloul, Chem. Mater. 1994, 6, 796 - 802
[16] R. Corriu, D. Leclercq, P. H. Mutin, A. Vioux, J. Mat. Sci. 1995, 30, 2313 -2318
[17] L. Bois, J. Maquet, F. Babonneau, D. Bahloul, Chem. Mater. 1995, 7, 975 - 981

III Reliability

Reliability of Ceramic Valve Plates for Common-Rail Injection Pumps

R. Speicher[1], G.A. Schneider[2], W. Dreßler[3], G. Lindemann[3], H. Böder[3], V. Knoblauch[1]
[1] Robert Bosch GmbH, Technical University Hamburg-Harburg/Advanced Ceramics Group, Germany
[2] Technical University Hamburg-Harburg/Advanced Ceramics Group, Germany
[3] Robert Bosch GmbH, Stuttgart, Germany

1 Introduction

In order to use ceramic materials in automotive applications, high reliability has to be ensured. Therefore, calculations of failure probabilities and lifetimes are required. For metal components, calculations of fatigue limits and guidelines for materials selection are well-known. Ceramic materials however, are characterized by brittle failure without plastic deformation. Therefore, voids caused by production which are predominantly present in the specimens volume, and voids from surface grinding are failure origins. These voids are different in shape, length and orientation to principal stresses, which results in a large scatter in strength. A statistical approach is required as well as convenient methods to determine materials data. For the evaluation of strength data it is important to know which specimen type has to be used. Often, data from standard specimens do not characterize the failure behavior of the ceramic component. In such cases the component itself has to be investigated. However, the determination of parameters of subcritical crack growth is influenced more by the stress state and the shape of loading.

In this paper, the dimensioning of ceramic components is demonstrated for a valve plate of Si_3N_4 which is used in high-pressure pumps for Common-Rail injection systems. Finite Elements (FE) calculations prior to lifetime considerations are done in order to investigate the stress states during cyclic loading. For calculations of failure probabilities and lifetimes the FE-post processor STAU [1] is used. 4-point-bending strengths as well as results from valve plate tests are presented. Static and cyclic subcritical crack growth parameters are investigated in bending techniques as well. Three different material suppliers of commercial Si_3N_4 are investigated.

2 Common-Rail Injection Pumps

In the last years there was an increasing interest in improving the performance of direct-injection automotive diesel engines, in order to reach low emissions of NO_x and low fuel consumption. Common-Rail systems offer a large variability to improve these parameters, e.g. mass flow, pilot injection, injection time.

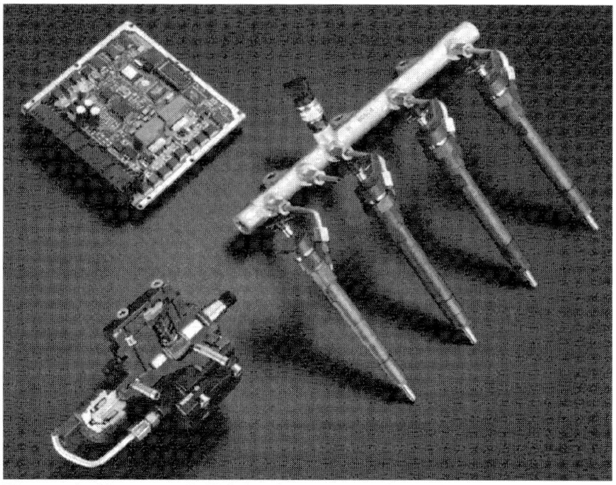

Figure 1. Common-Rail injection system, including high-pressure pump, rail, injectors and electronic control unit

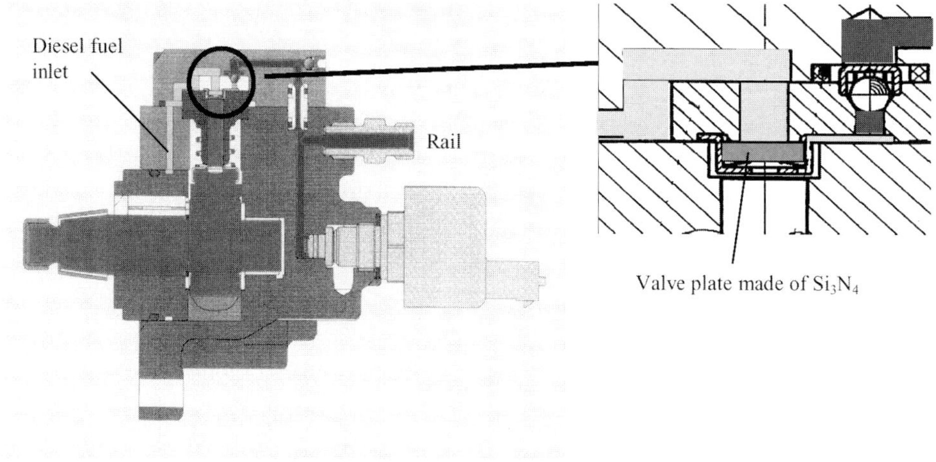

Figure 2. High-pressure pump with detailed view of the cylinder head including valve plate

Fig. 1 shows the Common-Rail injection system including high-pressure pump, rail, injectors and control device. The pump generates a maximum pressure of about 1350 bar, which is led via the rail to the injectors. The rail offers the advantage of acting as a pressure reservoir which provides a constant pressure during the whole injection period. The control device governs the pump and the injectors. The high-pressure pump with the valve plate made of Si_3N_4 which operates as an inlet valve is shown in more detail in fig. 2. When the piston goes downwards, the existing depression opens the valve plate until it is constrained by a cage, which can not be modified because of the compression cycle. For high performance it is very important that the cylinder is filled totally with fuel while the piston moves downwards. Therefore, the thickness of the valve plate has to be as small as possible. The aim of all investigations presented here is to determine the smallest thickness for reliable operation.

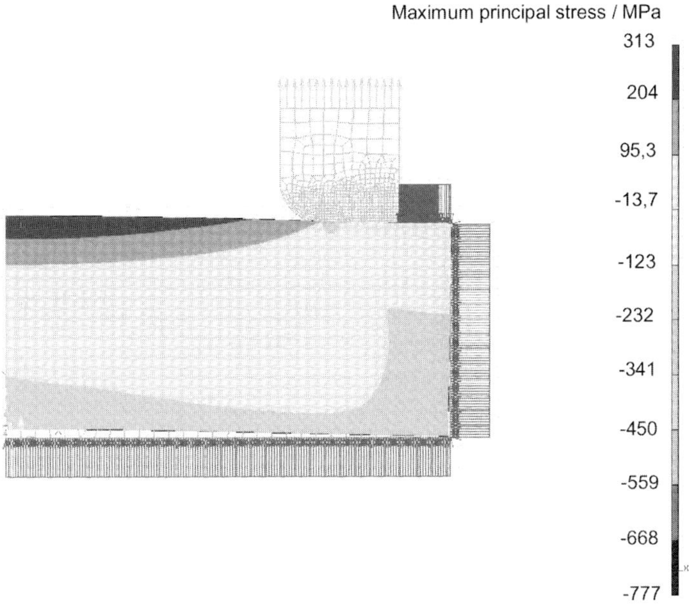

Figure 3. FE analysis of the valve plate at 1420 bar. Used materials data: valve plate E = 300 GPa, ν = 0.26 housing: elast.-plast. material behavior E = 207 GPa, R_{es} = 800 MPa, R_m = 1200 MPa

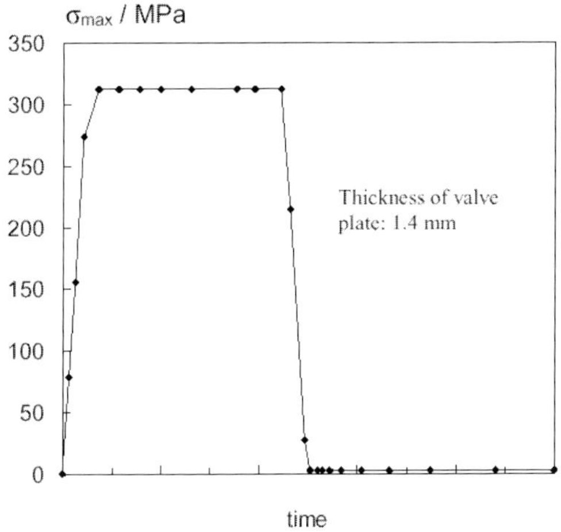

Figure 4. Maximum stress at the valve plate as a function of time

Fig. 3 shows the FE model of the valve plate including housing for a plate thickness of 1.4 mm and a plate diameter of 6 mm. The applied pressure is 1420 bar due to the maximum error of 70 bar in the pressure sensor. The analysis is done with the FE code ABAQUS using the surface-based contact algorithm. Axisymmetric elements are used with an element size of 0.02 mm at the contact areas. A friction coefficient of 0.2 is used. A variation to 0.5 did not result in a difference in the resulting stresses. Fig. 3 shows the resulting biaxial stress distribution with a maximum stress of 313 MPa at the center of the tensile surface of the valve plate. The applied pressure is a function of time as shown in fig. 4. The maximum speed of the high pressure pump is 3000 min^{-1}.

3 Materials Data

In order to carry out calculations of reliability, strength and subcritical crack growth data have to be determined. For measurements of strength, both 4-point bending tests and valve plate tests have been done, because the influence of specimen geometry and surface condition has to be examined. However, static and cyclic fatigue tests are performed using only 4-point bending tests, because these are not subject to be influenced by the specimen shape.

3.1 Strength measurements

4-point bending measurements were carried out according to DIN EN 843-1 [2]. The cross-section of the specimens was 3 x 4 mm. An inner load span of 20 mm and an outer load span of 40 mm were used. The velocity of loading was 5 mm/min, which resulted in fracture of the specimens after less than 10 s.

The valve plate tests have been performed using a special kind of ring-on-ring test adapted to the dimensions of the valve plate. The supports are made of WC/Co-hard material (1800 HV30) with a Young's modulus of 640 GPa and a Poisson's ratio of 0.22. The bending strength is about 3100 MPa. A FE analysis of the stress distribution is shown in fig. 5.

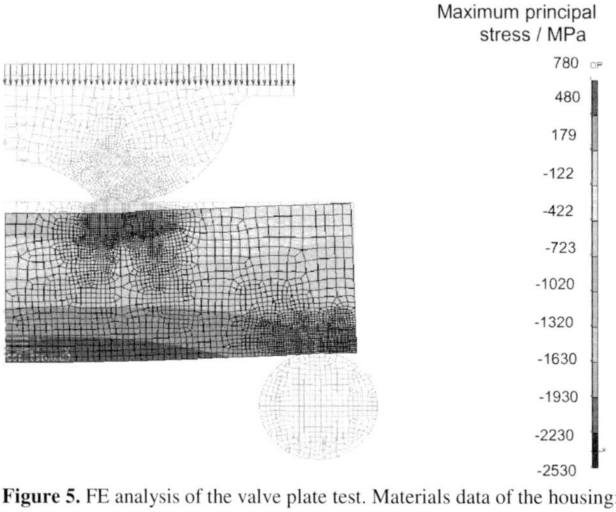

Figure 5. FE analysis of the valve plate test. Materials data of the housing: E = 640 GPa, ν = 0.22

Three different material suppliers, who use different shaping and sintering procedures, have been examined. Two material qualities were axially pressed, the other supplier used an injection molding process. The sintering process is done by gas pressure or hot isostatic pressing procedures. All these materials are commercially available.

In order to compare 4-point bending data and valve plate data, the effective volume of the specimens have to be taken into account. Therefore, the size effect has to be considered, according to:

$$\sigma_{0,corr} = \sigma_{0,\text{valve plate}} \cdot \left(\frac{V_{\text{eff,valve plate}}}{V_{\text{eff,4-point}}} \right)^{\frac{1}{m}} \tag{1}$$

In equation (1) $V_{\text{eff,valve plate}}$ is the effective volume of the valve plate in the valve plate test, and $V_{\text{eff,4-point}}$ the effective volume of the 4-point bending bar. $\sigma_{0,\text{valve plate}}$ and m are the Weibull parameters obtained from a Weibull evaluation using the Maximum likelihood method [3]. $\sigma_{0,corr}$ is the strength level of the valve plate calculated for the effective volume of 4-point bending bars. This value can be compared with the Weibull parameter σ_0 of 4-point bending tests. $V_{\text{eff,valve plate}}$ and $V_{\text{eff,4-point}}$ are calculated using STAU, where the orientation of flaws is considered. Therefore, analytical solutions for $V_{\text{eff,4-point}}$, e.g. given in [3], cannot be used. The results are shown in fig. 6. It is obvious that for all materials the characteristic strength σ_0 of 4-point bending tests is higher than the strength derived from valve plates. However, for material C the smallest difference is obtained. That means that the material supplier C fits best the strength level of bending bars and valve plates.

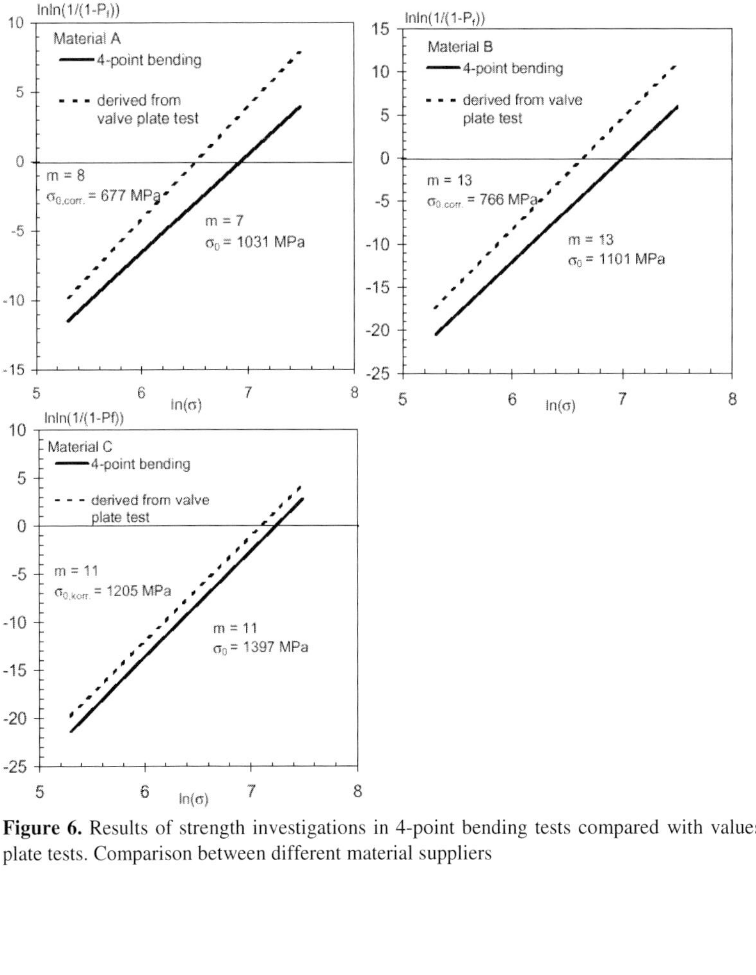

Figure 6. Results of strength investigations in 4-point bending tests compared with values derived from valve plate tests. Comparison between different material suppliers

3.2 Subcritical crack growth measurements

Static and cyclic fatigue tests were performed using 4-point bending techniques. To evaluate v-K-curves the modified lifetime method proposed by Fett [4, 5] has been used. For the calculation of the reliability of valve plates during cyclic loading in the high-pressure pump, no static experiments would be necessary. However, to check cyclic fatigue effects both tests have been done.

For static tests a maximum loading time of 500 h was chosen. The probability of spontaneous failure was set to a maximum of 10 % except for material A, where the low Weibull modulus led to a higher failure probability. The cyclic fatigue tests were done at R = 0.2. In the application a stress ratio of R = 0 will occur. But the chosen ratio of R = 0.2 enables an easier experimental set-up. The frequency in the high-pressure pump reaches a maximum value of 50 Hz. However, the resonance-frequency testing machine used necessitated a frequency of 78 Hz. The relation for static subcritical crack growth has to be written as a power law according to:

$$v = \frac{da}{dt} = A^* \cdot \left(\frac{K_I}{K_{Ic}}\right)^n \tag{2}$$

and respectively for cyclic loading:

$$v = \frac{da}{dt} = f \cdot \frac{da}{dN} = A^* \cdot \left(\frac{K_{I,max}}{K_{Ic}}\right)^n . \tag{3}$$

The results are shown in table 1. Material A shows very low n values for static and cyclic fatigue which are of interest for lifetime calculations because lifetimes are strongly influenced by this parameter. However, the difference between static and cyclic loading is very small. For Material B a large difference between static and cyclic n values was obtained, which means a cyclic fatigue effect. The result of material C is quite surprising, because the cyclic fatigue parameter n is larger than the static one. It has to be mentioned that the evaluation of parameters of subcritical crack growth has some uncertainties as revealed in [6].

Table 1. Parameters of subcritical crack growth from static and cyclic bending tests

Material	n_{stat}	A^*_{stat} m/s	n_{cycl}	A^*_{cycl} m/s
A	27	$3.2 \cdot 10^{-8}$	22	$1.6 \cdot 10^{-7}$
B	47	$8 \cdot 10^{-4}$	24	$2 \cdot 10^{-5}$
C	35	$4 \cdot 10^{-8}$	38	$1.2 \cdot 10^{-7}$

In order to evaluate if the determination of subcritical crack growth in 4-point bending techniques is valid for valve plates, cyclic fatigue investigations of valve plates have been performed using a similar experimental set-up as shown in fig 5. A radial fixation of the specimens prevented a sliding motion during cyclic loading. The tests were carried out at 84 Hz using a resonance-frequency machine which counted the cycles until failure. A statistical evaluation of these numbers of cycles together with the Weibull modulus of the inert strength distribution reveals the parameters of subcritical crack growth. Using a least square fit, n = 27 and $A^* = 1.2\cdot10^{-5}$ m/s is found for material B which is in very good agreement with data from cyclic 4-point bending tests.

4 Calculations of Reliability

4.1 Failure probability

Engineering applications of ceramic components need a FE-based calculation before production of prototypes, because they are too expensive and very time-consuming. Stress analysis of the components using FE in combination with reasonable materials data provide the basis for calculations of reliability. Ziegler [7] has expanded the existing models for the calculations of failure probability for cyclic periodic loadings. It leads to the failure probability P_f as a function of cycles N:

$$P_f(N) = 1 - \exp\left[-\frac{1}{V_0}\int_V dV \frac{1}{4\pi}\int_\Omega d\Omega \left(\frac{\sigma_0^2}{B}\cdot N\cdot\int_0^T\left(\frac{\sigma_{eq}(t)}{\sigma_0}\right)^n dt\right)^{\frac{m}{n-2}}\right] \tag{4}$$

In equation (4), V is the component volume, V_0 a reference volume, $d\Omega$ represents the consideration of flaw orientation, σ_0 and m are Weibull parameters of the inert strength distribution and n and B are the parameters of subcritical crack growth at cyclic loading. B and A^* are linked with equation (5)

$$B = \frac{2}{A^*\cdot Y^2\cdot(n-2)}K_{Ic}^2 \tag{5}$$

Y was chosen to be 1,99 according to very small cracks in 4-point bending [8]. T is the period time of one cycle. It has to be mentioned that σ_{eq} is a function of time and equation (4) is only correct for periodic stresses [7]:

$$\sigma_{eq}(t+T) = \sigma_{eq}(t) \tag{6}$$

The results for all three materials are shown in table 2. It is worth mentioning that the Weibull parameter derived from investigations of valve plates have been included in the calculations. As shown, only material C does not exceed the maximum acceptable failure probability of $1\cdot10^{-6}$.

To investigate the influence of materials data, m, n_{cycl} and A^*_{cycl} from material A and σ_0 from B was used for a calculation (combination 1, see table 2). The difference between combination 1 and material C is larger than 10^2. Using the inert strength data of material C and the subcritical crack growth data from B (combination 2) results in the difference of approx. 10^2 related to material C. Hence, for this cyclic calculation, both strength data and subcritical crack growth data have similar influence on the failure probability.

Table 2. Failure probability of valve plates after $4 \cdot 10^8$ cycles for materials A, B and C including used material data. Comparison with two combinations.

Material	m	σ_0 MPa	n_{cycl}	A^*_{cycl} m/s	P_f ($N=4 \cdot 10^8$)
A	8	677	22	$1.6 \cdot 10^{-7}$	$> 1 \cdot 10^{-2}$
B	13	766	24	$2 \cdot 10^{-5}$	$> 1 \cdot 10^{-3}$
C	11	1205	38	$1.2 \cdot 10^{-7}$	$< 1 \cdot 10^{-6}$
Combination 1	11	766	38	$1.2 \cdot 10^{-7}$	$1.27 \cdot 10^{-4}$
Combination 2	11	1205	24	$2 \cdot 10^{-5}$	$8.5 \cdot 10^{-5}$

4.2 Proof test

As shown in table 2, materials A and B do not fulfill the required survival probability. Thus, a proof test of the valve plates has been introduced. In a proof test all components are stressed with a load higher than the load during operation. Components which would fail during real operation, will fail spontaneously in the proof test.

The stress state in the valve plate should be as identical as possible in the proof test and in the high-pressure pump. For proof testing, the modified ring-on-ring supports shown in fig. 5 were used. A comparison of the stress distribution at the tensile surface in both loading conditions as a function of valve plate radius is illustrated in fig. 7.

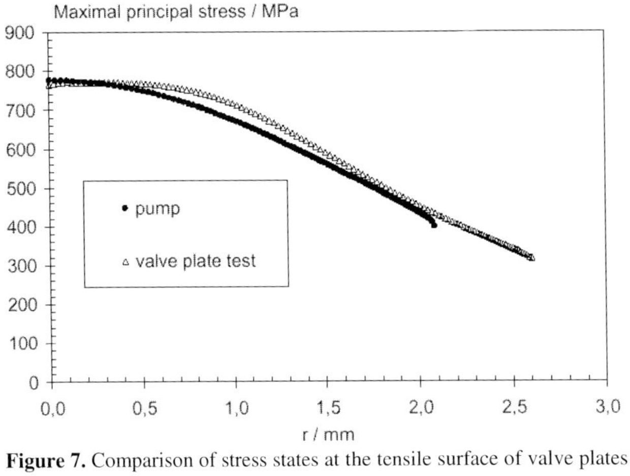

Figure 7. Comparison of stress states at the tensile surface of valve plates in high-pressure pump and valve plate test

Applying the same stress in the center of the valve plate, the stress in the valve plate during proof testing is slightly higher than in the high-pressure pump over the entire radius until the upper support. This means a conservative construction of the proof test. For the calculation of proof test ratios equation (7) is used:

$$\left(\frac{\sigma_P}{\sigma}\right)=\left(\frac{\sigma^2 \cdot t_{B,min}}{B}\right)^{\frac{1}{n-2}} \tag{7}$$

n and B are parameters of subcritical crack growth under cyclic loading conditions, σ the stress of the valve plate in the high-pressure pump, σ_P the maximum stress of the valve plate during proof testing and $t_{B,min}$ the operation time of the pump without failure of the valve plate. The result for a valve plate thickness of 1.4 mm is shown in fig. 8. It can easily been seen that for lower values of n the proof test ratio is more influenced by the parameter A^* than for high values of n.

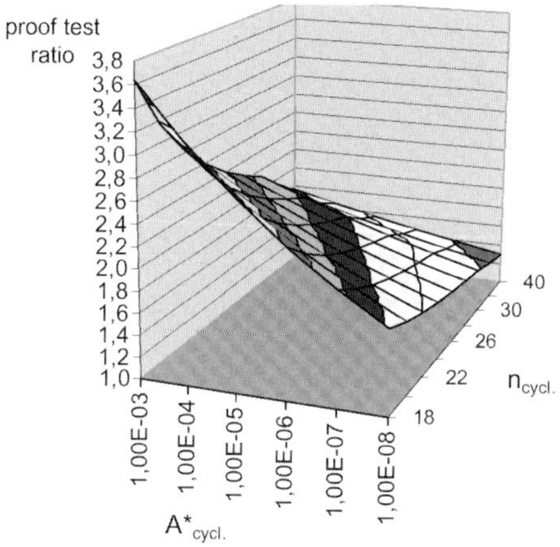

Figure 8. Proof test ratio for valve plates with thickness of 1.4 mm using cyclic subcritical crack growth parameters

5 Summary and Discussion

Calculations of failure probabilities using the FE post processor STAU are demonstrated for a cyclic loaded valve plate made of Si_3N_4 which is used in Common-Rail high-pressure injection pumps. Three different materials of commercial suppliers have been investigated.

It is shown that inert strength measurements of valve plates have to be used, because 4-point bending results do not match the strength distribution of valve plates. The Weibull modulus was between 7 and 13, the characteristic strength derived from valve plate tests and calculated for the effective volume of 4-point bending bars was between 677 and 1205 MPa.

Parameters of subcritical crack growth are determined under static cyclic conditions using 4-point bending techniques. Static n values ranged from 27 to 47, cyclic values from 22 to 38. For one material, cyclic fatigue tests of valve plates were performed. There was a very good agreement with the results of 4-point bending bars.

Calculations of failure probabilities after $4 \cdot 10^8$ cycles showed that two materials exceed the tolerable failure probability of $1 \cdot 10^{-6}$. They reached failure probabilities after operation time of about $1\cdot10^{-2}$. Therefore, proof testing procedures were investigated. Proof test ratios until a value of 2 are necessary for reliable operation of the valve plates in high-pressure pumps.

Further investigations have to be done examining the influence of R-curve behavior. Toughening mechanisms necessitate to modify K_I in equation (2) according to:

$$K_{I,tip} = K_{I,appl} + K_{I,sh} \quad (K_{I,sh} < 0) \tag{8}$$

A hint of existing R-curve behavior may be the difference between static and cyclic n value of subcritical crack growth of material B. Crack bridging of elongated grains is subject to cause R-curve behavior. These crack bridges are destroyed during cyclic loading, which results in the low n-values obtained.

6 Acknowledgements

The authors would like to acknowledge the Institute of ceramic engineering, University of Karlsruhe, Germany, for the measurements of materials data as well as the material suppliers for their interfirm cooperation.

7 References

1. Heger, in Fortschritt-Berichte VDI, Reihe 18, Nr. 132, VDI-Verlag Düsseldorf, Germany, 1993
2. DIN EN 843-1, Monolithische Keramik, Mechanische Eigenschaften bei Raumtemperatur, Teil 1: Bestimmung der Biegefestigkeit, Deutsches Institut für Normung, Berlin, Germany, 1995
3. Munz, T. Fett: Ceramics, Mechanical properties, failure behaviour, Materials Selection, Springer Verlag Berlin, Germany, 1999
4. T. Fett, D. Munz, J. Am. Ceram. Soc., Vol 68, 8, C-213 – C-215, 1985
5. T. Fett, G. Martin, D. Munz, G. Thun, J. of Materials Science, 26, 3320 – 3328, 1991
6. H. A. Lindner, B. Caspers in Mechanische Eigenschaften keramischer Konstruktionswerkstoffe (Ed.: G. Grathwohl), Verlag DGM Informationsgesellschaft,, Oberursel, Germany, 1993, Chapter V, 191 – 195
7. Ch. Ziegler, in Fortschritt-Berichte VDI, Reihe 18, Nr. 238, VDI-Verlag Düsseldorf, Germany, 1998
8. Draft of DIN 51109, Testing of advanced technical ceramics, determination of fracture toughness K_{Ic}, Deutsches Institut für Normung, Berlin, Germany, 1991

Failure Mechanisms During Thermal Shock Testing of Advanced Ceramics by Means of Lamp Irradiation

V. Knoblauch[1], G.A. Schneider[2], W. Dreßler[3], H. Böder[3], G. Schneider[4], R. Speicher[1], L. Jeannel[5]

[1] Robert Bosch GmbH, Technical University Hamburg-Harburg/Advanced Ceramics Group, Germany
[2] Technical University Hamburg-Harburg/Advanced Ceramics Group, Germany
[3] Robert Bosch GmbH, Stuttgart, Germany
[4] Polytechnic Aalen, Aalen, Germany
[5] Institut Français de Mécanique Avancée, Frankreich

Abstract

Thermal shock behavior of tape-cast Y-PSZ substrates is characterized by lamp irradiation, i.e. the generation of strong lateral temperature gradients in thin circular discs by rapid heating with a high power halogen lamp while in situ measuring the temperature fields with a high speed thermal imaging system. Special attention is paid on failure mechanisms of the samples. Numerical simulation of the specimens deformation indicates buckling of the discs due to radial compressive stresses. This causes significant bending components superposing the in-plane thermal stresses. Hence, depending on the geometry of the samples marked mistakes in the characterization of the thermal shock behavior are made if buckling is not considered. In situ off-plane deformation measurements with a Laser-Doppler-Interferometer confirm the theory and verify the Finite-Element simulation. An approach is proposed how buckling has to be taken into account both, in the field of experiment and simulation.

1 Introduction

The evaluation of thermal shock behavior of ceramics and ceramic substrates still presents itself as a many-sided problem. A lot of investigations has been done during the last years and several procedures have been proposed [1,2]. Irradiation methods whether by laser or halogen lamps seem to be useful and well defined tools which enable quantitatively the evaluation of the thermal shock behavior [2-5]. Regarding the samples geometry, i.e. thin circular discs, the methods should be very attractive for thermal shock testing of ceramic substrates [6].

The lamp irradiation method which is applied in this work uses a halogen lamp to generate the thermal loading. Thereby, the center of thin circular discs is irradiated one-sided while the transient and spatial-dependent temperature distribution is measured at the unheated side time-and spatial-resolved with a thermal imaging system and an IR-pyrometer. The higher thermal expansion in the disc center is constrained by the cooler edge causing tangential and radial compressive stresses in the hot zone and tensile hoop stresses at the outer rim. When the thermal stresses in the specimen reach material's strength typically the sample will be destroyed by unstable crack propagation. The knowledge of the almost axisymmetric temperature distribution, T(r), at the moment of failure together with the material properties, i.e. Young's-modulus, E, Poisson ratio, ν, and the thermal expansion coefficient, α, enables the calculation of the in-plane thermal fracture stresses according to [7]

$$\sigma_t(r) = \alpha E\left[\frac{1}{a^2}\int_0^a T(\tilde{r})\tilde{r}d\tilde{r} + \frac{1}{r^2}\int_0^r T(\tilde{r})\tilde{r}d\tilde{r} - T(r)\right] \quad (1)$$

for the tangential stress, and

$$\sigma_r(r) = \alpha E\left[\frac{1}{a^2}\int_0^a T(\tilde{r})\tilde{r}d\tilde{r} - \frac{1}{r^2}\int_0^r T(\tilde{r})\tilde{r}d\tilde{r}\right] \quad (2)$$

for the radial components, where a determines the radius of the sample and r the radial position. The Young's modulus, E, and the thermal expansion coefficient, α, are taken independent on temperature in this case.

In this work the authors will present results of lamp irradiation thermal shock tests of partially stabilized zirconia (Y-PSZ) substrates. Thereby, special attention will be focussed on the failure mechanisms. It will be shown that the radial stress components can cause buckling of the discs so that - in addition to the in-plane thermal stresses - substantial bending components have to be considered.

Solving the differential equation in cartesian coordinates x, y, z [8]

$$\frac{E \cdot h^3}{12(1-\nu^2)} \cdot \Delta\Delta w + h\left(\sigma_x \frac{\partial^2 w}{\partial x^2} + \sigma_y \frac{\partial^2 w}{\partial y^2} + \tau_{xy} \frac{\partial^2 w}{\partial x \partial y}\right) = 0 \tag{3}$$

for a completely hinged plate under hydrostatic radial compressive loading the critical buckling load σ_{bu} that causes spontaneous buckling, is given by

$$\sigma_{bu} = 4{,}2 \frac{E \cdot h^2}{12(1-\upsilon^2) \cdot a^2} \tag{4}$$

with the plates thickness, *h*, its radius, *a*, and the deflection described by *w*.

Comparing the critical buckling load, σ_{bu}, according to equation (4) with the radial compressive stresses, $\sigma_r(r)$, at the time of failure it becomes obvious that buckling has to be expected for some critical samples geometry i.e. an unfavorable ratio of sample's thickness to its diameter [9]. Thus, fracture is not exclusively caused by the in-plane tangential thermal stresses. Rather superposed bending components have to be taken into account to calculate the thermal stresses with satisfactory accuracy. Because the assumptions that form the basis of the solution of equation (3) do not represent the real test conditions equation (4) can only give us a rough estimate whether buckling has to be expected during thermal shock testing or not. However, equation (4) overestimates the samples stiffness and its resistance against buckling for different reasons.

- The radial compressive stresses $\sigma_r(r)$ under thermal loading are not radius independent but they increase towards the disc center.
- The calculations are carried out under the assumption of perfect plane-parallel circular plates without geometric imperfections. Even slight geometric imperfections – generated either during the samples preparation or possibly arising during thermal loading due to bending stresses owing to axial temperature gradients - reduce the stiffness of structure and therefore the critical buckling load [10].

Both, finite-element calculations and experimental investigations have been carried out to examine influences on this effect and its extent. Thereto, the geometry of samples was varied in a wide range. Because the samples thickness is almost determined by the tape-casting process parameters the discs diameter is the only parameter which can usefully be varied in a certain range.

2 Experimental

2.1 Samples Preparation

Circular discs of Y-PSZ with various diameters and 0,5 mm thick have been punched out of green tape-cast substrates (because of industrial applications of the materials by the Robert Bosch GmbH, Germany, a more detailed specification is not possible). After sintering the samples are nearly plane-parallel with diameters of 20,0 mm and 14,0 mm and a final thickness of 0,42 +/- 0,02 mm. They are used as sintered, there is no grinding or polishing process after sintering. Thus, the discs are slightly warped.

In order to obtain sufficient absorption of the infrared radiation on the one hand and, on the other hand to ensure a defined and specimen independent emission for the IR-thermometry the samples have been coated both-sided with a graphite spray (*Graphit 33 I CRC Industries Deutschland GmbH*).

2.2 Experimental Set-up and Performance

A detailed description of the experimental set-up is given in [6]. Fig. 1 shows a schematic diagram of the entire thermal shock system. To minimize a slight elliptical distortion of the temperature field due to the plane spiral-wound tungsten filament the IR-radiation is focused by an aperture situated in front of the sample in a distance of about 3 mm.

Two independent thermometry systems are used to record the time-dependent temperature distribution with high precision. The measurement is effected in situ at the unheated graphite-coated side. By use of a thermocouple (*Pt-Rh/Pt, Typ S*) laminated between two thin substrates the emission of the graphite-coating was determined to 0,80 for the thermal imaging system (*Thermovision 880, Agema Infrared Systems AG, Sweden*) and 0,75 for the IR-pyrometer (*TMR 95/105-d-G, Dr. G. Maurer GmbH, Germany*), according to the specific arrangement of the components.

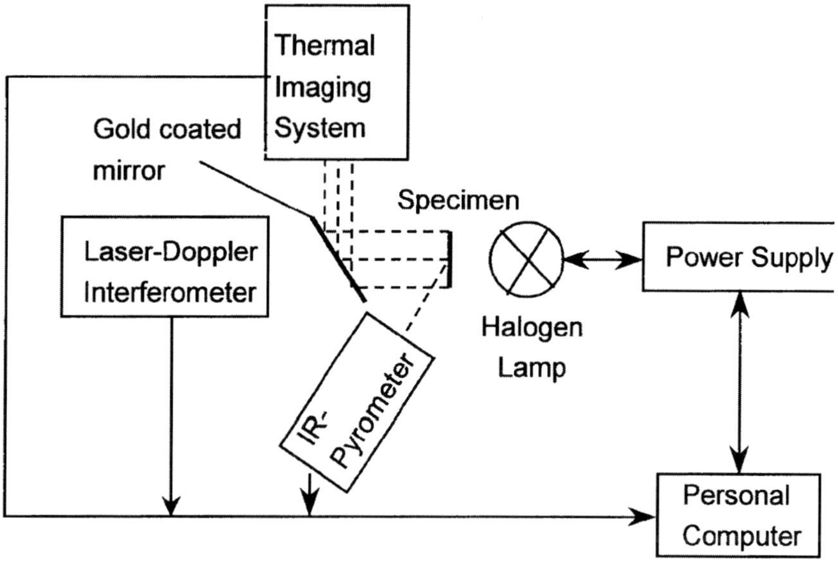

Figure1. Schematic diagram of the entire thermal shock equipment.

For the thermal shock tests the specimens have been irradiated until fracture. The temperature distribution at the moment of failure was measured one dimensionally along the horizontal axis of symmetry in the linescan-modus with the thermal imaging system. The good local resolution and the high scanning rate (up to 1250 Hz) enables a very precise temperature measurement of the transient heating process and an exact determination of the time of fracture. Out of that stress calculations have been performed both, analytically and numerically.

To determine the off-plane displacement time- and spatial-resolved a Laser-Doppler-Interferometer (LDI) (*Polytec GmbH, Germany*) was used measuring perpendicular to the samples surface. Hence, it was not possible to simultaneously record temperature and off-plane displacement. Therefore, a reference temperature profile, $T_{ref}(r)$, after a defined time of irradiation at a voltage of 15 V was recorded once that forms the basis of the finite-element calculations. Afterwards, the specimen was repeatedly irradiated under the same conditions while the off-plane displacement was measured pointwise along the x- and y-axis. Furthermore, buckling of the disc center was gauged during thermal loading to investigate its time-dependent progression until fracture.

3 Results and Discussion

3.1 Thermal shock testing

Using samples with a diameter d = 20,0 mm and various thickness 0,20 mm < h < 0,55 mm we determined temperature profiles and corresponding tangential stress distributions at the time of failure as shown in Fig. 2a and b. ###Fig. 2a and b### The stress calculations are carried out analytically based on the assumption of a state of plane stress because the samples are very thin compared to their diameter. Bending components owing to an axial temperature gradient should be negligible [11]. Hence, the thickness of the discs does not enter into the magnitude of the thermal stresses.

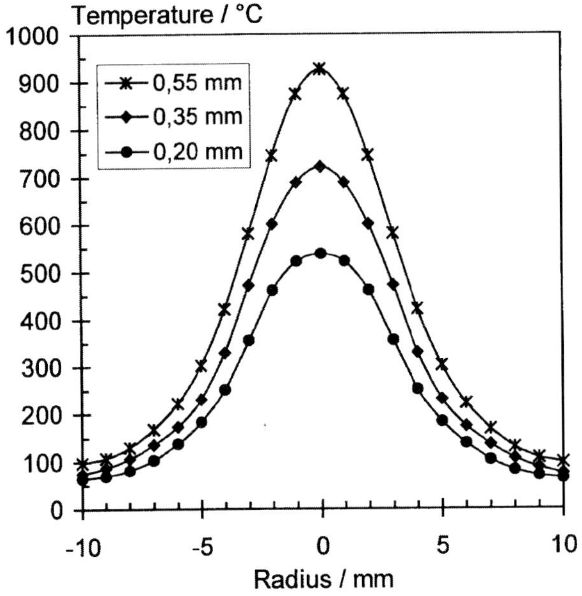

Figure 2a. Temperature profiles of Y-PSZ-specimens with different thickness at the moment of fracture.

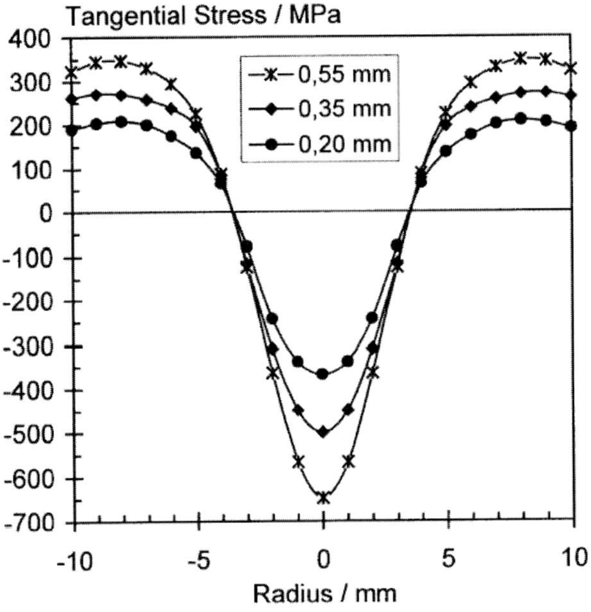

Figure 2b. Corresponding tangential stresses of specimens with different thickness.

On the other hand the stress loaded volume strongly depends on the samples thickness, h, and according to the Weibull theory it shall significantly influence the thermal fracture stress if volume defects are the failure origins. The relation between the stress loaded volume, V_{eff}, and the corresponding failure stress is given by

$$\frac{\sigma_1}{\sigma_2} = \left(\frac{V_{eff,2}}{V_{eff,1}} \right)^{\frac{1}{m}} \tag{5}$$

with the Weibull modulus, m. The stress loaded volume of the disc, V_{eff}, is

$$V_{eff} = 2\pi h \int_{r_0}^{a} \left(\frac{\sigma_t(r)}{\sigma_{t,max}} \right)^m r dr \tag{6}$$

where we assumed axisymmetric loading without temperature gradients in thickness direction. r_0 describes the change over from compressive to tensile tangential stresses and $\sigma_{t,max}$ represents the reference loading. Regarding the temperature- and stress-profiles (see Fig. 2a and b) it becomes obvious that the temperature as well as the stress distributions have self similar shapes and can be described by a time dependent function f(t) multiplied by a thickness independent shape function g(r)

$$\sigma_t(r,t) = g(r) \cdot f(t) \tag{7a}$$

The reference loading, $\sigma_{t,max}$, may be chosen to be the maximum tensile stress in the disc at the time t:

$$\sigma_{t,max}(t) = g_{max} \cdot f(t) \tag{7b}$$

Consequently, equation (6) can be rewritten as

$$V_{eff} = 2\pi h \int_{r_0}^{a} \left(\frac{g(r)}{g_{max}} \right)^m r dr = const \cdot 2\pi h \tag{8}$$

and V_{eff} is increasing with increasing samples thickness. According to equation (5) one would expect a lower thermal shock fracture stress for thicker samples but seeming higher values up to about 150 % are observed in contradiction to the Weibull theory comparing the 0,20 mm and 0,55 mm specimens.

To ensure that there are no additional stresses due to a martensitic tetragonal-monoclinic phase transformation with volume expansion [12] which would not have been considered in the thermal stress calculations X-ray diffraction (XRD) examinations have been carried out. Thereto, the amount of phases before and after the thermal loading up to 900 °C in the disc center was determined. 42% cubic and 58 % tetragonal modification was found in both cases [13].

3.2 Buckling of the disc

In the following chapters it will be shown that the radial compressive thermal stresses can cause buckling of the discs. Therefore, the in-plane thermal stresses are superposed by partially substantial bending components. Thus, the seeming low thermal shock fracture stress and its dependence on the samples thickness becomes understandable.

3.2.1 Experimental investigations

To get an idea of the extent and the kind of buckling the off-plane displacement was investigated using 20,0 mm samples (h = 0,40 mm). It was measured time-resolved in the disc center at three different samples. Samples 1 and 2 were measured two times with irradiation times, t_{irr}, of 2,1,s. Sample 3 was at first thermally loaded 1,6 s, afterwards its was irradiated until fracture. The voltage of the halogen lamp amounted to15 V. Furthermore, one specimen was repeatedly loaded by a short flash at a voltage of 15 V and a period of 2,1 s while the deformation was measured pointwise along the horizontal (x) and vertical (y) axis of symmetry.

In Fig. 3 the deformation of the disc center is plotted versus irradiation time. ###Fig. 3### We discover a large off-plane distortion of the disc center up to about 300 µm at the moment of fracture. Besides, even low thermal loading leads to a significant deformation of about 80 µm. Apparently in this case the off-plane deformation is not about a spontaneous buckling-process coming along with a bifurcation of the loading-displacement curve when exceeding the critical buckling load of the structure, rather we detect a continuos deformation from the beginning of the loading. This behavior is attributed to the geometric imperfections of the samples [14], i.e. a slight warping of about 0,02-0,05 mm due to the preparation process. Looking at the magnitude of buckling at the time of fracture of about 0,75 times the samples thickness it becomes evident that huge bending stresses will occur which have to be considered in the stress calculations.

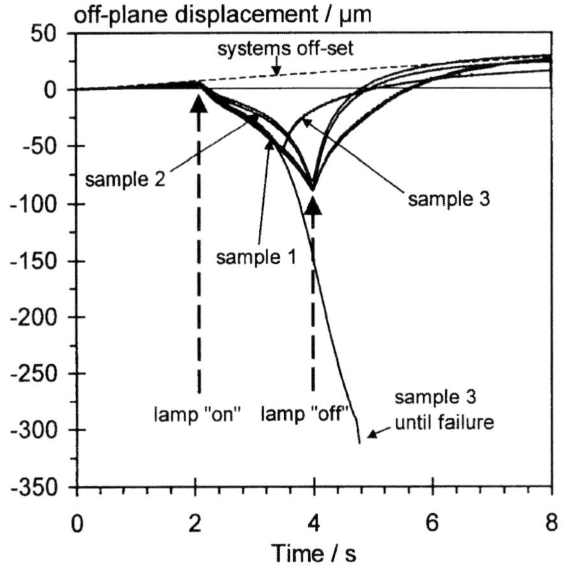

Figure 3. Off-plane displacement of the disc center (samples′ diameter: 20,0 mm) versus time. Samples 1 and 2 were measured two times with irradiation times, t_{irr}, of 2,1,s (see almost perfect reproducibility of the behavior of the discs). Sample 3 was at first thermally loaded 1,6 s, afterwards its was irradiated until fracture (regard the huge buckling of about 320 µm). The voltage of the halogen lamp amounted to 15 V.

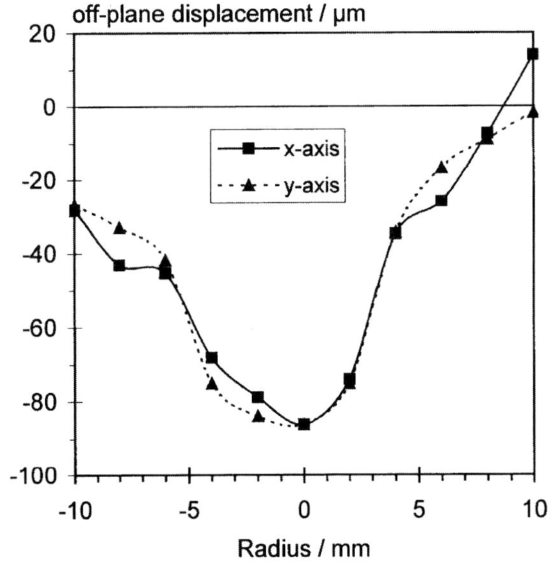

Figure 4. Off-plane displacement along the horizontal (x) and vertical (y) axis of symmetry (sample′s diameter: 20,0 mm). The specimen was repeatedly thermally loaded 2,1 s at a voltage of 15 V and the off-plane displacement was measured pointwise.

Observing the deformation along the x- and y-axis during short thermal loadings a more or less symmetric off-plane displacement of the sample superposed by a slight rigid-body displacement of the disc becomes visible (see Fig. 4). ###Fig. 4###

According to equation (4) the discs stiffness decisively depends on the ratio of thickness/diameter. Therefore, analogous investigations have been performed using samples with a reduced diameter of 14,0 mm. To measure the spatial-dependent off-plane displacement the specimen (h = 0,43 mm) was repeatedly thermally loaded 1,5 s at a voltage of 15 V (see Fig. 5a and b). ###Fig. 5a and 5b### Once again we observe an almost symmetric buckling about the x-and y-axis, superposed by a slight rigid body displacement. However, the magnitude is substantial lower than for the 20,0 mm specimen as to be expected. Additionally, the result of a finite-element calculation is inserted in hatched lines (open symbols), whereby the lateral buckling of the specimen is added computationally to make the result comparable to the experimental findings. We will go into this point later.

The off-plane displacement of the disc center versus irradiation time (see Fig. 5b) shows a continuos deformation instantaneously from the beginning of the thermal loading as observed at the 20,0 mm samples, too. However, in contrast to them a plateau is reached after a short irradiation period and there is no further increase of the off-plane displacement with increasing thermal loading. We conclude that the deformed structure is stiffer concerning this kind of loading and distortion, so that the radial compressive stresses do not cause further buckling. Furthermore, the almost dead match of the graphs of different samples is remarkable (see also the results of the 20,0 mm specimens, Fig. 3a). On the one hand it reveals the high reproducibility of the measuring technique, on the other hand it proves the nearly identical deformation behavior of different specimens.

3.2.2 Finite-Element Modeling

In order to estimate the effect of buckling on the thermal shock fracture stresses and their spatial distribution FE-calculations have been carried out using the FE-code ABAQUS. To model the more or less axisymmetric wave-like warping of the discs, w, of commonly 0,02-0,05 mm at first a simple axisymmetric model was performed described by a cosine-function A(r):

$$A(r)=\left(\frac{w}{2}\right)\cos\left(r\cdot\frac{\pi}{a}\right) \tag{9}$$

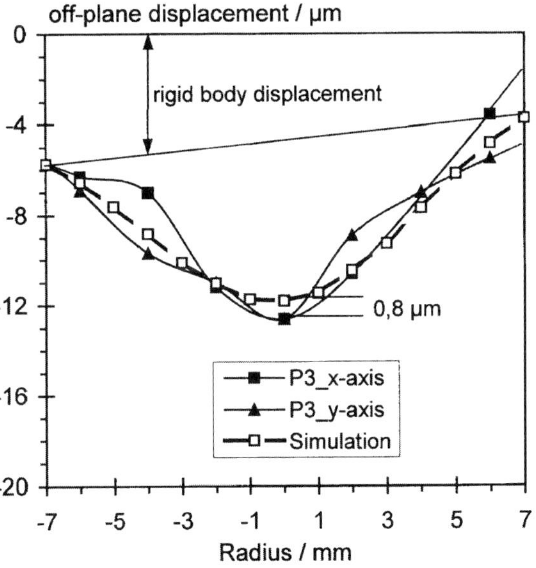

Figure 5a Almost symmetric buckling (superposed by a rigid body displacement) along the horizontal (x) and vertical (y) axis of symmetry (sample´s diameter: 14,0 mm). The sample was repeatedly irradiated 1,5 s at a voltage of 15 V and the off-plane displacement was measured pointwise.

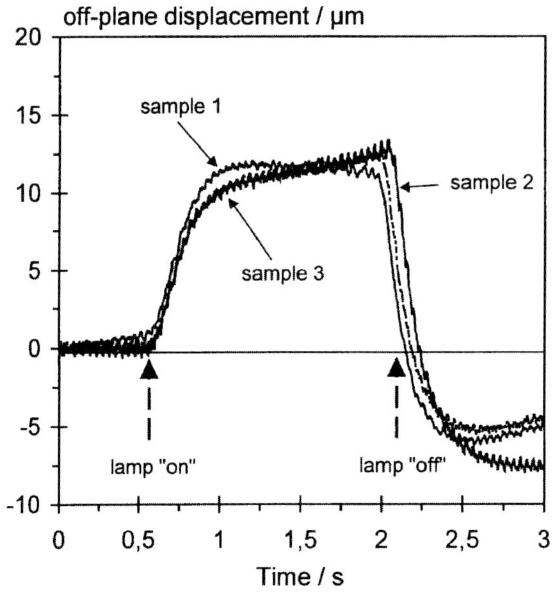

Figure 5b. Off-plane displacement of the disc center (samples´ diameter: 14,0 mm) versus time. All samples were irradiated 1,5 s at a voltage of 15V (see almost identical deformation behavior of the different samples).

Calculating the off-plane deformation based on corresponding temperature profiles after identical thermal loadings as for the LDI-measurements we obtain a maximum off-plane displacement in the disc center of about 7 µm. A comparison of the simulation and the experimental findings is given in Fig. 5a as described in chapter 2.1. Apart from some slight deviations we discover a high grade of coincidence between simulation and experiment. Hence, the model represents the real conditions with high precision. Furthermore, the almost identical deformation behavior of different samples (see Fig. 5b) justifies the generalization of slight variation of the discs shapes by modeling only one geometry type according to equation (9).

3.2.3 Thermal fracture stresses

Calculations of the thermal shock fracture stresses have been carried out, both, analytically according to equation (1) and numerically on the basis of the "real" geometry, i.e. the FE-model according to equation (9). In Fig. 6a the stress distributions of a 14,0 mm sample modeling the "real" geometry at the time of fracture is plotted. As it can clearly be seen the tensile components are shifted towards the surface ###Fig. 6### whereas in case of a plane-parallel disc the tangential stresses are not z-depending [7]. Besides, we obtain a significant higher thermal shock fracture stress of about 500 MPa (the analytical calculation reveals about 430 MPa), now in a scale expected from results of ring-on-ring tests. In case of 20,0 mm samples we obtain numerically calculated "real" thermal shock fracture stresses of about 1,5 to 2,0 times higher than analytically calculated.

Both facts are based on the superposed bending components induced by buckling. Therefore, it becomes obvious that possibly marked mistakes can be made neglecting buckling effects.

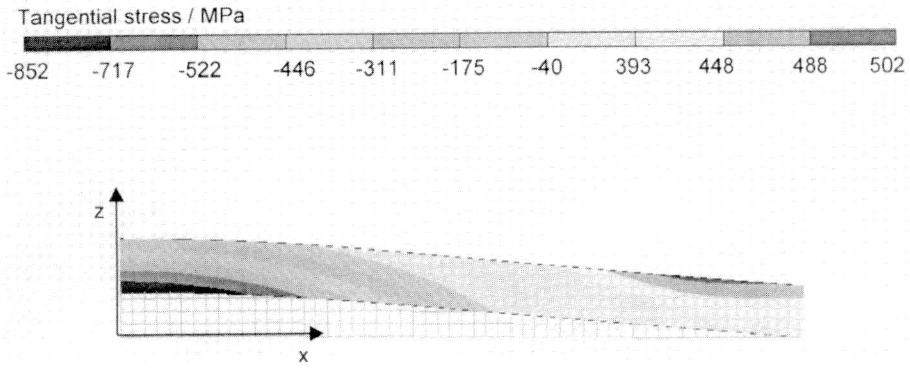

Figure 6. FE-calculation of the tangential stresses at the time of failure (sample´s diameter: 14,0 mm, thickness: 0,43 mm) of the "real" geometry considering buckling. The tangential tensile components are shifted towards the surface of the disc and are about 1.2 times higher than in the plane case.

4 Summary

To investigate the thermal shock behavior of tape cast Y-PSZ substrates a lamp irradiation thermal shock system was used. The in situ temperature measurement was carried out using a thermal imaging system that enables a precise determination of the fracture time and the temperature distribution.

First, analytical calculations of the thermal shock fracture stresses have been carried out. However, the seeming increase of the thermal shock fracture stress with increasing samples thickness, i.e. an increasing of the stress loaded volume could not be understood regarding only in-plane displacements and tangential tensile stresses. Rather, the radial compressive components and their effect on buckling of the discs have to bee taken into account.

Measuring the off-plane displacement in situ during thermal loadings with a Laser-Doppler-Interferometer significant buckling up to 300 µm was observed, depending on the diameter of the specimens, i.e. significant lower displacements with decreasing diameter as to be expected from theoretical considerations. Furthermore, we determined an almost axially symmetric deformation behavior. In addition, it could be shown that geometric imperfections due to the preparation process of the samples cause a continuos buckling process even at low thermal loadings in contrast to a spontaneous buckling of perfect structures when exceeding the critical buckling load.

Finite-element calculations of buckling yield a high grade of correspondence compared to experimental results. Two effects could be observed: A shift of the tensile stresses towards the sample´s surface and - depending on the diameter of the samples - partly substantially higher tensile stress values owing to the superposing bending components. That means that possibly a marked underestimation of the material´s thermal shock fracture stress can occur neglecting such buckling effects.

It becomes obvious that either optimization of the geometry of the specimen or simulation of buckling and consequently taking into account the calculated additional bending stresses is necessary to measure correctly the thermal shock fracture strength of the investigated Y-PSZ material.

References

1. Bradt, R.C., Hasselman, D.P.H., Munz, D., Sakai, M., and Shevchenko, V.Ya. (eds.), *Fracture Mechanics of Ceramics 11*, Plenum Publishing Corporation, New York, 1996.
2. Schneider, G.A. and Petzow, G. (eds.), *NATO ASI Series*, Series E, Vol. 241, Kluwer Academic Publishers, Dordrecht, The Netherlands, 1993.
3. Schneider, G.A. and Petzow, G., *J. Am. Ceram. Soc.*, 1991, **74** (1), 98-102.
4. Kirchhoff, G., Weiss, H.-J., Holzherr, M., Bahr, H.-A., Pflugbeil, I. and Pompe, W., in *Proceedings of the First International Symposium on Thermal Stresses and Related Topics*, Shizuoka, Japan, 1995, 119-122.
5. Amada, S., Nong, W.Y., Min, Q.Z. and Akiyama, S., *Ceramics International*, 1999, **25**, 61-67.
6. Knoblauch, V., Schneider, G.A., Schneider, G., Böder, H., accepted for publication in *J. Europ. Ceram. Soc.*
7. Boley, B.A., Weiner, J.H., *Theory of Thermal Stresses*, Robert E. Krieger Publishing Company, Florida, 1985.
8. *Taschenbuch für den Maschinenbau (Mechanical Engineering Handbook)*, (Ed.: Beitz, W. and Hüttner, K.-H.), 17. Auflage, Spinger-Verlag, Berlin, 1990.
9. Knoblauch, V., Diploma Thesis, Polytechnique Aalen, Germany, 1996.
10. Palassopoulos, G.V., in *Computational Stochastic Mechanics*, (Ed.: Spanos, P.D. and Brebbia, C.A.), 1991, 211-222.
11. Rettig, U., in *Fortschritt-Berichte VDI*, Reihe **18**, **237**, VDI Verlag GmbH, Düsseldorf, 1998.
12. Bansal, G.K. and Heurer, A.H., *Acta Metallurgica*, 1972, **20**, 1281-1288.
13. Knoblauch, V., Schneider, G.A., Dreßler, W., Böder, H., Schneider, G., submitted to Acta Materialia.
14. Palassopoulos, G.V., in *Uncertainly Modeling in Finite Element, Fatigue and Stability of Systems*, (Ed.: Haldar, A., Guran, A. and Ayyub, B.M.), 1997, 311-356.

Design and Production of High Reliability Soda-lime Silicate Glass

Vincenzo M. Sglavo and Marco Bonafini
Dipartimento di Ingegneria dei Materiali, Università di Trento, Trento (Italy)

1 Abstract

The introduction of a compressive residual stress with a maximum below the surface in a glass article can promote the stable growth of surface flaws and result in high strength value with limited scatter. In this work, ion exchange techniques are used for the production of an optimal residual stress profile in soda-lime silicate glass. Both single and double ion exchange processes are used. In the optimization procedure the residual stress is related to the obtained strength data and to the stable propagation of surface flaws.

2 Introduction

Glass has been used in very numerous applications since time immemorial and its strengthening has represented a fundamental goal for scientists and engineers at least since the seventeenth century when Prince Rupert of Bavaria introduced in England tadpole-shaped glass drops with exceptional strength.[1] The so called Prince Rupert's Drops owe their strength to the surface compressive residual stresses generated by pouring drops of molten glass into cold water.

Several techniques are nowadays available to increase glass strength, ranging from physical tempering to ion-exchange, from surface crystallization to lamination.[2] All these process allow to produce glass components with an average strength sometimes higher than 600-800 MPa. Nevertheless, the strengthening process is often associated to larger strength variability.

Conventional strengthening technique are based on the principle that, if a residual compressive stress σ_c is generated within a surface layer, the strength will be $\sigma_f = \sigma_f^0+\sigma_c$ where σ_f^0 is the strength of the pristine glass. Unfortunately, this approach is associated to two problems. First of all, the strength of the pristine glass is randomly distributed as the defect sizes are on the surface of the specimen. Secondly, typical residual stress fields are not constant but decrease with the distance from the surface. Therefore, larger defects, associated to small values of σ_f^0, are subjected to smaller σ_c values and this results into limited strength increase. Conversely, small defects, corresponding to high σ_f^0 values, are more strongly reinforced.

In a recent paper Sglavo *et al.*[3] proposed a different approach for the design of high reliability brittle material based on fracture mechanics principles. The main idea is to promote a stable growth of surface defects up to a certain depth before final catastrophycal failure. In this way, regardless the initial defect size, a constant failure stress could be obtained.

Stable growth of surface defects can be observed in materials presenting R-curve effect.[4] For example, if the material under interest has a fracture toughness which varies with crack length as shown schematically in Fig. 1(a), an interval of crack stable growth can be defined. For the situation depicted in Fig. 1(a), defects ranging from c'_0 to c'_{max} are subjected to stable growth all reaching the final length c'_{max} (the notation $c' = (\pi c)^{0.5}$ is here used). The corresponding failure stress is defined as $\sigma_f = (K_c+\Delta K_c)/c'_{max}$ where K_c is the fracture toughness of the pristine glass and ΔK_c is the toughening effect developed for crack lengths in excess of c'_{max}. Theoretically, one can chose the appropriate R-curve for obtaining a pre-defined strength and stable growth of defects with length included in a certain interval. This would mean to fix points B and C in the diagram of Fig. 1(a), point A being automatically defined by the fracture toughness of the material.[3] In a previous paper Sglavo *et al.* demonstrated that an apparent R-curve like that represented in Fig. 1(a) can be obtained by introducing a particular residual stress profile within the surface layers of the specimen.[3] The R-curve is here defined as apparent because the effect of the residual stress is combined to the fracture toughness of the material resulting in an apparent crack resistance force for external applied loads. If the apparent K_c curve is approximated by a polynomial function, the profile of the residual stress field can be analytically evaluated.[3] For example, the R-curve shown in Fig. 1(a) can be produced by the residual stress profile reported in Fig. 1(b).

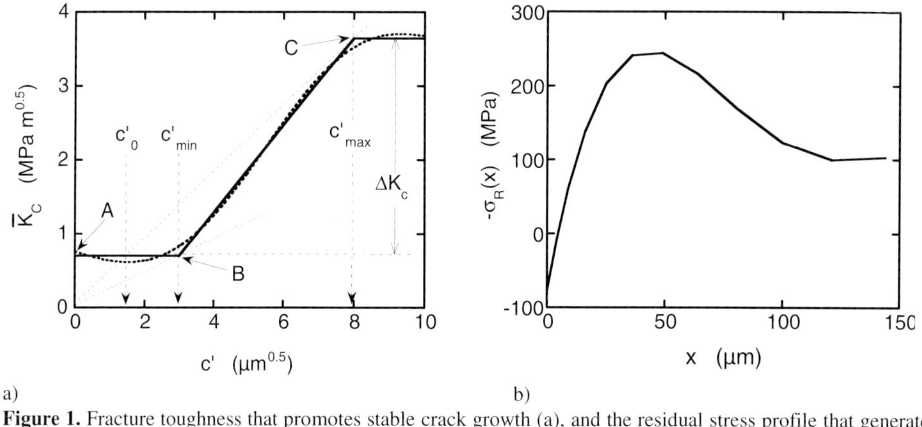

Figure 1. Fracture toughness that promotes stable crack growth (a), and the residual stress profile that generates it (b).[3] Thick dashed line represents the polinomial approximation of K_c ($c' = (\pi c)^{0.5}$

Using a double ion-exchange process a residual stress profiles similar to that shown in Fig. 1(b), i.e. characterized by a compressive maximum at a certain depth below the surface, can be produced in soda-alumina silicate glass.[5,6] The obtained glass samples showed evident stable growth of surface cracks and were characterized by bending strength as high as 500 MPa with a standard deviation smaller than 5%.

In this work, the approach used by Sglavo *et al.*[5] in soda-alumina silicate glass is transferred to more common glass like soda-lime silicate glass with the objective to find experimental conditions for the production of high-strength high-reliability articles.

3 Experimental Procedure

Commercial soda-lime silicate glass plates with nominal thickness of 3 mm were used in this work. The composition (wt%) of the glass is: SiO_2 71, Na_2O 13, CaO 10, MgO 4, Al_2O_3 1, other 1. The transition temperature measured by differential scanning calorimeter (DSC) method[7] is equal to 557°C.

The original plates were cut into 10 mm x 50 mm samples. The edges were polished using silicon carbide paper and diamond paste with grain size of 3 μm in order to remove macroscopic defects introduced during the cutting procedure. For the removal of any residual stress samples were then annealed at 520°C for 8 h using a cooling rate of 40°C/h.

Batches of 25 specimens were subjected to ion exchange process (in a semi-automatic furnace - LEMA TC 20A, Parma, Italy) for the production of a residual stress field similar to Fig. 1(b). Some groups of samples were subjected to single ion exchange treatment in fused KNO_3 at temperatures ranging from 425°C to 500°C with duration from 20 h to 96 h. Other batches were initially treated in pure KNO_3 at 450°C for 48-96 h and then in a solution of KNO_3 and $NaNO_3$ at 350-450°C for 30 min. KNO_3 / $NaNO_3$ weight ratios ranging from 2 to 8 were considered. In the following section, specimens subjected to double treatment will be identified by the parameters t_1, T_2 and r where t_1 is the duration of the first treatment, T_2 the temperature of the second ion exchange process and r the KNO_3 / $NaNO_3$ weight ratio. Samples were carefully cleaned with distilled water at the end of each ion exchange treatment and observed by an optical microscope.

Both annealed and ion exchanged samples were subjected to four-point bending test in deionized water, which represnts a constant environment with regard to fatigue behaviour. An actuator speed of 10 mm/min was used.

The residual stress profile introduced by the ion exchange process was measured using a technique in which one surface is etched away [8]. This procedure induces curvature in the specimen, which was measured using a profilometer. The residual stress $\sigma(t)$ at a certain depth t from the surface is related to the curvature $\varphi(t)$ produced by etching away a layer of thickness $\delta = s - t$ (where s is the thickness of the specimen) as:[8]

$$\sigma(t) = -2kt^2 \frac{d\varphi}{dt} - 8kt\,\varphi(t) - 4k\int_0^t \varphi\, dt + g \qquad (1)$$

where $k = E\,s^3 / [12\,(1-\nu^2)]$ (E = Young modulus = 71 GPa, ν = Poisson's ratio = 0.23) and g is a constant depending on the initial curvature and the thickness of the sample.

4 Results and Discussion

The observation of the samples subjected to ion exchange process sometimes revealed the presence of a fine network of cracks on the surface. This effect was observed in samples subjected to single ion exchange treated at 500°C. Among samples double treated, cracks were observed in specimens identified by (t_1 = 96 h, r < 4, T_2 = any value), (t_1 = 96 h, r = 4, T_2 = 350°C) and (t_1 = 48 h, r = 2, T_2 = 350°C or 450°C). The other specimens were perfectly transparent.

The residual stress profiles obtained by ion exchange treatment are shown in Fig. 2. One can easily observe that the presence of surface cracks after the tempering process is associated to residual tensile stresses on the surface. These latter can be accounted for either strong stress relaxation within the surface layers at high temperature (500°C) or considerable back-exchange of K^+ with Na^+ ions on surface layers when the second treatment is performed using high $NaNO_3$ loads.

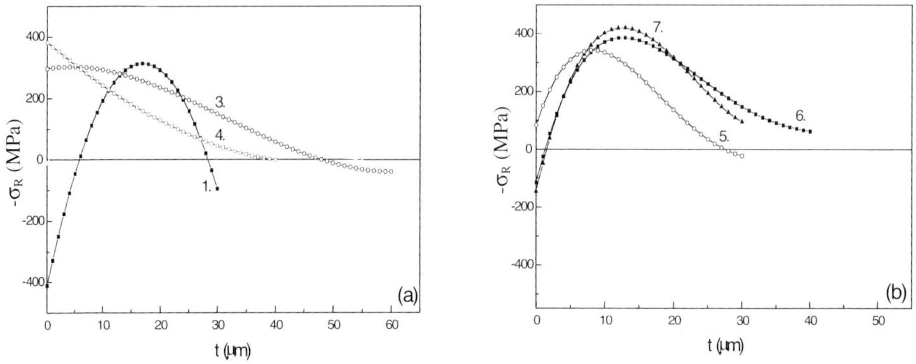

Figure 2. Residual stress profiles for single ion exchanged (a) and double exchanged (b) specimens: 1. 500°C for 20 h, 2. 475°C for 96 h, 3. 450°C for 96 h, 4. 450°C for 48 h, 5. (t_1 = 48 h, r = 4, T_2 = 350°C), 6. (t_1 = 96 h, r = 4, T_2 = 400°C), 7. (t_1 = 48 h, r = 2, T_2 = 400°C).

The presence of surface cracks influences not only the transparency of the glass but also its strength. As an example, Fig. 3 shows the strength for samples subjected to single ion exchange. The presence of a network of surface cracks is usually associated to limited strength values which, in some cases, are smaller than data recorded for annealed specimens.

Failure tests performed in silicone oil on annealed samples, this latter corresponding to an inert environment with regard to fatigue allowed an estimate of the flaw size characterizing the glass used in this work. Starting from a strength of 136±24 MPa, one can evaluate that the flaw size ranges from ≈5 μm to ≈20 μm. On the basis of the arguments proposed in the Introduction of this work, the residual stress field produced by ion exchange will have to promote the stable growth of defect in this depth range.

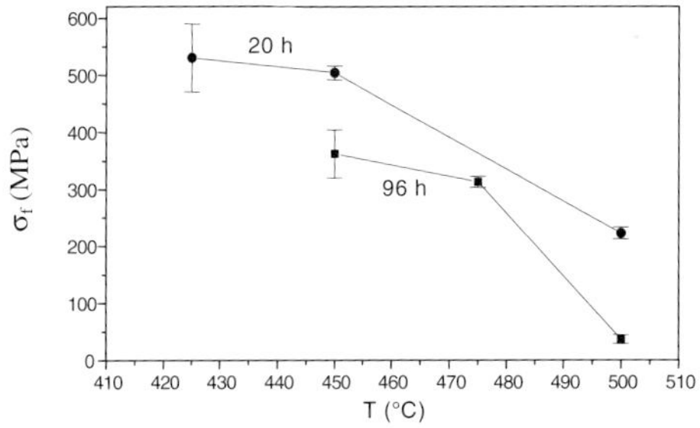

Figure 3. Flexural strength of specimens subjected to single ion exchange treatment.

Among the numerous treatments used in work, those fulfilling the conditions for surface crack stable growth only (Fig. 1) were considered. Three treatments were therefore chosen: single exchange at 475°C for 96 h in KNO_3 and double exchange characterized by (t_1 = 96 h, r = 4, T_2 = 400°C) and (t_1 = 48 h, r = 2, T_2 = 400°C). For simplicity the corresponding samples will be labelled as A, B and C. It is interesting to observe that in the first specimen the optimal residual stress field is obtained using a single ion exchange treatment only where the surface stress relaxation is the solely responsible for the maximum compressive stress at a depth of about 18 μm. For the other two conditions, the surface stress is reduced by the second treatment in the $KNO_3/NaNO_3$ solution.

Strength values of 294±23 MPa, 378±42 MPa and 401±37 MPa were measured for specimens A, B and C, respectively. These data can be compared to the resistance of the annealed glass which was evaluated to be equal to 123±17 MPa. An important feature is that the standard deviation for these samples is lower than 11%, smaller than that measured for annealed glass which is around 14%.

The apparent fracture toughness curve for samples A, B and C associated to the residual stress field can be evaluated as:[3]

$$\overline{K}_C = K_C - \frac{\psi}{\sqrt{\pi c}} \int_0^c \sigma_R(x)\, g(x)\, dx \qquad (2)$$

where c is the defect length, ψ the shape factor and g the Green's function associated to the defect/sample geometry.[4] If a through thickness surface defect is considered for simplicity, $\psi = 1.12$ and $g = (c^2 - x^2)^{-0.5}$.

Figure 4 shows the apparent R-curves for samples A, B and C. The treatment at 475°C for 96 h determines a stable growth interval from 5 and 21 μm and results in a theoretical strength of 320 MPa. For samples B and C a stable growth interval from 3 μm to 16 μm can be evaluated. Theoretical strength is equal to 324 MPa, 368 MPa and 384 MPa for specimens A, B and C, respectively.

Theoretical calculations well agree with strength measurements. Discrepancies can be partially accounted for the fatigue effect which occurs in humid environment while theoretical calculations correspond to inert conditions.

The stable growth of surface crack in specimens A, B and C was verified experimentally. Samples were loaded in bending up to stresses lower than failure resistance and then observed by an optical microscope. Surface cracks appeared on the surface of B and C samples at stress level of about 320 MPa and 260 MPa, respectively, in good agreement with predictions which can be advanced on the basis of the apparent fracture toughness curve (Fig. 4). At increasing loads these cracks became more numerous and propagated perpendicularly to the direction of the applied tensile stress. Examples of the observed arrested cracks are shown in Fig. 5. Surprisingly, no crack formation was observed in samples A. This effect can probably be related to the profile of the apparent K_c curve. As shown in Fig. 4, the slope of the K_c curve in the stable growth interval is limited and the approximation of the theoretical approach and small variation of the residual stress field (related also, for examples, to experimental errors) could determine the reduction or even the disappearance of the interval.

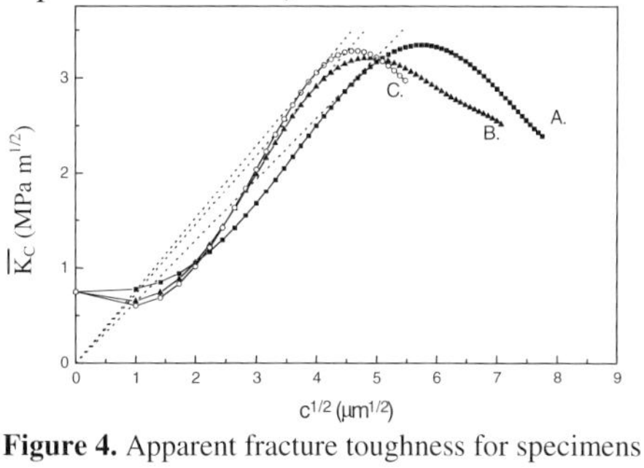

Figure 4. Apparent fracture toughness for specimens A, B, C.

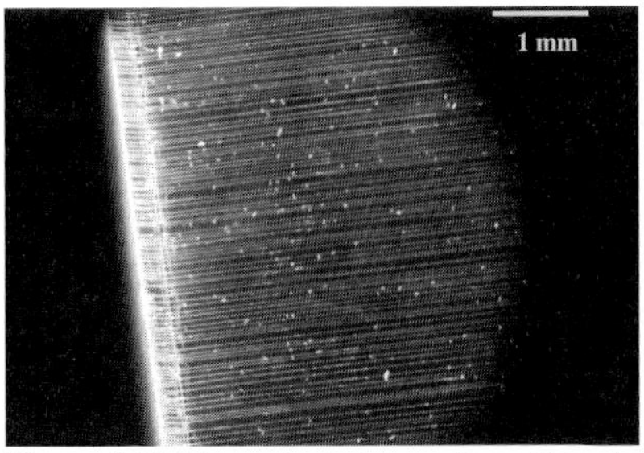

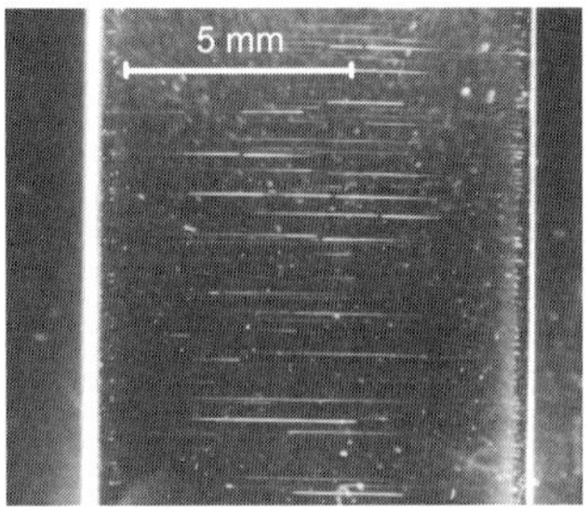

Figure 5. Stable cracks in (a) specimen B loaded to 340 MPa and (b) C loaded at 360 MPa.

6 Conclusions

A residual stress field which can promote stable growth of surface defects can be produced in soda-lime silicate glass by ion-exchange technique. Two optimal processes have been identified: the first one requires an initial treatment in KNO_3 at 450°C for 96 h and a second exchange in $KNO_3/NaNO_3$ (weight ratio = 4) at 400°C for 30 min; in the second process the sample is treated in KNO_3 at 450°C for 48 h and then in $KNO_3/NaNO_3$ (weight ratio = 2) at 400°C for 30 min. Obtained samples show surface crack arrest effect at stress higher than 300 MPa and are characterized by a strength of 378±42 MPa and 401±37 MPa. Similar strength and limited standard deviation (294±23 MPa) were observed also in samples subjected to single ion exchange at 475°C for 96 h in KNO_3. This latter treatment guarantees a residual stress which potentially can result in stable growth of surface defects though this effect could not be verified experimentally.

7 References

[1] M. Faraday, Lectures On The Matter, 1859.

[2] I. W. Donald, J. Mater. Sci. 1989, 24, 4177-4208.

[3] V. M. Sglavo, L. Larentis and D. J. Green, J. Am. Ceram. Soc. 1999, submitted for publication.

[4] B. Lawn, Fracture of Brittle Solids, 2nd ed., Cambridge University Press, Cambridge, UK, 1993.

[5] V. M. Sglavo, C. Camin and D. J. Green, J. Am. Ceram. Soc. 1999, submitted for publication.

[6] D. J. Green, R. Tandon and V. M. Sglavo, Science 1999, 283, 1295-1297.

[7] ASTM norm E 1346-91, in Annual Book of ASTM Standards, Vol. 15.01, American Society for Testing and Materials, Philadelphia, PA, USA, (1995).

[8] V. M. Sglavo and M. Bonafini, 1999, in preparation.

High Temperature Mechanical Properties of Si-B-C-N Precursor-Derived Amorphous Ceramics

Martin Christ, Günter Thurn, Joachim Bill, and Fritz Aldinger
Max-Planck-Institut für Metallforschung
Institut für Nichtmetallische Anorganische Materialien, Universität Stuttgart
Pulvermetallurgisches Laboratorium, Heisenbergstrasse 5, 70569 Stuttgart, Germany

1 Abstract

The high temperature mechanical properties of Si-B-C-N amorphous ceramics and their characterization by isothermal compression creep tests in the temperature range between 1350 °C and 1500 °C and at stresses up to 300 MPa are reported. The deformation rate was found to be composed of two components. One is proportional and the other one independent on the applied stress. The stress-dependent component is supposed to be due to viscous flow and the stress independent one to thermal activated shrinkage. The deformation behavior can be described by the free volume model developed for metallic glasses. According to this model, an activation energy $Q_f = 0.6$ MJ/mol for the movement and production/annihilation of flow defects was found. In the initial period of the creep tests some anelastic behavior was observed.

2 Introduction

Thermolysis of organometallic polymers provides the possibility for the fabrication of ceramics without the use of sintering aids, i.e. without grain boundary phase softening at high temperature. Of special interest are materials of the system Si-B-C-N since they possess excellent high thermal stability [1, 2] and oxidation resistance [3]. Compression creep tests with ternary precursor-derived Si-C-N materials have shown substantial creep resistance. Similar investigations of quarternary compositions [4] indicate that the deformation behavior of such amorphous ceramics is similar to that observed in metallic glasses which have been under investigation for some decades (e.g. [5, 6, 7, 8]). It is the objective of this paper to discuss the results of compression creep tests of Si-B-C-N precursor-derived amorphous ceramics in terms of the free volume model developed for metallic glasses [5, 8].

3 The Free Volume Model

According to this model, the deformation mechanism of amorphous materials is based on flow defects formed by deviations of the material structure from the "ideal" amorphous structure [5], i. e. fluctuations in the distribution of free volume. For this mechanism the following expression for the viscosity, based on simple rate theory, was derived [5, 8]:

$$\eta := \frac{\sigma}{3\dot{\varepsilon}} = \frac{kT\Omega}{c_f k_f (\gamma_0 v_0)^2} \quad (1)$$

where σ represents the applied compression stress, $\dot{\varepsilon}$ the creep rate, k the Boltzmann constant, T the temperature, Ω the atomic volume, c_f the concentration of flow defects, v_0 the volume of one defect, γ_0 the shear strain per jump of a single defect and $k_f = \upsilon_f \exp(-Q_f/kT)$ the jump frequency of a flow defect with activation energy Q_f and pre-exponential factor υ_f. It has been shown that the evolution of defect concentration c_f towards a temperature dependent metastable equilibrium value c_{eq} can be described by the rate equation [9]:

$$\dot{c}_f = -k_r c_f (c_f - c_{eq}) \quad (2)$$

The jump frequency k_f may be used as rate constant k_r [8]. Thus the activation energy of formation/annihilation of flow defects is equal to that of movement Q_f. In the limit case $c_f » c_{eq}$, valid for short times or intermediate temperatures, the equations (1) and (2) imply a linear increase of the viscosity with time:

$$\eta(T,t) = a(T) + C \cdot C_0 \cdot T \cdot t \quad (3)$$

with the abbreviation

$$a(T) = C \cdot T \cdot c_{f_0}^{-1} \cdot \exp\left(\frac{Q_f}{kT}\right) \quad (4)$$

where c_{f_0} is the initial value of the flow defect concentration and C and C_0 are constants. For long creep times or at high temperatures, where $c_f » c_{eq}$ is not valid, the viscosity increase deviates from linearity and the viscosity reaches a temperature dependent saturation value:

$$\eta_{eq} = \frac{kT\Omega}{c_{eq} k_f (\gamma_0 v_0)^2}$$

4 Experimental Procedure

High temperature mechanical properties of Si-B-C-N ceramics derived from a boron modified polyvinylsilazane with the chemical composition $Si_3B_{1.0}C_{8.2}N_{2.6}H_{20}$ (T2-1) were investigated. Details of the polymer synthesis are presented elsewhere [10]. The brittle and unmeltable polymer was ground by use of a vibrating-disk mill and sieved to particle sizes smaller than 100 µm. For compaction it was is heated in a graphite die up to a temperature, where the polymer softens (150 °C...350 °C), and subsequently densified at a uniaxial pressure of 48 MPa. The obtained green bodies were converted into amorphous ceramic monoliths by thermolysis in argon atmosphere by heating up to 1400 °C and holding this temperature for 2 h. The chemical composition of the T2-1 derived ceramic was determined to be $Si_3B_{1.0}C_{4.3}N_{2.0}$ [3]. Optimizing the powder compacting conditions, the obtained ceramic monoliths contained a total porosity of about 15 %. Approximately 1/3 of the porosity in the obtained ceramic is closed. The procedure of the porosity measurement is described in [4].

From the as-thermolyzed samples creep specimen with a height of 3 mm and a cross sectional area of 1.8x1.8 mm^2 were prepared. The high temperature deformation behavior were investigated by compression creep tests in air at temperatures between 1350 °C and

1500 °C and applied stresses up to 300 MPa. The mechanical testing equipment and the testing procedure are described elsewhere [11]. After a testing time of 100 h at 1400 °C, the total deformation was in the range of about 2...5 %, depending on the applied stress. The elastic deformation will be not taken into account for the following analysis. XRD analysis and TEM investigations were used to investigate whether the specimens crystallize during the creep experiments.

5 Results and Discussion

As has been shown in [4], the deformation of amorphous precursor-derived ceramics consists of two components, one with a linear stress dependence and a stress independent one that can be ascribed to shrinkage. Thus to determine the Newtonian viscosity, the two components were separated and only the stress dependent component was further investigated.

Figure 1 shows viscosity data measured at 1400 °C at various stresses applied. In compliance with the definition of the Newtonian viscosity, after shrinkage adjustment no stress dependence of the viscosity can be observed. The viscosity changes within 10^6 s by more than two orders of magnitude. The full line represents the least mean square error fit of equation (3) to the data for times longer than 3×10^4 s. Obviously, for large values of t this affine linear relation describes the change of the viscosity with time very well. At times shorter than 2×10^4 s significant deviations form the behavior predicted by equation (3) occur. This effect may be explained by anelastic behavior and will be discussed later on.

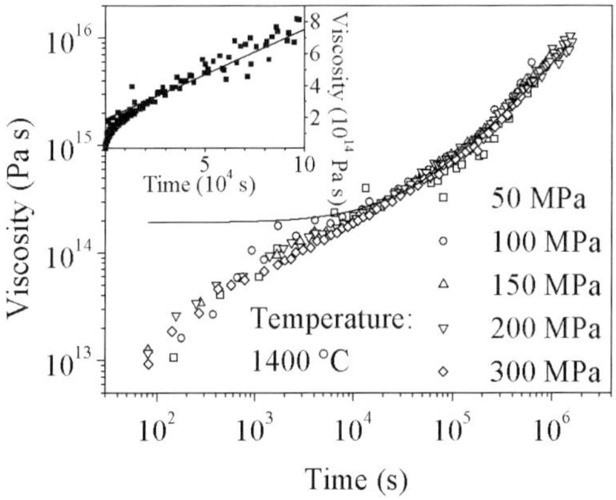

Figure 1. Dependence of the viscosity of precursor-derived amorpous ceramic on time at 1400 °C. The inset shows the deviation from the behavior expected according to equation (3) during the initial creep stage.

At a temperature of 1500 °C the viscosity also varies linearly with time (Fig. 2). In contrast to the measurements at 1400 °C, nearly no deviation from the expected affine linear behavior can be observed at short times. On the other hand, for times longer than 3×10^5s the observed viscosity is lower than expected according to the linear increase. A possible reason for this may be that the limit case $c_f \gg c_{eq}$ is no longer valid. The flow defect concentration c_f approaches the value c_f and thus the viscosity saturates. As the annihilation of flow defects is thermally activated, this deviation from linear behavior is expected to occur earlier at higher temperatures. This might explain why it was not observed at the creep tests performed at 1400 °C within testing times up to $1.6x10^6$ s.

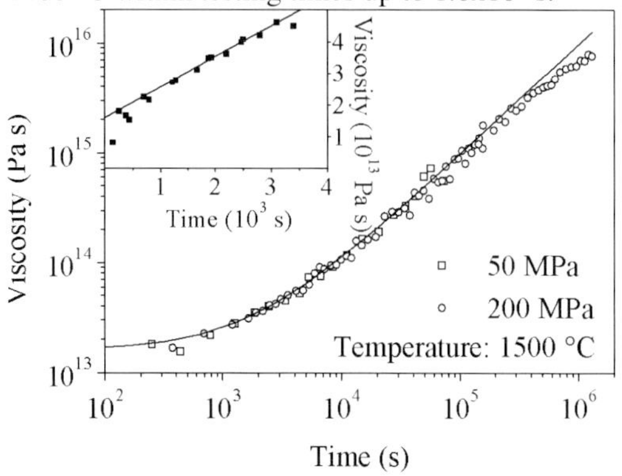

Figure 2. Dependence of the viscosity on time at 1500 °C. In the inset only at short times a slight deviation from the behavior predicted by equation (3) can be observed.

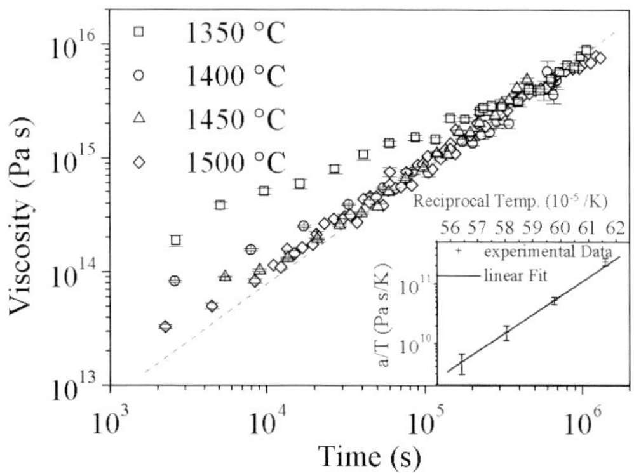

Figure 3. Time and temperature dependence of the viscosity. The inset shows a linear relation between $\log(a(T)/T)$ and $1/T$, where $a(T)$ is the initial value of the viscosity determined by fitting an affine linear function to the viscosity data after a correction allowing for shrinkage and anelastic behavior.

Figure 3 shows the dependency of the viscosity on time and on temperature in the range between 1350 °C and 1500 °C. The starting value of the viscosity shows a strong temperature dependence, whereas after about $3x10^5$ s nearly no variation of the viscosity with temperature can be observed. The inset in Fig. 3 shows, that equation (4) predicts the temperature dependence of the starting viscosity quite well. From the slope of the fitted straight line in the inset, the activation energy $Q_f = 0.6$ MJ/mol for the formation/annihilation and the movement of flow defects was estimated.

As mentioned above, for short times the measured viscosity deviates from affine linear behavior predicted by equation (3) (Fig. 1). This may be explained by anelastic behavior of the amorphous ceramic. The anelastic strain rate can be calculated from the difference between the measured viscosity and the viscosity predicted from fitting equation (3) to the data measured at long times. Figure 4 shows the dependence of the anelastic strain rate calculated in this way on time in double logarithmic scale for different stresses applied. The straight lines representing the slope –1 fit the data rather well. This indicates that the anelastic strain rate is indirect proportional to the time:

$$\dot{\varepsilon}_a = \frac{A}{t}$$

As can be seen in the inset in Fig. 4, the value of the proportionality constant A is proportional to the applied stress.

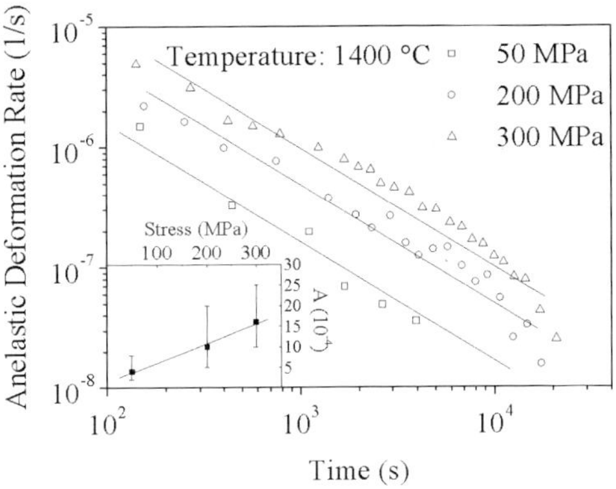

Figure 4. Anelastic component of the deformation rate.

6 Conclusions

Isothermal creep tests of precursor-derived amorphous Si-B-C-N ceramics in the temperature range between 1350 °C and 1500 °C and at stresses up to 300 MPa revealed a Newtonian deformation behavior after a correction of the data allowing for the thermally activated shrinkage. An increase of the viscosity of more than two orders of magnitude within testing times of about 300 h was observed. The dependence of the viscosity on time and temperature can be explained in terms of the free volume model. Applying this model, an activation energy of 0.6 MJ/mol for the movement and production/annihilation of flow defects was found. In an early creep stage at 1400 °C deviations from the behavior expected from the free volume model were observed. They may be explained by anelasticity. At a temperature of 1500 °C the influence of anelastic behavior is much smaller. The anelastic deformation rate at 1400 °C was found to be proportional to the applied stress and indirect proportional to the time.

7 References

[1] J. Bill, F. Aldinger, Adv. Mater. 1995, 7, p. 775.
[2] R. Riedel, A. Kienzle, W. Dressler, L. M. Ruwisch, J. Bill, F. Aldinger, Nature, 1996, 382, p. 796.

[3] M. Weinmann, J. Schuhmacher, H. Kummer, S. Prinz, J. Peng, H. J. Seifert, K. Müller, J. Bill, and F. Aldinger, submitted to Chem. Mater.
[4] M. Christ, G. Thurn, M. Weinmann, J. Bill, and F. Aldinger, submitted to J. Am. Ceram. Soc.
[5] F. Spaepen in Physics of Defects, (Ed.:R. Balian, M: Kléman and J.-P. Poirier), Les Houches Lectures XXXV, 1980, 134-174.
[6] A. L. Mulder, S. van der Zwaag, E. Huizer, A. van den Beukel, Scripta Met. 1984, 18, 515-519.
[7] M. H. Cohen and D. Turnbull, J. Chem. Phys. 1959, 31, p. 1164.
[8] A. van den Beukel, E. Huizer, A. L. Mulder and S. van den Zwaag, Acta metall. 1986, 34 (3), 483-492.
[9] P. de Hey, J. Sietsma and A. van den Beukel, Acta mater. 1998, 46 (16), 5873-5882.
[10] A. Kienzle, PhD thesis, Universität Stuttgart 1994.
[11] G. Thurn, J. Canel, J. Bill, and F. Aldinger, J. Europ. Ceram. Soc. 1999, 19 2317-2325.

Testing of Ceramics and Glasses by Indentation Techniques

Vincenzo M. Sglavo
Dipartimento di Ingegneria dei Materiali, Università di Trento, Trento (Italy)

1 Abstract

Indentation undoubtedly represents the most widely used technique for the fracture and fatigue characterization of glasses, ceramics and, in general, brittle materials. In this work, traditional indentation methodologies and recent developments are presented and compared. Advantages and limitations of each technique are critically analyzed.

2 Introduction

Since the pioneering works by Hertz[1] and Auerbach[2] at the end of the 19th century, indentation fracture, i.e. the fracture produced by the contact of a hard indenter on a brittle surface, has risen the interest of numerous researchers. Especially after the fundamental studies performed in the last thirty years, indentation has become the most widely used of all fracture techniques in the analysis of mechanical properties of glasses and ceramics.

Indentation allows the analysis of fracture behaviour of brittle solids and yields information on deformation processes at concentrated contacts; the production of an indentation on the surface of a ceramic material immediately furnishes data regarding the interaction between crack and microstructure; in addition, indentation allows the production of small sharp and controlled flaws similar to natural defects which can be used for the determination of material fracture characteristics such as fracture toughness and fatigue behaviour coefficients;[3, 4] it can be also be used for the study of crack initiation and propagation and for the simulation of strength degradation during surface finishing, wear and erosion;[3] finally, indentation produced at the interface between two different materials represents an easy tool for the determination of adhesion or interface resistance.[5]

The versatility of indentation fracture is accompanied by the limited cost and easy of testing related to the simple preparation of the sample and to the sole use of a routine hardness testing machine and of a microscope.

Indentations are usually classified as sharp or blunt depending on whether or not there is irreversible deformation at contact. Typical blunt indentation is obtained when a hard spherical indenter is loaded normally on the surface of a body. This process, at a critical load, usually results in the formation of the Hertzian cone crack.[1,3] Conversely, when sharp indenters like the Vickers or Knoop diamond pyramids used in hardness testing are used, inelastic and irreversible deformations are produced beneath the contact point and several types of crack propagates around the indentations.[3,6] Although sharp indentation is complicated from a mechanical point of view by the presence of a "plastic" zone, Vickers and

Knoop indentations have represented the most extensively studied subject in the field of indentation fracture and, probably, in fracture testing of brittle materials.

Deformation and fracture process produced by Hertzian contact re-entered the stage in the last ten years especially thanks to the studies of Lawn and co-workers devoted to the analysis of the damage produced on heterogeneous ceramics from brittle to "quasi-plastic" behaviour.[7]

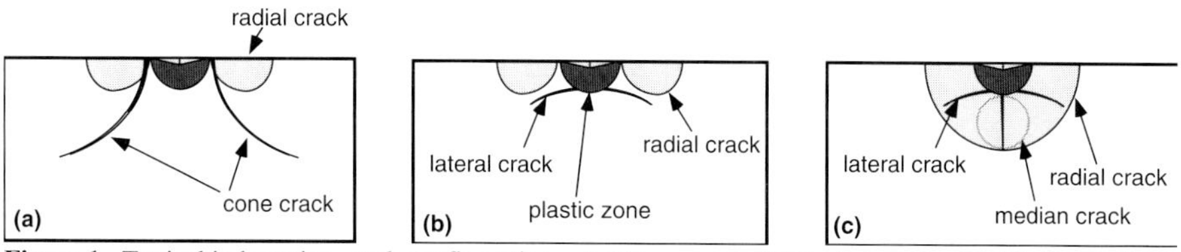

Figure 1. Typical indentation crack configurations. (a) Cone crack; (b) Radial cracks; (c) Median-radial cracks.

In the present paper, testing methodologies of glasses and ceramics related to sharp indentations will be reviewed, leaving to other papers arguments regarding blunt contact.[3,7] Initially, indentation mechanics and fracture mechanics principles will be considered. Then, methodologies for fracture toughness measurement and fatigue behaviour analysis will be considered.

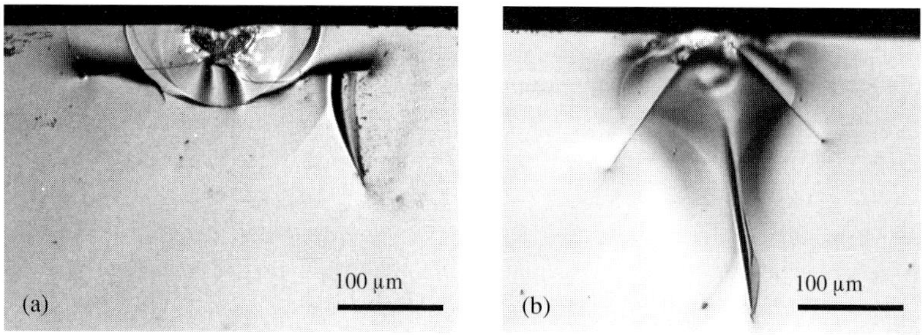

Figure 2. (a) Median-radial crack in soda-lime silicate glass; (b) cone crack in silica glass. $P = 9.8$ N.

3 Indentation Mechanics and Fracture Mechanics

The deformation and fracture pattern considered in the present section correspond to that produced by a standard Vickers diamond pyramid. Nevertheless, the proposed arguments can be easily extended to Knoop indentation.

Indentation on a brittle surface produces a semi-spherical plastic zone and a diversity of well-developed crack patterns illustrated in idealized form in Fig. 1.[3,6] Cracks which emanate from the corners of the indentation site are usually defined as median-radial and are often considered as semi-circular around the contact point. In some cases, especially at low indentation loads and in materials with high E/H ratio (E = elastic modulus, H = hardness) only radial Palmqvist cracks can be observed. Radial cracks usually are responsible for the strength of the sample. A second system, defined as lateral cracking, which propagates almost parallel to the surface of the sample from the bottom of the plastic zone is not usually responsible for failure but becomes important in the material removal associated to wear and erosion processes.

In some cases, a cone crack is produced around the indentation site as observed in the so called "anomalous" glasses where the densification processes which occurs beneath the indentation site blunt the contact.[6,8,9]

Examples of median-radial and cone cracks obtained in glasses are shown in Fig. 2.

The stress intensity factor for a semi-circular crack centered at the contact point and subjected to the stress field produced by the elastic contact is:[10-12]

$$K_e = \chi_e \frac{P}{c^{1.5}} \tag{1}$$

where c is the crack radius and χ_e is a factor which depends on the angular position on the crack front (ϕ), c and the ratio E/H.

On the other side, the stress intensity factor corresponding to the residual stress field generated by the plastic zone is:[10-12]

$$K_r = \chi \frac{P}{c^{1.5}} \tag{2}$$

where χ is another factor depending on ϕ and E/H. In this latter equation the load P' can be represented as a function of time but is equal to the maximum load P during unloading and after the indenter is removed.

The total stress intensity factor becomes:[10-12]

$$K = (\chi_e + \chi)\frac{P'}{c^{1.5}} = \chi_t \frac{P'}{c^{1.5}} \quad \text{on loading} \tag{3a}$$

$$K = \chi_e \frac{P'}{c^{1.5}} + \chi \frac{P}{c^{1.5}} \quad \text{on unloading} \tag{3b}$$

Due to the analytical definition of K crack propagates in a stable fashion (as $dK/dc<0$) upon loading,[3] reaching an equilibrium size given by:[10-12]

$$c = \left(\frac{\chi_t P'}{K_c}\right)^{2/3} \tag{4}$$

where K_c is the fracture toughness of the material. Equation (4) can be used for the determination of the length of median or cone cracks produced on loading.

On unloading the different stress distribution determines the propagation of crack along different path. The process is still stable and when the load is completely removed the crack length will be:[10-12]

$$c = \left(\frac{\chi P}{K_c}\right)^{2/3} \tag{5}$$

This latter relation represents, for examples, the surface crack length of radial cracks produces upon unloading.

It is important to point out that Eqs (4) and (5) were obtained assuming inert conditions. When kinetic effects produce sub-critical growth, the crack length will be larger.

4 Indentation Cracks for Fracture Toughness Determination

Vickers indentation probably represents the easiest tool for the evaluation of fracture toughness in glasses and ceramics.

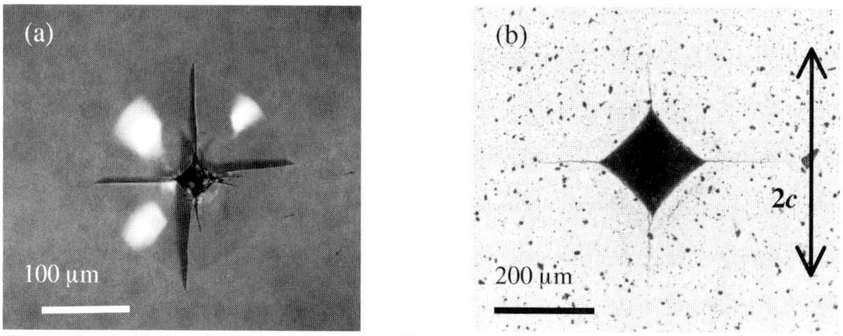

Figure 3. Radial cracks in (a) soda-lime silicate glass and (b) silicon nitride. The lateral crack halo can be observed in glass.

As obtained radial cracks (Fig. 3) can be directly used for K_c determination. From Eqs (4) or (5) one obtains:

$$K_c = \chi \frac{P}{c^{1.5}} \tag{6}$$

where χ represents χ_t for cracks produced on loading and c is the surface length of radial cracks (Fig. 3). Equation (6) allows fracture toughness evaluation from the measurement of crack length once χ is known.

For well developed radial cracks the factor χ is usually defined as:[10,13]

$$\chi = \xi (E/H)^{0.5} \tag{7}$$

where ξ is a geometrical factor. Experimental determination of ξ has been proposed by Anstis *et al.* who calculated a value of 0.016 ± 0.004.[13]

The described procedure, usually named as indentation fracture (IF) method, appears extremely simple, economical and fast. In addition, only limited amount of material is needed and this makes the IF method extremely useful during early stages of material development or optimization. Nevertheless, crack length is sometimes difficult to be measured even on polished surfaces especially in coarse grain materials. The choice of the proper χ value is also very important. A certain number of relations alternative to Eq. (6) have been proposed and often their use results in quite different values of K_c. In general, due to the a-priori calibration of the factor χ, which refers to the average behaviour of a certain number of different materials, can results in K_c values in disagreement with data obtained using alternative techniques. The sub-critical growth of radial cracks under the influence of the residual stress field, which can occur in glasses and in glass-containing ceramics at room temperature, can determine substantial underestimate of fracture toughness. Another limitation of the present procedure is that it can be easily used only at room temperature, the hardness test resulting quite complicated at high temperature.

An alternative approach for fracture toughness measurement is the so called indentation strength (IS) method.

If an external stress field, σ, is applied to the indented specimen, the stress intensity factor on radial cracks becomes:[3,14,15]

$$K = \chi \frac{P}{c^{1.5}} + \psi \sigma \sqrt{c} \tag{8}$$

where ψ is the shape factor depending on the geometry of the crack and the dimensions of the sample. Shape factor for semi-circular crack in a semi-infinite solid is equal to 1.29.[16] Nevertheless, due to finite dimensions of the specimens and to crack shape evolution upon bending ψ can assume values considerably different.[10,16-18] An important feature is that, when

σ increases and the equilibrium condition ($K = K_c$) is reached, crack undergoes some stable growth before the final failure.[3] In fact, due to the particular form of Eq. (8), a crack length interval can be found in which dK/dc < 0. If both ψ and χ are constant, the final instability condition in absence of fatigue phenomena, corresponds to a crack length equal to:[3,14,15]

$$c_f = \left(\frac{4\chi P}{K_c}\right)^{2/3} \tag{9}$$

and to a strength of[3,14,15]

$$\sigma_f = \frac{3}{4}\left(\frac{K_c^4}{4\psi^3 \chi P}\right)^{1/3} \tag{10}$$

This latter equation can be rewritten as:[3,15]

$$K_c = \delta\left(\frac{E}{H}\right)^{1/8}\left(\sigma_f P^{1/3}\right)^{3/4} \tag{11}$$

where

$$\delta = \left(\frac{256\,\psi^3 \xi^{2/3}}{27}\right)^{1/4} \tag{12}$$

Fracture toughness can be therefore evaluated by Eq. (11) from inert strength measurement of indented beams or disks, provided δ value is defined and the ratio *E/H* is known. In previous works the value of δ was calibrated to be equal to 0.59±0.12.[3,15]

The IS method does not require the measurement of crack length and this makes the procedure easier. Nevertheless, the application of the method requires the fabrication of flexure specimens (beams or disks) and only one data point can be obtained per specimen. As well as for IF method, tests can be performed only at room temperature due to the relaxation of the residual stress field at higher temperature.

The two approaches presented so far are extremely simple from an experimental and analytical point of view but require the a-priori assumption of the residual stress factor χ. In addition, both ψ and χ are considered as constant during the stable growth process. This fact represents the major reason of the disagreement of fracture toughness values obtained by using the proposed indentation techniques and data derived using typical fracture mechanics tests.

In order to eliminate the problem of χ determination one can try to remove the residual stress field. Two different approaches can be followed. In a first case the surface layer containing the plastic zone can be removed by polishing thus obtaining a stress-free semi-circular crack.[19] This latter is then used as starting notch for successive bending test. This method has been shown to be applicable also in material where Palmqvist cracks are formed in place of typical median-radial flaws. If the sample is indented using a tilted and canted Vickers pyramid one of the Palmqvist cracks is larger than the opposite one. The successive polishing procedure removes both the plastic zone and the smaller Palmqvist flaw leaving one semi-circular crack which acts as critical flaw in successive bending test. The crack profile can be decorated by penetrant dye and this makes the measurement easier after the bending test. On the basis of the flaw and sample sizes the shape factor can be evaluated. For semi-circular and semi-elliptical surface cracks ψ can be obtained on the basis of the arguments proposed by Newman and Raju.[20] It is important to remember that the shape factor varies along the pre-crack boundary and that the maximum value for ψ shall be used. This maximum value can occur either at the deepest point of the crack or at the intersection with the surface. Fracture initiating from the former location is preferred and can be promoted by sufficient polishing which makes the pre-crack a shallow semi-ellipse. The fracture toughness is then calculated from the bending strength as:

$$K_c = \sigma_f \, \psi \sqrt{c} \qquad (13)$$

where c corresponds to one half of the surface crack length.

The present procedure, named as Surface Crack in Flexure (SCF), has been shown to be extremely reliable for a large variety of ceramics and is part of a recent ASTM norm.[21]

A second possibility for the elimination of the factor χ is to remove the residual stress field by annealing.[22-24] The indented sample is heated at a particular temperature which guarantees the complete relaxation of the stress field. The maximum temperature must be carefully controlled to avoid crack healing or blunting. If the annealing is carried out correctly, the resulting specimen can again be used in bending test and the fracture toughness can be calculated following the procedure reported for the SCF method.

One common comment regarding the two latter procedures is that they can be suitable also for the measurement of fracture toughness at high temperature.

In the last decade, a development of the indentation strength method which takes into account the stable growth of indentation cracks has been proposed. This procedure consists in the observation and measurement of radial crack extension during *in situ* loading.[25] This approach requires either a microscope with miniaturized loading fixture or alternatively the attachment of a microscope to the mechanical testing machine. Indentations are produced using different loads and the samples are then bent up to failure. The couples (σ, c) measured during stable growth for different indentation loads are used to obtain the values of K_c as a function of crack size by Eq. (8). The present approach has been used also for the determination of R-curve behaviour in ceramic composites and coarse grain materials.[4,25] It is important to point out that the successful application of the methodology requires the knowledge of the factors χ and ψ. These latter can be calibrated using control specimens of known constant toughness with structure and composition similar to the material of interest: the residual stress factor can be calculated by Eq. (9) from the fracture toughness of the control material; analogously, the shape factor is evaluated by using Eq. (10). The hypothesis is made that χ and ψ are constant during the stable growth process and that they are invariant for similar materials. For the material under interest K_c is then calculated by Eq. (8). The complete set of fracture toughness values is generally plotted as a function of the crack length. The presence of possible toughening mechanisms is shown by increasing K_c values for larger crack length.[4,25]

In the procedures presented so far, with the exception of the SCF technique, fracture toughness can be determined once factors χ and ψ are *a priori* defined. In addition, these parameters are usually accepted as constant and indentation cracks are often considered shape-invariant during propagation process. These requirements are not always fulfilled. Depending on the material and the conditions under which the indentation is performed and stored, crack shape and growth can be complicated by several factors. Environmental effects acting both during and after the indentation process can affect crack shape and residual stress field. Prior to length measurement or bending tests, cracks can grow sub-critically to dimensions sensibly larger than those defined by equilibrium condition (Eqs (4) and (5).[12,16,26,27] In the same manner, the residual stress field can be partially relieved.[14,26] Moreover, the crack shape can change during bending and often flaw size becomes comparable with specimen dimensions. These facts can lead to strong variations in the shape factor.[17,20]

Testing procedures for the measurement of K_c by indentation cracks which relies to the direct calibration of the two parameters χ and ψ have been recently proposed by Steinbrech and co-workers[28,29] and Sglavo and co-workers.[23] The method is based on a detailed analysis of the crack propagation upon bending. This can be performed either by the observation of the crack stable growth during the test by an *in situ* optical microscope or, better, by successive bending tests interrupted at loads lower than fracture strength. In this latter case, the crack can be decorated by penetrant dye which allows an easier observation after the successive manual breakage. In this way the real shape of the crack during the complete stable growth process can be detected (Fig. 4).[23]

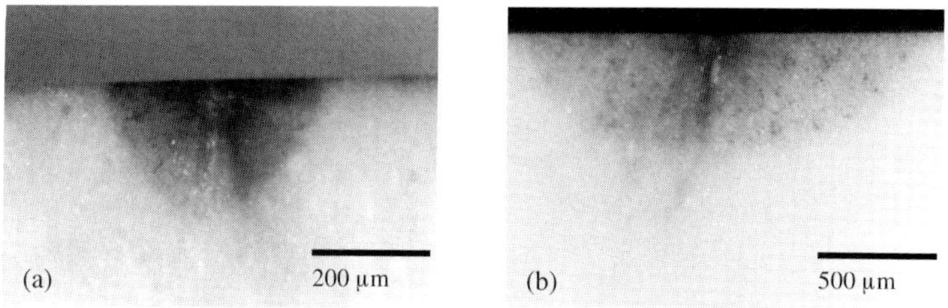

Figure 4. Sub-surface profile of as-obtained indentation crack (a) and crack after stable growth (b) in alumina (P = 294 N).

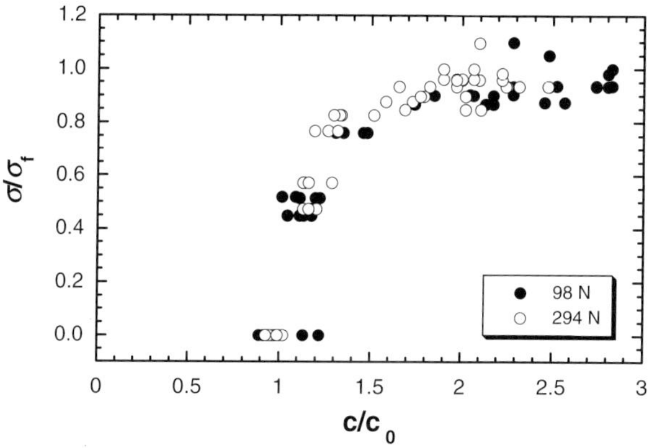

Figure 5. Evolution of surface crack length during stable growth in alumina for two different indentation loads. (From Ref. 23)

For each applied stress value a set of crack length and shape data is therefore obtained (Fig. 5). Depending on the crack geometry the shape factor can also be evaluated. The calculus is precise when semicircular or semi-elliptical cracks are observed as usually occurs in alumina, silicon nitride and other technical ceramics. The methods can therefore hardly be applied to glass where the interaction of radial and lateral cracks produces a complicated geometry in median-radial cracks.[18] For semi-elliptical cracks the shape factor can be calculated as a function of the surface crack length, c, by using the relations proposed by Newman and Raju.[20] Equation (8) can be transformed into:

$$A = K_c\, B - \chi \tag{14}$$

where

$$A = \frac{\psi \sigma c^2}{P} \tag{15}$$

and

$$B = \frac{c^{1.5}}{P} \tag{16}$$

A linear trend is usually obtained between A and B (Fig. 6). Values of K_c and χ are then obtained by interpolation of A *vs.* B data.

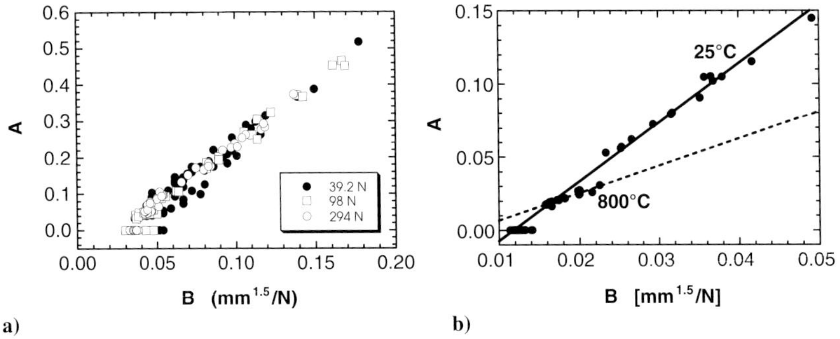

Figure 6. Plots of A vs. B for (a) alumina (room temperature) and (b) SiC_w - Si_3N_4 composite (P = 294 N). (From Refs 23 and 31)

This methodology can be applied also at relatively large temperatures provided the residual stress factor is larger than zero,[30,31] otherwise stable growth does not occur and results from linear fitting can be affected by large errors. Nevertheless, in this situation, the procedure previously described and related to the residual stress removal by annealing can be applied.[23]

The present approach has been applied also to ceramics characterized by toughening effects. Results were shown to point out the R-curve effect for ceramics presenting frontal-wake shielding mechanism such as partially stabilized zirconia.[32] Conversely, the method furnished anomalous and constant K_c values for different crack sizes in ceramics characterized by bridging mechanisms.[28,31] This anomaly can be associated to the large COD associated to indentation cracks which modifies the development of the bridging stresses.[28]

This latter observation probably represents the most important limitation of the indentation methods for fracture toughness measurement. Even when the more sophisticated methods are used and all the parameters (included ψ and χ) are experimentally and independently calibrated, indentation produces values which differs from data obtained by using typical fracture mechanics methodologies for materials characterized by bridging cracks. Nevertheless, in this situation, indentation techniques can still be used for an easy qualitative comparison of similar materials. Conversely, for constant fracture toughness ceramics, indentation method can definitely be used with high reliability.

5 Indentation Cracks for Fatigue Testing

Indentation cracks represent a fundamental tool also in the determination of fatigue behaviour of glasses and ceramics. Due to the presence of the residual stress field which can be removed at high temperature, indentation flaws have been widely used in the determination of sub-critical crack growth behaviour at room temperature.

Fatigue testing needs the analysis of the crack propagation at $K<K_c$ in and active environment under the effect of a specified load. The value of crack length as a function of time is related to the applied stress intensity factor and this is usually expressed by the power law:[3]

$$v = \frac{dc}{dt} = v_0 \left(\frac{K}{K_c} \right)^n \tag{17}$$

where v is the crack velocity and v_0 and n two parameter which depend on the couple material/environment.

Indentation crack can be used either directly, under the influence of the indentation or residual load, or as controlled starting defect during bending tests for fatigue testing of ceramic materials.

The first approach has been particularly used in glass. The residual stress field itself can represent the driving force for the sub-critical propagation of radial cracks.[27] It is common experience that if, after the indenter removal, the indentation is kept in the active environment crack size increases. This happens for example in silicate glasses when the indentation are stored in humid environment. The radial crack length as a function of time can be derived from Eqs (5) and (17):[12,27]

$$\frac{dc}{dt} = v_0 \left(\frac{\chi P}{K_c} \right)^n \frac{1}{c^{3n/2}} \tag{18}$$

Equation (18) can be solved for χ = constant to give:

$$c = c_0 \left(1 + h\, v_0 \frac{t}{c_0} \right)^{1/h} \tag{19}$$

where c_0 is the crack length produced under inert conditions and $h=(3n+2)/2$. In most cases Eq. (19) can be simplified as:[12]

$$c \cong c_0 \left(h\, v_0 \frac{t}{c_0} \right)^{1/h} \tag{20}$$

which can be linearized as:

$$\log c = \frac{1}{h} \log \left(h\, v_0 c_0^{h-1} \right) + \frac{1}{h} \log t \tag{21}$$

The fatigue parameters, n and v_0, can be therefore evaluated from the measurement of radial crack length at different times after the indenter removal.[27,32]

The linear fitting of log c *vs.* log t data allows the detemination of n from the slope and of v_0 from the intercept. One problem associated with the method is the change in the shape of the median-radial crack during the sub-critical process. This fact is complicated by the interaction of radial and lateral cracks which makes both χ and ψ strongly dependent from crack length and which arrests radial cracks at K values larger than the fatigue limit.[18]

An alternative methods has been proposed by Sglavo and Green[9,33] where the sub-critical growth is measured on cracks subjected to the maximum load during the indentation test. The load is kept on the sample for a pre-defined time and the sub-surface crack length is then measured. The depth of the median crack is measured in "normal" glasses (e.g. soda-lime silica, aluminosilicate)[33] while the length of the cone crack is measured in "anomalous" glasses (silica and borosilicate) (Fig. 7).[9] The sub-critical growth of both median and cone cracks is ruled by a relation similar to Eq. (19). Experimental data are fitted by using Eq. (21) and fatigue parameters can be obtained as well.

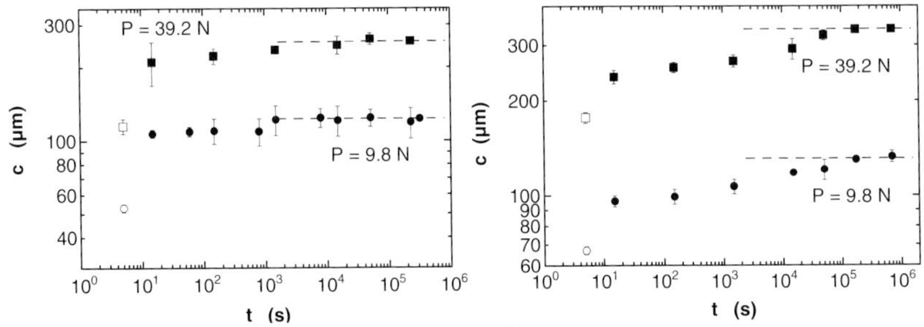

a) b)

Figure 7. (a) Median crack depth in soda-lime silicate glass and (b) cone crack length in silica glass as a function of the duration of the indentation test. Empty symbols correspond to measurements performed in inert environment. (From Refs 23 and 31)

This latter approach has been shown to allow an easy determination of another important fatigue parameter, i.e. the fatigue limit, K_{th}.[9,33] In fact, when the duration of the indentation test is long enough the stress intensity factor applied to the indentation crack can reach the fatigue limit and this is revealed by an arrest of the crack. The invariant length is used to back – calculate K_{th} by Eq. (3a).

Indentation cracks can be also used as starting and controlled defect for fatigue testing when an external load is applied to the specimen.

The stress intensity factor-crack velocity (K-v) relationship for sub-critical crack growth can be determined from *in situ* observation of indentation flaws.[34] The indented specimen is subjected to an external load using especially designed fixtures which allow the observation of the surface crack propagation during the test. Tests are usually performed on radial cracks which have undergone to significant sub-critical growth in order to minimize the effect of the residual stress and the interaction with lateral cracks. Successive measurements of crack increase for given applied stress and time interval are made. The sub-critical propagation usually occurs with a semi-elliptical geometry and the observation of the *post mortem* crack geometry allows the determination of the shape factor. This latter is used to evaluate the stress intensity factor as a function of the crack length and, therefore, of time, as $K = \psi \sigma c^{0.5}$.

Indirect methods for fatigue testing of glasses and ceramics involve dynamic or static fatigue tests on indented samples. In these cases, indentation crack represents the controlled defect and allows a drastic reduction of the scatter of strength or lifetime data usually revealed when natural flaws are considered.

Dynamic fatigue test consists in the failure of different samples at various stressing rates. When indented samples are used, under conditions of dynamic fatigue, provided ψ and χ are constant, the strength in corrosive environment can be expressed as a function of the stress rate:[3,35-39]

$$\sigma_f = \zeta\, \dot{\sigma}_G^{1/(n'+1)}, \tag{22}$$

where

$$\zeta = \left(\frac{2.84 n'^{0.462} \sigma_f^{n'} c_f}{v_0} \right)^{1/(n'+1)} \tag{23a}$$

$$n' = 0.763 n; \tag{23b}$$

$\dot{\sigma}_G$ is the stressing rate and σ_f and c_f are the strength and crack length corresponding to unstable fracture under inert conditions. Linear regression analysis of log σ_f against log $\dot{\sigma}_G$ is used for the calculation of n and v_0 (Fig. 8). The slope of the fitted line is equal to 1/(n'+1) while the intercept is equal to log ζ.

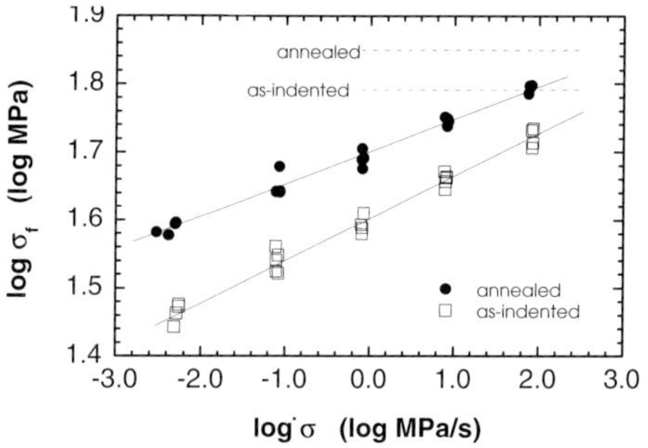

Figure 8. Dynamic fatigue plot for as-indented and annealed specimens tested in deionized water (P = 9.8 N). Dashed lines correspond to strength measured in inert environment. (From Ref. 39)

If the residual stress term is eliminated as previously described either by annealing or by surface polishing, Eq. (22) reduces to[35]

$$\sigma_f = \zeta' \dot{\sigma}_G^{1/(n+1)} \tag{24}$$

where

$$\zeta = \left(\frac{2\,(n+1)\,\sigma_i^n\, c_0}{(n-2)\, v_0} \right)^{1/(n+1)} \tag{25}$$

c_0 being the initial crack length.

This latter situation can be used for fatigue testing at high temperatures as it does not require the analysis of the influence of the stabilizing residual stress term, K.

Indentation cracks in which the residual stress field has been removed can also be used for fatigue analysis in static fatigue tests. In this case, the time-to-failure, T_f, is related to the constant applied stress, σ_a, by the expression:[35,39]

$$T_f = \frac{\gamma}{\sigma_a^n} \tag{26}$$

where

$$\gamma = \frac{2}{n-2} \left(\frac{K_c}{\psi} \right)^n \frac{1}{v_0} \frac{1}{c_0^{(n-2)/2}} \tag{27}$$

c_0 being the initial crack length.

Linear regression analysis of log T_f *vs.* log σ_a allows the evaluation of n from the slope (equal to -n) and of v_0 from the intercept.

On the basis of the arguments reported in the previous section, as-indented samples can be used for fatigue characterization of glasses and ceramics at room temperature. Conversely, measurements at higher temperature are better performed if indented samples where the residual stress field has been removed either by polishing or by annealing are used.

6 Conclusions

Indentation techniques represent an extremely versatile tool for testing ceramics and glasses. Both fracture characteristics, *in primis* fracture toughness, and sub-critical crack growth parameters can be determined by alternative approaches based on the production of indentation cracks. Indentation techniques have several advantages compared with other experimental procedures: low cost, easy of testing and limited requirement of material. On the other side, care must be taken in the choice of the residual stress factor, χ, and shape factor, ψ, associated to the produced crack and the more specific is their determination the more reliable and comparable to other procedures is the technique. The presence of the residual stress field limits the use of as-obtained indentation cracks at room temperature testing. High temperature tests can be performed after the residual stress field has been removed by polishing away the surface layer or by annealing. In fracture toughness measurement, problems arise when toughened materials are considered as the large COD associated with indentation flaws limits the development of crack bridging mechanisms. In fatigue testing, indentation cracks allow to limit sensibly the typical strength scatter of brittle materials thus allowing more precise determination of sub-critical crack growth parameters.

7 References

[1] H. Hertz, Hertz's Miscellaneous Papers, Ch. 5 & 6, Macmillan, London, UK, 1896.
[2] F. Auerbach, Ann. Phys. Chem. 1891, 43, 61.
[3] B. R. Lawn, Fracture of Brittle Solids, 2nd Edition, Ch. 8, Cambridge University Press, Cambridge, UK, 1993.
[4] S. J. Bennison in Mechanical Testing Methodology for Ceramic Design and Reliability (Eds: D. C. Cranmer and D. W. Richerson), Marcel Dekker Inc., New York, USA, 1998.
[5] R. Dal Maschio, V. M. Sglavo, L. Mattivi, L. Bertamini and S. Sturlese, J. Thermal Spray Techn. 1994, 3(1), 51-56.
[6] R. F. Cook and G. M. Pharr, J. Am. Ceram. Soc. 1990, 73(4), 787-817.
[7] B. R. Lawn, J. Am. Ceram. Soc. 1998, 81(8), 1977-94.
[8] A. Arora, D. B. Marshall and B. R. Lawn, J. Non-Cryst. Solids 1979, 31, 415-28.
[9] V.M. Sglavo and D. J. Green, J. Appl. Phys., submitted for publication.
[10] D. B. Marshall and B. R. Lawn, J. Mater. Sci. 1979, 14, 2001-12.
[11] B. R. Lawn, A. G. Evans and D. B. Marshall, J. Am. Ceram. Soc. 1980, 63(9-10), 574-81.
[12] V. M. Sglavo and D. J. Green, J. Am. Ceram. Soc. 1995, 73(3), 650-56.
[13] G.R. Anstis, P. Chantikul, B. R, Lawn and D. B. Marshall, J. Am. Ceram. Soc. 1981, 64(9), 533-38.
[14] D. B. Marshall, B. R. Lawn and P. Chantikul, J. Mater. Sci. 1979, 14, 2225-35.
[15] P. Chantikul, G. R. Anstis, B. R. Lawn and D. B. Marshall, J. Am. Ceram. Soc. 1981, 64(9), 539-43.
[16] S. M. Smith and R. O. Scattergood, J. Am. Ceram. Soc. 1992, 75(2), 305-15.
[17] R. F. Krause, Jr., J. Am. Ceram. Soc. 1994, 77(1), 172-78.
[18] V. M. Sglavo and D. J. Green, Eng. Fract. Mech. 1996, 55(1), 35-46.

[19] G. D. Quinn, R. J. Gettings and J. J. Kübler in Fracture Mechanics of Ceramics (Eds: R. C. Bradt, D. P. H. Hasselman, D. Munz, M. Sakai and V. Ya. Shevchenko) Plenum Press, New York, USA, 1996, pp. 203-218.
[20] J. C. Newman and I. S. Raju, Eng. Fract. Mech. 1981, 15, 185-92.
[21] ASTM norm C1421, in Annual Book of ASTM Standards, Vol. 15.01, American Society for Testing and Materials, Philadelphia, PA, USA.
[22] M. Leoni, P. Scardi and V. M. Sglavo, J. Eur. Ceram. Soc. 1998, 18, 1663-68.
[23] V. M. Sglavo and P. Pancheri, J. Eur. Ceram. Soc. 1997, 17, 1697-1706.
[24] A. M. Thompson, H. M. Chan, M. P. Harmer and R. F. Cook, J. Am. Ceram. Soc. 1995, 78(3), 567-71.
[25] L. M. Braun, S. J. Bennison and B. R. Lawn, J. Am. Ceram. Soc. 1992, 75(11), 3049-57.
[26] W. T. Han, P. Hrma and A. R. Cooper, Phys. Chem. Glasses 1989, 30(1), 30-33.
[27] P. K. Gupta and J. N. Jubb, J. Am. Ceram. Soc. 1981, 64, C112-14.
[28] D. Bleise and R. W. Steinbrech, J. Am. Ceram. Soc. 1994, 77(2), 315-22.
[30] V. M. Sglavo, E. Trentini and M. Boniecki, J. Mater. Sci. Lett. 1999, 18, 1127-30.
[31] V. M. Sglavo and P. Bosetti in Ceramics: Getting into the 2000's, Part A (Ed.: P. Vincenzini), Techna, Faenza, Italy, 1999.
[32] G. W. Dransmann, R. W. Steinbrech, A. Pajares, F. Guiberteau, A. Dominguez-Rodrigues, A. Heuer, J. Am. Ceram. Soc. 1994, 77(5), 1194-201.
[32] R. F. Cook and E. G. Liniger, J. Am. Ceram. Soc. 1993, 76(5),1096-105.
[33] V. M. Sglavo and D. J. Green, J. Am. Ceram. Soc. 1999, 82(5), 1269-74.
[34] P. J. Dwivedi and D. J. Green, J. Am. Ceram. Soc. 1995, 78(8), 2122-28.
[35] V. M. Sglavo and D. J. Green, Acta Metall. Mater. 1995, 43(3), 965-72.
[36] B. R. Lawn, D. B. Marshall, G. R. Anstis and T. P. Dabbs, J. Mater. Sci. 1981, 16, 2846-54.
[37] E. R. Fuller, B. R. Lawn and R. F. Cook, J. Am. Ceram. Soc. 1982, 66(5), 314-20.
[38] V. M. Sglavo M. Gadotti and T. Micheletti, Fatigue Fract. Eng. Mater. Struct. 1997, 20(8), 1225-34.
[39] V. M. Sglavo and D. J. Green, J. Mater. Sci. 1999, 34, 579-85.

Indentation Induced t- and m-Phase Orientation in ZrO_2

Michael Dorn, Klaus Georg Nickel
Universität Tübingen, Institut für Mineralogie, Petrologie und Geochemie, Wilhelmstraße 56, 72074 Tübingen.

1 Abstract

The structural changes in the vicinity of indentations in Y-PSZ were investigated by micro-raman spectroscopy and micro-focus sychroton radiation diffraction. It is shown that some Raman band intensities are strongly susceptible to the orientation relative to a polarised laser beam, both for monoclinic and tetragonal phase bands. This feature was used to demonstrate a preferred orientation of both t- and m-grains around indentations. The orientation direction rotates along the edges of an indent, which is related to the curvature of stress field lines.

This feature is also important for the evaluation of relative proportions of *t*- and *m*-phase in partially transformed ZrO_2 and is a potential source for large errors.

2 Introduction

The *t-m* phase transformation of yttria-PSZ is of great technological interest because of its toughening mechanism, which improves the mechanical properties of ZrO_2 ceramics. Toughening arises as a result of the metastability of *t*-precipitates in partially stabilised zirconia. The metastabile *t*-ZrO_2 grain transforms to the *m*-phase under localised tensile stress whereby the volume increases by about 4%. The resulting volume strain reduces the stress intensity at a crack tip [1].

The appearance of hydrostatic and deviatoric stresses in yttria-PSZ during Vickers hardness indentation at room temperature does not only cause cracking and plastic deformation, but also phase transformation in and around the indentation.

To determine the lateral extend of the transformed region micro-Raman spectroscopy is a useful tool. However, the quantitative or semi-quantitative estimation of the amount of monoclinic phase in a partially transformed region by band ratios [2; 3; 4] is hampered by the dependence of the Raman signal of a polarised laser beam on the crystallographic orientation for some bands, which has been demonstrated for the monoclinic phase [5].

In this paper we show that this dependence is not restricted to the monoclinic phase and can be used to determine a indentation induced orientation. Only ratios for some well-chosen bands should be used to indicate the relative proportions of t- and m-ZrO_2 by micro-Raman spectroscopy with a polarised laser beam, if grains are oriented.

3 Experiments

Experiments were carried out using a commercially available tetragonal 3Y-ZrO_2 with high purity, obtained from the Zentrum für Zahn-, Mund- und Kieferheilkunde of the Eberhardt-Karls-Universität Tübingen. The samples had an average grain size of about 0.6 µm and a density >6.05 g/cm³ without open porosity. They were polished to a 1 µm finish using diamond pastes. The indentations were made using an Instron 4502 universal test machine with different loads and constant loading and unloading speed (0.01mm/min) using Vickers (tetragonal diamond pyramid, 136° included angle) indenter.

To monitor the phase transformation, a LabRam-2 micro-Raman spectrometer (Dilor, France) was used with excitation wave length of 488.0 nm (argon ion laser) . To obtain comparable results, the same parameters were chosen for all investigations, including laser intensity, objective and focus on the sample. A measurable re-transformation to the t-phase caused by laser annealing was not observed even after the longest record time of a spectrum (5 min). The lateral resolution depended on the objective used and varied between 1-5 µm.

Fine-focus x-ray investigations were carried out at the European Synchrotron Research Facility (ESRF) in Grenoble to examine the microstructure of the indented samples. Because of the relatively large beam diameter of 50 µm it was necessary to investigate large indentations (about 200µm edge width) obtained at a load of 500N. The indentations were scanned in x- and y-direction in steps of 20 µm.

4 Results

4.1 Raman spectroscopic investigations

It is well known, that phase transformations do not only occur in the direct contact area of indenter and sample, but also in a region next to the indentation, where a strong uplift of the surface takes place. This uplift appears to be primarily due to the volume increase during the phase transformation and to a minor extend to the transport of material out of the contact area [6].

The determination of the extend of microstructural changes around indentations is of technological interest, because it will allow to evaluate the behaviour of transformation toughened ceramics under non-uniform stresses, which are typical for wear or machining processes.

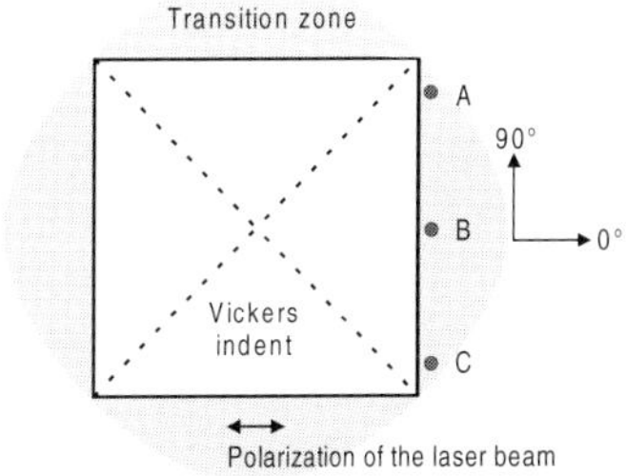

Fig. 1: Positions A, B, and C at the edge of the Vickers indentations, from which spectra were taken

The formation of monoclinic phase in tetragonal zirconia is characterised by the appearance of typical bands of different modes in the raman spectrum [5]. The effect of the intensity dependence of some bands on the direction of the polarisation of the laser beam in relation to the crystal orientation [7] is of concern for all cases of preferred orientation. Therefore, experiments were carried out, in which the sample was rotated under the laser beam. Measurements were taken next to the indentation at the points A, B and C in Fig.1. The distance to the edge was about 3-5 μm.

The sample was rotated around the points indicated in Fig. 1, taking care to retain the exact focus and position. The spectra were taken with a 100x objective, which yields a spot diameter of app. 1μm. The variation of the band intensities with of the rotation angle is exemplified for point B in Fig. 2, which shows that in particular the bands at 264 and 650 cm^{-1}, which belong to the tetragonal phase, show a strong dependence of intensity on orientation. The 188 cm^{-1} band intensity of the monoclinic phase is likewise orientation dependent. The bands least influenced are 147 cm^{-1} of the tetragonal and 378 cm^{-1} of the monoclinic phase.

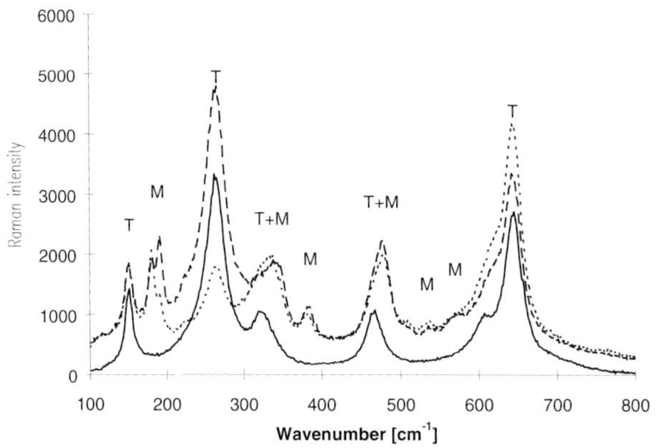

Fig. 2: Raman spectra of pristine surface (solid), and partially transformed zone in direction 0° (dotted) and 90° (dashed) at Point B of Fig. 1.

To exclude artefacts from possibly changing background values, the ratio between the fix 147 cm^{-1} and the orientation susceptible 264 cm^{-1} band is used to evaluate orientations.

The pristine surface of the Y-PSZ does not show any dependence of the band intensities on the rotation angle (Fig 3). This indicates a statistical distribution of grain orientations in the original material after machining and is evidence that enough grains are detected within the excitation volume of the laser spot to guarantee a statistical result for the spot.

In contrast to the pristine sample the spectra of positions A, B and C do vary significantly with sample orientation in relation to laser beam polarisation (Fig. 4).

Furthermore, the rotation angles, at which the maxima appear, change dependent on the position relative to the Vickers indent (Fig. 4). Thus an orientation of the grains in the vicinity of the indentation edges has occurred, but it is not uniform around the impression nor along the edge of the indent.

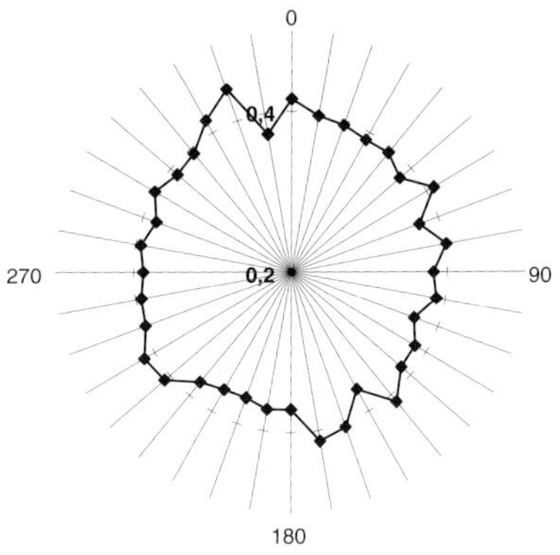

Fig. 3: Intensity ratio of the *t*-bands 147/264cm^{-1} as a function of sample orientation in the pristine sample

Similar results would be obtained by plotting ratios for appropriate monoclinic bands, supporting the findings of Behrens et al. in 1993 [4]. However, from Fig. 2 the best suited band with sample orientation susceptibility is 188 cm^{-1}. This has an overlap with the band at 175 cm^{-1}, which seems not to vary. A proper evaluation will thus need a fitting of the spectra and this will be presented elsewhere.

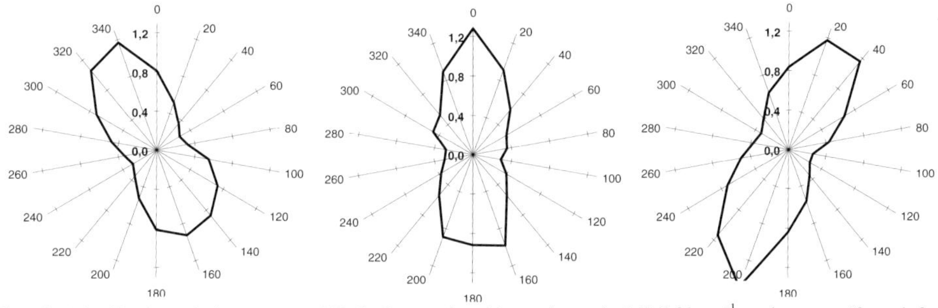

Fig. 4: Distribution of the tetragonal ZrO_2-Raman band intensity ratio 147/264 cm^{-1} at the spots (from left to right) A, B and C of Fig. 1.

4.2 X-ray microfocus investigations

X-ray microfocus investigations were carried out using a large indentation (500N load), which was fractured along indentation induced cracks at the corners of the indent. A relatively long edge length of about 200 µm was obtained. Using a 50µm X-ray beam, the indentation area was scanned in steps of 20 µm in x-direction . The sample was placed in an angle of 5° relative to the incident beam [Fig 5].

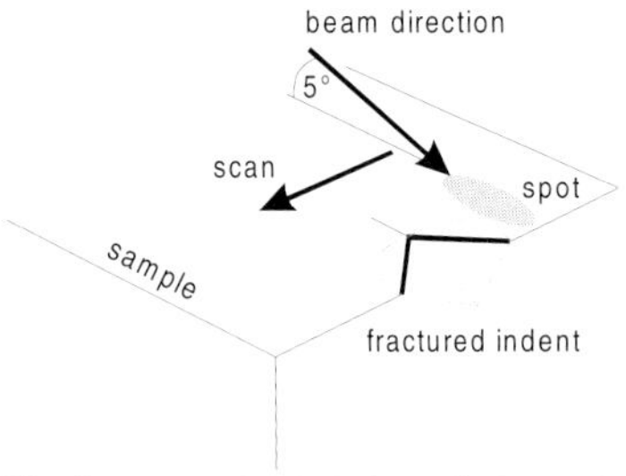

Fig. 5: Schematic of the beam/indent orientation in synchrotron experiments

Scanning the indentation and the areas beneath, Debye Scherrer rings with an inhomogeneous intensity distribution were found [Fig 6]. The non-uniform monoclinic -111m and 111m reflexes rotate while scanning over the indentation and completely disappear outside the indented area. The tetragonal 101 reflexes cannot show orientation features for crystallographic reasons.

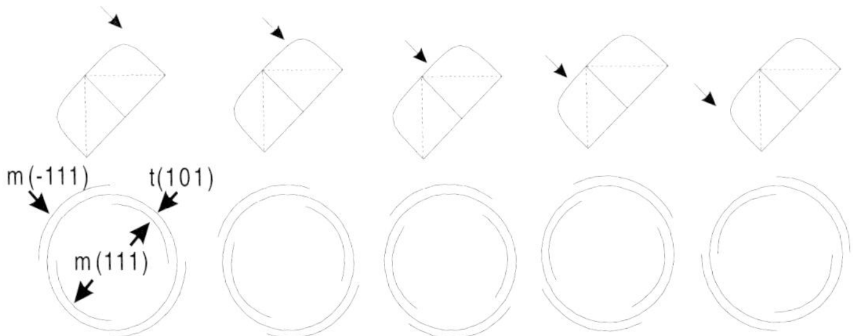

Fig. 6: X-ray microfocus investigations at the ESRF, Grenoble. Scan over a cracked 500N HV indentation. The schematic pictures show the rotation sequence of the nonuniform scattered monoclinic Debye-Scherrer rings. Arrows indicate measured spot relative to the indentation

This is again an evidence for preferred orientation, here for the monoclinic phase. A similar effect was also reported in cold drawn wire [8].

5 Discussion

The orientation susceptibility of Raman band intensities of tetragonal ZrO_2 and X-ray micro-focus diffraction were used to demonstrate the orientation of both mother- and daughter-phases (t and m) in the vicinity of hardness impressions.

At present we do not know, which crystallographic symmetry element is responsible for the susceptibility and hence we do not know, which geometrical orientation the grains have in detail. Nonetheless it is plain that the angle of the maxima and minima in Fig. 4 must indicate rotating crystallographic orientation.

This can be rationalised by looking at the stress distribution in Vickers impressions and residual compressive stress in outer regions. The calculations of Galanov et al. [9] for the distribution of compressive stresses under such indenters and residual for outer regions [10] are shown in Fig. 7 in relation to the direction of the maxima of Fig. 4. It seems feasible to interpret the direction change as mixture attributed to the curvature of iso-stress lines in the indentation and in outer regions.

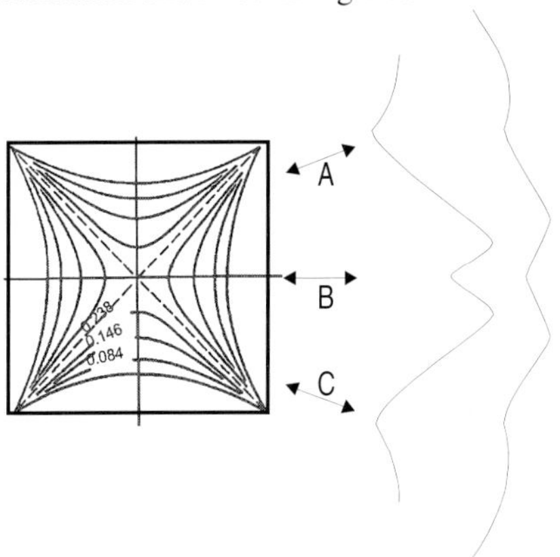

Fig. 7: Maxima directions from Fig. 4 at spots A, B, and C of Fig. 1 in relation to compressive iso-stress lines after Galanov et al. [9] and residual compressive stress after Zeng et al. [10]

This does not reveal the mechanism of orientation. A twinning in the course of phase transformation may be just as able to induce the orientation as ordinary plastic deformation under high stresses.

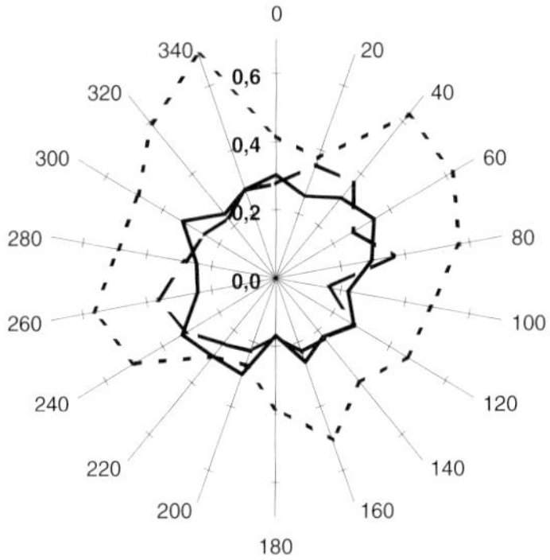

Fig. 8: Variation of the band intensity ratio 147/380 cm^{-1} (t/m-ZrO_2) with sample orientation for the spots A (dashed), B (solid) and C (dotted) of Fig. 1

From Fig. 2 we used the 147 cm^{-1} band of tetragonal ZrO_2 as a fix band. From the same figure a band of similar stability for the monoclinic ZrO_2 seems to be the 380 cm^{-1} band. Accordingly these would be candidates to give an estimation of the relative change in proportions of both phases at a given spot. This is evaluated in Fig. 8.

Fig. 8 indicates that this ratio is a useful one, which gives equal ratios regardless of orientation for most conditions. However, some orientation features remain for high ratios (please note the scale: it is a variation by a factor of about 2 instead of the variation of a factor of 5 in Fig. 4!).

Thus presently we caution against the use of band intensity ratios for the quantification of relative proportions of ZrO_2 phases unless statistical orientation is proven or a depolarisation of the laser beam is established.

6 Literature

[1] Lawn B.R.: „Microfracture beneath point indentations in brittle solids“, *J. of Mat. Sc.*, vol. 10, pp. 113-122, 1975.

[2] Clarke D. R. and Adar F.: „Measurement of the Crystallographically Transformed Zone Produced by Fracture in Ceramics Containing Tetragonal Zirconia“, *J. Am. Ceram. Soc.*, vol. 65, pp. 284-288, 1982

[3] Kaliszewski M. et al.: „Indentation Studies on Y_2O_3-Stabilized ZrO_2: I, Development of indentation-Induced Cracks“, *J. Am. Ceram. Soc.*, vol. 77, pp. 1185-1193, 1994

[4] Behrens G. et al.: „On the isothermal Martensitic Transformation in 3Y-TZP“, *J. Am. Ceram. Soc.*, vol. 76, pp. 1025-1030, 1993

[5] Ishigame M. and Sakurai T.: „Temperature Dependence of the Raman Spectra of ZrO2“, *J. Am. Ceram. Soc.*, vol. 60, pp. 367-369, 1977

[6] Celli A. et al.: „AFM, a Tool for Investigating Indentation Damage in ZrO_2“, *Am. Ceram. Soc. Bul.*, vol. 78, No. 8, pp. 87-91, 1999

[7] Pelletier M. J.: „Analytical Applications of Raman Spectroscopy“, *Blackwell Science*, pp. 444-445, 1999, ISBN 0632053054

[8] Klug H. and Alexander L.: „X-Ray Diffraction Procedures“, *Wiley Interscience Publication*, pp. 550 ff, 1974, ISBN 0471493694

[9] Galanov, B.A. & Grigoriev, O.N.: „Deformation and fracture of superhard materials under contact loading“, *Problemy Prochnosty (Strength of Materials)*, vol. 10, pp. 36-42, 1986 (in Russian)

[10] Zeng, K. et al.: „Residual Stress Fields at the Surface of Sharp Pyramid Indentations“, *J. Am. Ceram. Soc.*, vol. 81, pp. 689-530, 1998

Influence of the Processing on the Mechanical Properties of Precursor-Derived Si-C-N Ceramics

Arndt Bauer, Günter Thurn, Joachim Bill, and Fritz Aldinger
Max-Planck-Institut für Metallforschung
Institut für Nichtmetallische Anorganische Materialien, Universität Stuttgart, Pulvermetallurgisches Laboratorium, Heisenbergstrasse 5, 70569 Stuttgart, Germany

1 Abstract

The commercial available polysilazane Ceraset was used to obtain amorphous Si-C-N-ceramics, thermolysed at a temperature of 1050°C. In addition to the excellent thermal resistance, these materials show high creep resistance. The creep properties are severely influenced by the processing conditions, such as polymer cross linking, particle size distributions, powder compaction and thermolysis temperatures. For the characterization of the mechanical properties of Si-C-N-ceramics, isothermal and non-isothermal compression creep tests were performed at temperatures above 1350 °C and stresses up to 500 MPa. The creep behavior is explained with a free volume model.

2 Introduction

Thermolysis of polymer precursors leads to amorphous ceramics, which are known to exhibit high temperature stability [1-3] and outstanding high temperature creep resistance [4-7].

The properties of such materials are severely influenced by the processing conditions such as polymer cross linking, milling, powder compaction and thermolysis. Subsequent heat treatments at higher temperatures can strongly influence the amorphous state or lead to crystallization of silicon carbide and silicon nitride phases [1].

The influence of different processing parameters on the mechanical properties is not well understood yet. Therefore an improvement of the ceramic materials with respect to their mechanical properties is very difficult. Systematic investigations on the influence of the processing parameters are necessary to optimize the properties of the ceramics. Such investigations are presented in this paper together with the mechanical properties of the ceramic materials obtained. The high-temperature deformation behavior was investigated in compression creep tests at temperatures above 1350 °C and stresses up to 500 MPa. The free volume model, which was first applied for metallic glasses by Spaepen [8] and proven to be applicable to amorphous ceramics [6] is used to understand the mechanical behavior of ceramics derived from a new commercially available preceramic polymer.

3 Experimental

Commercially available polysilazane Ceraset (Allied Signal Composites Inc., [9]) was used as a preceramic polymer. The slightly yellowish liquid was handled under inert atmosphere in an argon-filled glovebox to avoid oxygen contamination. To receive a non-meltable solid, the polymer was thermally crosslinked for 1-13 h at temperatures between 250 °C and 400 °C or at 110 °C with a free radical generator. In this case a peroxide was used. The hard glassy body was grinded with a tungsten carbide ball mill and sieved to obtain particle size distributions between 32 μ and 200 μ. The polymer powder was densified in graphite dies under uniaxial compression stress of 48 MPa for 30 min at temperatures between 250 °C and 350 °C, taking advantage of the plastic deformability of the polymer [10]. The green body was then thermolysed under argon atmosphere at 1050 °C for four hours, using a heating ramp of 25 K/h.

The degree of cross linking and the optimum pressing conditions were investigated by thermomechanical analysis (TMA) for materials obtained under different cross linking conditions. In these experiments, the shrinkage or extension was measured during the heating with a constant heating rate. Additionally the temperature range where the sample softens was determined by the application of a dynamic load.

Total porosity of the samples, which generally consists of closed and open porosity, was measured with helium pygnometry and mercury porosimetry as described in [11]. In order to obtain the density of the amorphous matrix surrounding the pores, the material was crushed into fine powder particles which were assumed not to contain pores. The density of the pore-free powder particles, defined as theoretical density, was determined by helium pygnometry. Porosity as a function of plastic forming temperature was investigated. The evolution of porosity and the change in density during thermolysis was determined using specimens with similar cross linking and pressing temperature which were thermolysed at different temperatures with same heating rate and holding time of 4 h.

Compression creep tests in air were performed at a constant temperature. The deformation of the samples was measured using scanning pins and an inductive strain gage. The construction of the creep apparatus is described in [5]. The dimensions of the creep samples were $1.5 \times 1.5 \times 4.5$ mm^3. To investigate the influence of the processing conditions on the mechanical properties, creep samples, produced under different processing conditions but with the same total porosity were measured at 100 MPa and 1400 °C. Creep tests have been performed under different stresses from 5 MPa to 300 MPa.

Non-isothermal creep tests were carried out in the temperature range between 1350 °C and 1550 °C at 500 MPa stress with heating rates of 4 K/min. This heating ramp is fast enough to avoid annealing effects and crystallization during heating. It was necessary to measure a correction curve at 5 MPa to consider shrinkage of the ceramic and thermal expansion of the creep apparatus.

4 Results and Discussion

4.1 Processing of Ceraset precursor

Figure 1 shows TMA curves of the material for different cross linking times and temperatures. The maximum of the volume expansion is an indication of the optimum warm pressing temperature, which varies with the degree of cross linkage. If the temperatures during the plastic forming process are higher, the sample tents to shrink in the warm pressing form. Frictional forces between the walls of the pressing die and the specimen perpendicular to the pressing direction induce tensile stresses in the specimen which provoke crack formation. At lower temperatures viscosity may be too low for a complete compaction of the specimen.

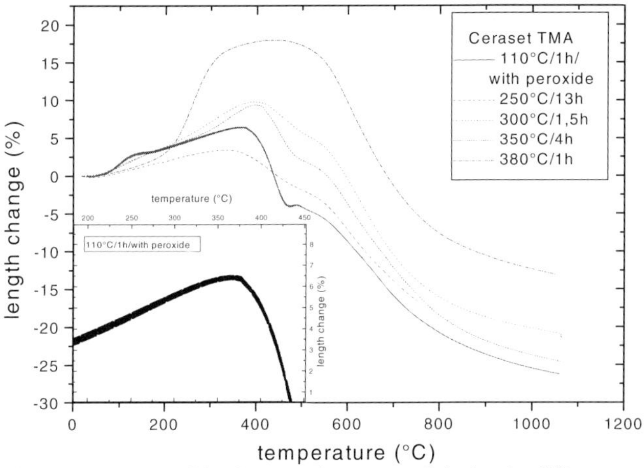

Fig. 1. TMA curves of the Ceraset polymer cross linked under different conditions

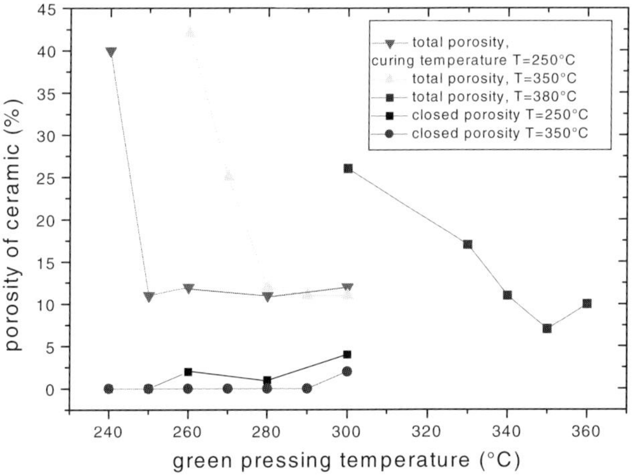

Fig. 2. Total and closed porosity of the thermolysed ceramic vs. plastic forming temperature.

The plastic forming has a great influence on total porosity of the resulting ceramic, which should be minimized to improve the mechanical properties. It is necessary that the green body has some open porosity to let the reaction gases out of the bulk during thermolysis. However, it can be seen in Fig. 2 that the thermolysed ceramic also contains closed pores. The minimum porosity which could be achieved in crack-free ceramics was 11 %. At lower porosity the reaction gases which are developed during the thermolysis tend to fracture the specimens.

The optimal pressing temperature rises with higher cross linking temperature because the viscosity of the polysilazanes increases with the degree of cross linkage which is controlled by the cross linking temperature. Therefore the minimum porosity that can be achieved does not depend on the cross linking conditions as long as the pressing stress is kept constant. However, it is assumed that the pore size distribution and the amount of closed pores could be reduced if a higher pressure would be applied to densify polymer powders with higher viscosity.

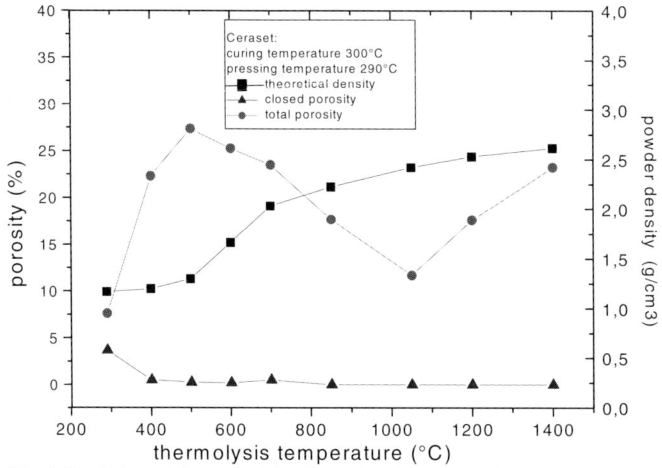

Fig. 3. Evolution of theoretical density and total and closed porosity during thermolysis

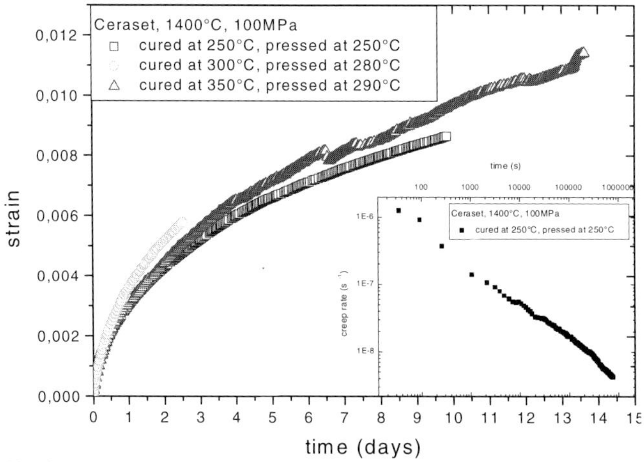

Fig. 4. Creep curves of cermics obtained with different cross linking conditions

When pressed at the optimum temperature, the green body has a porosity of about 7 %. Figure 3 shows porosity and theoretical density during thermolysis. The porosity increases between 300 °C and 500 °C due to gas evolution and decreases above some 600 °C because of shrinkage (Fig. 2). The optimum temperature to get dense ceramics during thermolysis of Ceraset, was about 1050 °C. This shows, that the ceramic is not in an equilibrium state and porosity rises because of further gas evaporation. However, higher porosity at higher temperatures does not mean to have worse mechanical properties, as discussed in the next chapter.

The variation of processing parameters like cross linking temperature, warm pressing temperature and thermolysis temperature, shows, that all processing steps have a strong interdependence. Therefore it is very difficult to find in a straight-forward optimization procedure for precursor-derived ceramics with well-defined properties.

Comparing the processing parameters for the transformation of Ceraset into a ceramic monolith with those of other polysilazanes like VT50 and NCP200 [10] or boron containing polysilazanes [11], it is evident, that each parameter has to be optimized individually for every precursor. This is due to different structures and compositions of polymers [1-3] and different phases and microstructures to be formed during thermolysis.

4.2 Mechanical properties of the Ceraset-derived ceramics

To investigate the influence of different processing conditions on the high-temperature deformation properties of the ceramics, creep tests were performed at a temperature of 1400° C and 100 MPa compression stress. The samples, were cross linked at temperatures of 250 °C, 300 °C and 350 °C and had same total porosity between 11-13 %.

The creep rate, which is the time gradient of strain, exhibits pronounced primary creep and no stationary creep (Inset Fig. 4). This is likely to be due to the metastable amorphous state of the ceramic material, which changes during annealing. This was previously shown for other precursor ceramics [5-7,12].

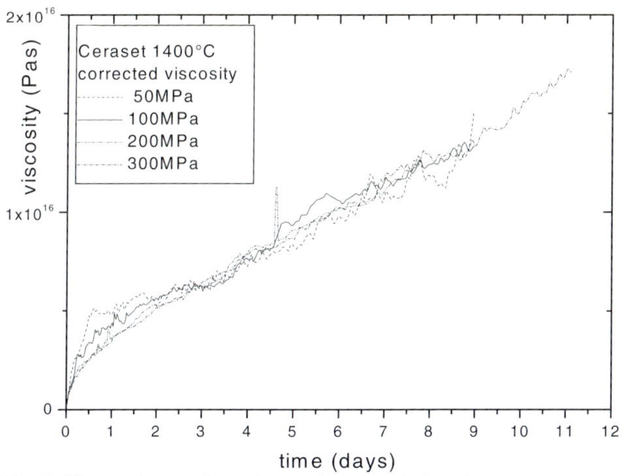

Fig. 5. Time and stress dependance of corrected viscosity

When thermolysed at 1050°C, the influence of the cross linking conditions on the creep behavior is rather small (Fig. 4). Therefore it seems that the influence of the processing conditions on the structure of the amorphous matrix is smaller than the influence of annealing.

The dependence of the creep behavior on the powder compaction conditions is that enormous that the creep properties of the ceramics cannot be measured because their strength is either not sufficient to support the creep stress, if the pressing temperature is too low or because of brittleness and cracks if the temperature too high.

Figure 5 shows the time dependence of viscosity , measured at different stresses from 50 MPa to 300 MPa. The viscosity is calculated from $\eta = \sigma / 3\dot{\varepsilon}$, where $\dot{\varepsilon}$ is the creep rate. Shrinkage of the material occurs parallel to creep. The rate of shrinkage was measured at a creep stress of 5 MPa, which is the stress for clamping force for the specimen during the creep tests. It has the same mathematical progression as the creep rate and was subtracted from creep curves mentioned above. Viscous flow increases linearly with time and can be defined as Newtonian, since it is independent of the applied stress. Therefore, creep rates rise linear with stress. The non-linear increase of viscosity at short times can be attributed to anelastic deformation of the ceramic.

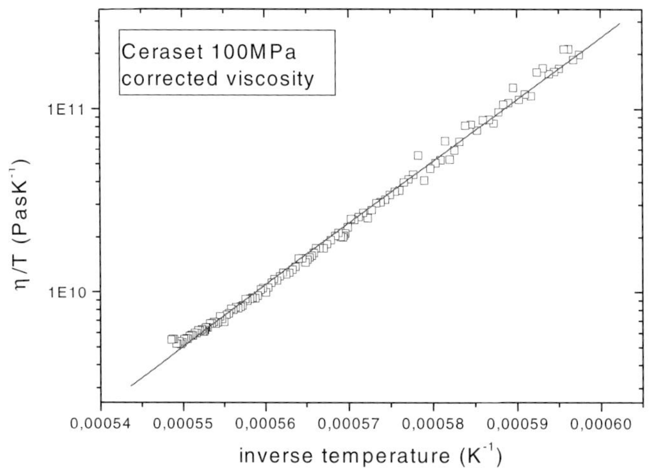

Fig. 6. Temperature dependance of corrected viscosity

Figure 6 shows a non-isothermal creep experiment. A linear correlation between the applied inverse temperature and the ratio between viscosity and temperature was found.

To describe the stress, time and temperature dependence of viscosity of amorphous ceramics, a free volume model developed for metallic glasses by Spaepen [8] and later by van den Beukel [13] has been used [6,12]. In this model the change of viscosity during annealing is attributed to a change of the defect concentration c_f. When c_f is much higher than a metastable equilibrium defect concentration c_{f0} , viscosity can be described as[6,8,12,13]:

$$\eta = \frac{\sigma}{3\dot{\varepsilon}} = C c_{f0}^{-1} T \exp(-Q/RT) + C C_1 T t \quad (1)$$

where T is the temperature, t the time, Q the activation energy and C, C_1 constants. The activation energy, which corresponds to the annihilation of flow defects, was calculated by linear regression carried out in Fig. 6 to be 0,6 MJ/mol. Frank et. al.[14] found activation energies in the order of 0,5 MJ/mol for the diffusivity for various Si-C-N-ceramics. This may be an indication, that deformation in these materials occurs by diffusion controlled viscous flow as already discussed by Spaepen for metallic glasses[8].

5 References

1. R. Riedel, A. Kienzle, W. Dressler, L. Ruwisch, J. Bill, and F. Aldinger, Nature, **382,** 796-798 (1996).
2. J. Bill, J. Seitz, G. Thurn, J. Dürr, J. Canel, B.Z. Janos, A. Jalowiecki , D. Sauter, S. Schempp, H. P. Lamparter, J. Mayer and F. Aldinger, Phys. Stat. Sol. **166**, 269 (1998).
3. F. Aldinger, M. Weinmann, J. Bill, Pure & Appl. Chem., **70**, 439-448 (1998).
4. L. An, R. Riedel, C. Konetschny, H.-J. Kleebe, and R. Raj, J. Am. Ceram. Soc., **81** 1349-52 (1998).
5. G. Thurn, J. Canel, J. Bill and F. Aldinger, J. Europ. Ceram. Soc., 19, 2317-2325 (1999).
6. M. Christ, G. Thurn, M. Weinmann, J. Bill and F. Aldinger, submitted to J. Am. Ceram. Soc.
7. B. Baufeld, H. Gu, J. Bill, F. Wakai and F. Aldinger, submitted to J. Europ. Ceram. Soc.
8. F. Spaepen, Les Houches Lectures XXV, North Holland, Amsterdam (1981).
9. U.S. Patent 4,929,704; 5,001,090; 5,021,533; 5,032,649; 5,155,181
10. R. Haug, M. Weinmann, J. Bill and F. Aldinger, J. Europ. Ceram. Soc. **19**, 1-6 (1999).
11. M. Christ, S. Katz, M. Weinmann, J. Bill and F. Aldinger, Keramische Zeitschrift, **11** (1999).
12. M. Christ, G. Thurn, J. Bill and F. Aldinger, in this proceedings.
13. van den Beukel, E. Huizer, A. L. Mulder and S. van der Zwaag, Acta metal. **34**, 483 (1986).
14. W. Frank, P. Scharwaechter, R. Blüher, W. Gustin, M. Horz, S. Matics, and T. Schuler, Jahrbuch der Max-Planck-Gesellschaft 1998, edited by the Max-Planck-Gesellschaft, Vandenhoeck und Ruprecht, Göttingen 1998, p.554-562.

Aqueous Corrosion of Advanced Ceramics

Ralf Westerheide, Thomas Hollstein, Manfred Füting,
Fraunhofer-Institut für Werkstoffmechanik, Freiburg und Halle, Germany

1 Introduction

Advanced ceramics are well known for their excellent strength, hardness, high temperature strength, resistance against abrasion and their resistance against corrosion. Therefore ceramics are in use in several applications where corrosion resistance in aqueous media is of great importance. For example in fields of mechanical engineering, in the medicine technology or they are in use as chemical components. The fluids, which act as the corrosive media, are hot water, body fluids, juices, acids, caustics, molten metals or salts.

The factors, that influence the corrosion of advanced ceramics, are the material parameters (composition, crystalline matrix grains, glassy or crystalline grain boundary phases and impurities), and they are greatly influenced by the processing, the machining influenced surface structure (roughness, damage and stress state), the medium (composition and concentration) and its flow velocity, the ambient temperature and the exposure time. In the case of indirect (incongruent) corrosion of a ceramic by a liquid a reaction product is formed, which can be either a gas, a fluid, a solid or a combination of them. If the reaction product remains in contact with the ceramic, it may form more or less protective layers, or - acting as a catalyst - even a more aggressive substance. In the most simple case of direct (congruent) corrosion the ceramic dissolves directly into the liquid.

In this work the corrosion behaviour of a dispersion ceramic, Al_2O_3 + 10 weight-% ZrO_2, is described. The loss of mass, the strength degradation and the phase changes are measured at temperatures between 25°C and 100°C in concentrated and 1 m acids of phosphor and sulphur and caustics of sodium.

To enable the study of early stages of corrosion the possibilities of environmental scanning electron microscopy (ESEM) were investigated. First results offering advantages for the investigation of corrosion behaviour and therefore for an optimisation of materials.

2 Material

The dispersion ceramic Al_2O_3 + 10% ZrO_2 (Friatec FZT) discussed here is based on a 99,7% pure alumina ceramic with some MgO and SiO_2 < 500ppm. The base material without the ZrO_2 has a good corrosion resistance and a strength of about 350 MPa. The 10% ZrO2 were added in order to increase the strength and wear resistance, but this led to a material, which was susceptible to aqueous corrosion [1]. The addition of Y_2O_3

instead of MgO as a stabiliser resulted in both good corrosion resistance and good wear properties at an increased strength level of about 450 MPa. The zirconia phases in the as sintered state are 18% monoclinic, 26% tetragonal and 56% cubic [1].

3 Experimental results

The specimens, which were clamped in a support, were tested in a 4 liter container made of glass or teflon for the tests in acids or caustics, respectively. The test container with magnetic -glass or teflon coated- stirrers was placed on a heating plate in an oil bath. The mass losses of 3x4x45 mm^3 longitudinally fine ground standard 4-point-bend bars in the corrosion media at room temperature and 90°C, respectively, is linear with the time, except at the beginning up to about 200 h, where a square root dependence with time can be used to describe the data (see Fig. 1).

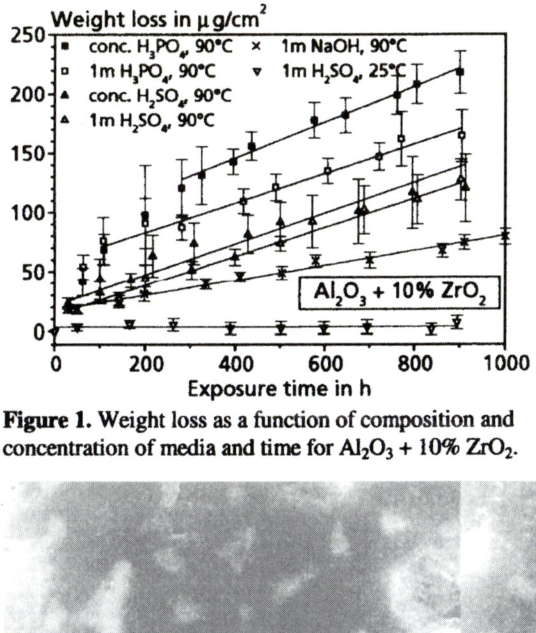

Figure 1. Weight loss as a function of composition and concentration of media and time for Al_2O_3 + 10% ZrO_2.

Whereas no attack is observed at 25°C, a small weight loss occurs at 90°C. The highest corrosion rate is observed for the phosphoric acids, and especially for the concentrated one, the lowest for NaOH. SEM-pictures of corroded specimen surfaces show, that the corrosion takes place at the grain boundaries (see Fig. 2).

The observed changes of the surface state due to the corrosive attack at the grain boundaries can be identified by

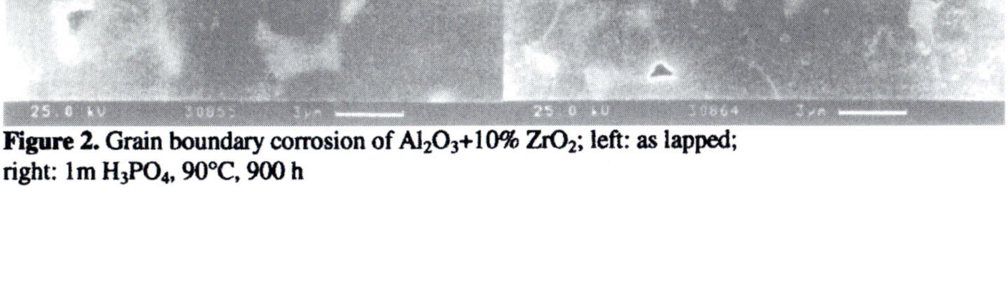

Figure 2. Grain boundary corrosion of Al_2O_3+10% ZrO_2; left: as lapped; right: 1m H_3PO_4, 90°C, 900 h

strength measurements using, for example, 4-point-bend specimens. In figure 3 the characteristic strength of the Al_2O_3 + 10% ZrO_2 ceramic after exposure to different solutions as a function of time is shown. The confidence intervals are based on the results of 10 specimens (otherwise the number of specimen is stated). Within the first 100 hours the strength decreases by about 10%. Then, up to 900 or 1000 hours no further decrease or even an increase of the strength is observed. This can be seen clearly for the exposure in NaOH.

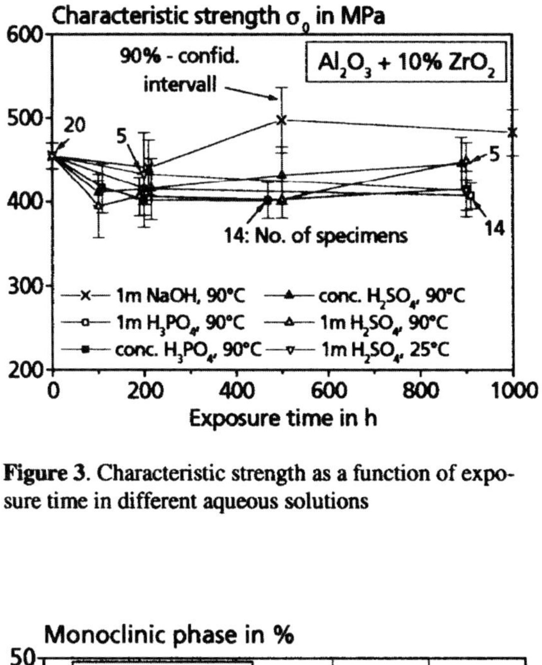

Figure 3. Characteristic strength as a function of exposure time in different aqueous solutions

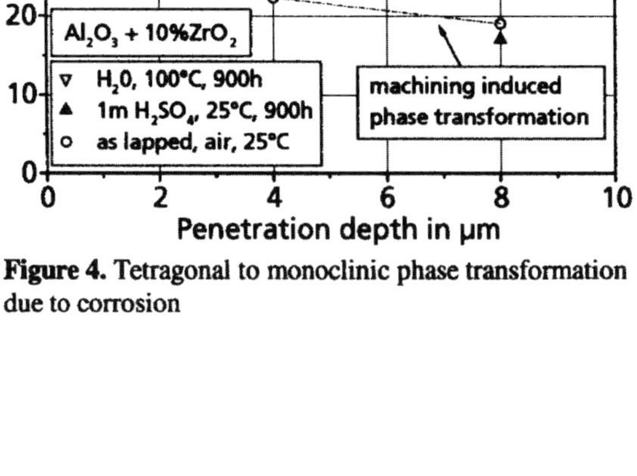

Figure 4. Tetragonal to monoclinic phase transformation due to corrosion

The increase of strength for the exposure in NaOH and the constant level of strength after a first decrease for the exposures in acids can be attributed to the transformation of the ZrO_2 particles from the tetragonal phase to the monoclinic phase after the stabilising additive Y_2O_3 has been dissolved. The phase transition is accompanied by an increase in volume and therefore induces compressive residual stresses. Consequently two different aspects have to be considered: a strength decreasing damage effect due to dissolution of secondary phase, i.e. crack growth, and a strength increasing crack closure effect due to compressive stresses. For a separation of these effects one has to determine both the residual stress distributions below the specimen surface and the crack depths.

The monoclinic and tetragonal phases in the surface region were determined by a X-ray-diffraction analysis. In figure 4 the monoclinic phase content in the subsurface region before and after corrosion are shown. The bulk values measured in a penetration depth of about 8μm value were measured as 18% monoclinic phase and 26%

tetragonal phase. The increase of the monoclinic phase content of the lapped unexposed material near the surface shows the machining induced phase transformation The X-ray measurements show that there is no phase transformation, if the material is exposed in 1 molar sulphuric acid at room temperature. In contrast, additional phase transformations occur for corrosion in 90°C hot caustic and acidic solutions. At the surface the whole amount of the tetragonal phase is transformed to the monoclinic phase and a maximum value of about 44% was measured.

The investigation of specimen after exposure in boiling water has shown that a phase transformation occurs also. It can be concluded that the Y_2O_3 is dissolved in this case as well. Since simultaneously a weight gain of about 15 µg/cm3 is measured after 900 hours [2], some complexes and/or hydroxides seem to be formed at the specimen surface.

4 The potential of ESEM testing

An environmental scanning electron microscope (ESEM) allows the investigation of specimens under different atmospheres without the necessity of a vacuum. It is also possible to investigate electrically non-conducting materials without a special preparation like C- or Au-coating. In this way *in situ* experiments are possible to study the wetting and drying behaviour on a microscopic scale by use of a microinjector. With the microinjector small droplets with diameters of 100µm to 300µm can be applied to the specimen surface [3]. In this study the early stages of corrosion of ceramic materials in contact with liquid media was investigated.

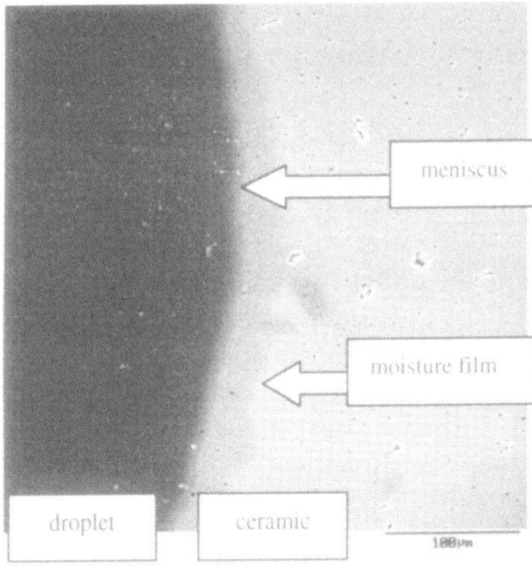

Figure 5. Formation of a thin moisture film in front of droplets from water or aqueous corrosion media

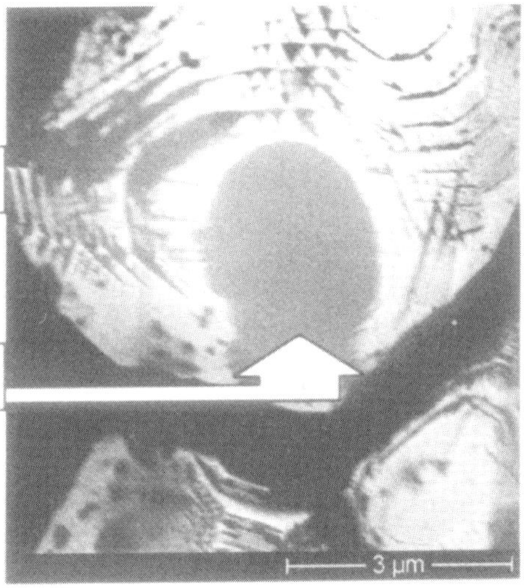

Figure 6. The formation of the thin moisture film depends on the crystallographic orientation

It is assumed that the early stages of corrosion are influenced by the wetting behaviour. An increase in wetting may result also in a stronger corrosive attack. It corresponds to an increasing penetration rate in the extension of microcracks and flaws and it generates favourable conditions for the transport of corrosion products.

The investigation of the wetting behaviour of an Al_2O_3 + 10% ZrO_2 ceramic has shown that it is quite different to that of Si_3N_4. This is caused by the higher surface energy of Al_2O_3 + 10% ZrO_2 compared to Si_3N_4. Condensed water droplets and droplets of sulphuric acid on the Al_2O_3 + 10%ZrO_2 surface show the usual droplet forming and the formation of a thin moisture film in front of the meniscus of the droplet (Fig. 5).

The moisture film formation was also observed on cracked surfaces. Here the formation depends on the crystallographic orientation (see Fig. 6) and small morphological features (growth steps). Further characterisation of this phenomenon showed that the moisture film is stable up to temperatures of 55°C and up to reduced water vapour pressures of about 100 Pa. The presence of the moisture film allows the in situ observation of corrosive attack without removing the corrosive media. However, the question arises, whether the conditions at the surface in contact with the moisture film are similar to those which are in contact with the droplet. This will be investigated by parallel observations of different surface regions.

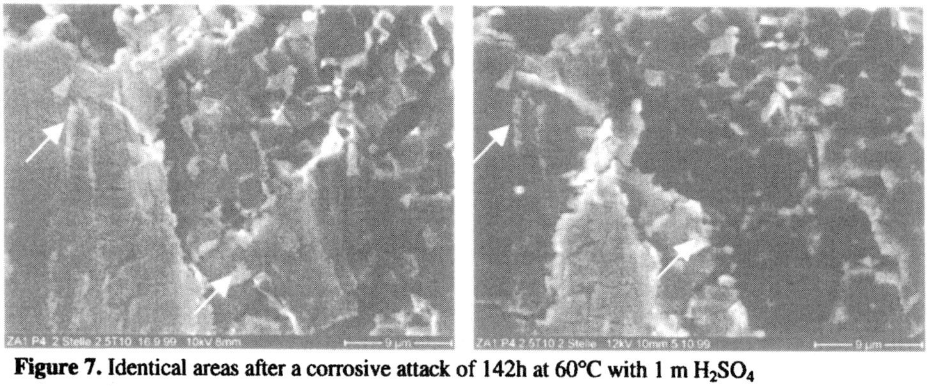

Figure 7. Identical areas after a corrosive attack of 142h at 60°C with 1 m H_2SO_4

Basic investigations of long term corrosion in the ESEM have shown that due to the renunciation of surface preparation and a careful arrangement and positioning of specimens it becomes possible to characterise identical areas of the sample surface at different times (see Fig. 7). For the Al_2O_3 + 10% ZrO_2 ceramic a shell-like cracking after 142h at 60°C in contact with 1 m H_2SO_4 was observed. This can be correlated to the compressive stresses induced due to the tetragonal-monoclinic phase transition of the ZrO_2 after the dissolution of the stabilising Y_2O_3 as it was observed in previous investigations [2].

5 Conclusion

It is shown for an Al_2O_3 + 10% ZrO_2 ceramic that the corrosion in caustic or acidic aqueous media below a temperature of about 100°C is dominated by the corrosion of the secondary phases. The dissolution of the stabilising Y_2O_3 leads to a phase transformation from the tetragonal to the monoclinic phase. The accompanied increase of volume induces compressive residual stresses. After a first decrease of strength due to the crack growth at the grain boundaries the phase transformation effect results in a prevention of further decrease and therefore in a constant level of strength. In some cases also a strength increase could be measured.

The use of an environmental scanning electron microscope (ESEM) for corrosion testing of advanced ceramics is possible. Identical areas of the sample surface at different times can be characterised for corrosion experiments without modifying the specimen by any preparation. Short-term corrosion studies can be carried out as in situ experiments inside the microscope. They provide information about the early stages of the corrosive attack as well as about the wetting behaviour and the formation of thin moisture films in front of droplets of the corrosive media. This complex microscopic characterisation of the ceramics can be an useful tools for material development.

Acknowledgement

The authors gratefully acknowledges the Bundesministerium für Bildung, Wissenschaft, Forschung und Technologie (BMBF) and the Deutsche Forschungsgemeinschaft (DFG) for financial support of parts of this work.

References

1. A. Reckziegel, A. Danzer, H.-J. Graf, DE-Patent 40 29 066 C2, 1992
2. T. Graas, T. Hollstein, W. Pfeiffer, A. Reckziegel, cfi/Ber. DKG 1997, 74, 723 – 726
3. J. Hopfe; M. Füting, Fundamentals and Applications of Environmental Scanning Electron Microscopy. - in: In situ Scanning Electron Microscopy in Materials Research, Wetzig/Schulze (Eds.), Akademie-Verlag Berlin 1995, 219-240

Defect Analysis in Ceramics by Microwaves

H.Dannheim, H. Hädrich
Institute of Material Science, Departement Ceramics University of Erlangen-Nürnberg

1 Introduction

Because failures of ceramics may cause costly production interruptions special attention has to be paid to their quality control.

During the last years nondestructive testing techniques have been developed and tested for ceramics. For fused cast refractories a mobile device for ultrasonic testing was constructed and tested very sucessfully in practise [1].

More recently, attention has been turned towards the development of NDT methods for the testing of sintered ceramics. Useful results can be obtained with an ultrasonic device but the usually high porosity of those materials is a major disadvantage for the use of this technique. For ultrasonic setups you need a coupling agent , normally water, and so the water penetrats the material while measuring. Alternatively, microwave equipment has been set up and the good results promise chances for the use in quality control.

Because of the electromagnetical nature of the microwaves it is not necessary to have a coupling medium like water, so it is possible to start the measurement immediately even by samples with high porosity.

A microwave setup with variable frequency of 20 - 40 GHz was built up and the microwave absorption dependent on the local position was measured. A broad variety of ceramics was examined.

To knew the propagation of the waves inside the materials it is necessary to measure the dielectric constants of the materials. They were determined with an angle method.

With lenses it is possible to improve the recognisation of faults and cracks. We examined different systems of lenses.

It is important to do the quality control as early as possible. So the unfired bodies should be examined with microwaves. If the water content of unfired ceramics is too high, there are difficulties with microwaves because of absorption. We measured the dependence of the absortion on frequency to find the less absorption, this depends on the sintered material.

2 Experimental Setup

The figure 1 shows the experimental setup. It is possible to change the frequency from 20 to 40 GHz. Because of interferences between waves from the sender and reflected waves from the surface of the ceramic sample it is necessary to modulate the frequency with about 2 - 3 GHz.

The samples can move between sender and receiver in two dimensions, the measurement of the absorption of microwaves in a plane is called a C-scan.

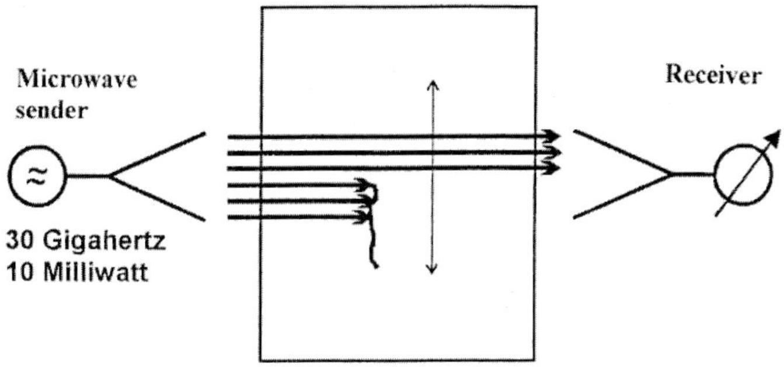

Figure 1 Schematic experimental setup

To knew the propagation of the waves inside the refractory materials it is necessary to measure the dielectric constant of the materials, because:

$n = \sqrt{\varepsilon}$

The figure 2 describes the method with which the dielectric constant can be measured. From the distance x behind the material one can calculate the refractory index n and so the dielectric constant.

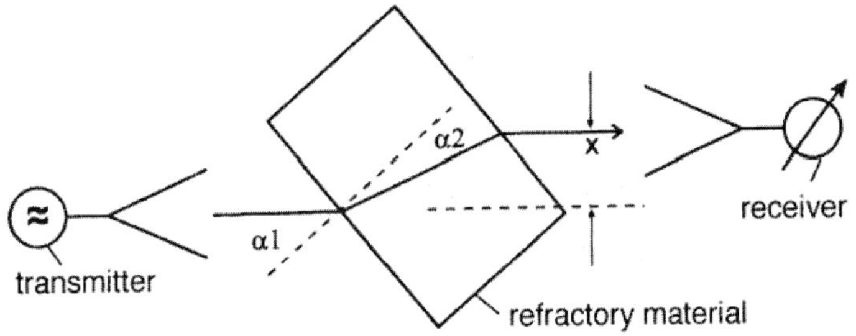

Figure 2 Experimental determination of the dielectric constant

The following table shows the results of the dielectric constant for different refractories. The variation is from 3 to 11 and so there is a very different refractive behavior inside the refractories. Also the identification of faults depend on the refractive index.

Table 1. Dielectric constant of different ceramics

Type of ceramics	Dielectric constant
Silica	2,8
Chromalumina	7,6
Alumina	9,2
Zirconmullite	6,8
Zirconsilica	10,2

Another important physical property is the absorption of the microwaves inside the materials depending on the frequency. Because of a limited sensitivity of the apparatus one have to measure the damping to calculate the maximal dimensions, which the microwaves are able to penetrate. The table 2 shows some selected damping coefficients for different refractories measured at 30 GHz.

As mentioned in the introduction it is important to do the quality control as early as possible. So the unfired bodies should be examined. Here the damping is extreme important because of the water, which is still inside the dried bodies. If the water content is less then 2 - 5% depending on the type of ceramics it is possible to penetrate sufficient large samples also depending on the frequency. The figure 3 shows the transmission of a unfired sillimanite refractory depending on frequency.

Table 2. Damping coefficient of different refractories

Type of refractory	Damping coefficient in dB/cm
Chromalumina	1,0 +/- 0,3 dB/cm
Zirconmullite	0,5 +/- 0,3 dB/cm
Zirconsilica	0,1 +/- 0,05 dB/cm

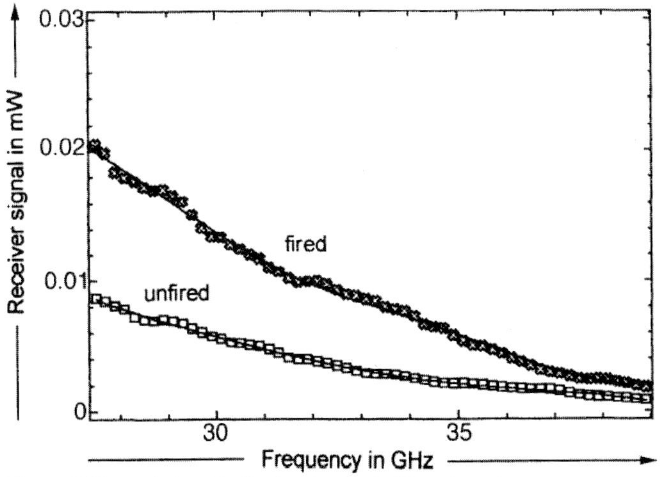

Figure 3 Transmission of a fired and unfired sillimanite refractory depending on frequency

As expected, the transmission decreases with increasing frequency. So for the transmission it is better to go to lower frequencies. On the other hand the dissolution of faults increases with higher frequencies because the less detected faults are in the region of the wavelength. Therefore the selection of the used frequency is always a compromise between the possible detection of cracks ore pores and the maximal dimension of the sample, which the microwaves are able to get through.

3 Increase of Fault Detection with Lenses

With polymer lenses it is possible to focuse the microwaves like glass lenses in the geometrical optic. This is a possibility to increase the recognization of a fault by a given frequency. Figure 4 shows schematically the experimental verification.

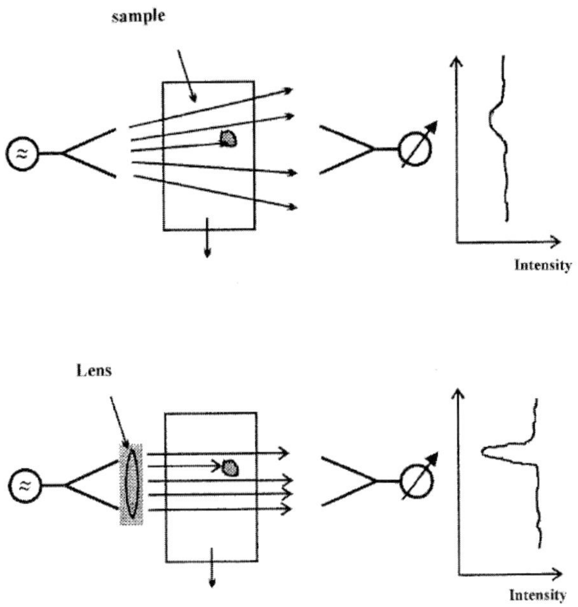

Figure 4 Increase of fault detection with lenses

Several lenses with different focus and several lens systemes were tested, but for practical purposes the use of one lens with a focus f = 86 mm is sufficient. To test the dissolution of the system we prepare different ceramic samples with definite holes in it, with different diameters. Along a line the attenuation of a sillimanite brick was measured; figure 5 shows the result.

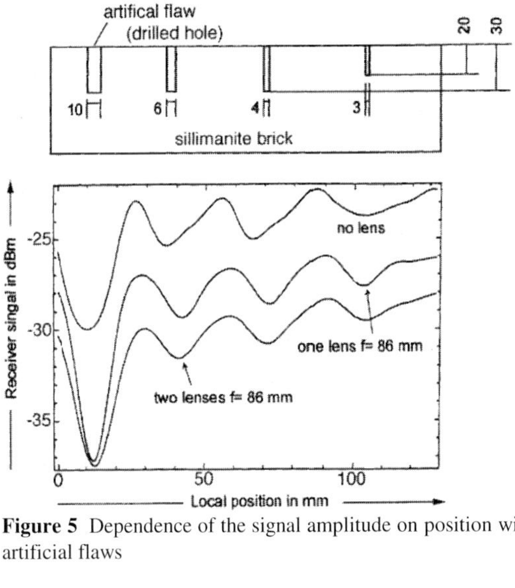

Figure 5 Dependence of the signal amplitude on position with and without lenses for a sillimanite material with artificial flaws

4 Device for Automatic Scanning

In order to detect flaws and cracks within a ceramic material, it is necessary to measure the attenuation step by step over the whole sample, even in two directions, to determine their exakt position and size. For this purpose a device was built to create complete C-scans of the samples, the principle setup is shown in [1].

As an example the C-scan of two sillimanite bricks placed together, one of them showing an inclusion, is given in figure 6.

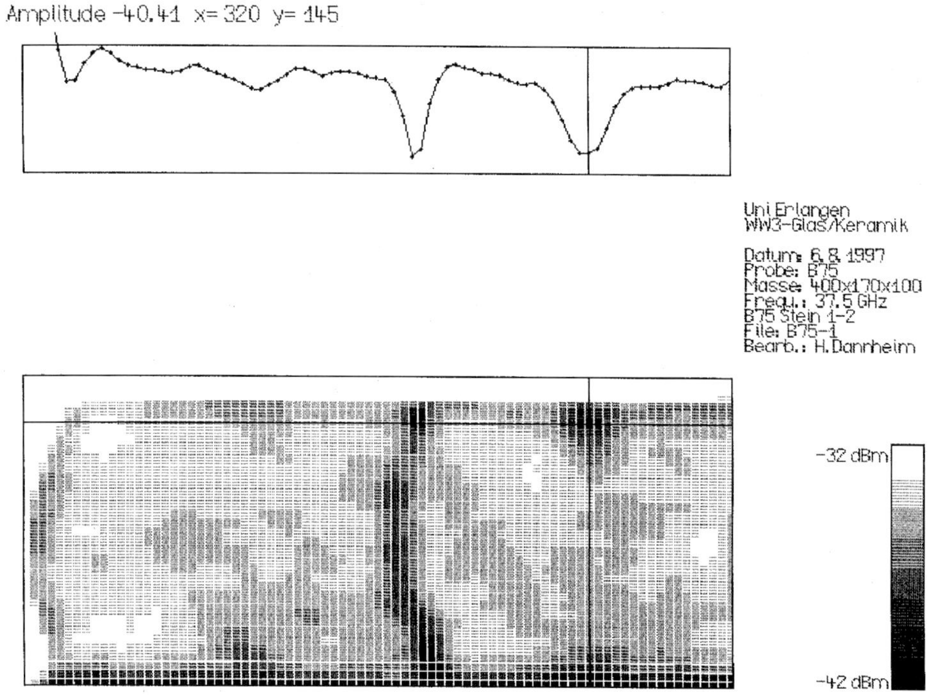

Figure 6 C-san of two sillimanite bricks placed together, one of them showing an inclusion

5 Conclusions

The microwaves have some advantages against other nondestructive techniques. There is no coupling medium necessary, through the improvement with polymer lenses the microwave images achieve a clarity and resolution better than ultrasonics. The flaw detection goes up to 1 mm and at lower frequencies the length of possible penetration is sufficient enough when the water content especially by unfired bricks is less.

6 References

1. Ulitzka, S.; Hädrich, H.; Riedling, M.: Nondestructive microwave testing of bonded refractories. Glastechn. Ber. 63K(1990) p. 403-409
2. Ulitzka, S.; Hädrich, H.: Non-destructive testing of ceramically bonded refractories - Application of ultrasound and microwaves. Glastechn. Ber. Glass Sci. Technol. 68(1995) p. 51-58
3. Dannheim, H.; Hädrich, H.; Ulitzka, S.: Möglichkeiten der zerstörungsfreien Untersuchung von gesinterten feuerfesten Steinen. Proceedings of the International Colloquium on Refractories Aachen 1993, p. 121- 125
4. Dannheim, H., Hädrich, H.: Quality assurance of sintered refractories through non-destructive testing with microwaves. Glastechn. Ber. Glass Sci. Technol. 71(1998) p. 223-229

Fracture Toughness of Thermally Shocked SiC-Fibre Reinforced Glass Matrix Composite

Ivo Dlouhý[1], Michal Reinisch[2], Aldo R. Boccaccini[3]
[1]Institute of Physics of Materials ASCR, Brno, Czech Republic
[2]Institute of Materials Engineering, Technical University Brno, Czech Republic
[3]Institute for Material Technology, University of Ilmenau, Germany

1 Introduction

The development of flaw-tolerant fiber reinforced glass, glass ceramics and ceramic matrix composites has resulted in a new family of structural materials exhibiting quasi-ductile fracture behaviour, oxidation resistance and high temperature capability [1]. A key requisite for the glass composites to be useful at high temperatures is the retention of a weak fibre/matrix interface, which is responsible for their flaw-tolerant behaviour [1,2].

In silicate matrix containing silicon oxycarbide fibres (Nicalon) as reinforcement [4], the weak interface is provided by a carbon-rich layer around the fibers which is grown in-situ during composite fabrication [5]. This carbon-rich interlayer can be severely degraded when the material is used in oxidising environments at temperatures as low as 400 °C [4,6]. Oxygen penetration can occur through matrix microcracks pre-existing in the material as a result of the fabrication process, or through crack formed under mechanical or thermomechanical (thermal shock) loading [7].

In ceramics reinforced by brittle fibres, an elastic fibre bridging and a pull-out bridging mainly cause the toughening. Both these mechanisms increase in some extent behind the crack tip along the process zone wake [2,3]. Crack growth resistance rises as the crack propagates and leaves the wake. It is difficult to define the intrinsic fracture toughness as a material parameter due to the increasing crack growth resistance curve [2].

The purpose of the contribution is the verification of the suitability of the chevron notch technique for changes monitoring in the fracture behaviour of fibre/glass matrix composites that may occur following ageing in non-oxidising atmosphere or thermal cycling. Material aspects of the microstructural integrity and mechanical property degradation due to ageing treatment are also addressed.

2 Material and Experimental Procedures

2.1 Material characterisation and thermal treatment

The material investigated was a commercially available unidirectional SiC Nicalon (NL202) fiber reinforced borosilicate (DURAN) glass matrix composite fabricated by Schott Glaswerke (Mainz, Germany). Information on the composite constituents is given in Table 1. The composite was prepared by the sol-gel-slurry method [8]. The density of the composites was 2.4 g/cm^3 and their fibre volume fraction 0.4. Fairly regular fiber distribution and the absence of porosity was found during microstructural investigations. Fiber distribution in section perpendicular to specimen axis is obvious from Fig. 1.

Table 1. Properties of the composite constituents [8]

	Density [g/cm^3]	Young's modulus [GPa]	Poisson's ratio	Thermal exp. coefficient [K^{-1}]	Tensile strength [MPa]
matrix DURAN	2.23	63	0.22	$3.25{*}10^{-6}$	60
fibre SiC Nicalon	2.55	198	0.2	$3{*}10^{-6}$	2750

Two sets of thermal exposition was applied: (i) *Thermal ageing in argon atmosphere*. This involved encapsulating the as-received bars in silica tubes after these had been evacuated and filled with argon of technical purity (the oxygen content in the capsules being of a level lower than 0.001 torr). The capsules were aged at 500, 600 and 700 °C for 100 hrs. For 600°C, ageing for 250 and 1000 hs was conducted. (ii) *Thermal shocks and cycling*. Thermal shocking was produced by cooling in water from temperature between 620 and 670 °C. Thermal cycling was carried out at temperatures varying from room temperature to 700 °C by two number of cycles, 466 and 821. The temperatures cover the upper temperatures at which the composites are likely to find applications.

The mass and dimensions of the samples were measured, and the outer surfaces inspected for the appearance of any damage, such as delamination, fibre debonding/protrusion etc. [9]. X-ray diffraction analyses were performed on polished sections of the samples after ageing in order to detect any crystallisation in the matrix.

2.2 Mechanical testing

The flexural strength and Young's modulus were determined in four-point bending using 32mm outer span and 16 mm inner span. A cross-head speed of 0.1 mm/s was used.

The chevron notch (CN) specimen technique was employed for fracture toughness determination. Chevron notches with angles of 90° were cut using a thin diamond wheel. A three point bending (with span of 16 mm) at a constant cross-head speed of 0.1 mm/min was employed. Graphs of load versus deflection were recorded and the maximum force was determined from each trace. The fracture toughness value was calculated from the maximum load (F_{max}) and the corresponding minimum value of geometrical compliance function (Y^*_{min}). The calculation of the geometric function Y^*_{min} for chevron notch bend bars was described in detail elsewhere [10]. The chevron-notch depth a_0 was measured after testing from SEM micrographs of fractured specimens.

Additionally, straight-notched specimen technique recommended as standard procedure has been also employed . The straight notch having width of about 150 μm and depth of about 1.5 mm has been ground by thin diamond wheel. Then the tip was sharpened by hand by polishing with a conventional razor blade and diamond paste of size of 0.1 μm. Testing conditions have been the same as in case of chevron notch specimen technique.

The acoustic emission technique (AE) was applied during the test. Traces of cumulative number of counts (AE events) were obtained in the same time scale as the load vs time plots. This technique allows for an accurate detection of the microcrack initiation onset at the chevron notch, which occurs when a sharp increase in the number of AE events is observed.

Valid measurements for computing K_{IC} are those in which this increase of AE events coincides with the end of the linear part of the force versus time trace.

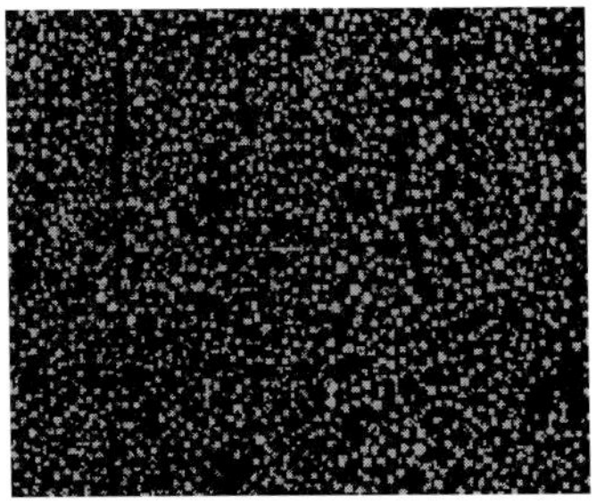

Figure 1. Composite microstructure in cross - sectional direction

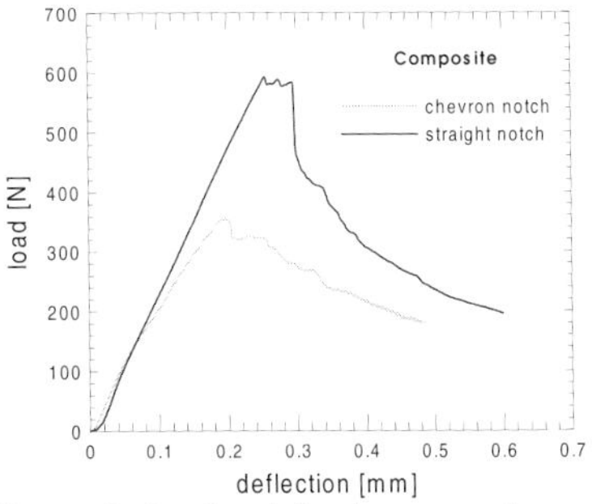

Figure 2. Load - deflection traces for samples with Chevron and straight notch

3 Results and Discussion

3.1 Applicability of CN technique to fiber reinforced glass matrix composites

The CN specimen technique has not received wide application to measure the fracture toughness of fiber reinforced ceramic composite materials. Other specimen configurations have been more extensively used, for example notched three-point bending specimens with notch depth-sample thickness ratios of 0.5 [11]. As long as the notch is large enough to cover many fibers at the critical crack length, the chevron-notch test should be a useful method to measure toughness in ceramic composite [12]. Moreover, the CN specimen technique is well adapted for test at elevated temperatures and, therefore it should be considered more frequently for K_{IC} determination in high-temperature fiber reinforced ceramics composites. In addition, literature results and own previous experiments have shown that chevron-notched specimens can provide highly reliable and accurate results for fracture toughness determination in particulate reinforced brittle matrix composites [4]. (ii) For unidirectionally reinforced materials in which the fracture behaviour is governed not only by the matrix brittleness but also by the fiber/matrix interface, the chevron-notch is as successful crack initiator.

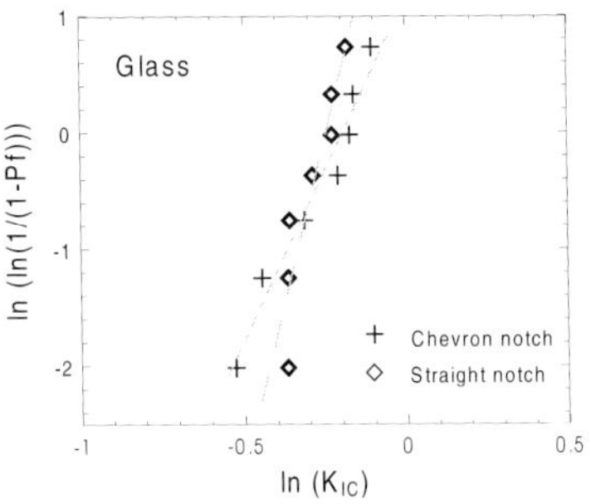

Figure 3. Weibull plot for measurements carried out with pure borosilicate glass

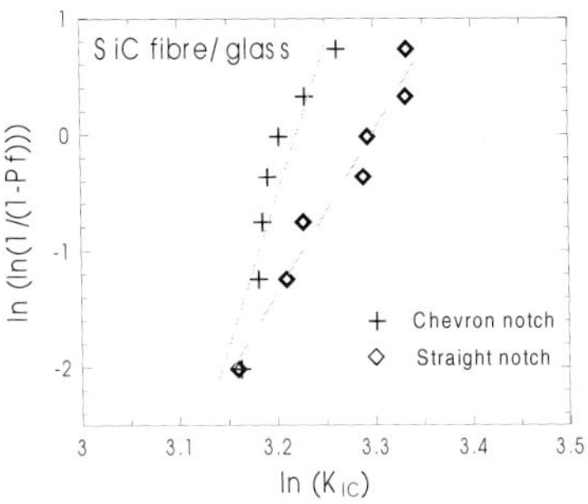

Figure 4. Weibull plot for measurements carried out with composites

In order to obtain exact evidences about the applicability of CN specimen technique for fracture toughness determination of fibre/glass matrix composites the specimens with straight notch have been included into analyses. The study has been carried out with pure borosilicate glass and fibre/glass matrix composite in as received state.

For pure glass the load deflection traces have been found to be qualitatively comparable. The typical examples of load - deflection trace for composite is shown in Fig. 2. For the composite, the record from specimens with straight notch is typical by pop-in effects that can be observed at maximum load. The pop-ins has been assigned to the stopping of major crack propagation at fiber/matrix interfaces followed by the interface delamination. Fractographic observations have supported this explanation. For chevron notch specimen the crack is developed at about 1/2 to 2/3 of maximum load and is smoothly running up to maximum load after which unstable fracture occurs.

Nearly comparable K_{IC} values have been gained from both specimen techniques when comparing pure glass (0.75 $MPam^{1/2}$ for CN specimens; 0.73 $MPam^{1/2}$ for specimens with straight notch). As shown in Fig. 3 the scatter characteristics are slightly different (due to different process zone sizes). Differences have been found for composite material however (Fig. 4). Not only the scatter characteristic but also the average K_{IC} values differ in remarkable extent showing overestimation of toughness characteristic for specimen with straight notch (24.59 $MPam^{1/2}$ for CN technique in comparison to 26.19 $MPam^{1/2}$ for straight notch technique). Only the data calculated for load deflection traces with no more than one pop-in are on level fully comparable with data from CN specimens.

3.2 Fracture toughness of thermally exposed composites

For composites thermally aged in non-oxidising atmosphere, the fracture toughness values determined by the CN specimen technique seem to be little dependent on the ageing condition. The material exhibits average K_{IC} values on level of about 24 $MPam^{1/2}$, which are similar to data quoted in the literature (16). Summary of fracture toughness data in dependence on ageing condition is shown in Fig 5 and detailed assessment has been described elsewhere [9]. The results are in broad agreement with data in the literature, in that thermal exposures at temperatures below 1000 °C in inert atmospheres do not have deleterious consequences on the properties of the SiC-Nicalon fibres and on the interface in silicate matrix composite [9]. Thus, the temperature capability of the matrix is the limiting factor for applications involving non-oxidising environments. For the borosilicate glass matrix investigated here the limiting temperature is in the range 600 - 700 °C, depending on the duration of the exposures, since at this temperature range, the matrix softens leading to microstructural damage that results in loss of mechanical properties. Short excursions at these temperatures may be possible, however.

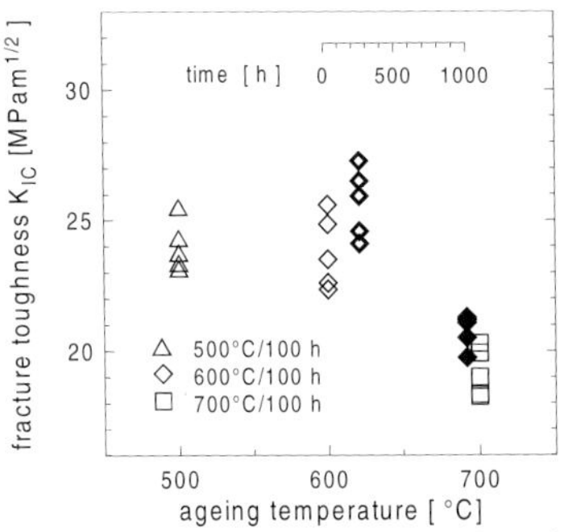

Figure 5. Fracture toughness data in dependence on ageing condition

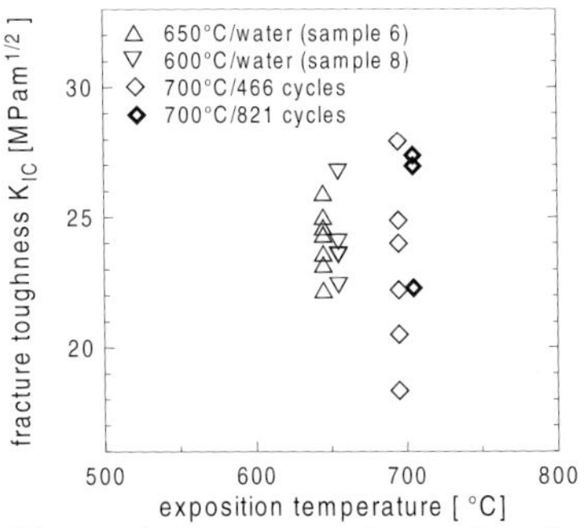

Figure 6. Fracture toughness data for thermally cycled specimen

3.3 Fracture characteristics of thermally shocked and cycled specimens

No substantial changes of K_{IC} have been observed when analysing the fracture behaviour of thermally shocked samples. Fracture toughness data for samples shocked in water from temperature interval 620 to 670 C are represented in Fig. 6 by triangles. Both the average value and the data scatter (24.4 ± 1.14) are comparable to data obtained with as received state. Fracture surfaces exhibit the extensive fibre pull-out. Assessment of the micrographs has shown that the average pull-out lengths in both samples are similar, independently of the ageing condition. This is consistent with the little variation of fracture toughness.

Decrease in K_{IC} values has been detected for thermal shocked samples. For specimens exposed to 466 cycles the drop by 1.8 $MPam^{1/2}$ has be observed (from 24.59 to 22.97 $MPam^{1/2}$, in Fig. 6 by rhombi). For specimen after 821 general delamination along fiber/matrix interfaces occurred and specimen giving invalid data has to be rejected from the assessment. been tested because of general delamination

4 Concluding Remarks

The present results have demonstrated that the chevron-notch technique can be a reliable method to assess fracture properties in brittle matrix composites reinforced by brittle fibres. The method has some advantages for estimating the fracture resistance of toughened materials, since it is possible to measure fracture toughness from the running crack. The K_{IC} values determined using the chevron-notch technique (in the range 24 - 26 $MPam^{1/2}$) are comparable to data obtained in similar materials using other techniques.

The CN test measurements for different thermal exposures were consistent with the results of interfacial properties determined by a push-out indentation method [9]. For the materials investigated, it was found that exposures to inert atmospheres at temperatures in the range 500 – 700 °C do not have deleterious consequences on the properties of the SiC-Nicalon fibres and on the interface. No damage has been detected when assessing the influence of thermal shock in water from temperatures 620 to 670 °C. Slight embrittlement has been observed for specimens thermally cycled between room temperature and 700 °C.

5 Acknowledgements

The research was financially supported by grant Nr. 101/96/K264 of the Grant Agency of the Czech Republic and project of binational Czech - German collaboration supported by Czech Ministry of Education (TSR-007-98). The authors gratefully acknowledge Dr. W. Beier and Mr. R. Liebald of Schott Glaswerke, Mainz, Germany, for supplying the composite samples.

6 References

[1] Chawla, K.K., Ceramic Matrix Composites, Chapman and Hall, London (1993);
[2] Akatsu, T., et al., Fracture Mechanics of Ceramics, Vol. 11, p. 245;
[3] Thouless, M.D., Evans, A.G., Acta Metall., 1988, p. 517;
[4] Sutherland, S., Plucknett, K.P., Lewis, M. H., Comp. Eng., 5, 1995, p. 1367;
[5] Benson, P.T., Spear, K.E., Pantano, G.C., Ceram. Eng. Sci. Proc., 9, 1988, pp. 663;
[6] Daniel, A. M. et al., Ceram. Eng. Sci. Proc., 17, 1996, pp. 280-287;
[7] Boccaccini, A.R., et al., J. Mat. Eng. Perfor., 6, 1997, pp. 344-348;
[8] Pannhorst, W., Ceram. Eng. Sci. Proc., 11, 1990, pp. 947-963;
[9] Boccacini,A.R., Janczak-Rusch, J.,Dlouhý, I., Mat.Chem. Phys., 53, 1998, p. 155;
[10] Dlouhý, I. et al., Metallic Mater., 32, 1994, p. 3;
[11] Brennan, J.J., Prewo, K.M., J. Mat. Sci., 17, 1982, pp. 2371-2383.
[12] J.J. Mecholsky, Ceram Bull. 65, 19896, p. 315;
[13] Dlouhý I. et al. Fatig. and Fracture of Engng Mat. and Structures, 20, 1997; p.1235;

Annealing Induced Changes in Fracture Behaviour of Bulk Plasma-sprayed Alumina

R. J. Damani
Department of Structural and Functional Ceramics, University of Leoben, A-8700 Leoben, Austria

1 Abstract

The microstructural development and fracture characteristics of a bulk plasma-sprayed alumina material is investigated in the as-sprayed and various annealed material conditions, using electron microscopy, and in-situ SEM fracture experiments.

Toughness increases considerably after heat treatment. The material exhibits pronounced *R*-curve behaviour. In the short-crack range broken splats act as frictional bridges resulting in steeply rising toughness. In the longer-crack range larger features cause a less steep, but steady rise in toughness. It is found that the fracture behaviour is governed by the alignment of splats and a intra-splat microcrack sub-structure which provides low energy crack paths. This sub-structure changes extensively on annealing.

2 Introduction

Plasma spraying is gaining increasing importance as a ceramics processing technology. Deposits of several micrometres for coatings to tens of millimetres for free-standing components can be produced. Both coatings and bulk components can have excellent mechanical properties, such as high wear and thermoshock resistance. [1].

The various properties of plasma sprayed ceramics result from their porous, anisotropic, layered *splat* microstructures, which are a consequence of the deposition process [2, 3]. Furthermore, as-sprayed ceramics often contain metastable phases [4, 5]. After spraying, components are frequently subjected to heat treatments to homogenise the phase system and reduce porosity. Such thermo-mechanical loading can lead to extensive changes in the mechanical properties [6-8].

An understanding of the relationships between changes in microstructure resulting from heat treatment and the related effects on properties would facilitate the ability to engineer plasma-sprayed ceramics with different property–microstructure configurations. It has been the the objective of this work is to investigate the fracture behaviour of plasma-sprayed alumina as a function of annealing induced microstructural evolution.

3 Material Evaluation and Experimental Procedure

The Al_2O_3 of this investigation was bulk APS deposited by LWK Plasmakeramik, Gummersbach, Germany. The as-sprayed material contained a mix of about 65 % γ and γ-near transition phases and 35 % α-Al_2O_3. Heat treatments were carried out with a hold time of 12 h at 900 °C, 1050 °C, 1180 °C and 1550 °C. These temperatures were chosen on the basis of the transformation and sintering temperatures of Al_2O_3 phases [9, 10]. Samples were heated in air at 5 °C/min and furnace cooled to RT.

Phase analysis was conducted by X-ray diffraction. Microstructural characterisation was conducted by TEM, SEM. Fracture behaviour was investigated in-situ on compact tension samples in an SEM with a purpose-built loading device [11]. The samples were 5 x 60 x 50 mm^3. They were prepared with a notch cut perpendicular to the splat layers and sputtered on one side with gold. Fracture toughness, K_{Ic}, was evaluated from load and respective crack length according to standard formulae [12].

4 Results

During heat treatment the deposited transition phases transformed through the intermediette stages δ and θ-Al_2O_3 to α-Al_2O_3. After heat treatment at 1180 °C, the material was almost completely α-Al_2O_3.The splats of the as-sprayed material had an internal, 0.2 to 1 µm wide columnar substructure. Significant cracking and agglomeration of porosity between columns was seen, Fig. 1a. Generally, the as-sprayed material contains a high concentration of internal defects.

After 900 and 1050 °C for 12 h intra-columnar porosity became a little more pronounced, otherwise there was little obvious change. After twelve hours at 1180 °C a columnar type substructure was still evident despite the reconstructive phase transformation to α-Al_2O_3 The amount of porosity also increased considerably.

After 12 h at 1550 °C the splat internal columnar substructures had been completely replaced by small, slightly elongated grains, Fig 1b. However, the quasi-laminated splat structure still prevailed. The splats had recrystallised.

A summary of the crack resistance tests is presented in Fig. 2. The material exhibited pronounced *R*-curve behaviour in all conditions where extended crack advance was achieved. The initial toughness of the as-sprayed material is about 0.55 MPa $m^{1/2}$ and rises to about 1.2 MPa$m^{1/2}$. After both the heat treatments at 900 °C and 1050 °C the material exhibits similar long-crack fracture toughness at between 1.5 to 2 times that of as-sprayed material. It is useful to note that the rate of increase of fracture toughness is similar for as-sprayed material and heat treated materials in this range.

After the phase transformation to α-Al_2O_3 at 1180 °C the initial toughness rose to about 2.3 MPa $m^{1/2}$. The toughness increased with crack extension at a higher rate than in the long crack range of materials before transformation. The highest valid toughness recorded was about 4.75 MPa $m^{1/2}$, an increase of about 100 % on the initial value and about 8.8 times higher than the initial fracture toughness of the as-sprayed material.

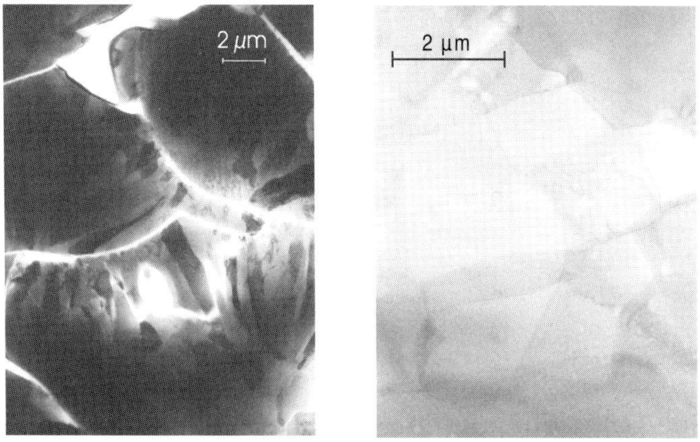

a) b)

Fig. 1. TEM micrographs of internal substructure of splats: a) columnar structure of as-sprayed splats, b) recrystallised grains within one splat after heat treatment at 1550 °C for 12 h

The samples heat treated at 1550 °C showed no stable crack growth, however, the initiation fracture toughness was up to 45 % higher than that of material heat treated at 1180 °C, and almost 6 times the initial fracture toughness of as-sprayed material.

Post-fracture examination of the fracture surfaces revealed a cog-tooth topography in the vicinity of the notch tip, typical of neighbouring splats which have broken across their thickness in different planes and have been pulled away from each other. In general, the main toughening mechanisms observed operating were crack flank bridging, crack hinging, crack deflection and crack splitting. A typical fracture path is shown in Fig. 3.

5 Discussion

In the as-sprayed condition the exeptionally low toughness compared to conventionally sintered aluminas may be attributed to cracks finding low energy paths between pre-existing crack columnar interfaces. TEM investigation has shown these to be areas of weakness with high defect populations [13]. Healing of these defects with heat treatment leads to an increase in the mechanical integrity of the splats by making easy crack paths less accessible. Transformation to α-Al_2O_3 causes further consolidation of the microstructure by replacing columnar grains initially by a system of laths and eventually leading to recrystallisation. The heat treatment at 1550°C also results in better cohesion between splats, making crack propagation through splats increasingly difficult. Thus, cracks are forced to greater deviation and the toughness of the material improves towards that expected of sintered, dense aluminas.

On crack initiation and advance, a crack tended to appear at a notch tip and advance through the first few splats with little deflection, after which it began to deflect around large blocks of material and deviate away from the plane of the notch. It would then either be deflected further from, or back into the plane of the notch (Fig. 3). Thus there was clearly a transition between relatively straightforward crack advance in the short crack range, and the formation of large crack bridging ligaments in the long crack range.

The initial toughening probably results from the mechanical interlocking and pull out of broken splats segments. Thus the larger the segment that has to be pulled out, the greater the toughening increment per splat. The low toughness of as-sprayed material fesults from the fact that cracks can easily advance through splats by using the many pre-existing low energy paths and only short segments are pulled out.. It follows that, any reduction in the number of weak crack paths available can result in a significant increase in toughness in this short crack range.

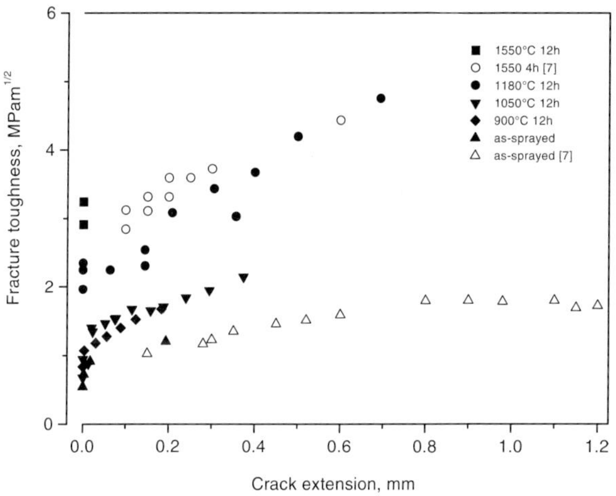

Fig. 2: Summary of *R*-curves for as-sprayed and annealed materials

After the crack has advanced sufficiently, other mechanisms of toughening can take effect: The crack may deflect around localised areas of better mechanical stability, or be offered easier paths by areas of relative weakness, resulting in crack splitting or, large blocks of material which have been circumvented may act as elastic hinges. All such mechanisms have a long range effect as they require the activation of larger scale microstructural features. Increasing microstructural integrity should result in larger blocks of material being circumvented or acting as ligaments and consequently a more tortuous crack path. Thus a greater toughening increment would be expected. Since, however, the heat treatments do not cause dramatic changes in the material at this larger scale of microstructure, the long range effect is similar in all material conditions.

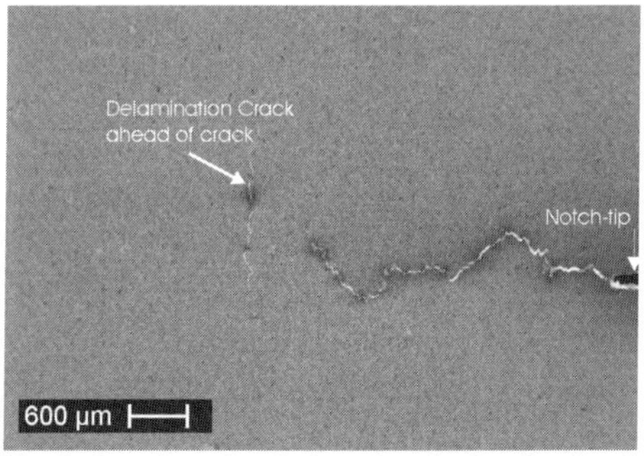

Fig. 3: Overview of typical crack path, also showing delamination beginning ahead of crack tip.

5 Synopsis and Conclusions

- The weak interfaces between the columns and other substructural defects in as-sprayed spalts provide low energy paths for crack advance and are largely responsible for the low toughness of the as-sprayed material.
- Annealing at 900 °C and 1050 °C for 12 h does not significantly change the microstructure of plasma sprayed alumina. Heat treatment at 1180 °C for 12 h results transformation of almost all metastable phases to stable α-Al_2O_3 and an increase in porosity. The increase in integrity of the splat internal structure enhances toughening mechanisms and results in excellent toughness of about 4.75 MPa $m^{½}$.
- After 12 h at 1550 °C the splats are internally recrystallised, but the macroscopic splat structure is largely retained. The intersplat cohesion is a little improved due to some sintering. This leads to much improved initial toughness, but results in an inability to accomodate stable crack propagation. The improvement is largely a consequence of the improved mechanical stability associated with recrystallisation of the splats.
- All the material conditions (with the exception of material heat treated at 1550 °C for 12 hours, for which no controlled crack extension was achieved) exhibit pronounced *R*-curve behaviour. The *R*-curves may be divided into two regions: very steep in the initial phase of crack propagation, and flatter, but persistent, in the long range. In the short crack range toughening is a consequence of the pull-out of splats; and in the long crack range, the toughening effect results from crack branching and hinging by larger microstructural features.
- The change in microstructure concurrent with annealing as-sprayed alumina leads to significant improvements in its fracture resistance.

6 Acknowledgement

This work was partially funded by the Austrian Ministry for Science and Transport under the contract no. GZ 140.539/3-V/A/6/98.

7 References

1. L PAWLOWSKI, The Science and Engineering of Thermal Spray Coatings. Wiley, Chichester, 1995.
2. E. H LUTZ,. Fachberichte der DKG, 1995, **72,** 713–716.
3. D WANG; C. C BERNDT ;Proc. 2nd Plasma-Technik-Symp., Vol. 2, Lucerne, 1991, p 295–304.
4. R MCPHERSON, J. Mat. Sci., 1973, **8,** 851–858.
5. A DAUGER, D FARGEOT, J. P LAVAL, In Mat. Res. Soc. Symp. Proc., Vol. 21., Elsevier, 1984, 207–211.
6. E. H LUTZ, J. Am. Ceram. Soc., 1994, **77,** 1274–1280.
7. R. J DAMANI, E. H LUTZ, J. Eur. Ceram. Soc., 1997, **17,** 1351–1359.

8. S. SAFAI, H HERMAN, In Treatise on Materials Science and Technology, Vol. 20. Academic Press, New York, 1981, 183–213.
9. K WEFERS, C. MISRA, Oxides and Hydroxides of Aluminium. Alcoa Techn. Paper No. 19, Revised, Alcoa Laboratories, 1987.
10. R. MORRELL, Handbook of Properties of Technical & Engineering Ceramics, Part 2: Data Reviews, Section 1: High-Alumina Ceramics. HM Stationary Office, London, 1987.
11. TASCHL, ., et al., In Sonderband der Praktischen Metallographie, (Eds.: A. KNEISSL, F. JEGLITSCH), Verlag Werkstoff-Informationsgesellschaft, Frankfurt, 1999 (in press).
12. ASTM Standard E 399-83: Standard Test Method for Plane-Strain Fracture Toughness of Metallic Materials.
13. R. J DAMANI, ., P MAKROCZY, ., Heat Treatment Induced Phase and Microstructural Development in Bulk Plasma Sprayed Alumina. Accepted for publication in J. Eur. Ceram. Soc.

IV Tribology and Wear

Wear Mechanisms and Influence of Humidity on the Tribological Behaviour of SSiC Ceramic under Unidirectional Sliding

R. Blattner, K.-H. Zum Gahr
Universität Karlsruhe (TH), Institute of Materials Science and Engineering II and Forschungszentrum Karlsruhe, Institute of Materials Research I D-76021 Karlsruhe

1 Introduction

SSiC ceramic is increasingly used for wear components such as mechanical face seal rings or sliding bearings in pumps for the chemical industry [1]. This development is based on the favourable sliding properties and high corrosion resistance of SSiC in many gases and liquids [2, 3]. Nevertheless, under specific operating conditions the sliding surfaces can be damaged which causes leakage of the sealed fluid or abrasive grooving owing to spalling of SiC grains. Critical operating conditions occur when the mating surfaces get into dry sliding contact due to lack of lubrication [4]. As a result the surface temperature can rise up to very high values in small contact areas. Following relubrication of the mated surfaces can lead to cracking and fracture due to thermal shock loading [1, 5]. Hence, the tribological behaviour of self-mated SSiC ceramics is very important for the performance of such components. The scanning electron micrographs (SEM) of Figure 1 show the surface of a SSiC seal ring covered with a partially delaminated SiO_x surface layer which was formed during running in a laboratory water pump test.

The aim of this study was to investigate the wear mechanisms of different commercial SSiC ceramics which differed in size of the grains, porosity and content of free carbon.

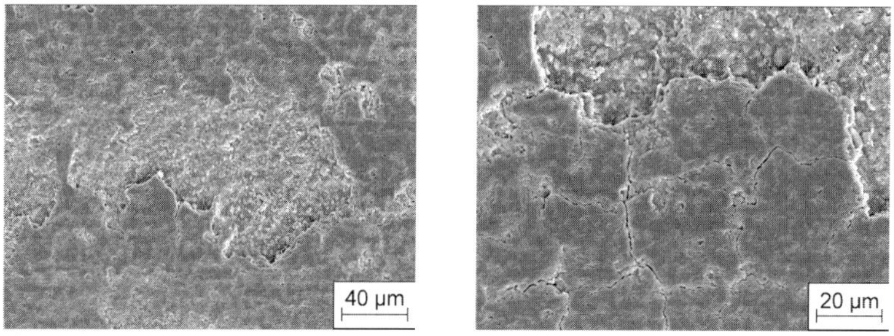

Figure 1. Surface of a SSiC mechanical face seal ring damaged during a laboratory water pump test.

2 Materials and Experimental Methods

Five different commercial SSiC ceramics were provided by the Elektroschmelzwerk Kempten GmbH (ESK). Microstructures of the ceramics varied in the size and size distribution of the grains, closed porosity and content of free carbon as shown in Table 1.

Table 1. Microstructural elements of the SSiC ceramics used.

Trade Name	EKasic® D	EKasic® TRIBO 2000	EKasic® TRIBO 2000-1	EKasic® W	EKasic® HW
(Abbrev.)	(D)	(T 2000)	(T 2000-1)	(W)	(HW)
Grain size	< 10 μm	< 10 μm	< 10 μm	100-1000 μm	100-1000 μm
Porosity	< 3.5%	6.0%	6.0%	< 3.5%	< 3.5%
Pore size	-	< 40 μm	< 60 μm	-	-
Carbon size	-	-	40-60 μm	-	40-60 μm
Carbon cont.	-	-	1 vol.-%	-	0.6 vol.-%

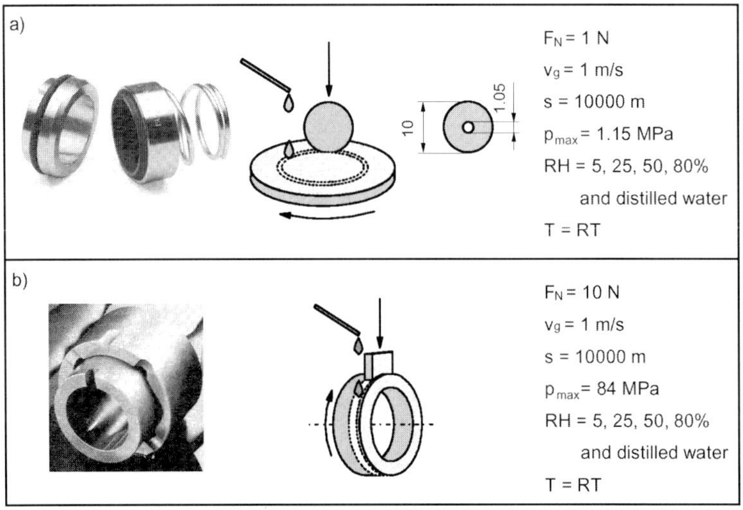

Figure 2. Face seal ring (a) and radial bearing (b) as well as the model tribosystems (a) ball-on-disc test and (b) block-on-ring test and operating conditions used.

Unidirectional sliding wear tests were carried out using a ball-on-disc tribometer and a block-on-ring tribometer (Figure 2). Figure 2a shows a mechanical face seal with conformal contact geometry which was simulated by the ball-on-disc geometry using a flattened ball. In Figure 2b a sliding bearing with conformal contact geometry is shown, which was simulated by using the block-on-ring geometry. In addition to the self-mated ceramics D, T 2000 and T 2000-1, tribological tests were run on sliding pairs of the ceramic D (ball and ring, respectively) mated to the ceramics W and HW. The relative humidity of the surrounding air was systematically varied from 5%, 25%, 50% up to 80% and supplementary distilled water was used. In both tribosystems the tests were run at the sliding speed of 1.0 ms^{-1} and the wear path length of 10000 m. In the ball-on-disc geometry the flattened ball with a contact area of 1.05 mm diameter and at a normal force of 1 N led to a contact pressure of 1.15 MPa at the beginning of the test. In the block-on-ring system the maximum contact pressure for line contact was 84 MPa at the applied normal force of 10 N. Friction force and total linear wear W_l (sum of the sliding pair) were measured continuously during each test. After the tests the linear amount of wear W_l^* was measured separately on the mated specimens. The surface roughness values R_a (c.l.a.) and R_z, respectively, of the wear track of the disc specimens were measured using an atomic force microscope (AFM) before and after the testing on the ceramic

D. The tests were run in normal laboratory air at temperatures between 20°C and 30°C. All reported values are the average of at least two identical tests.

3 Microstructure and Tribological Properties

Figure 3a shows a SEM micrograph of a polished and chemically etched surface of the ceramic T 2000 with an average grain size < 10μm and an average closed porosity of 6% with a pore size up to 40 μm. The ceramic HW contained a bimodal grain structure and average grain sizes of 100 μm (fine crystallites) and 1000 μm (large crystallites), respectively, and carbon particles between 40 μm and 60 μm of size (Figure 3b).

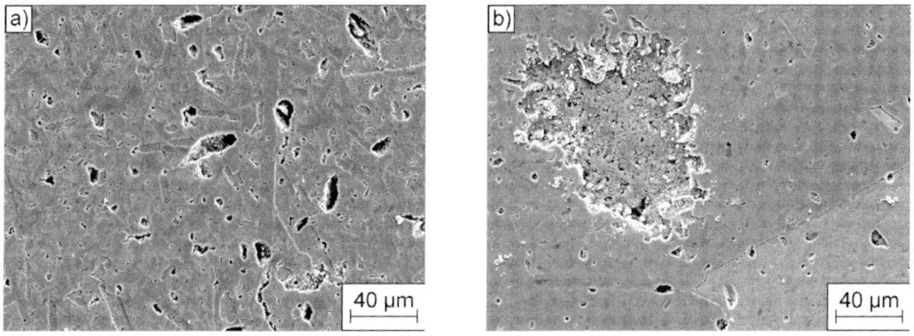

Figure 3. SEM micrographs of polished and chemically etched surfaces of the ceramics: (a) T 2000 with 6% closed porosity and (b) HW with bimodal grain structure and 0.6% vol.-% free carbon.

Figure 4 shows the quasistationary friction coefficient μ and the linear amount of ball wear W_l^* measured after the tribological tests over a sliding path of 10000 m as function of the relative humidity. The linear amount of wear of the sliding pair was equal to that of the ball because the amount of wear measured on the disc specimens could be neglected. The friction coefficient of the different sliding pairs (Figure 4a) increased from μ = 0.4 to 0.62 at 5% RH up to μ = 1.18 to 1.25 at 80% RH. At the presence of distilled water values of the friction coefficient were less than μ = 0.05 for all sliding pairs tested. At a relative humidity of 50% the value of friction coefficient decreased from about μ = 1.03 to 0.73 in the order of the sliding pairs D/D, T 2000/T 2000, T 2000-1/T 2000-1, D/W to D/HW. Lowest friction coefficient was measured in the range between 5% to 50% relative humidity when mating ceramic D to HW. Linear amount of wear after 10000 m of running decreased from W_l^* = 18.3 to 20.4 μm at 5% RH to W_l^* = 1.1 to 2.7 μm at the presence of distilled water (Figure 4b).

Quasistationary values of friction coefficient μ and linear amount of wear W_l^* after 10000 m of sliding path of the block are shown in Figure 5 as function of the relative humidity using the tribosystem block-on-ring. Linear amount of wear of the sliding pair was equal to that of the block because the wear measured on the ring specimens could be neglected.

Values of the friction coefficient of the differently mated materials are presented in Figure 5a. Depending on the pair the friction coefficient ranged from μ = 0.55 to 0.6 at 5% RH and increased to μ = 0.75 to 0.97 at 80% RH. Values of friction coefficient less than μ = 0.05 were measured at the presence of distilled water. The linear amount of wear after the path length of 10000 m ranged from W_l^* = 32.0 to 38.7 μm at 5% RH depending on the pairing and decreased to W_l^* = 3.2 to 5.6 μm at the presence of distilled water (Figure 5b). Amount of wear W_l^* of the self-mated ceramic T 2000-1 was approximately twice as large as that of the other sliding pairs at relative humidities between 25% and 80% RH.

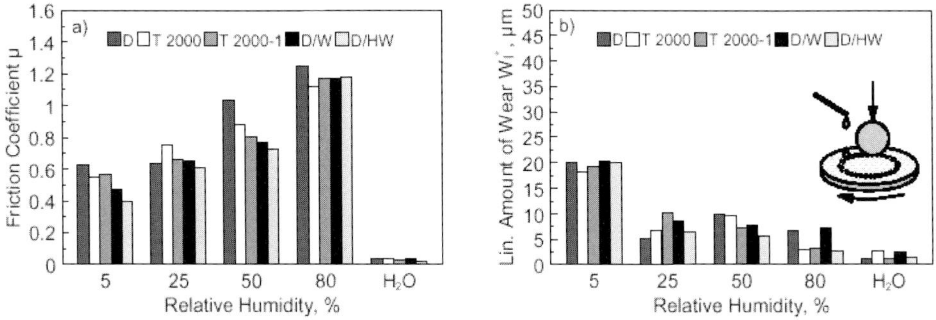

Figure 4. (a) Friction coefficient and (b) linear amount of ball wear for self-mated D, T 2000, T 2000-1 and the sliding pairs D mated to W and D mated to HW, respectively, vs. the relative humidity of the surrounding air.

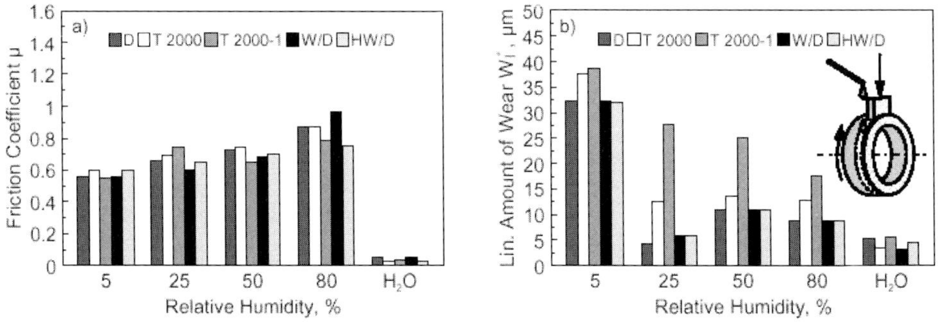

Figure 5. (a) Friction coefficient and (b) linear amount of block wear for self-mated D, T 2000, T 2000-1 and the sliding pairs W and HW, respectively, mated to D vs. the relative humidity of the surrounding air.

Table 2. Surface roughness R_a and R_z for discs of ceramic D before and after testing in the tribosystem ball-on-disc at 5, 25, 50, 80% RH and at presence of distilled water.

	before test	after testing at 5% RH	25% RH	50% RH	80% RH	H_2O
Roughness R_a, μm	0.115	0.091	0.027	0.027	0.012	0.072
Roughness R_z, μm	0.8	0.645	0.228	0.30	0.092	0.052

Table 2 shows the influence of the relative humidity on surface roughness (measured using AFM) values R_a and R_z, respectively, of discs of ceramic D before and after tests in the tribosystem ball-on-disc. The surface roughness values R_a and R_z of the disc specimens were 0.115 μm and 0.8 μm, respectively, before the test. Running at relative humidities of 5% to 80% led to a decrease of R_a values of 0.091 to 0.012 μm and R_z values of 0.645 to 0.092 μm. After running at the presence of distilled water, values of R_a = 0.072 μm and of R_z = 0.052 μm were measured.

4 Discussion and Conclusions

Scanning electron micrographs of worn disc and ring surfaces after a wear path length of 10000 m at a relative humidity of 50% RH are shown in Figure 6. According to the SEM micrographs the surface of the worn disc was in-situ polished during sliding contact and wear particles were embedded in pores. The worn surface of the ring specimen was partially covered with a SiO_x-surface layer, which was produced by compressing small wear particles mainly generated on the block specimen. This surface layer reduced the contact area for adhesion between block and ring during dry sliding and resulted in lower friction coefficient when using the block-on-ring tribosystem compared to the results of the ball-on-disc tribosystem. With increasing the relative humidity the wearing surfaces became smoother due to an in-situ polishing process at the presence of water molecules on the surface and the effect of meniscus forces at interacting asperities was enhanced. Both the increasing true area of contact and increasing attracting forces between the surfaces resulted in higher values of friction coefficient.

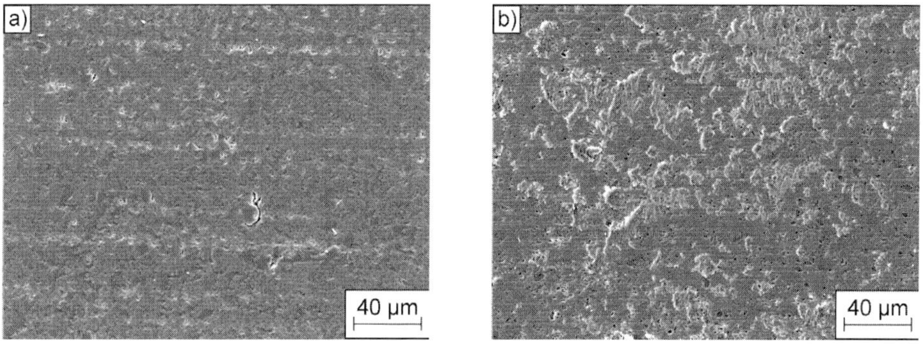

Figure 6. SEM micrographs of worn surfaces of ceramic D after tests at 50% RH and a wear path length of 10000 m: (a) disc from the ball-on-disc test and (b) ring from the block-on-ring test.

Carrying out wear tests in two different tribosystems with different specimen geometry, contact geometry and normal loads showed equal dependence of tribological properties on relative humidity of the surrounding air. In both systems the friction coefficient increased with increasing relative humidity between 5% and 80%. This was attributed to a decrease of surface roughness values R_a and R_z, respectively, on the wear tracks during time of running which was shown on the discs of the ceramic D. Friction coefficient was quite low at the presence of distilled water. In general, the linear amount of wear tended to decrease with increasing relative humidity. The amount of wear of the self-mated ceramic T 2000-1 measured in the block-on-ring geometry was substantially larger than all other pairs at relative humidities between 25% and 80% (Figure 5b). This was explained by the higher loading or contact pressure of this system compared to that of the ball-on-disc geometry which caused severe wear on the ceramic T 2000-1 of lower mechanical strength owing to the internal defects (pores and free carbon).

5 Acknowledgements

The authors wish to thank the Ministry of Science, Research and the Arts of Baden-Württemberg for supporting this work through the KKS programme. The material tested was provided by the Elektroschmelzwerk Kempten GmbH (ESK).

6 References

[1] J. Greim, in the VDI-Berichte Nr. 1331, VDI-Verlag, Düsseldorf, Germany, 1997, pp. 153-159.
[2] B. Löffelbein, M. Woydt, K.-H. Habig, Wear 1993, 162-164, 220-228.
[3] B. Zimmerlin, K.-H. Zum Gahr, Proc. 6th International Congress on Tribology, Budapest, Vol. 3, 1993, pp. 46-51.
[4] J. Greim, M. Fundus, Δp Das moderne Pumpenmagazin 1995, 2, 54-57.
[5] H. Knoch, M. Fundus, in Ceramic Technology International (Ed.: I. Berkby), Stirling Publications Ltd., London, UK, 1995, pp. 59-63.

Surface Modified Alumina Ceramic for Unlubricated Tribological Applications

J. Schneider, K. Przemeck, I. Lenke *, K.-H. Zum Gahr
Universität Karlsruhe (TH), Institute of Materials Science and Engineering II and Forschungszentrum Karlsruhe, Institute of Materials Research I, * CeramTec AG, Lauf

1 Introduction

Advanced Al_2O_3 ceramics offer a high potential for technical components with high mechanical or tribological requirements such as cutting tools, seal rings or bearing parts. Major problems of Al_2O_3 are its low fracture toughness and low resistance to thermal shock, which can result in low wear resistance and relative high friction coefficient of self-mated Al_2O_3 under unlubricated sliding wear [1 - 3]. The tribological behaviour of alumina is characterized by a transition from mild to severe wear, owing to a change of mechanism from predominantly tribochemical reactions to surface damage by microfracture. Beside environmental and operating conditions, microstructural parameters can influence this transition. Improved tribological properties can be expected by microstructural modifications resulting in enhanced fracture toughness or by multiphase materials [3]. Considering, that all tribologically induced interactions are concentrated on a relative thin surface zone, it may be sufficient and economical to modify the microstructure of this surface zone only. Recent studies [4 - 8] showed that CO_2-laser irradiation can be successfully used for surface modification of alumina by embedding second phases.

The aim of the present paper is to show the high tribological potential of laser surface modified Al_2O_3. Therefore, alumina was modified using CO_2-laser irradiation by adding HfO_2, TiN or W and ZrO_2. Friction and wear of the modified ceramics were studied under oscillating sliding contact against alumina balls in a laboratory tribometer.

2 Materials and Experimental Methods

Surface modification using a 4 kW CO_2-laser was carried out on an Al_2O_3 ceramic (Al24, Fa. Friatec) with an open porosity of about 5 vol.%. Properties of this ceramic, the laser modified ceramics Al24L, Al24Hf, Al24TiN and Al24WZ are given in Table 1. The Al_2O_3 (Al23, Fa. Friatec) used for reference in the tribological tests is also listed. For surface modification the ceramic Al24 was precoated with a suspension consisting of HfO_2, TiN or ZrO_2 powder mixed with isopropyl alcohol. The precoated specimens were heated up to 1200 to 1500°C and then remelted with an average laser power of about 200 W using a laser beam cross-section of 1x6 mm^2. A commercial powder feeder conveyed the W powder into the melt bath by a specially adjusted injection nozzle (Fig. 1a). To avoid oxidation the specimens were protected by an argon gas shield. The laser treatment is discribed in more details elsewhere [6, 8]. Microstructures and worn surfaces were analysed with standard ceramographic techniques, scanning electron microscopy (SEM) and energy dispersive x-ray spectroscopy

(EDX). Mechanical properties were characterized by Vickers hardness tests and determination of indentation fracture toughness K_{Ic}*. Tribological properties were measured in a ball-on-block tribometer (Fig. 1b, Optimol SRV) under unlubricated oscillating sliding contact against Al_2O_3 balls (F99.7, Fa. Friatec, 1700 HV_{500}, average grain size 9 μm). Before testing, the blocks were ground and polished to a surface roughness of $R_a < 0.1$ μm comparable to that of the Al_2O_3 balls. All tribological tests were run over 144 m at a normal load of 40 N, a frequency of oscillating of 20 Hz, and a stroke of 0.5 mm (v = 0.02 m/s). The tests were carried out inside a climate chamber with varied relative humidity (3 - 80%) or temperature (RT - 500°C). During each test, friction force and amount of linear wear W_l were measured and recorded continuously. Linear wear W_l* was separately determined for balls and blocks from surface profiles recorded with a profilometer after the tests. Friction coefficients reported were calculated from the stationary values after a short period of running-in. All results presented are average values of at least two tests.

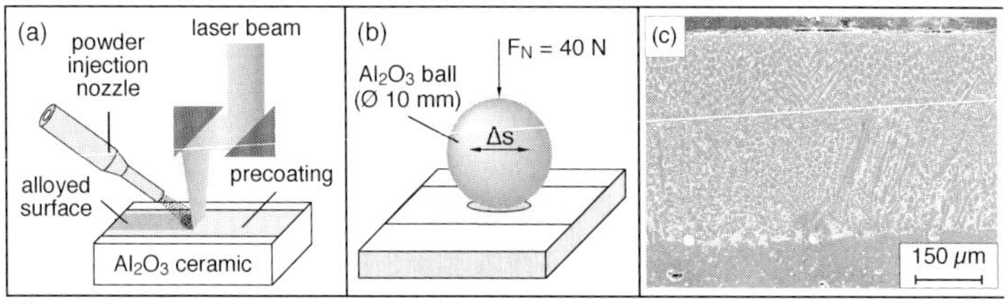

Figure 1. (a) Experimental setup for laser treatment, (b) tribological system used in oscillating sliding wear tests and (c) cross section through the surface modified ceramic Al24Hf.

Table 1. Materials, additives and microstructural parameters of the surface modified ceramics and the monolithic Al_2O_3 ceramics in the as-received condition.

Materials Designation		Al24	Al24L	Al24TiN	Al24Hf	Al24WZ	Al23
Added Compounds		---	---	TiN	HfO_2	W + ZrO_2	---
Average Powder Size	μm	---	---	1.5	1.5	W: 6.8 ZrO_2: 1.3	---
Al_2O_3 Matrix	vol.%	100	100	72	75	45	100
Grain Boundary Phase	vol.%	---	---	16	25	35	---
Reinforcing Particles	vol.%	---	---	12 (TiN)	---	20 (W)	---
Average Grain Size	μm	30	81	4	11	10	6.5
Hardness	HV_{500}	1900	2050	1900	1750	1250	2030
Fracture toughness	$MPam^{1/2}$	< 3.0	2.5	4.5	4.6	5.0	3.7

3 Results

A scanning electron micrograph of a cross section through the modified ceramic Al24Hf is shown in Fig. 1c. The thickness of the alloyed surface layers could be varied from about 100 to more than 400 μm, depending on the chemical composition of the additives and the parameters of the laser process. Microstructures of the different ceramics are presented in Fig. 2. The dense alumina Al23 (Fig. 2a) showed an average grain size of 6.5 μm and a hardness of 2030 HV_{500}. Remelting of the alumina ceramic Al24 led to surface sealing, but was accompanied by an increase of the average grain size from 30 μm to more the 80 μm

(Fig. 2b, c). Laser modification of Al24 with TiN particles resulted in a composite microstructure with 12 vol.% TiN particles distributed homogeneously in a fine grained Al_2O_3 matrix (Fig. 2d). A reaction phase consisting of Ti, O and Al was formed (16 vol.%) due to partial oxidation of TiN particles during laser processing. The HfO_2 alloyed ceramic Al24Hf (Fig. 2e) showed about 25 vol.% of a fine lamellar eutectic phase along the grain boundaries of the Al_2O_3 crystallites, which reduced grain size and hardness compaired with the Al24 substrate. Laser remelting of ZrO_2 precoated Al24 and simultaneous injection of W particles resulted in a multiphase microstructure containing Al_2O_3 crystallites of about 10 μm average size, a grain boundary phase (35 vol.%) and about 20 vol.% W particles (Fig. 2f).

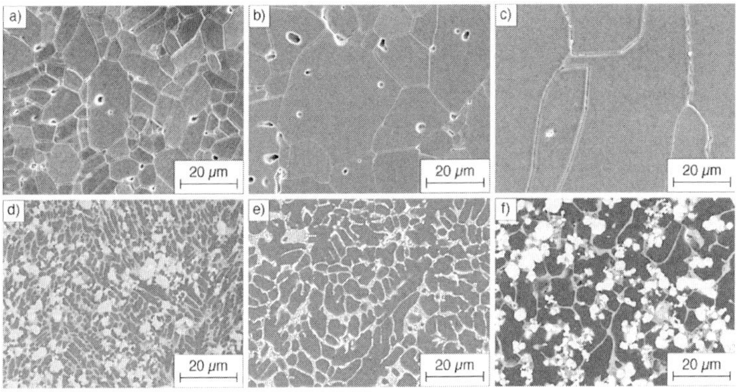

Figure 2. Scanning electron micrographs of the commercial alumina ceramics (a) Al23, (b) Al24 and the laser surface modified ceramics (c) Al24L, (d) Al24TiN, (e) Al24Hf and (f) Al24WZ.

Friction coefficient and amount of linear wear for the different ceramics in oscillating sliding contact against Al_2O_3 balls as a function of length of wear path at room temperature and 50% RH are shown in Fig. 3. The friction coefficient of the self-mated alumina pair (Al_2O_3/Al23) increased sharply to a value of about 1.0 after a short running-in period. With prolonged time the friction coefficient decreased to a quasi-stationary value of about 0.7 (Fig. 3a). The sliding pairs of the modified ceramics showed substantially lower values of friction coefficient and reduced periods of running-in. Lowest values of quasi-stationary friction coefficient between 0.36 and 0.4 were measured on the pair Al_2O_3/Al24WZ. High values of the friction coefficient observed at the beginning of the tests corresponded with high wear intensities on the self-mated alumina (Fig. 3b). Owing to the ball-on-block geometry, surface pressure decreased with increasing amount of wear on the ball, which led to lower wear intensity at prolonged time. Compared with the self-mated alumina pair the amount of wear for the pairs with the modified ceramics was up to a factor of 9 lower after a sliding distance of 144 m.

Fig. 4a - d show scanning electron micrographs of surfaces of Al23 and the modified ceramics worn during oscillating sliding contact against Al_2O_3 balls over a sliding distance of 144 m at 50% RH. The worn surface of the reference ceramic Al23 (Fig. 4a) was covered with a partially cracked layer consisting of desified wear debris and loose wear particles. In contrast, the surface of the laser modified ceramics Al24TiN, Al24Hf and Al24WZ (Fig. 4b - d) displayed smooth or rather polished surfaces after a total length of wear path of 144 m. Small rolls on the worn surface of the ceramic Al24TiN indicated the detachment of thin films produced by tribochemical reactions (Fig. 4b).

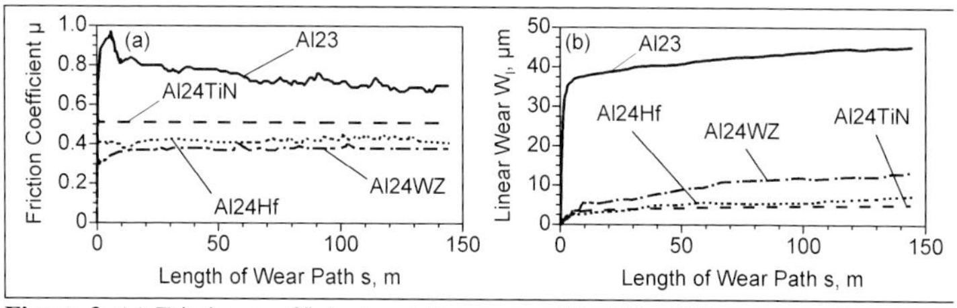

Figure 3. (a) Friction coefficient and linear wear W_l of the different ceramics versus length of wear path during oscillating sliding contact against Al_2O_3 balls (F_N = 40 N, RH = 50%, RT).

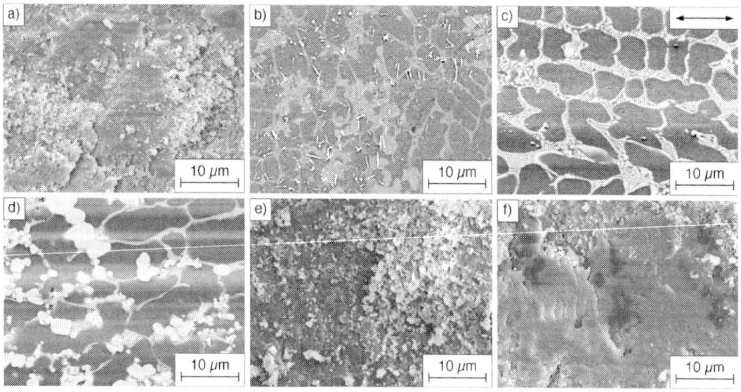

Figure 4. Scanning electron micrographs of surfaces of (a, e) Al23, (b) Al24TiN, (c) Al24Hf and (d, f) Al24WZ worn during oscillating sliding contact against Al_2O_3 balls over a sliding distance of 144 m at room temperature and (a - d) 50% or (e, f) 3% RH (arrow indicates sliding direction).

Fig. 5 shows the effect of relative humidity on friction and wear of the ceramics Al23 and Al24WZ during oscillating sliding contact against Al_2O_3 balls. The highest friction coefficient was measured for both pairs in dry air (3% RH) and the lowest values for high humidity (80% RH). The ceramic Al24WZ showed lower values of friction coefficient and amount of wear than the sliding pair Al_2O_3/Al23 in both cases. After the tribological tests in dry air the worn surface of Al23 was covered with a rough layer consisting of densified and loos wear particles (Fig. 4e). Smooth areas occurred on the worn surface of the ceramic Al24WZ, where W was analysed by EDX (Fig. 4f).

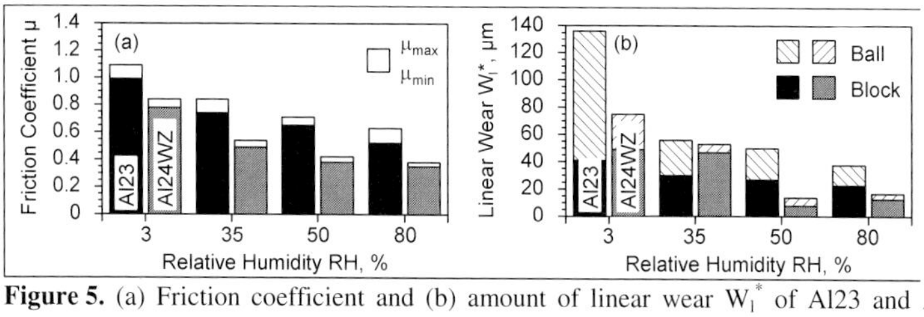

Figure 5. (a) Friction coefficient and (b) amount of linear wear W_l^* of Al23 and Al24WZ under oscillating sliding contact against Al_2O_3 balls at various relative humidities (F_N = 40 N, RT).

Fig. 6 shows the effect of temperature on friction and wear of the ceramics Al23 and Al24TiN during oscillating sliding contact against Al_2O_3 balls. The pair Al_2O_3/Al23 showed an increase of both, friction coefficient and amount of wear with increasing temperature from

28 to 110°C. Between 110 to 500°C friction and wear were nearly independent of test temperature. For the pair Al_2O_3/Al24TiN, friction coefficient increased from 0.5 at 28°C to 1.0 at 250°C followed by a decrease with further increasing temperature. At 500°C friction coefficient was about equal to that at 28°C. Corresponding with the friction coefficient highest wear occurred at 250°C.

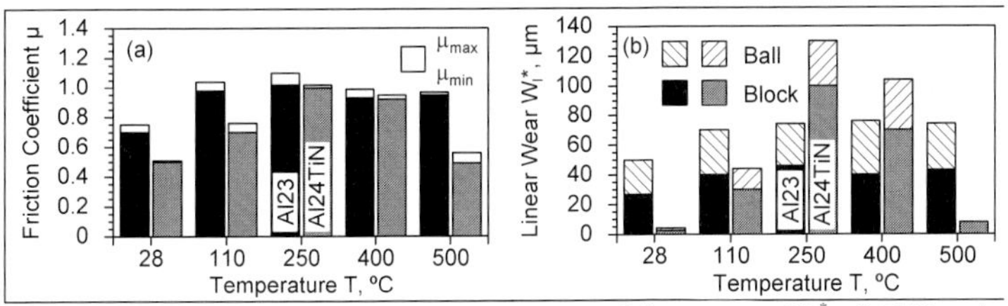

Figure 6. (a) Friction coefficient and (b) amount of linear wear W_l^* of Al23 and Al24TiN under oscillating sliding contact against Al_2O_3 balls at various test temperatures ($F_N = 40$ N).

4 Discussion and Conclusions

The experimental results showed that CO_2-laser irradiation can be effectivelly used for surface modifying Al_2O_3 by embedding HfO_2, TiN or W and ZrO_2. The thickness of the composite surface layers ranged from about 100 to 400 μm depending on the precoating and the process parameters. The multiphase microstructures contained up to 55 vol.% of second phases homogeneously distributed in an Al_2O_3 matrix.

Compared with the monolithic ceramic (Al23) the multiphase ceramics showed 25 to 40% lower friction coefficient and up to a factor of 9 lower linear wear under unlubricated oscillating sliding contact against Al_2O_3 balls at 50% RH. High surface pressure at the beginning of the tests (hertzian pressure ca. 1500 MPa) resulted in intergranular microcracking and spalling of individual grains in the contact zone of the self-mated Al_2O_3 sliding pair (Al_2O_3/Al23). This could be explained by the low fracture toughness of monolithic alumina at short cracks along the grain boundaries [9]. Increased fracture toughness and soft grain boundary phases could hinder crack formation and growth in the multiphase ceramics resulting in rather polished surfaces at the end of the wear tests [3]. In the modified ceramics spalling of individual grains due to intergranular fracture could also be hindered by the columnar shape of the Al_2O_3 crystallites and their orientation perpendicular to the worn surfaces [7]. Furthermore additions like HfO_2, ZrO_2 or W resulted in lower hardness and Young´s modulus and thereby reduced local contact pressure by increasing the true area of contact [8]. Increased wear resistance owing to a grain refinement was particular evident for the TiN particle reinforced ceramic Al24TiN [7].

The laser modified ceramic Al24WZ showed reduced friction and wear in dry air (3% RH) compared with the ceramic Al23. This might be due to plastic deformation of the metallic W particles in the contact zone, which resulted in a relatively smooth and stable surface layer [8]. Surface layers of densified wear debris generated during the fracture controlled period of running-in dominated the tribological behaviour of the self-mated alumina sliding pair (Al_2O_3/Al23) independent of relative humidity and resulted in relative high friction coefficient. Low friction coefficient of the multiphase ceramics at relative humidities > 35% could be explained by the formation of thin films of soft aluminium hydroxides, which reduced friction due to their low shear strength [10]. The low friction coefficient of the pair Al_2O_3/Al24TiN at 500°C indicated the positive influence of the embedded TiN particles and the Ti-O-Al reaction phase, which can led to the formation of lubricious oxides (e.g. TiO_{2-x}) [11].

The authors are grateful to A. Reif, P. Riegel and F. Keller for helpful assistance and wish to thank the *Ministry of Science, Research and the Arts of Baden-Württemberg* and the *Deutsche Forschungsgemeinschaft (DFG)* for supporting the work in the frame of the *KKS programme* and the *Graduiertenkolleg "Technische Keramik"*, respectively.

5 References

1. S. Hsu, M. Shen: Wear, 200 (1996) 154 - 157.
2. H. Czichos, D. Klaffke, E. Santner, M. Woydt: Wear, 190 (1995) 155 - 161.
3. K.-H. Zum Gahr: Wear, 200 (1996) 215 - 224.
4. R. Rowlings, et al.: Proc. Laser-6. Paris, (1990) pp. 41 - 46.
5. F. Glasser, X. Jing: Br. Ceram. Trans. J., 91 (1992) 195 - 198.
6. S. Z. Lee, K.-H. Zum Gahr: Ceram. Intern., 20 (1994) 147 - 157.
7. K.-H. Zum Gahr, C. Bogdanow, J. Schneider: Wear, 181-183 (1995) 118 - 128.
8. K. Przemeck, K.-H. Zum Gahr: J. Mat. Sci., 33 (1998) 4531 - 4541.
9. S.-J. Cho, et al.: J. Am. Ceram. Soc., 79 (1989) 1249 - 1252.
10. Blomberg, M. Olsson, S. Hogmark: Wear, 171 (1994) 77 - 89.
11. M. Gardos: Tribo. Trans., 32 (1988) 427 - 436.

Surface Engineering to Improve Tribological Performance of Gamma Based Titanium Aluminides

Ahmad Reza Rastkar and Tom Bell
School of Metallurgy and Materials, The University of Birmingham, Birmingham, U.K.

1 Abstract

Titanium aluminide with the composition of Ti-48Al-2Nb-2Mn was plasma nitrided to modify the surface conditions. The effects of nitriding temperature, nitriding time and H_2/N_2 gas ratio were investigated. The microstructure and phases of the nitrided layers on the surface were studied by optical and electron microscopy, X-ray diffraction (XRD) and glow discharge spectroscopy (GDS). The samples were wear tested using a ball-on-disc configuration. Experimental results indicated that nitriding temperature and time control the thickness, and the H_2/N_2 ratio is responsible for the crystal structure of the nitrided layers. The nitrided compound layers consist of TiN and Ti_2AlN phases. The diffusion zone, in most cases is comprised of recrystallised gamma grains beneath the compound layer.

Keywords: Plasma Nitriding; Titanium Aluminides; TiN; Ti_2AlN

2 Introduction

Gamma based titanium aluminides have been widely studied with the objective of attaining their full potential for high temperature applications such as required by the aerospace industry [1-4]. These materials have been recently recognized to have wider applications, for example in land-based turbines, power generation plants and as a substitute for steel in car exhaust valves, allowing a reduction in weight and higher operating temperatures [5-7]. The main obstructions to the proper performance of these materials, are their low hardness and poor wear resistance. There is a dearth of information concerning the tribological behavior of the gamma titanium aluminides but a preliminary study [8] on the hardness and wear properties of these alloys has shown them to have a low hardness and poor wear resistance, compared to that of commercial titanium alloys against steel.

Plasma nitriding has been successfully used to improve the tribological properties of steels and Ti alloys, but it has not been employed widely on titanium aluminides. In the present work, the plasma nitriding of Ti-48Al-2Nb-2Mn alloy is investigated. The characteristics of the nitrided layers and their tribological properties are discussed.

3 Experimental Work

The titanium aluminide Ti-48Al-2Nb-2Mn alloy was prepared through plasma melting in a twin torch Retech plasma furnace in The University of Birmingham from elemental compacts, which was then semi-continuously cast into 100 mm diameter ingot. The as cast alloy was sectioned by Electrical Discharge Machining (EDM) into 50 mm diameter discs and 10 mm in thickness for wear testing which were then ground to a surface finish of 0.02 μm (Ra). Before plasma nitriding, all specimens were cleaned by acetone to remove surface contaminants.

Plasma nitriding was carried out in a Klockner Ionon GmbH plasma unit with a DC 40 kW power supply. Plasma nitriding tests were carried out at different temperatures, times and gas mixture from which, two typical treatments have been shown in Table 1. The microstructures and phases of the nitride layers were studied by X-ray diffraction (XRD) and scanning electron microscopy (SEM) with an energy-dispersive X-ray analysis (EDX) facility.

The wear tests were performed using a ball-on-disc wear testing machine. In the present work, steel balls of 10 mm in diameter were loaded against horizontal rotating discs. During the sliding process the disc was driven by a servo-controlled d.c. motor which provided a continuously variable range of rotating speeds. Loads were applied to the ball through a load arm by means of a spring loaded weight system, designed to reduce inertial loading. The experimental data were collected and displayed graphically using a microcomputer connected via an analogue-digital converter to a strain gauge.

Table 1 Plasma nitriding conditions applied to the Ti-48Al-2Nb-2Mn alloy sample

Sample	Treatment Temperature, °C	Gas composition ratio %	Time, h
PN3	700	20 H2 / 80 N2	100
PN6	800	80 H2 / 20 N2	100

4 Results and Discussion

4.1 Microstructure of Plasma Nitrided Ti-48Al-2Nb-2Mn Alloy

After plasma nitriding, the surfaces of specimens showed a range of brown to golden yellow color, depending on the nitriding parameters. The cross sectional micrograph of sample PN6 plasma nitrided at 800 C for 100 h is typically shown in Figure 1. By examination of Figure 1, two layers can be observed on the lamellar Ti-48Al-2Nb-2Mn alloy substrate: A dense thin layer is observed at the top of the sample similar to the compound layer seen in nitrided pure titanium and titanium alloys [9], and beneath this layer, a zone of nearly equiaxed grains has formed which is different from thin lamellae of the alloy. This zone is formed due to nitrogen diffusion into the lamellae which recrystallises (coarsens) the TiAl lamellae into the equiaxed grains [10, 11].

The microstructure of the compound layer depends strongly on the H_2/N_2 treatment gas ratio and temperature. Fracture sections shown in Fig. 2 represent two typical structures of the samples plasma nitrided for 100 h. The PN3 treatment (at 700 C in 20 H_2/80 N_2 gas ratio), produced a columnar compound layer (Fig. 2a) without an observable diffusion layer, while the PN6 treatment (at 800 C in 80 H_2/20 N_2 gas ratio) formed a fine-grained compound layer (Fig. 2b). The compound layer of PN6 treatment is in turn composed of at least two layers, a very fine-grained layer at the top and a nearly equiaxed coarse-grained layer underneath the top layer. The different crystal structures are thought to correspond to the sputtering and redeposition rates during plasma nitriding treatment, which increase with increasing hydrogen gas ratio [12]. In fact this hydrogen increase causes more nuclie and hence more grains to be formed or in other words finer grain size is created. Therefore the higher nitrogen ratio

(20 H_2/80 N_2) leads to a columnar or coarse-grained compound layer, while treatment in a high hydrogen ratio (80 H_2/20 N_2) produces a fine-grained structure.

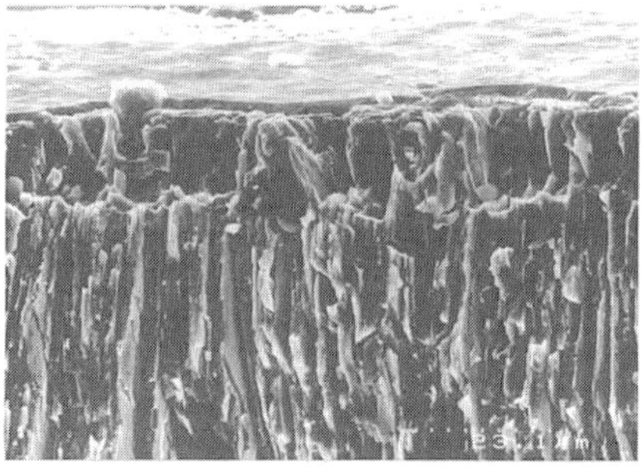

Figure 1. FESEM micrograph of the fractured cross section of the sample plasma nitrided for 100 h at 800 °C in 80 H_2/20 N_2 gas ratio showing the compound layer and diffusion zone

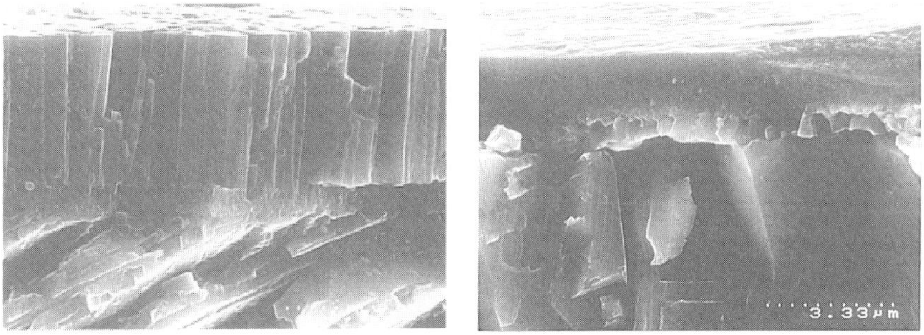

a) b)

Figure 2.FESEM micrographs showing the fractured cross sections of the compound layers formed on Ti-48Al-2Nb-2Mn alloy after plasma nitriding for 100 hrs at (a) 700 C n 20 H_2/80 N_2 gas ratio and (b) at 800 C in 80 H_2/20 N_2 gas ratio.

The microstructure of the diffusion (recrystallised) zone is strongly dependent on the temperature. In fact, the diffusion zone is not observed at 700 C treatments even after 100 h, whilst treatment for 20 h at 800 C produces the diffusion layer in a considerable depth beneath the compound layer.

4.2 Compositional Analysis of Plasma Nitrided Ti-48Al-2Nb-2Mn Alloy

XRD results shown in Fig. 3 indicate that the nitrided layers are essentially composed of TiN and Ti_2AlN. Ti_2N was not observed in plasma nitrided samples. According to Fig. 3, the relative amount of TiN to Ti_2AlN varies depending on the treatment conditions. For instance, in sample PN3, Ti_2AlN is not only the main constituent of the nitrided layer but also has a strong (110) peak, which is regarded as the preferential orientation of columnar grains in that direction. However, most peaks of TiN and Ti_2AlN are comparatively large in sample PN6, corresponding to the fine-grained structure of the compound layer in that sample.

GDS analysis of plasma nitrided samples show generally a gradual decrease in the amount of diffused nitrogen into the depth of sample. This characteristic is shown for two samples PN3 and PN6 in Fig 4. The lower amount and depth of nitrogen diffusion into sample PN3 than those of sample PN6 is due to the lower treatment temperature and

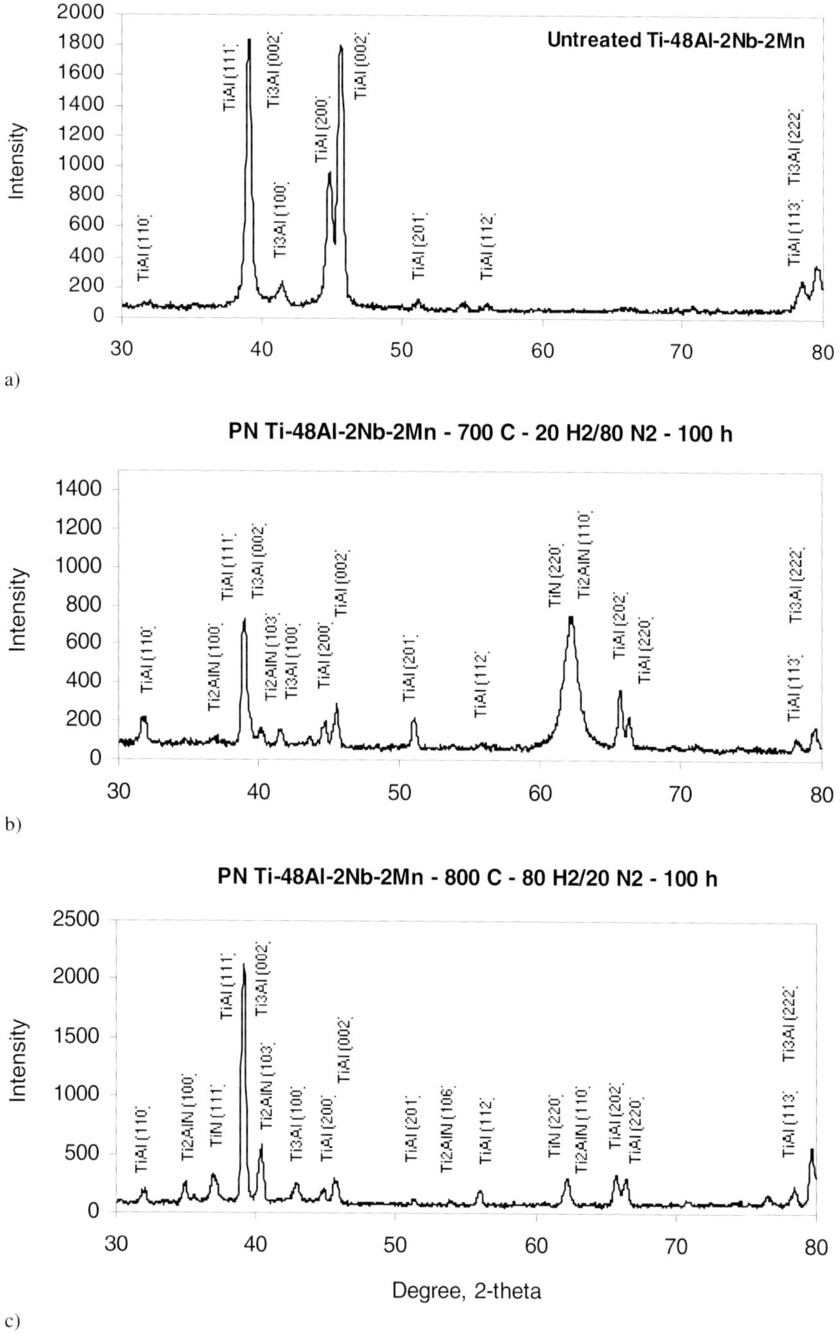

Figure 3. XRD patterns of (a) untreated Ti-48Al-2Nb-2Mn titanium aluminide and the samples plasma nitrided for 100 h (b) at 700 C in 20 H_2/80 N_2 gas ratio and (c) at 800 C in 80 H_2/20 N2 gas ratio

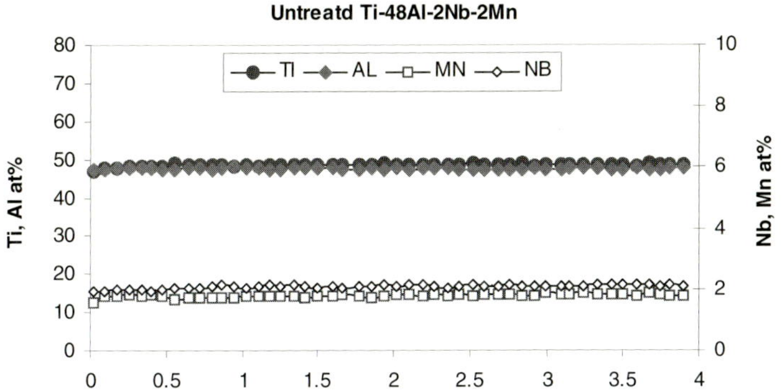

a)

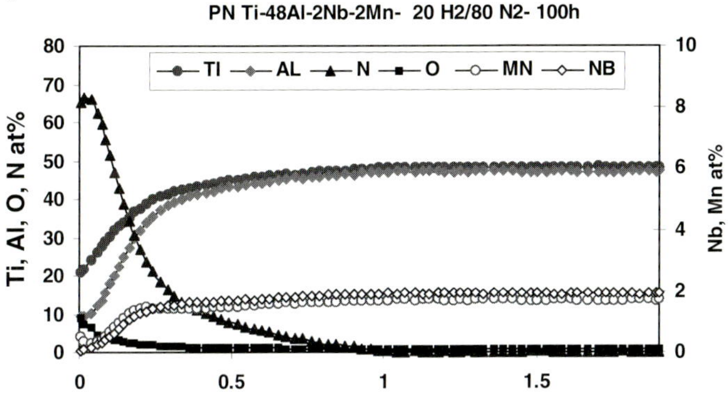

b)

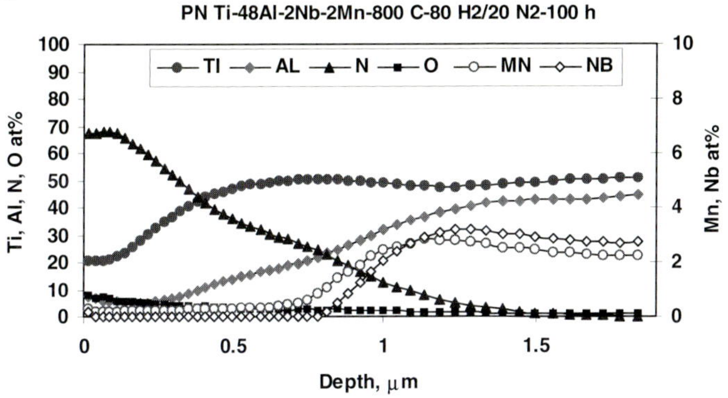

c)

Figure 4. GDS depth profiles of (a) untreated Ti-48Al-2Nb-2Mn titanium aluminide and the samples plasma nitrided for 100 h (b) at 700 C in 20 H_2/80 N_2 gas ratio and (c) at 800 C in 80 H_2/20 N_2 gas ratio

4.3 Mechanical Properties of Plasma Nitrided Ti-48Al-2Nb-2Mn Alloy

4.3.1 Effect of Plasma Nitriding on Surface Hardness

The two-phase (gamma + alpha 2) lamellar Ti-48Al-2Nb-2Mn alloy has a hardness value of about 230-320 HV. This value is about one-tenth of the hardness value of wear resistant materials or coatings such as TiN, TiC and Al_2O_3. Plasma nitriding of Ti-48Al-2Nb-2Mn forms TiN and Ti_2AlN compounds which increases the surface hardness significantly. The high hardness was revealed at testing loads lower than 25 g through Vickers indentation (Fig. 5a). Since even at the lowest testing load of 15 g, the penetration depth of Vickers indenter is higher than one-tenth of the impression diagonal, it should be noted that the hardness values are a combination of the substrate and compound layer hardness values. With respect to the low hardness of the untreated alloy, the high hardness values should in major be due to the contribution of the compound layer. The higher Knoop hardness values for long (100 h) treatments (Fig. 5b) may be more representative of the compound layer.

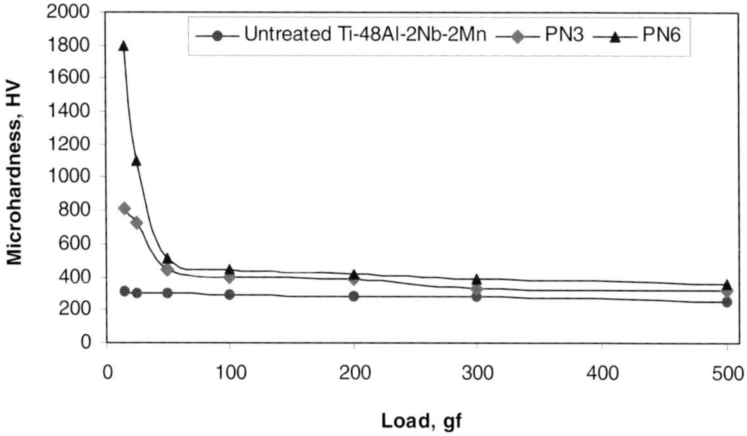

a)

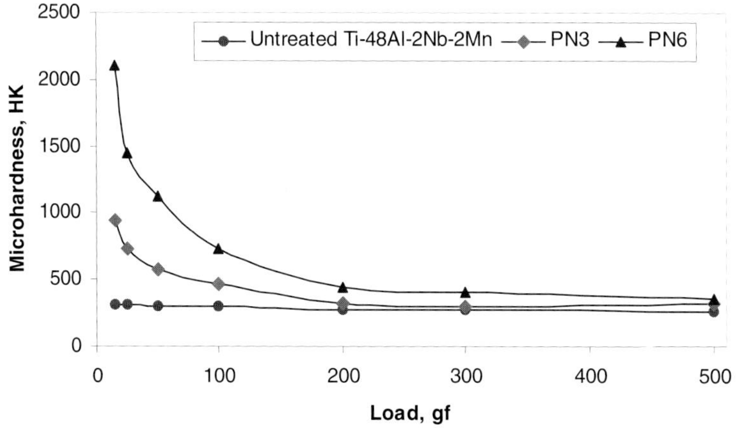

b)

Figure 5. Microhardness values on the surface of untreated and plasma nitrided Ti-48Al-2Nb-2Mn titanium aluminides using (a) Vickers indenter, and (b) Knoop indenter.

4.3.2 Tribological Behavior of Plasma Nitrided Ti-48Al-2Nb-2Mn Alloy

(a) Friction Behavior

Fig. 6 shows typical curves of coefficient of friction vs. sliding distance, which demonstrate the influence of, temperature and gas composition on the friction behavior of plasma nitrided Ti-48Al-2Nb-2Mn titanium aluminide. The untreated Ti-48Al-2Nb-2Mn alloy has a highly varying coefficient of friction (Fig. 6a), which is reduced by plasma nitriding to a lower mean value and variation, depending on the treatment conditions. From Fig. 6b and 6c, it is seen that after 100 h treatment at 700 C in 20 H_2/80 N_2 gas ratio results in a less varying coefficient of friction than that of the treatment at 800 C in 80 H_2/20 N_2 gas ratio. These changes show the effect of temperature and gas mixture on the coefficient of friction. The higher temperature and higher hydrogen ratio results in more variation in the coefficient of friction. However the mean values of the coefficient of friction at all conditions is less than that of the untreated material. These higher variations are directly correlated to the higher roughness of the sample plasma nitrided at higher temperature and higher hydrogen ratio.

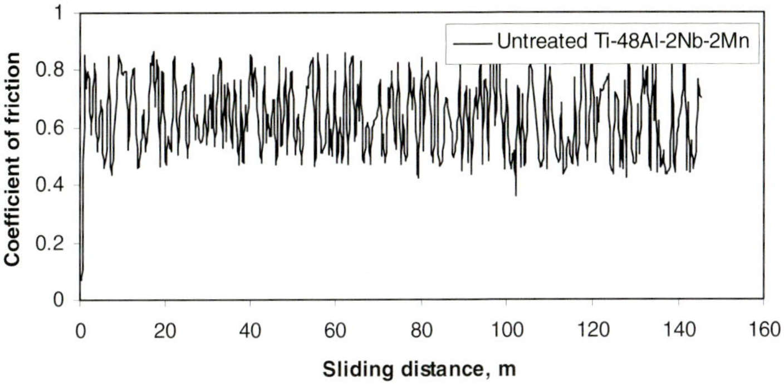

a)

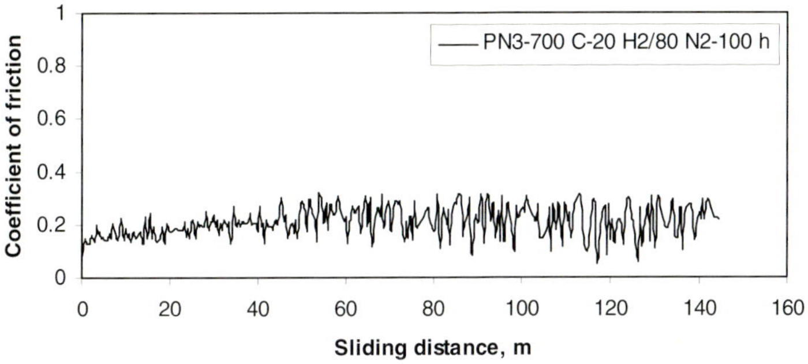

b)

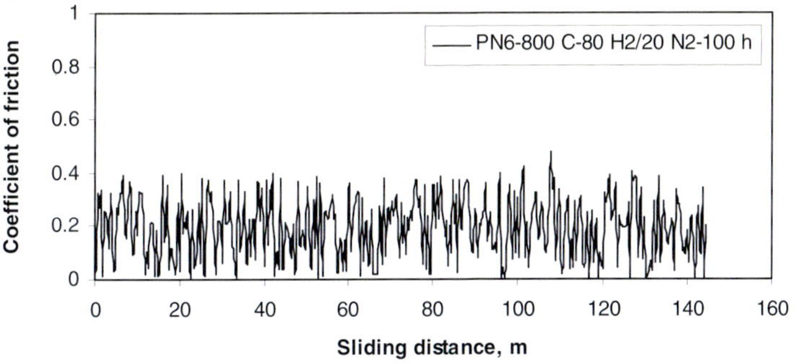

c)

Figure 6. Coefficient of friction vs. sliding distance curves produced under 5 N and 0.1 m/s, on (a) untreated Ti-48Al-2Nb-2Mn alloy and the samples plasma nitrided for 100 h (b) at 700 C in 20 H_2/80 N_2 and (c) at 800 C in 80 H_2/20 N_2 gas ratio

(b) Wear Loss in Dry Sliding Wear

The wear volume loss of untreated and Plasma nitrided samples are shown in Fig 7. Under loads up to 10 N load (maximum normal contact stress 1139 MPa) [13] and sliding speeds of 0.1 m/s all coatings had a very high wear resistance in comparison with the untreated material. However the maximum normal contact stress of 746 MPa which is produced under 5 N load was higher than the yield strength of TiAl alloys [2]. As a measure of load bearing capacity of the treated surfaces, most nitrided surfaces were wear tested under 20 N load (maximum normal contact stress 1436 MPa) and sliding speed of 0.3 m/s. After 1080 m sliding distance all samples were worn nearly as much as the untreated material except two PN3 and PN6 samples. The volume loss measured from sample PN6 was nearly 10 times less than that of the untreated materials. The wear resistance of plasma nitrided surfaces depends on the thickness and crystal structure of the compound layers, which were revealed through high resolution SEM (FESEM) examinations of the worn surfaces.

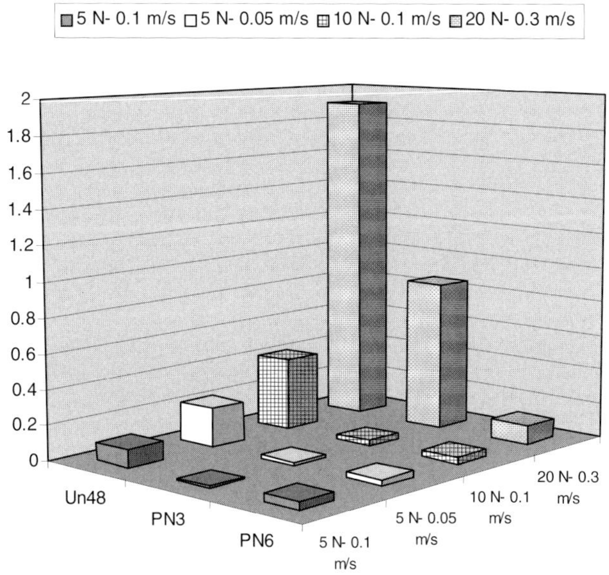

Figure 7. Wear volume loss of untreated Ti-48Al-2Nb-2Mn alloy and two plasma nitrided samples under different loads and speeds.

Microscopic examinations on the untreated surface showed mainly plastic deformation and plowing (Fig. 8a). It was found that the lamellae of Ti-48Al-2Nb-2Mn are deformed or slightly bent under sliding and hence they fracture due to the intrinsic brittleness of the intermetallic lamellae. It is believed that the lamellae are fractured through interlamellar boundaries and translamellar slip lines [14] (Fig. 8b) which are extruded step-like in the areas adjacent to the wear tracks. The separated agglomerates of debris are fragmented under the subsequent abrasive action of the sliding parts into flake-like debris (Fig. 9). The mechanism of deformation and subsequent fracture of lamellae have been explained in another work [15]. The accumulated flake-like debris on the untreated material adhered strongly on the wear tracks, leading to the cohesive failure of the substrate rather than the failure of the junctions formed through the adhesion of the surfaces. This resulted in the considerable wear loss of the material. EDX analysis of these debris revealed a low amount of oxygen under most loads, which indicated a dominantly adhesive metallic wear of the counterparts rather than the oxidative wear.

a) b)

Figure 8. Images showing the (a) Optical micrograph of the wear track produced on untreated Ti-48Al-2Nb-2Mn alloy under 10 N load and 0.1 m/s sliding speed after 180 m and (b) SEM micrograph of the area adjacent to the wear track produced on untreated Ti-48Al-2Nb-2Mn under 20 N load and 0.3 m/s sliding speed after 1080 m sliding distance

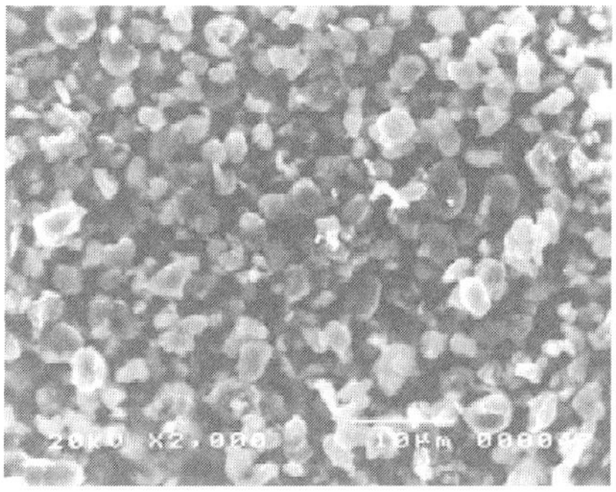

Figure 9. SEM micrograph showing the size and shape distribution of the wear debris obtained from untreated Ti-48Al-2Nb-2Mn alloy worn under 10 N load for 360 m sliding distance.

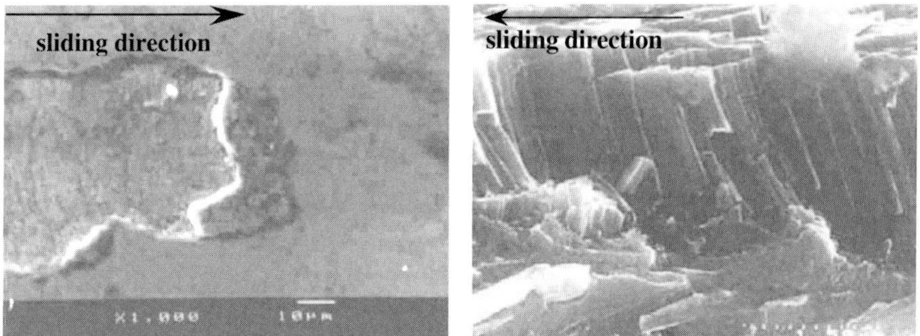

a) b)

Figure 10. SEM micrographs of the wear tracks produced on sample PN3 under (a) 10 N load after 180 m and (b) 20 N load after 1080 m sliding distance.

The worn surfaces of two samples PN3 and PN6 are typically discussed to clarify the role of the compound layer structure in the wear performance of plasma nitrided Ti-48Al-2Nb-2Mn alloy. Fig. 10a represents the spalled (delaminated) areas of sample PN3, worn under 10 N load. Examination on the fracture cross section of the area delaminated under 20 N load (Fig. 10b) did not show bending or shear deformation of the columns [16] but their detachment from the substrate resulting in total failure of the compound layer. This microstructural response to wear is correlated to the weak bonding of the columns to the substrate due to the lower treatment temperature which has not been capable to produce a significant diffusion layer and adhesion. Furthermore, the short columns have transferred the tractional force due to sliding through the hard and tough columns to the weak interface rather than the substrate or within the columns, resulting in the delamination of the compound layer.

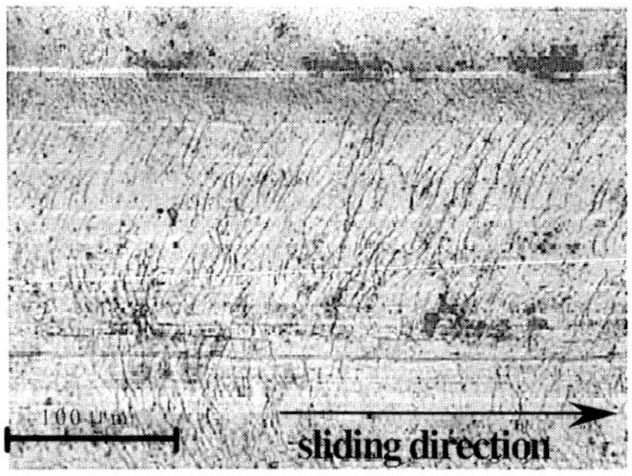

Figure 11. SEM micrograph showing a typical semicircular crack pattern produced on sample PN6 after sliding under 5N load and 0.1 m/s sliding speed for 180 m sliding distance

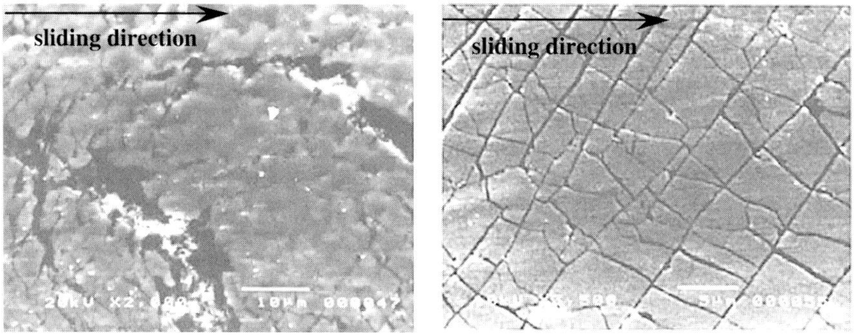

a) b)

Figure 12. Wear tracks of sample PN6 produced under (a) 10 N and 0.1 m/s and (b) 20 N load and 0.3 m/s sliding speed.

The failure of the worn surface of sample PN6 occurred through semicircular cracks appearing along the wear track which was geometrically similar to the trailing end curvature of the sliding ball (Fig. 11), indicating the tensile failure of the compound layer [17, 18]. At 10 N load level and 0.1 m/s sliding speed, the cracks were associated with a slight shear deformation of the compound layer (Fig. 12a), implying the more plastic behavior of the compound layer than that of sample PN3. Under 20 N load and 0.3 m/s sliding speed, the crack pattern was also semicircular but contained more cracks in the form of side crack branches (Fig. 12b). However sample PN6 cracked under all loading conditions, it is believed that the strong bonding of the compound layer to the substrate and its fine-grained structure were responsible for its resistance to the total failure or spallation of the compound layer as occurred with sample PN3. Due to the formation of the nearly equiaxed diffusion layer in sample PN6 rather than a lamellar substrate, it is suggested that there is more gradation in properties between the compound layer and the substrate through the presence of the diffusion layer. The fine-grained compound layer is more wear resistant and the low wear rate occurred through removal of the fine grains from the layer as opposed to total detachment of the compound layer. Furthermore any inter-granular crack initiating within the coating or at the boundaries of different layers will have a longer, tortuous path through the fine grains in order to cause complete compound layer failure. The crack patterns on sample PN6 are evidences of this speculation, since they are produced under the same sliding conditions as sample PN3 without causing the complete removal of the compound layer.

5 Conclusions

1) The friction behavior of the untreated titanium aluminides is characterized by a highly fluctuating and a high mean value of friction coefficient. Adhesive wear of the oxidation resistant lamellae play an important role in the typical stick slip behavior of this material.
2) During the sliding wear of the untreated Ti-48Al-2Nb-2Mn titanium aluminide, the cracks are initiated at the interlamellar boundaries and extended into the gamma lamellae through slip and slip lines, resulting in the cleavage fracture of the lamellae. The fractured lamellae are further fragmented under the abrasive action of the sliding parts into fine-grained debris.
3) Plasma nitriding of lamellar Ti-48Al-2Nb-2Mn titanium aluminide results in a thin compound layer composed of TiN and Ti_2AlN and a diffusion zone composed of nearly equiaxed grains of recrystallised lamellae. The diffusion layer was observed at 800 C.
4) plasma nitriding of Ti-48Al-2Nb-2Mn results in the reduction of the level of fluctuation and mean value of friction coefficient
5) Plasma nitriding of Ti-48Al-2Nb-2Mn results in a high wear resistant compound layer which reduces the wear loss of the alloy up to one order of magnitude.

6 References

1. Kim Y W (1989), " Intermetallic Alloys Based on Gamma Titanium Aluminide", JOM, pp. 24-30.
2. Kim Y W and Froes F H (1990) “Physical Metallurgy of Titanium Aluminides”, High Temperature Aluminides and Intermetallics, ed. S. H. Whang, C. T. Liu, D. 112 Minerals, Metals & Materials society, pp. 465-492.
3. Kim Y W (1991), " Microstructural Evolution and Mechanical Properties in Gamma Titanium Aluminides ", Microstructure / Property Relationships in Titanium aluminides and Alloys, ed. Y. W. Kim and R. R. Boyer, The Minerals, Metals & Materials Society, pp. 91- 103.
4. Kim Y. W. and Dimiduk D. M., 1991, "Progress in the Understanding of gamma Titanium Aluminides", JOM, Vol. 43, No. 8, August , pp. 40-47.

5. Hurta S, Climens H, Frommeyer G, Nicolai H P, Sibum H (1995), "Valves of Intermetallics γ-TiAl Base alloys: Processing and Properties", Titanium' 95: Science and Technology (Proceedings of the Eighth World Conference on Titanium), pp. 97-104.
6. Jones P. E., Porter W. J. III, Keller M. M. and Eylon D (1995), "Microstructure/property studies supporting development of low cost processes for TiAl automotive valves", Mat. Res.Soc. Symp. Proc., Vol. 364, pp. 769-774.
7. Dowling W E, Donlon W T and Alison E J (1995), "Development of TiAl-Based Automotive Engine Valves", Mat. Res.Soc. Symp. Proc. , Vol. 364, pp. 757-767.
8. A. Bloyce (1994), "Environmental Sensitivity of Surface Engineered Titanium Aluminides": EssenTiAl University Network, Meeting report, .
9. J. Lanagan (1989), "Plasma Surface Engineering of Titanium Alloys", Ph.D. Thesis, The University of Birmingham.
10. Yuki I, Amano N, Uozumi M, Inui H and Yamaguchi M (1994), " Effect of Nitrogen Addition on the Microstructure and Mechanical Properties of Cast γ Titanium Aluminide Alloys ", J. Japan Inst. Metals, Vol. 58, No. 5, pp. 564-570.
11. Lee DS, Stucke MA and Dimiduk (1995) "Effects of Thermal Exposure on the Properties of Two Phase gamma Alloys", Material Science and Engineering, A192/193, pp. 824-829.
12. B. Edenhofer (1974), " Physical and Metallurgical Aspects of Ionitriding (part 1)", Heat Treatment of Metals, No. 1, pp. 23-28
13. Hutchings I M (1992), " Tribology: Friction and Wear of Engineering Materials", Edward Arnold, London, pp. 4, 46, 47.
14. Umakoshi Y, Yasuda H Y and Nakano T (1996). " Plastic Anisotropy and Fatigue of TiAl PST crystals: a review ", Intermetallics, 4, pp. S65-S75.
15. A. R. Rastkar, " Surface Engineering of Gamma Based Titanium Aluminides to combat wear", Ph.D. Thesis, The University of Birmingham.
16. K. J. Ma , A. Bloyce and T. Bell (1995), " Examination of Mechanical properties and Failure Mechanisms of TiN and Ti-TiN multilayer coatings", Surf. Coat. Technol., Vol. 76-77, pp. 297-302.
17. Holmberg K (1992),"A concept of friction mechanisms of coated surfaces", Surface and Coating Technology, 56, pp. 1-10.
18. Burnett P J and Rickerby D S (1988), "The Relationship Between Hardness and Scratch Adhesion ", Thin Solid Films, 154, 7, pp. 403-416.

Assessment and Optimization of Tribological Systems

Wulf Pfeiffer, Bernhard Blug, Martin Zimmermann[1]
Fraunhofer IWM, Freiburg;

1 Introduction

Tribological systems in components are characterized by a complex interaction of mechanical, thermal and chemical use. The assessment of the tribological situation and the optimization of component performance needs the identification of the main wear effects and the improvement of the related near-surface characteristics of the components.

The wide field of different applications can roughly be divided into three main classes with respect to the lubrication conditions: systems operating under boundary lubrication, under mixed lubrication or under hydrodynamic lubrication. The main wear effects in components are closely related to these lubrication conditions.

This paper deals with efficient experimental and numerical methods used for the optimization of tribological systems with ceramic components. Different tribological systems are examined and strategies for the optimization of the component performance are presented.

2 Sliding contact under mixed friction conditions

The first tribological system presented is related to mechanical face seals, operating in the mixed friction mode.

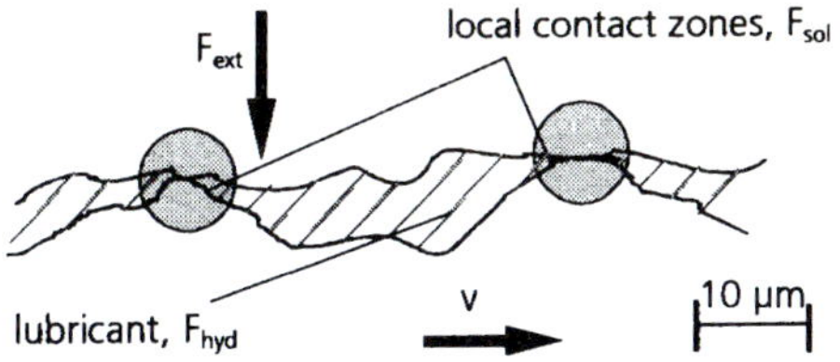

Fig. 1: Real rough surfaces under mixed lubrication

When the load in sliding systems is carried by an elasto-hydrodynamic lubricating film, the rate of wear is negligible. When this state is disturbed, for example by a sudden impact, or in the starting or stopping process, or when elasto-hydrodynamic lubrication cannot occur because the load is too high or the velocity is too low, the surfaces get into contact, a situation which is called

---

[1] now at Schott Glas, Mainz, Germany

mixed lubrication (Fig. 1). These contacts cause very high local stress concentrations and as a consequence high friction and wear of the surfaces [1]

The optimization of such systems can be performed by a semi-numerical model which consists of two basic steps:

- the calculation of the load and the deformation due to solid-solid contacts,
- the calculation of the internal pressure in the lubricant due to micro-hydrodynamic effects.

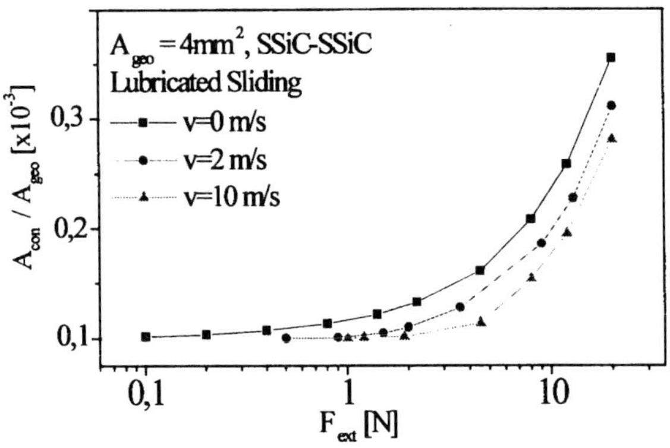

Fig. 2. Ratio of real to geometric contact area of lapped SSiC-surfaces for different sliding speeds .

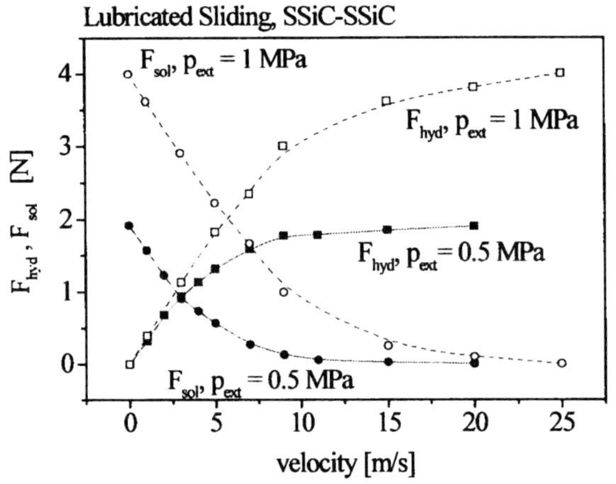

Fig. 3. Load carried by solid contacts and hydrodynamic effects for different external pressures

Input parameters are the complete data of the measured surface roughness without statistical simplifications, the material parameters of the contacting bodies and the lubricant and the parameters of the contact situation (load, velocity). The procedure is described in detail elsewhere [2]. The following investigations is focused on ceramic materials (SSiC) which have a high potential for sliding system applications. The surfaces are lapped, the lubrication is a mineral oil with a viscosity η=1 Pa s.

As a first result of the calculations Fig. 2 shows the ratio of the real contact area to the geometric contact area as a function of the external force F_{ext}. The results are plotted for three different sliding speeds. The curve for v = 2 m/s starts at an external force F_{ext} = 0.5 N (v = 10 m/s at F_{ext} = 1 N) because the contact for lower forces is completely hydrodynamic which means that no contact of the summits takes place.

In Fig. 3 the loads carried by solid-solid contacts (F_{sol}) and hydrodynamic effects (F_{hyd}) are calculated as a function of the sliding speed for two different external pressures. The sum of both forces is equal to the external force.

The calculations show that in a lubricated sliding contact the real contact area is very small compared to the geometric contact area of a component. For ceramic materials it will not exceed 1/1000 of the geometric contact area. For a constant sliding speed the real contact area is approximately linearly proportional to the applied load. The load carrying micro-hydrodynamic effects is enhanced when the sliding speed and the external pressure on the lubricant increases.

3 Sliding wear under boundary friction condition

In systems operating under boundary friction conditions (like water-lubricated face-seals during start/stop operations) micro-hydrodynamic effects are negligible and the wear behavior will be related to the surface integrity of the components. This is illustrated by Fig. 4 , where the depth distributions of residual stresses (determined by XRD), the near surface strength (determined by ball-on-plate tests), and the material removal due to sliding wear (2 MPa normal pressure, water lubricated) of differently machined sintered silicon carbide samples are shown (details of the experimental procedure are given in [3]).

Although rough machining conditions like lapping with coarse grained abrasives may introduce more damage than polishing processes, the near-surface strength is increased due to the additionally machining induced high and deep-reaching compressive residual stresses. Obviously, these compressive residual stresses are able to reduce the wear rate after the roughness peaks have been removed. The polished samples show very high wear rates at early stages of wear due to adhesive failure followed by a continuous material removal.

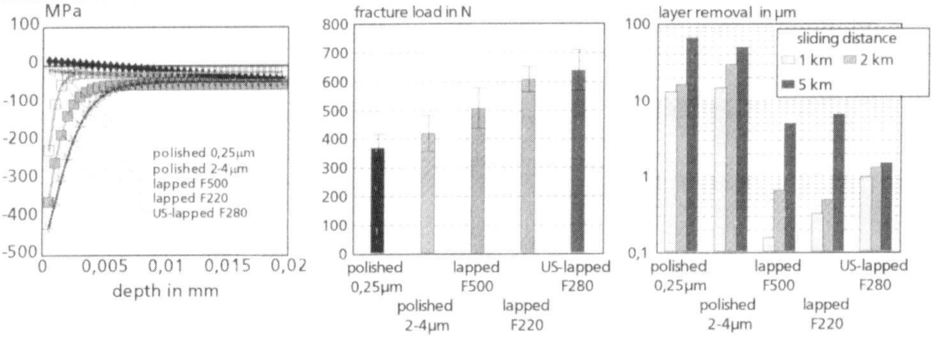

Fig. 4. Machining induced residual stresses, near–surface strength and sliding wear behavior of differently machined silicon carbide samples.

4 Rolling wear without lubrication

This tribological system is related to full or hybrid ceramic roller bearings designed to operate without lubrication, e.g., in vacuum applications. Theoretical and experimental investigations show that the failure behavior of these bearings is controlled by tensile stresses near the border of the contact zone [1, 4]. Thus, the load capacity of a specific ceramic material may be increased by compressive residual stresses introduced by machining [3], or, more effectively, by a newly developed shot-peening process [5] or by surface coatings which reduce the friction coefficient.

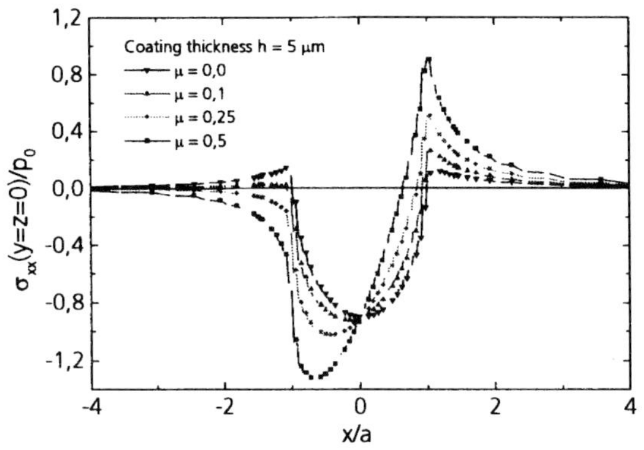

Fig. 5. The σ_{xx} stresses at the surface with different friction coefficients μ under a sliding load

Anti - friction surface coatings combine a low wear rate with a low friction coefficient which enables them to cope with a wide range of wear situations. But the use of low friction coatings as a design element in tribological systems requires an accurate knowledge of the wear and failure mechanisms of the coating and the substrate under different loading and environmental conditions. The properties of the coating and the substrate can then be optimized with respect to the needs of the desired application. The process of optimization is started by the numerical analysis of the loading conditions and the resulting stress distributions. Then, the best fitting mechanical parameters of the coating and the

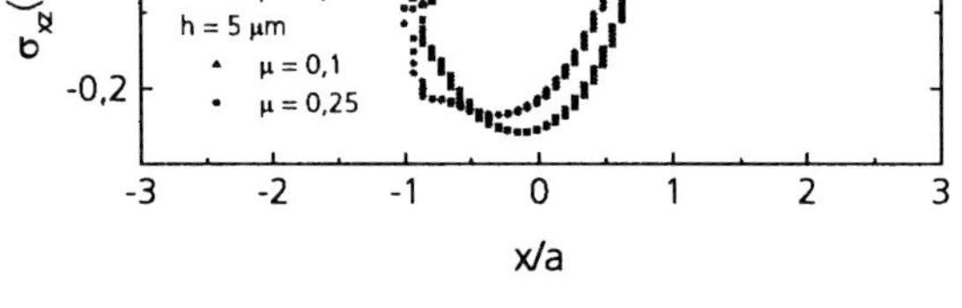

Fig. 6. Shear stresses σ_{xz} at the coating - substrate interface under a sliding load

substrate are calculated in an iterative process (such as coating thickness, friction coefficient, elastic and plastic properties, hardness, residual stresses, bonding strength of the coating, grading) and an optimal coating can be chosen from a large variety of low friction coatings. It is often quiet difficult to gain information on the failure properties of commercial coatings.

Another possibility is to develop a new coating which has nearly the optimal mechanical properties calculated before. Basis for such an optimization must be a type of coating where the mechanical properties can be altered within a wide range. Diamond – like carbon (DLC) coatings combine the requirements of a low friction coatings with the possibility to vary the mechanical properties with the help of different dopants [7].

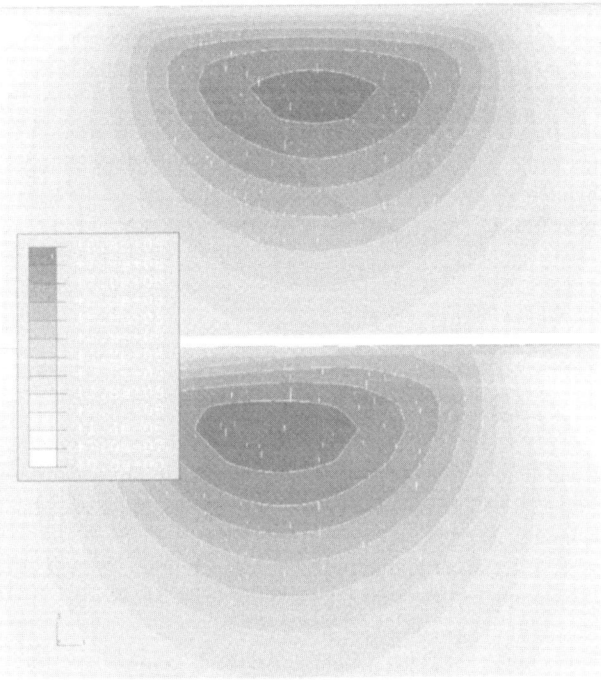

Fig. 7. Von Mises stresses under rolling load with different slide- roll – ratios (SRR = 0,75 %, 2,5 %)

Hydrogenated amorphous DLC coatings (a - C:H) are used in high speed hybrid roller bearings to enable an usage without lubricants. Fig. 5 – 7 show a parameter study within a 3 dimensional finite element simulation of the a – C:H coating in order to find the optimal coating properties. The stress distribution in the coating and the substrate under a rolling (including slip) and a sliding load are calculated. The stresses depend strongly on the friction coefficient (Fig. 5, sliding). A higher coefficient of friction results in higher stresses at the surface, but not inevitably at the coating interface. The shearing stresses, which are important for the delaminating of the coating, can be reduced by choosing a proper coating thickness (Fig. 6, sliding). Under high loading plastic deformation of the substrate is another failure mechanism. Fig 7 shows the von Mises stresses in the coating and the substrate under different slide - roll - ratios (SRR). The maximum of the von Mises stresses moves with higher slide roll ratios to the trailing edge of the contact zone and simultaneously nearer to the surface.

5 Conclusions

The compilation of tribological systems here presented show, that the characterization and assessment of such systems based on the main failure and wear mechanisms in most cases allow for a straight-forward optimization of the service behavior of the components.

The life time of components operating with sliding contact under mixed friction conditions is increased at most by all those parameters, which enhance micro-hydrodynamic effects. A consequent design of the topography and an increased pressure on the lubricant may be most effective.

If systems operate at boundary friction conditions or completely dry, the function of the missing lubricant may be (partially) replaced by friction reducing surface coatings and/or by improving the surface integrity by compressive residual stresses. Generating compressive residual during fabrication is easier, but the low friction coatings have a wider range of application.

6 References

[1] M. Rombach, Experimental and Theoretical Investigations on Plastic Deformation and Brittle Cracking in a Ball-on-Plate Contact of Ceramic Materials, in: Ceramics Charting the Future (Edit. P. Vincenzini, Techna Srl) 1995, 1055-1064.

[2] M. Zimmermann, M. Rombach, Microhydrodynamic Effects in Sliding Contacts of Real Surfaces: A Numerical Approach, in Austrib 98 – Tribology at Work, The Institution of Engineers Australia, 1998, 201 – 206.

[3] W. Pfeiffer, T. Hollstein, B. Schunk, R. Telle, Charakterisierung bearbeiteter und tribologisch beanspruchter Randschichten von Hochleistungskeramiken anhand werkstoffmechanischer Kenngrößen. In: Werkstoffe für die Fertigungstechnik / Werkstoffwoche 1998, Symposium 5. Wiley-VCH Verlag GmbH, Weinheim, Germany, 181-186

[4] T. Hollstein, M. Rombach, W. Pfeiffer, M. Popp, Vollkeramische Wälzlager aus Siliciumnitrid: Anwendung, Auslegung und Optimierung, VDI Berichte 1995, 1151, 3-11.

[5] W. Pfeiffer, M. Rombach, Residual Stresses and Damage in Ceramics Due to Contact Loading. Proc. of the Fifth Int. Conf. on Resid. Stresses, Linköping Univ., Sweden, 1998, 1, 302-307.

[6] Cheng, H.S., »Lubrication Regimes«, in ASM Handbook, Vol. 18, Friction, Lubrication and Wear Technology, 1995, 90-97.

[7] K. Holmberg, A. Matthews, Coatings Tribology, 2nd ed., Elsevier, 1998, 213 - 236

Hybrid- and Ceramic Rolling Bearings with Modified Surface and Low Friction Rolling Contact

M. Popp, R. Sternagel
CEROBEAR GmbH, Herzogenrath (D)
W. Pfeiffer, B. Blug, S. Meier
Fraunhofer-IWM, Freiburg (D)
G. Wötting, L. Frasseck
CFI, Rödental (D)

1 Abstract

Conventional steel bearings often fail under extreme strains. All ceramic bearings and bearings with ceramic rolling bodies, the so called hybrid ceramic bearings, provide a good alternative for such applications. However, important fields of application of the hybrid- and all ceramic bearings can be accessed only, if long lasting load capacity and easy motion of dry running bearings can be ensured even under increased loads. A main objective is to reduce friction within the rolling contact and to significantly enhance the rolling strength particularly in unlubricated contact through modification of the working surfaces of the bearing rings and roll bodies. The approaches are: optimised DLC coatings on steel for hybrid bearings and the incorporating of friction minimising additions of BN and TiN into the Silicon Nitride bulk material for hybrid and ceramic bearings. Coatings yielded good results in lab scale while the friction reducing Si3N4-composites already showed their potential to significantly reduce the rolling friction in underlubricated bearings. A new method of surface treatment of Si3N4 was developed based on a shot peening process. Under optimized conditions load capacity of a ball on plate contact can be increased remarkable. Additionally, in an innovative approach the project focuses on intelligent adaptation of the rolling contact macrogeometry of ball bearings. This development, intended to allow a load related control of the bearing characteristics, now is in the status of numeric and practical evaluation. The work described is funded by BMBF within the MaTech programme (Projects 03N2002 and 03N2018).

2 Reduction of Friction

The reduction of friction is one of the keys for a successful application of hybrid or full ceramic roller bearings under dry or nearly dry environmental conditions [1]. In case of hybrid bearings diamond-like thin film coatings deposited on the raceway seem to be an appropriate method for improvement of the performance. For full ceramics roller bearings, newly developed silicon nitride materials with self-lubricating phases promise to be the successful way. The main goals have been defined to be: a significant reduction of the friction coefficient (<< 0.6) and a load capacity of more than 1 GPa, both under dry conditions.

2.0.1 Coatings

The optimization of diamond-like coatings (doped a-C:H) deposited by a PA-CVD-process to steel raceways needs three steps: (a) optimization of the coating-substrate interface, (b) reduction of the friction coefficient, (c) improvement of life time.

The results of scratch tests (figure 1) show that an improved control of the residual stress state of the coating results in a significant gain of adhesion. Once the adhesion between substrate and coating is optimized, the friction coefficient and the thickness of the coating are the most relevant characteristics determining the performance of the hybrid bearing. An important detail in the subsequent optimization of the coating techniques is the configuration of the standard pin-on plate-test with the relevant pin material. Although for different ceramic materials the friction between ball and coated substrate could be dropped well below a factor of 0.25, it turnes out, that silicion nitride reveals a slightly higher friction and shorter lifetime.

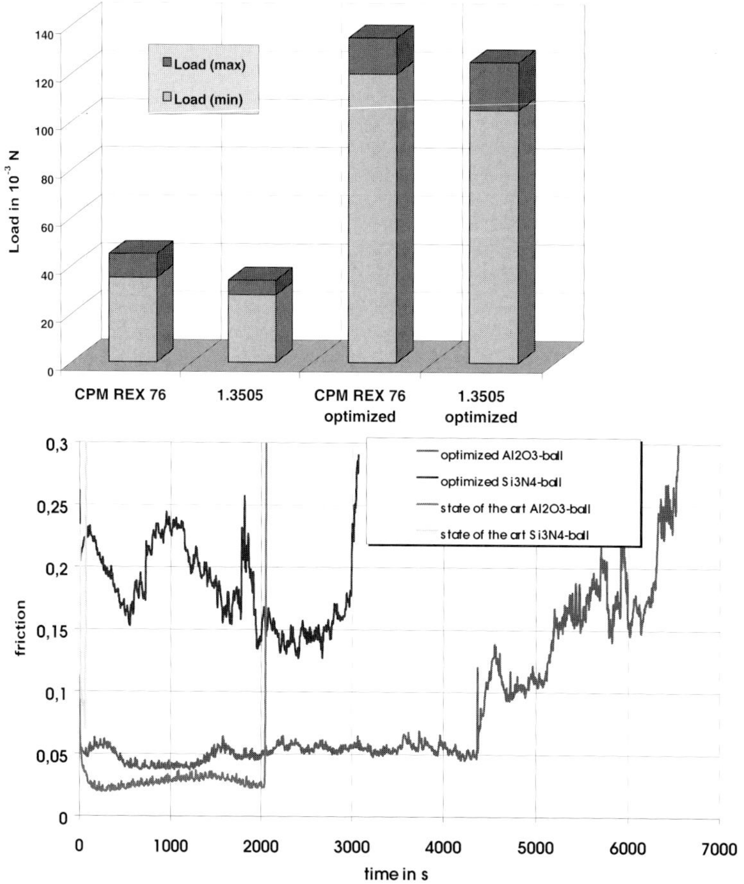

Figure 1: upper: Optimization of the adhesion of diamond-like coatings on different steel substrates by control of the residual stress states (micro-indenter scratch tests, 10 μm indenter tip radius). Lower: The optimization of the coefficient of friction and the life time of coatings have to be related to the ball material (oscillating pin-on-isk sliding wear test, ball diameter = 10 mm, initial normal pressure=1.4 Gpa, frequency = 20Hz, amplitude = 1mm, dry)

2.0.2 Self-Lubricating SN-Based Composites

Besides the coating concept, the introduction of self-lubricating phases into the basic silicon nitride material is a second approach to reduce friction under dry-running conditions. According to literature, primarily BN and TiN are recommended for this aim. Bearing applications, however, require a certain level of load capacity of the bearing components. Thus, the introduction of these phases may not weaken the basic material markedly. As it is well-known that the introduction of crystalline BN-powder into silicon nitride is associated with processing problems and loss of strength and hardness, within this project the synthesis of BN during processing was developed using a SiBN-precursor as BN-source. This precursor can be distributed very homogeneously within the basic composition during its preparation and decomposes to BN and Si_3N_4 during thermal consolidation. Varying the concentration of the added precursor resulted in an essential reduction of mechanical properties with about 5 vol% BN and more in comparison with the BN-free reference material. With half of this concentration, bending strength and hardness are only reduced by about 10 – 15 % with respect to the reference material, qualifying such a composition for bearing applications. The microstructure is characterized by finely and homogenously distributed BN-plates between the silicon nitride grains.

There resulted problems to machine this material to bearing balls with a quality grade G10. With these balls as well as bearing races out of this material, friction tests were performed under dry-running conditions. Results demonstrate that this material has a significantly reduced the rolling friction coefficient compared to the reference SN under these conditions. As the load-carrying capacity of bearings is exponentially influenced by the friction coefficient, its reduction allows an increase of load of about 60 % in comparison with the reference. This demonstrates that also this development is successful to reduce friction under dry-running conditions, opening up a variety of new applications and designs.

2.1 Bearing Performance with new Silicon Nitride Materials

Test bearings have been manufactured from four different materials as given in table 1. The main interest focussed on the performance in dry run and the reduction of the bearing friction torque at dry run.

Table 1: New ceramic bearing materials tested in all ceramic bearings.

Material	Si3N4-MgO	α' / β ' SiAlON	Si3N4-TiN (0,95%) composite	Si3N4-BN (2%) composite
Development Objective	advanced corrosion resistance in acid liquids	very high level of hardness	reduced friction in rolling contact	reduced friction in rolling contact

The experimental approach for the investigation of bearing friction was a test set-up in which an axial spring loaded angular contact bearing couple of 7006 E Type (30x55x13mm, 25° contact angle) is mounted. Tests were performed at ambient temperature with minimal grease lubrication, with water and in dry run, respectively. Tests performed with reference bearings made of a modified standard Silicon Nitride (Si3N4-MgO, CFI No. N3209) showed the completely different bearing friction at various lubrication levels. Obviously the bearing friction of the tested angular contact bearings is closely associated with the friction coefficient between balls and rings in the rolling contact as figure 2 demonstrates.

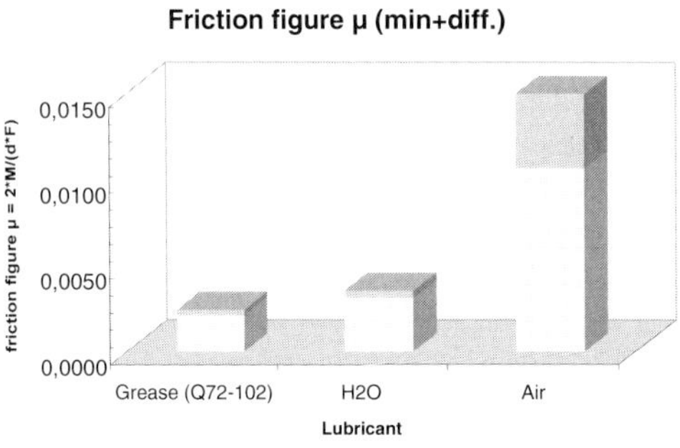

Figure 2: Normalized friction figures of an angular contact bearing 7006 E at operation with varied lubrication condition. Load 500 N axial, p_0 = 1.500 MPa.

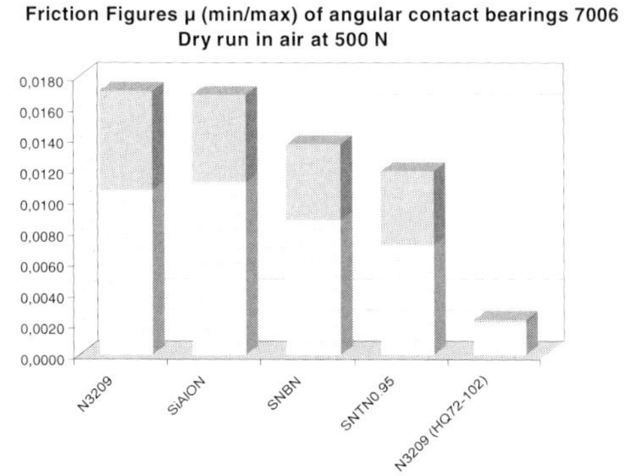

Figure 3: Normalized friction figures of angular contact bearings 7006 E. Material: new developed Si3N4-basesd composites with improved friction behaviour. Dry run, 500 N axial load, contact pressure p_0 = 1.500 MPa.

Dry run in air even with ceramic bearings result in a 5-8 times higher bearing friction as measured for proper lubrication. When operated in water, the friction torque is a function of bearing speed and load. At low speeds almost dry-run friction can be observed whereas with increasing speed a lubricant film is established and friction drops down to values similar to a full lubricated operation. In a test series carried out with all ceramic angular contact bearings the performance of the four developmental ceramics from table 1 was assessed. Bearing friction figures for dry-run in air and in water at room temperature are given in figure 3. Both friction reduced Si3N4-base composites exhibit a reduction of the overall friction around 20% (Si3N4-BN) and 35%(Si3N4-TiN) compared to the reference. Although these values do not reach that of lubricated service, this is already a remarkable improvement. Tests in water again demonstrated the improved rolling contact friction of both materials, that had a friction behaviour (level and ripple) like grease lubricated reference bearings. Here the improvement was around 35% against the reference. Thus, the approach of minimising the contact friction by way of self lubricating phases in a Si3N4 material provides a good potential for ceramic bearing applications in dry running or underlubricated hybrid bearings.

2.2 Improvement of surface integrity

Residual stresses due to mechanical loading have been thought to be no significant effect in brittle materials like ceramics. Recent contact loading tests (ball-on plate tests) show that it is possible to introduce macroscopic plastic deformation without creating damage, if the diameter of the indenter sphere is below a certain critical value [2]. The aim of the investigations was therefore to develop a method for the improvement of the near-surface load capacity of ceramics based on a shot peening process [3]. In figure 4 the load capacities determined in the ball-on-plate test point out that a significant gain of strength (up to 50 %) with respect to the polished reference samples can be established by the developed shot peening process.

The increase of the load capacity depends mainly on the peening time for a treatment with fine (75 µm - 125 µm) ZrO_2-beads and on the peening pressure for the coarse (250 µm - 425 µm) beads. The highest load capacity is found for the long time shot peening at high pressure using coarse grained beads.

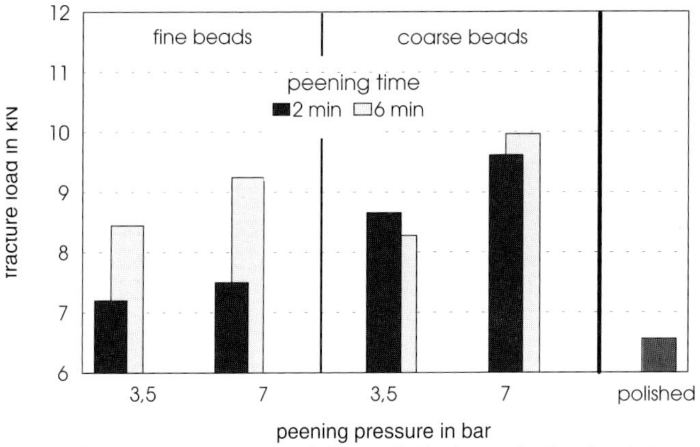

Figure 4: Gain of load capacity of silicon nitride due to the developed shot peening process (determined in the ball on plate test with a 11.11 mm silicon nitride ball).

2.3 Optimization of contact geometry

A quite novel strategy to optimize the performance of hybrid roller bearings is the on-line control of raceway geometry by piezo actor and sensor elements integrated into the bearing. The aim of this challenging project is to enable a bearing to a) run with a loose Osculation in case of a moderate loading conditions resulting in less friction, or b) to increase load capacity by a close Osculation in case of high loading. The results of a first FEM-based design-study on the principal possibilities of a geometry-controlled angular contact ball bearing (Figure 5) show the change in raceway geometry due to a force applied by an piezo-ceramic actor element.

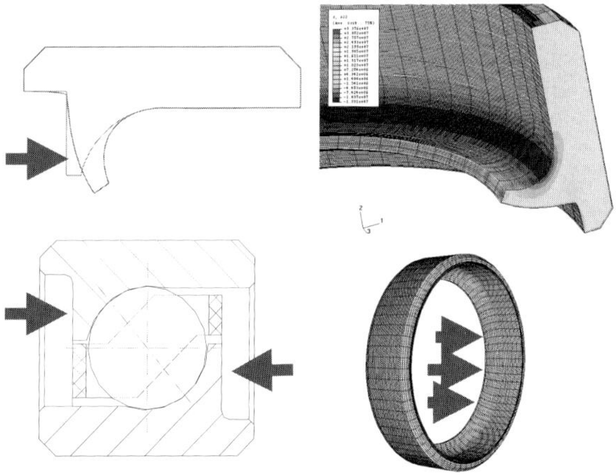

Figure 5: General principle and FEM-based design study of a roller bearing with piezo-actor controlled raceway geometry.

3 Summary and Examples of application

One target application of the above mentioned developments are lubricant free screw and turbo compressors for energy supply, for clean semiconductor production processes and in food industry. Reduced rolling contact friction is a key for realizing lubricant free operation. Another essential fact is that only ceramic or corrosion resistant bearing steels can be utilized in unlubricated bearings. The developments for optimised rolling contact, DLC coating and friction reduced Si3N4-components, both showed a good potential for applications with poor lubrication conditions like hermetic sealed chemical pumps, agitators, compressors, vacuum pumps and fans for process gases. The improvement of surface strength and surface integrity through shot peening can be considered for Si3N4 bearing components with strong requirements of reliability. The shot peening treatment is expected to reduce sensitiveness of surface defects due to the induced compressive stresses near surface and the macroscopic plastic deformation. This opens up new opportunities i.e. for bearing components with a high stress on reliability i.e. for aircraft engines or space applications. Examples for the potential of an intelligent bearing with load adapted friction and stiffness are HSC-spindles. Large cutting tools with moderate shaft speed require in first place high stiffness and load capacity while small tools typically need bearings that are capable for running at high speeds and moderate loads. The variation of contact geometry here can provide a very smart solution.

4 References

[1] T. Hollstein, M. Rombach, W. Pfeiffer, M. Popp: Vollkeramische Wälzlager aus Siliciumnitrid: Anwendung, Auslegung und Optimierung. VDI Berichte 1995, 1151, 3-11.

[2] M. Rombach: Experimental and theoretical investigations on plastic deformation and brittle cracking in a ball-on-plate contact of ceramic materials. In Ceramics Charting the Future (Edit. P. Vincenzini, Techna Srl) 1995, 1055-1064.

[3] W. Pfeiffer, M. Rombach: Residual stresses and damage in ceramics due to contact loading. Proc. of the Fifth Int. Conf. on Resid. Stresses, Linköping Univ., Sweden, 1998, 1, 302-307.

Development of C/C-SiC Materials for Friction Applications

B. Heidenreich, W. Krenkel
German Aerospace Center (DLR)
Institute of Structures and Design
Pfaffenwaldring 38 - 40
70569 Stuttgart

1 Introduction

New C/C-SiC materials, developed for the thermal protection of spacecraft at the DLR, Stuttgart, have demonstrated their high potential for the application in brake discs [1]. The aim of this work was to investigate the tribological behaviour of the C/C-SiC composite material and its individual main components, carbon fibre reinforced carbon (C/C) and silicon carbide (SiC), and the influence of the materials thermal conductivity, modified by graphite components. This led to significantly improved, new C/C-SiC friction materials with high wear resistance and a high coefficient of friction (COF) even under high energy test conditions.

The fibre reinforced ceramic materials are produced by a short three-step process via Liquid Silicon Infiltration (LSI-process) [2]. A conventional CFRP (carbon fibre reinforced polymer) green body is pyrolysed at 900°C in nitrogen atmosphere. The resulting C/C-preform is siliconized at 1600°C in vacuum. The liquid silicon is infiltrated in the open porosity of the C/C-preform by capillary forces and reacts with the carbon, forming a coating of SiC around the C/C bundles [3].

2 Tribological Tests

Within the basic research programme of the joined ceramic research group Karlsruhe-Stuttgart-Freiburg (KKS) the friction and wear behaviour of different C/C-SiC materials were investigated under low and high energy test conditions (Table 1).

Test condition	low energy	high energy
Test parameters		
sliding velocity: v_0 [m/s]	6	16
deceleration: a (= constant) [m/s^2]	0.1	0.36
contact pressure: p (= constant) [MPa]	0.1	0.35
Maximum total energy [kJ]	20	80

Table 1. Low and high energy test conditions for ring specimens of different materials

The ring specimens with an outer diameter of 68 mm (Fig. 1) were tested in pairs of identical material and geometry.

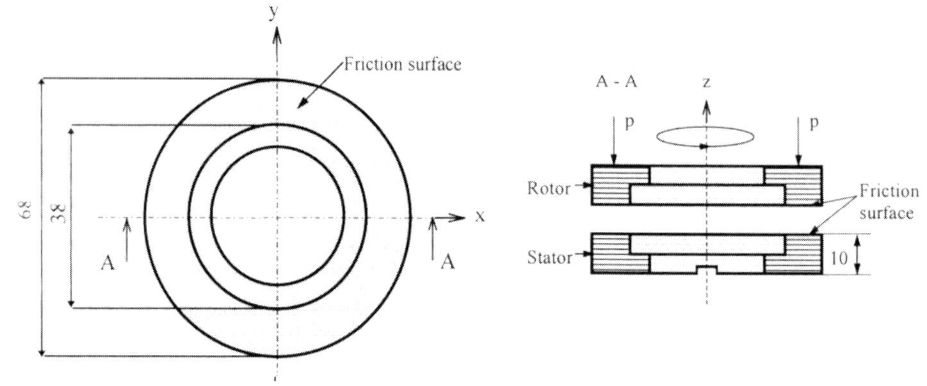

Figure 1. Geometry of C/C-SiC-specimens

In a high energy test facility the rotating specimen (rotor) was pressed onto the fixed specimen (stator) with constant force. The velocity of the rotor was regulated by the driving motor and was reduced with constant deceleration until stop. The influence of oxidation to the tribological behaviour was minimized by testing under purging nitrogen atmosphere.

3 Sample Preparation

Seven pairs of specimens were manufactured to investigate the tribological behaviour of the basic C/C-SiC material and its main components (C/C, SiC) and to analyse the influence of the materials thermal conductivity.

The standard C/C-SiC material was based on orthotropic woven fabrics of high tenacity (HT) carbon fibres and on a high yield carbonaceous precursor (XP 60). The C/C-SiC material consisted of about three quarters of carbon, respectively carbon reinforced carbon (C/C), and about one quarter of silicon carbide (SiC) by weight. The content of metallic silicon (Si) was about 4 % by weight.

The samples were cut out of rectangular plates (100 x 300 x 13 mm^3). Due to the resulting, anisotropic structure (Figure 2), the mechanical and thermal properties in plane, which are determined by the qualities of the fibre reinforcement, were diffferent to the properties in the vertical or z-direction (Table 2).

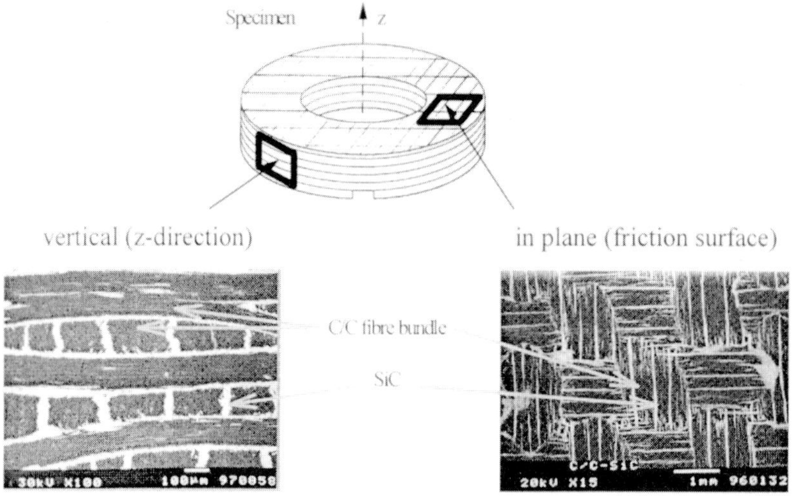

Figure 2. Microstructure of basic C/C-SiC material with laminated, orthotropic fabrics of carbon fibres

The tribological behaviour of the main components of the basic C/C-SiC material were tested with a C/C preform material used in the LSI process, and with C/C-SiC specimens coated with a thin layer of pure SiC (99.5 % β-SiC; 150 μm) via chemical vapour deposition (CVD-SiC).

Four pairs of specimens with different contents of graphite were manufactured to investigate the tribological influence of the specimens thermal conductivity vertical to the friction surface.

The tribological behaviour of monolithic graphite (V1633; Ringsdorff, Bonn) with high thermal conductivity (λ = 43 W/mK at 1200 °C) was investigated with two pairs of specimens from basic C/C-SiC material with joined graphite bolts (10 bolts Ø 12 mm). Pure graphite (100 % graphite) was tested with projected bolts (Fig. 3a). The surface content of the flushed graphite bolts was 45 % and led to a thermal conductivity of $\lambda_{z,C/C\text{-}SiC\,+\,45\,\%\,Graphite} = 23.5$ W/mK for the entire friction area.

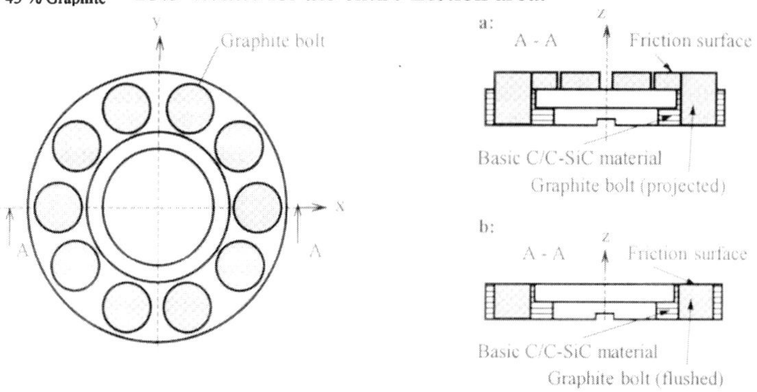

Figure 3. Tested specimens from basic C/C-SiC material with joined graphite bolts of monolithic graphite with different portion to the friction surface (a: 100 % Graphite; b: 45 % Graphite + 55 % C/C-SiC)

Two samples of C/C-SiC materials with modified C/C components were tested. In one pair of specimens the normally used HT fibres were replaced by graphitized high modulus (HM) carbon fibres with higher thermal conductivity in the fibre direction ($\lambda_{HM,\parallel}$ = 80 W/mK, $\lambda_{HT,\parallel}$ = 17 W/mK). The second pair of specimens was based on a graphitized C/C-preform which was thermally treated at 2700 °C prior to siliconizing (C/C-SiC_{graph}) [4]. Due to the laminated structure with its fibre orientation parallel to the friction surface the thermal conductivity vertical to the friction surface was only 60 % higher compared to the basic C/C-SiC material (T = 1200 °C: $\lambda_{z,HM}$ = 12 W/mK ; $\lambda_{z,C/C\text{-}SiCgraph.}$ = 12.4 W/mK ; $\lambda_{z,C/C\text{-}SiC.}$ = 7,5 W/mK).

Properties	C/C-SiC	SiC(1)	C/C preform	Graphite(2)
Density [g/cm^3]	1.9	3.21	1.3	1.87
Open porosity [%]	3.5	0	20	10
ILSS [MPa]	28		5	
Flexural strength [MPa]	160			80
Young`s modulus [GPa]	60	400		14
Thermal conductivity [W/mK] $\lambda_{\parallel}$ (200°C / 1200°C)	18.5 / 17	120 / 20	1.2 / 1.6	100 / 43
$\lambda_{\perp} = \lambda_z$ (200°C / 1200°C)	9 / 7.5		-	
CTE [10^{-6} K^{-1}] $\alpha_{\parallel}$ (100°C / 1500°C)	-1 / 2.5	6		4.3
$\alpha_{\perp}$ (100°C / 1500°C)	2.5 / 6.5			

(1) β-SiC (CVD-coating) (2) V 1633; Ringsdorff, Bonn

Table 2. Material properties of basic C/C-SiC material, C/C preform and monolithic components

4 Test Results

The main components of the basic C/C-SiC material (SiC, C/C) showed principal differences in their tribological behaviour under low and high energy test conditions.

The COF of the C/C component (C/C preform) was on a high level ($\mu_{min} > 0.28$) and was almost independent from the test conditions (Fig. 4). In contrast to this, the SiC component was extremely influenced by the test conditions. Under low energy input, the COF was very high ($0.6 < \mu < 0.9$) but decreased to a minimum of $\mu_{min} = 0.2$ under the high energy test conditions (Fig. 4).

The tribological behaviour of the basic C/C-SiC material was influenced by both of its main components. Under low energy test conditions, the COF of the basic C/C-SiC material was a combination of the high COF of the SiC-coated specimen and the lower

values of the C/C-preform. Under high energy inputs the COF decreased to a minimum of μ_{min} = 0.12, due to the dominant influence of the SiC component.
The decrease of the COF under high energy test conditions was explained by the low thermal conductivities λ_z of the SiC-coated and uncoated C/C-SiC specimens. The resulting high temperatures in the contact areas of the specimens led to the building up of friction layers with low COF and to locally overheated areas, so called „hot spots", which reduced the effective friction surface and therefore the COF [5].

All the specimens with graphite components showed a significantly higher COF under high energy test conditions compared to the basic C/C-SiC material (Fig. 5). Generally, high thermal conductivities vertical to the friction surface led to high COF.

The wear behaviour of the basic C/C-SiC material was almost independent from the test condition (low energy: 127 mm^3 / MJ; high energy: 123 mm^3 / MJ). Under high energy test condition the wear rate of the basic C/C-SiC material could be improved by the SiC-coated specimens (34.3 mm^3 / MJ), the specimens on the basis of the graphitized C/C preforms (70.6 mm^3 / MJ) and by the C/C-SiC material with flushed graphite bolts (87.5 mm^3 / MJ).
The porous C/C preform and pure graphite showed very high wear rates (2400 mm^3 / MJ and 6030 mm^3 / MJ) even in a non-oxidizing atmosphere and are therefore not suitable for brake applications.

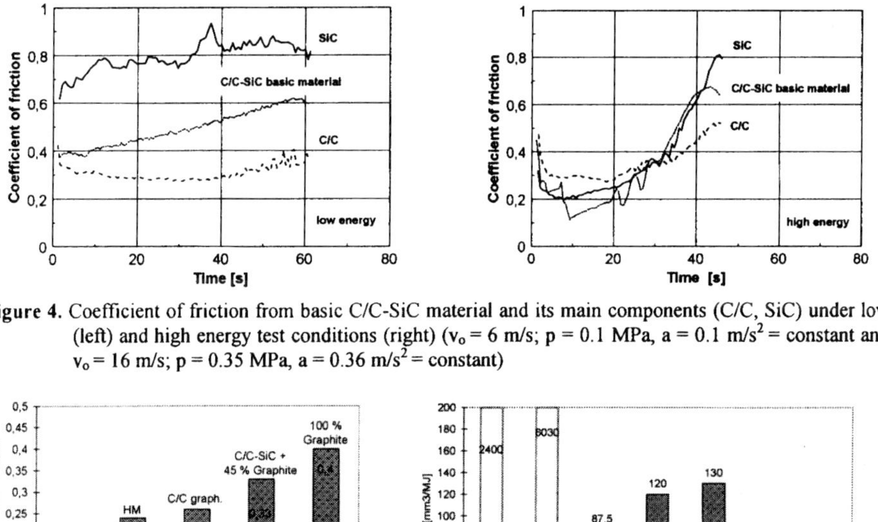

Figure 4. Coefficient of friction from basic C/C-SiC material and its main components (C/C, SiC) under low (left) and high energy test conditions (right) (v_o = 6 m/s; p = 0.1 MPa, a = 0.1 m/s^2 = constant and v_o = 16 m/s; p = 0.35 MPa, a = 0.36 m/s^2 = constant)

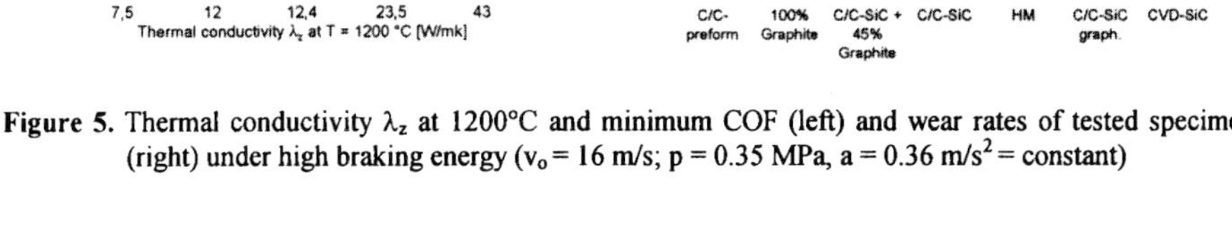

Figure 5. Thermal conductivity λ_z at 1200°C and minimum COF (left) and wear rates of tested specimens (right) under high braking energy (v_o = 16 m/s; p = 0.35 MPa, a = 0.36 m/s^2 = constant)

5 Summary

The tribological investigations of specimens ∅ 68 mm confirmed the great potential of liquid siliconized C/C-SiC materials for high performance brake applications. The basic C/C-SiC material, originally developed for the heat protection of spacecraft, showed sufficient wear resistance but too low coefficients of friction especially under high energy test conditions.

In order to improve this, the friction and wear behaviour of the basic C/C-SiC material was investigated by studying its main components C/C and SiC seperately. These investigations showed a relatively high COF for the C/C component under low as well as high energy test conditions. The decrease in the COF under high energy test conditions could be traced back to the dominant influence of the SiC component, in direct connection with an insufficient thermal conductivity of the basic C/C-SiC material.

The tribological behaviour under high energy test conditions was connected to the specimens thermal conductivity λ_z and could be significantly increased (100 % - 300 %) by modified C/C-SiC materials with graphitized carbon components. The homogeneous SiC- coating reduced the wear rate of the basic C/C-SiC material from 123 mm^3/MJ to 34.3 mm^3/MJ. C/C-SiC materials based on graphitized C/C-preforms or with graphite additives showed 30 % to 40 % lower wear rates compared to the basic C/C-SiC material.

The further development of C/C-SiC materials for brake discs is concentrated on wear resistant and economical manufactured SiC-coatings and basic materials with high thermal conductivity vertical to the friction surface. This seems to be possible by coatings with a minimal thickness of 1 mm, manufactured via the LSI process, and modified basic C/C-SiC materials with high contents of suitable additives or with graphitized C/C preforms.

References

[1] Krenkel, W., CMC Materials for High Performance Brakes, ISATA Conference on Supercars, Aachen, 1994

[2] Kochendörfer, R., Liquid Silicon Infiltration - a Fast and Low Cost CMC-Manufacturing Process, ICCM 8, Honolulu, 1991

[3] Krenkel, W., Gern, F., Microstructure and Characteristics of CMC manufactured via the Liquid Phase Route, ICCM 9, Madrid, 1993

[4] Frieß, M., Krenkel, W., Influence of the Graphitization Temperature on the Properties of C/C-SiC Manufactured via LSI-Processing, CIMTEC 9, Florence, June 14-19, 1998

[5] Heidenreich, B., Krenkel, W., Development of CMC for Friction Applications, IEKC 6, Stuttgart (MPI), September 13-15, 1998

Tribological Phenomena in the Outer Layer of Friction Linings Containing Composite Waste Materials

Jerzy Myalski, Józef Śleziona
Silesian Technical University, Katowice

1 Introduction

Management of waste materials from plastics is an unusually significant problem from the point of view of environmental protection and production economy. Economic factors show that the solution of this problem should prefer these methods and technologies which will allow to recover the material (raw material) contained in the waste element [1]. The management of waste materials from polymer plastics can be performed by different methods, including material recycling (reintroduction into processing), pyrolysis (raw material recycling), incineration (energy recycling) and storage [2]. Currently, the reutilization of polymer thermoplastic wastes is widely used, whereas the methods of recycling of waste materials containing thermosetting polymers (hardening plastics) are not applied on a large scale yet (e.g. company Ercom [3] offers a complete system to close the loop between used parts from automotive service and disassembly plants and reuse of fibrous reinforcing materials in new composite compounds). This is connected with the fact that hardening plastics do not melt at an elevated temperature and their spatial network structure does not change. Another difficulty of recycling, e.g. energy recycling or pyrolysis, is the presence of a big amount of mineral fillers (chalk) or fibres lowering the calorific value and causing the production of a big amount of carbonized wastes in polymers. Therefore, the recycling process of thermosetting polymers or composites on their matrix should be based, first of all, on raw materials recycling. This will also allow to preserve good properties of the reinforcing fibres contained in composites with a matrix made of hardening plastics and replace at least part of the reinforcing fibres deliberately introduced with fibres contained in the recycle material. The amount of the recycle material added must be precisely defined, otherwise it may result in a change of the final product properties or in changes of the processing properties.

This paper is an introduction to the investigation of the influence of composite waste materials additives on tribological properties of friction composite materials. These materials contain reinforcing fibres and various fillers which are widely used in polymer composites. Therefore, the introduction of this kind of wastes and a replacement of part of friction material components with them should, at the worst, lower the costs of friction composites production and contribute to a considerable decrease of the load to the natural environment.

2 Experimental Part

2.1. Materials used for investigation

The basic (initial) material used for the investigation of tribological properties was an asbestos free friction material intended for the production of clutch linings in tractors and agricultural machines. This material is a composition of organic and inorganic fibres, ceramic and metallic components combined by means of a resinous-rubber matrix. The initial material was selected on purpose, since it is characterized by quite low thermal resistance at an elevated working temperature and an improvement of its functional properties is economically justified.

Additives of recycle material of epoxide composite reinforced with glass fibres were introduced into this material. Since the initial material has relatively low thermal resistance, a decision was made to admix also epoxide-glass recycle material carbonized at a temperature of 800°C (i.e. at a temperature at which no changes occur in the material [4,5]. The recycle material (carbonized or uncarbonized) introduced into the initial material replaced the reinforcing fibres, in one case – ceramic fibres and in the other – organic ones. The recycle granulated product was admixed to the basic composition of the friction material by means of dry mixing technology [6] and from the admixtures prepared some test pieces were pressed for the investigation, applying standard processing parameters.

2.2 Results of investigation

The estimation of usefulness of different forms of the recycle material was carried out on the basis of the friction coefficient measurement at an elevated temperature. The measurement was taken with the use of a disc braking system of Bendix type at a constant pressure of 1.4 MPa and rubbing speed of 7 m/s. The registration of the temperature was carried out by means of 3 thermocouples located in the friction material, at its friction face. The results of the friction coefficient investigation are presented in figure 1.

For the investigation of the course of thermal destruction of friction materials derivatographic and chromatographic analyses were applied [7]. A Setaram GDTD16 thermoanalyser was used, to which a gas-chromatograph apparatus was connected. Simultaneously, the following parameters were registered: test piece temperature T, mass change TG, mass change speed DTG, sorts of gaseous products of the reaction EGA.

The measurements were taken in the conditions of a continuous temperature increase at a heating speed of 10°C/min, in helium atmosphere and with a weighed portion of 30 mg. The chromatographic analysis was performed at columns temperature of 65°C, were helium was the carrier gas. The changes taking place in the friction material structure and accompanying the transformation of the outer layer of the friction material under the influence of temperature, were carried out on a Hitachi S-4200 scanning electron microscope with field emission, equipped with a Voyager 3100 X-ray spectrometer.

The investigation made allowed to determine in what way the recycle granulated composite additive will influence the friction characteristics and if its additive can substitute ceramic or organic reinforcing fibres (fig.1 and 2).

A carbonized recycle granulated additive has a favourable influence on thermal stability of the friction coefficient. Such recycle glass fibre composite should be regarded as a component distinctly improving friction characteristics at an elevated temperature.

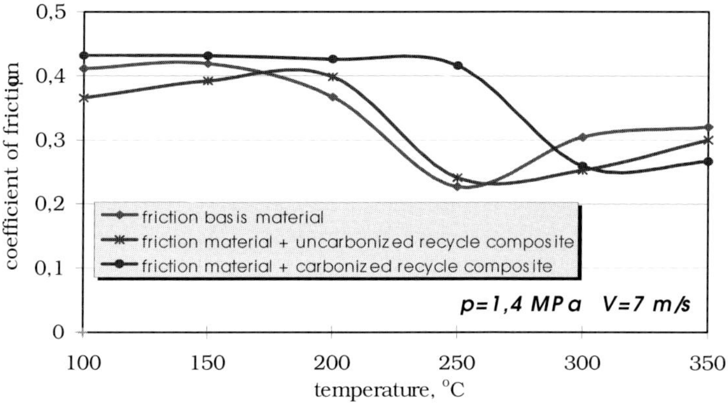

Figure 1. Results of the friction coefficient investigation at an elevated temperature of the friction material containing an additive of composite recycle granulated product which replaces glass fibres.

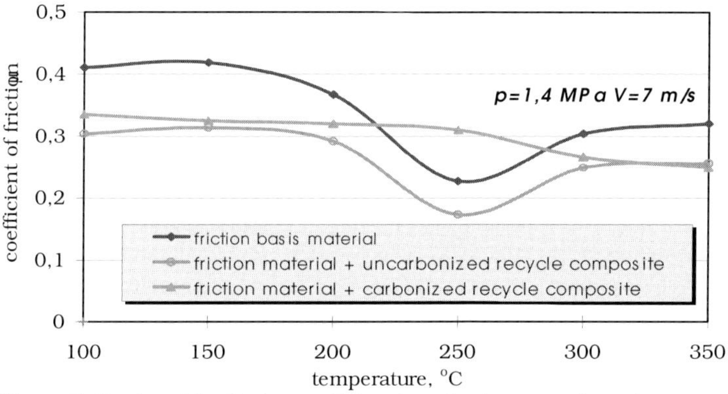

Figure 2. Results of the friction coefficient investigation at an elevated temperature of the friction material containing an additive of composite recycle granulated product which replaces organic fibres

This is particularly well visible in the case of carbonized recycle granulated product replacing organic fibres (fig. 2). In relation to the friction characteristic of the initial material, a constant level of the friction coefficient value is obtained to the temperature of about 50-80°C, after adding a carbonized component. Moreover, the decrease of the friction coefficient value at an elevated temperature is smaller than in the initial material not containing a carbonized component.

Modification of the friction material with uncarbonized recycle product does not cause such significant changes of the friction characteristic. This refers first of all to the thermal stability of the friction material. After adding epoxide-glass recycle material instead of glass fibres the level of the friction coefficient value does not change and the decrease of braking effectiveness in a critical temperature range of 200-250°C is even bigger than in the initial material. It is disadvantageous to replace organic fibres with the recycle material. As figure 4 shows, the material modified in this way has a friction coefficient lower by 25-30% than the initial material. The reason for such a behaviour of the friction material was probably an increase of the friction material hardness caused by the replacement of soft organic fibres with hard particles of the epoxide-glass recycle material. As a result, the real surface of contact in the friction material was decreased.

The increase of thermal stability of friction materials after adding the carbonized recycle material is the result of destruction limitation at an elevated temperature. As the thermogravimetric research has shown, the addition of carbonized recycle material caused an increase of the breakdown temperature and a decrease of decomposition intensity as well as a decrease of the test piece mass decrement at the time of measurement. For instance, the maximum decomposition intensity for the initial material occurred at a temperature of 416°C and in the friction material where glass fibres were replaced with carbonized recycle material, at a temperature of 508°C. The total mass decrement amounts respectively to 34% and 30%. Even smaller mass decrement and a different nature of decomposition were observed in a friction material containing carbonized recycle material which replaced organic fibres. The temperature of maximum destruction was about 426°C and the test piece mass decrement was 22%. However, the nature of the decomposition curve changed comparing to the initial material and the material containing the recycle granulated product which replaced the glass fibre. The elimination of glass fibres made the decomposition to be a one-stage phenomenon, in a rather narrow temperature range. Such a behaviour of the material resulted in higher stability of the friction coefficient in the whole range of the temperature under investigation coefficient of friction (fig. 2).

The destruction processes occurring at an elevated temperature decide about phenomena which take place in the outer layer. As a result of an elevated temperature, the matrix undergoes destruction. As the chromatographic investigation has shown, the destruction occurs at a temperature of about 300°C and manifests itself in educing liquid and gaseous products of decomposition, such as H_2, simple hydrocarbons – CH_4 and C_2H_6, steam, CO and CO_2. At this temperature phenols and cresols are also educed. Educing of gaseous products is also coupled with a formation of gaps, pores and microcracks which weaken the strength of the bonding between the matrix material and the friction material components. Gaps and cracks occur around the components, which is conducive to components chipping.

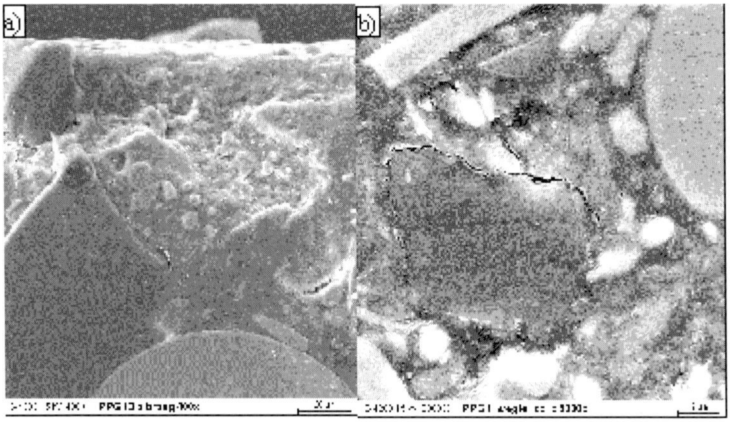

Figure 3. Structure of friction material containing carbonized recycle material replacing glass fibre, after friction coefficient investigation at an elevated temperature; a- near friction surface, b- in distance about 1000μm from friction surface (craks around carbonized particle).

An additive of carbonized recycle material causes a decrease of the area in which destruction takes place (fig. 3). One can expect that carbonized particles, due to their good conductivity, take over bigger amounts of heat generated during friction [8]. This is shown by the occurrence of cracks and gaps around carbonized particles (fig. 3b). As a result, the outer layer zone which undergoes thermal degradation is decreased.

Whereas in the case of friction materials containing additives of uncarbonized recycle material the destruction areas reach the size of 1500-2000 μm (fig.4), the additive of carbonized recycle material limits the destruction to about 1200-1500 μm.

Figure 4. Structure of friction material containing carbonized recycle material replacing organic fibre, after friction coefficient investigation at an elevated temperature; a- near friction surface, b- in distance about 2000μm from friction surface (cracks around recycled particle).

4 Conclusion

The investigation carried out shows that there is a possibility of composite waste materials utilisation as friction materials components. The waste material may be a component which replaces expensive reinforcing fibres and, simultaneously, a component which modifies the tribological properties of friction lining. An additive of carbonized composite recycle material has an advantageous influence on the friction coefficient value stabilization at an elevated temperature and limits the phenomena of thermal destruction of the friction material, which allows to increase the temperature of using the friction material.

5 Acknowledgements

The research was financed within the project of Polish State Committee for Scientific Research (KBN) (Contract No 7 TO8 E 025)

6 References

1. Błędzki A.K.:Recykling materiałów polimerowych, WNT, Warszawa 1997, p. 82.
2. Błędzki A.K., Gorący K.: Polimery, 1994, 39, no 9, p. 206.
3. Peters S.T.: Reuse and Disposal, Handbook of Composites, 7th ed., Chapman&Hall, London,1998, p.884.
4. Watremez M., Bricout J., Marguet B.,Oudin J.: Journal of Tribology, 118, pp.457-467, 1996;
5. Ostrovskij V.S., Brigileev C.: Isskustviennyj grafit, Mietallurgia, Moskva, 1986.
6. Hyla I., Myalski J, Śleziona J.:Patent P-294 492 PL., 1996;
7. Hyla I., Myalski J.: Polimery, , 43, nr 10, s.630-636; 1998
8. Hyla I., Myalski J., Śleziona J.:5th European Conference of Advanced Materials „EUROMAT'97", Maastricht, 1997, p. 478.

Wear Behaviour of Ceramics-Metal Couples for Water Hydraulics Applications

E. Bombonato°, L. Ceschini°°, G.L. Garagnani°, G. Magnani*, V. Miozza**
° Dept. of Engineering, University of Ferrara, (Ferrara) Italy
°° Institute of Metallurgy, University of Bologna, (Bologna), Italy
* ENEA, Dept. of Innovation, Faenza (Ravenna), Italy
** CERMET s.r.l, S.Lazzaro di Savena (Bologna), Italy

1 Introduction

The denomination " Water Hydraulics" identifies the sector of hydraulics that studies and produce components working with water, or water-based fluids, such as transmission power fluids.

The re-emergence of water-based hydraulic systems is related to some consideration: tap water is not inflammable, it does not pollute the environment, it is cheaper and it does not create problems for discharge. On the other hand, water creates some problems because it has little range of process temperature, it deposits limestone and allows micro-organism growth. Water does not have a good lubricant power and is highly aggressive for many metallic materials.

Problems related to water use are listed in table 1.

Table 1. Problems related to water use.

Problem	Cause	Remedy
Cavitation and Erosion	High superficial pressure of water High temperature High density of water High solubility of air in water High velocity of fluid Turbulent flux Particles in the fluid	Use of ceramics, anodized aluminium, stainless steels with high hardness, superficial treatments. Use of water with additives in order to increase viscosity or to reduce superficial pressure.
Sliding Wear and Lubrication	Low viscosity of water Poor lubrication power of water	Use of opportune couples of materials (i.e.: ceramic/ceramic, ceramic/polymer, stainless steel/bronze, stainless steel superficial treatment) Use of additives
Corrosion	pH Ion Cl^- O_2 Microbiological corrosion Stress corrosion Velocity of water	Use of corrosion resistant materials (i.e.: stainless steels, ceramics, bronzes, polymers, ecc.) Avoid galvanic couples. Add corrosion inhibitors.

Problem	Cause	Remedy
	Other (galvanic couples...)	
Fluid Losses	Low viscosity of water	Use of materials resistant to cavitation, erosion and abrasion (i.e.: ceramics, stainless steels with high hardness)
Microorganisms Content	The presence of micro-organism depends on nutritional substances, pH, and temperature of water.	Take care cleaning filters and other components. Use disinfection systems for water

It is clear that using water creates many problems, which a planner must take into account.

In order to produce hydraulic components working with water, a series of water lubricated tribological test have been carried out with the aim of identifing pairs of materials with good tribological behaviour. Experiments were made on both traditional and innovative materials like austenitic stainless steel, bronzes, anodized aluminium and two types of ceramics, such as pure Al_2O_3 and TZ-ZrO_2 (80% vol.)-Al_2O_3 (20% vol.).

This work aims at identifing the tribological behaviour of ceramics coupled with themselves and with traditional materials.

2 Experimental and Discussion

Ceramics specimens have been produced by ENEA, Faenza, Italy, with powders having the characteristics listed in table 2.

Table 2. Characteristics of ceramic powders.

	Al_2O_3	TZ ZRO_2-Al_2O_3
Powder manufacturer	Baikowsky International Corporation	Tosoh
Commercial name	Baikalox SM8	3Y20AB (Super-Z)
Purity	99.99%	--
Chemical composition	Al_2O_3 al 99,99%	80% ZrO_2 20% Al_2O_3 3% Y_2O_3
Specific superficial area (m^2/m)	10±1	15±3

The specimens have the following shape(Figure 1).

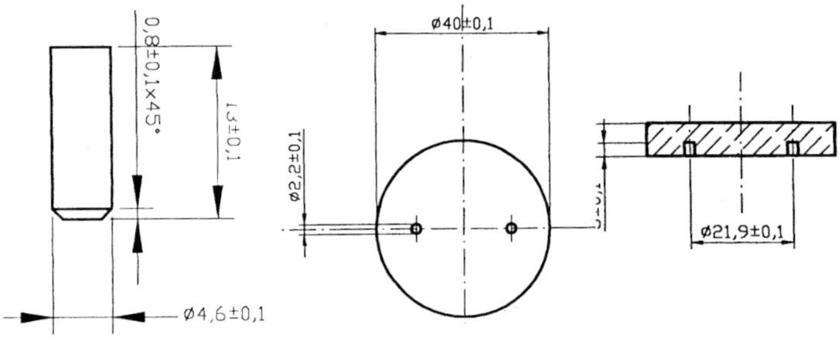

Figure 1. Speciment design and dimensions.

Ceramic Specimen was made by hot isostatic pressing process, worked by tools and then completely sinterized. Some properties are listed in table 3:

Table 3. Properties of ceramics.

Material	Roughness (Ra)	Density (kg/dm^3)	Vickers hardness (kg/mm^2)
Zirconia-Alumina	0,12	11,3	1644
Alumina	0,23	3,9	2200

Wear tests were carried out on a pin-on-disk tribometer, with samples completely submerged in water. An hour long initial screening, was made and, for the sliding pairs which showed better performance, six hour tests were carried out.

Before each tribological test, conducted with FALEX mod. FX-1506 tribometer, analyses of micro-hardness, density, dimensional and superficial measurements, and metallographic analysis were made; after the test, samples underwent measurements of weight and dimensional losses as well as relief on wear truck, metallographic observation by stereo, optical, and SEM microscopy, XDR on wear truck and debris. The tribometer gaves coefficient of friction value during the test; the wear rate was calculated by loss in weight.

In order to describe ceramics tribological performances, tests with the following sliding pairs were conduced (table 4).

Table 4. Sliding pairs tested

PIN Materials	Disk Materials
Alumina	Alumina
Stainless Steel X15T.N	Alumina
Bronze B14	Alumina
Bronze SM7	Alumina
AISI 316 stainless steel	Alumina
Zirconia-Alumina	Zirconia-Alumina
Stainless Steel X15T.N	Zirconia-Alumina
Bronze B14	Zirconia-Alumina
Bronze SM7	Zirconia-Alumina
AISI 316 stainless steel	Zirconia-Alumina

A good tribological behaviour of the alumina-alumina sliding pairs in water was observed; whereas, the behaviour of zirconia-alumina sliding on itself showed poorer results; both ceramics sliding against traditional materials under water lubrication showed good friction and wear behaviour.

2.1 Alumina against traditional materials

All tests have been carried out with 80N applied load and 0,75 m/s sliding velocity in tap water at room temperature; the test duration was one hour.

Disk material is alumina, pins are respectively: AISI 316 Stainless Steel, Alumina, Bronze SM7 (UNI 5275), Bronze B14 (UNI 4855, UNI 7013/10), Stainless steel X15T.N (X40CrMoVN16.2)

Results are reported in table 5:

Table 5. Coefficient of friction values.

PIN	Coefficient of friction (mean value)
Alumina	0,17
AISI 316 stainless steel	0,35
Bronze B14	0,29
Bronze SM7	0,40
Stainless steel X15T.N	0,21

Figure 2 shows comparative wear rates of sliding pairs:

It is clear that, in this condition, AISI 316 stainless steel has a bad tribological behaviour; the best behaviour resulted from alumina against itself which showed low wear rate, low coefficient of friction and low damage to specimen.

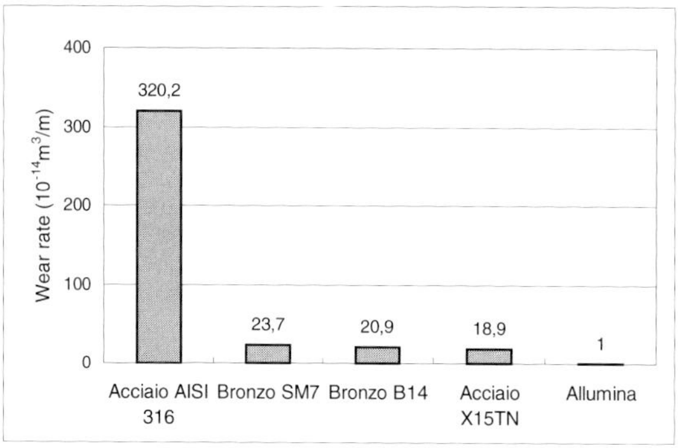

Figure 2. Wear rate.

2.2 Zirconia-Alumina against traditional materials

All of the tests were carried out with 80N applied load and 0,75 m/s sliding velocity in tap water at room temperature; the test duration was one hour.

Disk material is zirconia-alumina, pins are respectively: Zirconia-Alumina, Bronze SM7, Bronze B14, Stainless steel X15T.N.

Results are in the table 6:

Table 6. Coefficient of friction values.

PIN	Coefficient of friction (mean value)
Alumina Zirconia	0,45
AISI 316 stainless-steel	0,40
Bronze B14	0,29
Bronze SM7	0,25
Stainless steel X15T.N	0,16

The wear rate is shown in figure 3:

It is possible to see that zirconia-alumina on itself and AISI 316 stainless steel on zirconia-alumina showed a bad tribological behaviour, whilst bronzes and X15T.N stainless steel showed better performances.

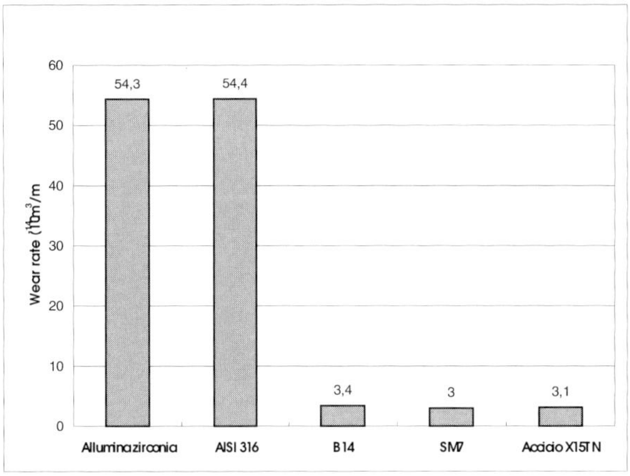

Figure 3. Wear rate.

After this initial hour long screening, for the sliding pairs which showed better performances, six hours tests were carried out. These tests were performed with the same applied load and sliding velocity conditions, always lubricated with water.

2.3 Alumina against traditional materials

Pins materials are alumina, bronze B14, bronze SM7, X15T.N stainless steel.

Results are listed in table 7:

Table 7. Coefficient of friction values.

PIN	Coefficient of friction (mean value)
Alumina	0,17
Bronze SM7	0,39
Bronze B14	0,28
Stainless steel X15TN	0,10

As evidenced, alumina against itself and X15T.N stainless steel showed better tribological performance, either in terms of coefficient of friction or in terms of wear rate.

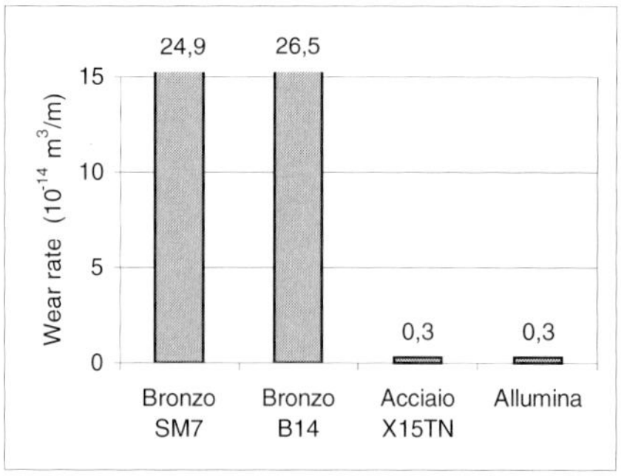

Figure 4. Wear rate.

2.4 Zirconia-Allumina against traditional materials

Pins materials are Bronze B14, Stainless steel X15T.N

Results are in table 8:

Table 8. Coefficient of friction values

PIN	Coefficient of friction (mean value)
Bronze B14	0,29
Stainless steel X15TN	0,28

It is possible to see that stainless steel X15T.N performs better than bronze; X15T.N performes worse when sliding on zirconia-alumina than on pure alumina.

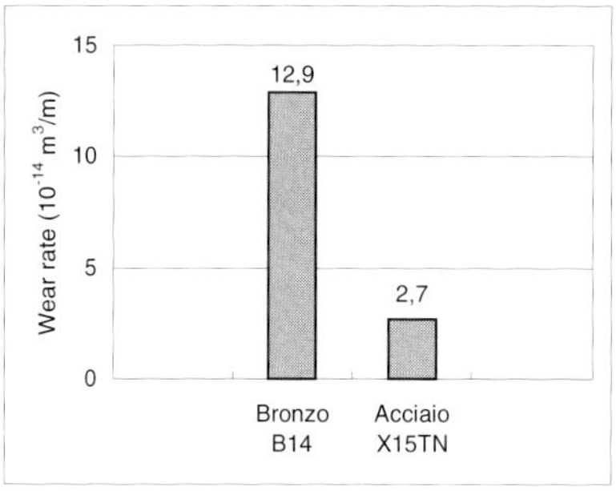

Figure 5. Wear rate.

3 Conclusion

These tests show the good tribological behaviour of ceramics, alumina in particular, bronzes and stainless steel X15T.N sliding on alumina, and zirconia-alumina coupled with bronzes and X15T.N.

Particularly interesting is the alumina-alumina sliding pair, which shows a low coefficient of friction and low wear rate; this is in line with other Authors [1],[2] and has been confirmed in long run tests. Like alumina-bronze and alumina-X15T.N stainless steel, the alumina-alumina sliding pair, showed an initial running in and a subsequent steady state behaviour during the test. These results can be explained by tribo-chemical phenomena which occur during the test between ceramics and water, modifing the contact surfaces.

Water seems to be a “lubricant” in the contact especially in the alumina-alumina sliding pair; a dry test failed immediately, the pin shattering after four minutes.

A result of more significance for our project, is the good performance of the ceramic-bronze sliding pair, technology allowed us to produce body-valves with foundry methods, while cursors or inner components of valves will be produced by sintering ceramic materials.

4 References

[1] Zum Gahr, K.H. "Sliding wear of ceramic/ceramic, ceramic/steel and steel/steel in lubricated and unlubricated contact.." Wear. n° 133. (1989). p. 1-22

[2] Ravikiran, A.; Pramila Bai, P.N. "Water-lubricated sliding of Al_2O_3 against steel." Wear. n° 171. (1993). p. 33-39

DYNAMIC FRICTION OF CARBON-CERAMIC COUPLES

P. VANTOMME, P. DEPREZ, J .F. COUTOULY, A. DEFFONTAINE
Centre d'Application des Lasers de Flandres-Artois, rue de l'Université, BP 819, 62408 Béthune - FRANCE

A. PLACET, D. GAILLOT
Cyclam S.A., 75 rue Robert Le Coq, 80038 Amiens cédex 1 - FRANCE

1. Introduction

In the majority of combustion engines, the cooling system and the outside environment are separated by a dynamic sealing element. This mechanical system is made up of two sub assemblies, which are, a static head and a dynamic seat. The assembled dynamic part press fitted to the shaft, includes the mating ring, an elastomer cup ring and a metallic sleeve. The mating ring is inter-dependant from the sleeve through the inside, and the cup ring ensures the static sealing between the two parts. The static head fitted into the body of the water pump is made up of a metallic retainer, an elastomer bellows, a primary ring and two ferrules which mutually ensure the clamping of the bellows to the primary ring and to the retainer in order to get a static sealing. The axial stress exerted by the spring keeps the friction faces, lubricated by a liquid, in contact during the movement. The primary ring and the mating ring are the two most important elementary parts of sealing of which the basic principle is to ensure the sealing between the two friction faces, in relative film. This film maintained by rotation, comes from liquid circulating through the water pump.

The sealing of the primary ring coupled to the mating ring must be ensured in all viability with a specified life expectancy. It is subject to pressure and temperature variations and undergoes wear due to friction.

The aim of this study is to experimentally determine the tribological properties of different carbon primary ring / porous silicon carbide mating ring couples in the scope of dynamic sealing applications.

2. Experimental method

2.1. Materials

The geometry and dimensions of the mating ring and the primary ring forming the studied friction couple are given in figure 1. The primary rings are moulded from a thermo-hardening material. This material is a mixture, in specific proportions of graphite, coke, phenolic resin, antimony and zinc stearate. These primary rings are made by hot compression moulding. From the same chemical composition, two materials are obtained from different thermal processes.

Material A : Carbon material of a thermo-hardening resin base, hot polymerised, undergoes sterilisation (250 - 300°C) to finish the polymerisation (Fig. 2.a).

Material B : From material A, a high temperature process (700 to 1100°C) is carried out in a neutral atmosphere to avoid oxidation. This enables the material to be compact and to form a carbon network in the resin (Fig. 2.b).

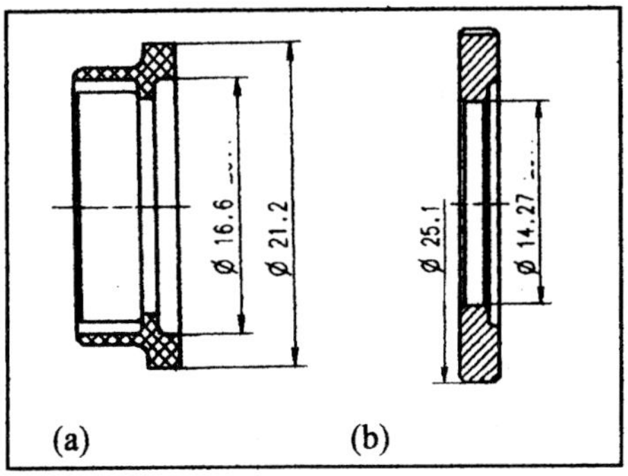

Fig. 1. Geometry and dimensions of : (a) primary ring, (b) mating ring

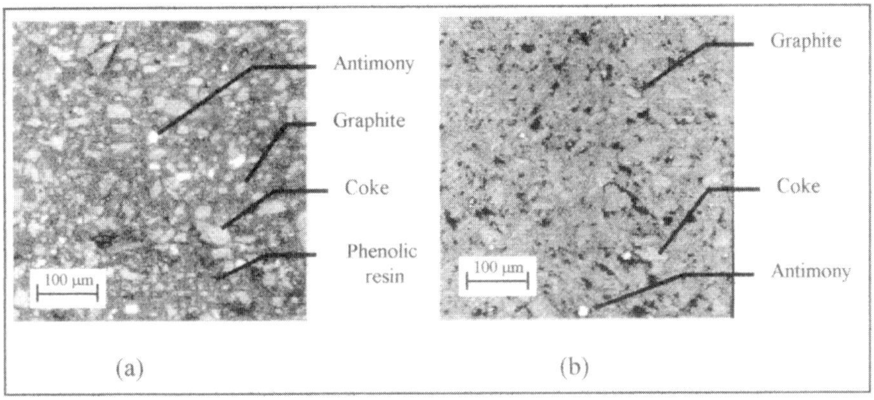

Fig. 2. Micrography of the primary ring : (a) material A, (b) material B.

Table 1 summarises the mechanical properties of the two different studied materials. It is interesting to note that the B material variant is a better thermal conductor and has a lower thermal expansion coefficient compared to the A material variant.

	material A	material B	porous silicon carbide
Young modulus (GPa)	16	25	350
Thermal conductivity ($W.m^{-1}.C^{-1}$)	5	20	130
Thermal expansion coefficient (10^{-6} / C)	14.8	4	4
Poisson coefficient	0.21	0.21	0.25
Density	1.77	1.85	2.82
Tensile strength (Mpa)	28	45	160
Compressive strength (Mpa)	100	180	1500
Brinell hardness (daN / mm^2)	40.08	48.44	
Vickers hardness (daN / mm^2)			2200 - 2800

Table 1. Physical properties of the studied materials

The tested mating rings are made from porous silicon carbide (Fig. 3) and are obtained by sintering. The part closed porosity is in the order of 10 to 14%, with pore diameters ranging from 10 to 200 µm. In order to remain sealed. the mating rings have no open porosity.

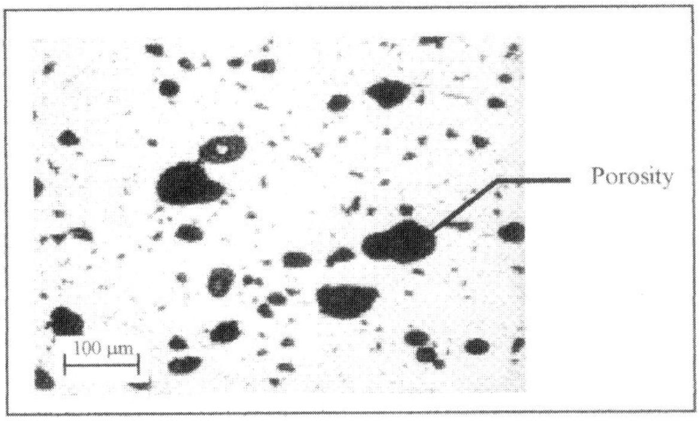

Fig.3. Micrography of the porous silicon carbide

The chemical composition of the silicon carbide is : SiC : 98.5 % mini, Si free : < 2000 ppm, SiO_2 : < 2000 ppm, Fe : <1000 ppm, other impurities (Al, Ca, Mg, K ...) : < 1000 ppm.

To remain sealed, the surface flatness and roughness of the parts tested by friction have to be of very high quality. The arithmetic roughness of the different tested parts is in function of the lapping applied to the surface of these parts. A lapping period followed by glazing (3µm) was carried out on the primary ring whilst the mating ring underwent another lapping period (boron 400, boron 800) or a glazing (3µm).

2.2. Tribological characteristics

Figure 4 describes the friction test implemented during this study.

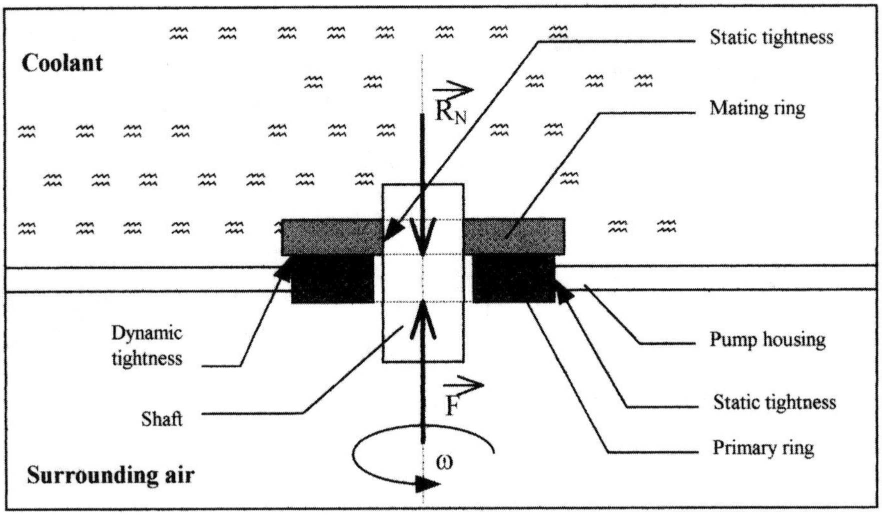

Fig. 4. Friction test

The tribometer used was studied in order to carry out friction tests on pairs of materials, without sensitive differential pressure, in the following conditions : liquid / surrounding air, surrounding air / surrounding air.
The tribometer tank was assembled turning between two bearings. A system made up of spring blades ensured that the tank was recalled to its initial position. The friction ring was assembled onto the bottom of the tank.
The tank contained about one litre of liquid. An electronic torque-meter, linked up between the tank and a fixed part of the machine allowed the measurement of the resisting torque. During the test, the rotational speed of the primary ring was raised through the use of a cog wheel and a sensor. These measurements were carried out at a frequency of 50 Hz thanks to an acquisition card. An average of the measurements was taken and recorded every second.

2.3 Experimental measurements

The parameters which influence static and dynamic friction coefficients are :

- the quality of the primary ring (material A or B),
- the surface finishing of the mating ring (boron 800, boron 400 and glazing 3μm),
- the normal load (1 to 5 daN),
- the surrounding (dry / lubricated),
- the speed in the case of dynamic friction (1000 to 8000 rd/min).

The static friction torque measured before testing can vary from 0 to 200 mm daN due to the influence of these different factors. The torque absorbed by two parts undergoing sealing friction is not generally constant in time. It is therefore difficult to fix a figure indicating the friction coefficient of a pair of materials since the dynamic friction torque can vary in proportions being able to reach 10 times the minimum value. However, to determine the area of possible variations, the start values, the maximum and minimum values, as well as values stabilised after 10 hours of testing materials have been retained. The dynamic friction torque, equally, can vary in function of the values taken by the influencing factors, from 0 to 200 mm daN. Thus, the primary ring will be carried out with static and dynamic friction torques contained between 0 and 200 mm daN.

3. Results

For the study of the dynamic friction coefficient, the dynamic torque was raised to determine the coefficient of dynamic friction, at the start, minimum, maximum, and at the end (10 hours) of the test without indication (Table 2). For tests 3, 4 and 20, due to the destruction of the carbon primary ring, the dynamic friction coefficient was deduced before the carbon ring breaks.
In a previous work [1], from the experimental results obtained in table 3, a linear model taking into account the studied factors and the interactions between those factors was determined. The average estimated effects of the main factors on the estimated coefficient of dynamic friction were established (Fig. 5). For this study, due to better stability, the coefficient of dynamic friction at 10 hours was retained.

test	load (daN)	speed (m/s)	primary ring quality	mating ring finishing	surrounding	coefficient of dynamic friction			
						start	mini	maxi	at 10 hours
1	1	7.956	A	boron 400	dry	0.29	0.09	0.35	0.18
2	3	4.475	A	boron 400	lubricated	0.22	0.15	0.22	0.15
3	5	4.475	A	glazing (3μm)	dry	0.65	0.03	0.65	0.30(1)
4	5	7.956	B	boron 400	dry	2.07	0.04	2.07	1.45(1)
5	3	7.956	A	boron 800	dry	0.14	0.10	0.22	0.14
6	5	0.995	B	boron 400	lubricated	0.51	0.29	0.57	0.31
7	1	0.995	A	glazing (3μm)	dry	0.60	0.11	0.69	0.69(2)
8	5	7.956	B	boron 800	lubricated	0.36	0.17	0.31	0.20
9	5	4.475	A	boron 800	lubricated	0.26	0.16	0.32	0.16
10	1	7.956	A	glazing (3μm)	lubricated	0.44	0.21	0.44	0.42
11	5	7.956	B	glazing (3μm)	lubricated	0.88	0.33	0.89	0.37(3)
12	1	7.956	A	boron 800	lubricated	0.56	0.50	0.68	0.62
13	1	0.995	B	glazing (3μm)	lubricated	0.47	0.30	0.55	0.41(4)
14	1	0.995	A	boron 400	lubricated	0.51	0.41	0.53	0.41
15	5	0.995	A	glazing (3μm)	lubricated	0.48	0.23	0.53	0.26
16	5	0.995	B	boron 800	dry	0.29	0.10	0.60	0.15
17	3	0.995	B	glazing (3μm)	dry	0.14	0.11	0.19	0.12
18	5	7.956	A	boron 400	lubricated	1.24	0.18	1.24	0.18(2)
19	1	0.995	B	boron 400	dry	0.34	0.27	0.41	0.39
20	5	0.995	A	boron 400	dry	3.10	0.25	3.19	1.19(1)
21	1	7.956	B	boron 800	dry	0.14	0.09	0.20	0.17
22	1	0.995	A	boron 800	dry	0.44	0.10	0.44	0.15
23	1	7.956	B	boron 400	lubricated	0.63	0.38	0.78	0.74
24	1	7.956	B	glazing (3μm)	dry	0.34	0.14	0.46	0.26
25	1	0.995	B	boron 800	lubricated	0.39	0.28	0.39	0.28

Table 2. Measurements of the dynamic friction coefficients. (1) breakage of the primary ring before the end of the test ; (2) sudden variation of the dynamic couple; (3) instability of the dynamic torque ; (4) electrical problem).

Fig. 5. Average estimated effects of the principal factors on the estimated dynamic friction coefficient.

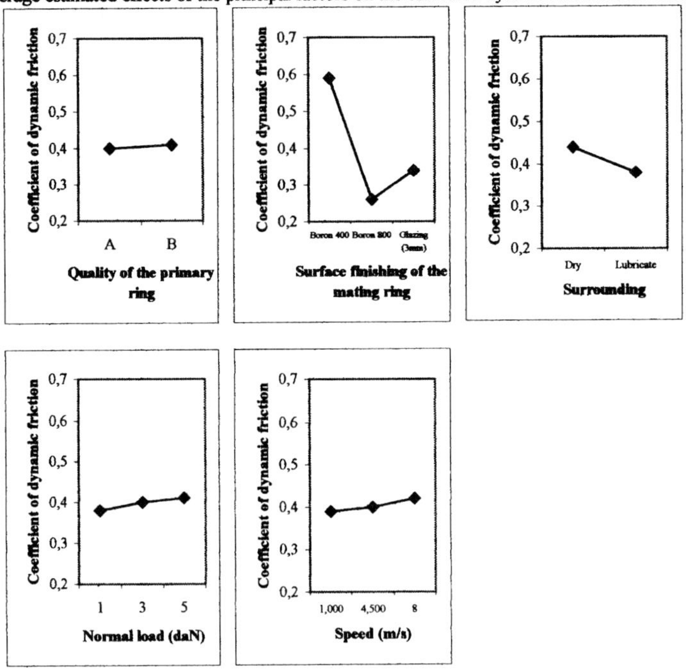

The quality of the primary ring, the speed and the load don't bring about a notable variation in the average estimated dynamic coefficient. These factors alone do not influence the evolution of the dynamic friction coefficient. However, figure 5 brings into evidence the average estimated effect of the principal finishing factor. A variation of 0.32 is observed when the boron 400 finish is replaced by the boron 800 finish for the mating ring. As with the coefficient of static friction, the glazed finishing slightly increases the dynamic friction coefficient.

The interdependence of the glazed finishing with the environmental and qualitative factors have little influence on the dynamic friction coefficient ($\pm$ 0.08), for the interaction finishing/quality ($\pm$0.06) and for the interaction finishing/environment. The boron 800 finishing has little interference with the load factor ($\pm$0.05).

It is also noted that the effect of the primary ring quality on the dynamic friction coefficient is compensated by the effect of the relative speed. The works of Golubiev [2] explain this phenomenon. During his attempts at measuring the pressure in the interface of the dynamic waterproof guards, he observed the presence of micro-humps due to thermal conductivities and thermal expansion coefficients as well as the locally different frictional temperatures. With intersecting contact pressure, due to the dilatation of the carbon produced by the heating crossed with the speed , the carbon face is pressed in the pores due to its weak coefficient of elasticity. In the case of a more elevated speed or load the carbon particles are cut.

With the two materials Carbon A/porous silicon carbide, in addition, deposits and traces of carbon are nearly always observed on the porous silicon carbide mating ring. In addition, the porosities tend to be blocked by the carbon. However, for the binomial B/porous silicon carbide, carbon traces in the porosities and on the friction face of the mating ring are rarely seen.

The lubricating quality of the graphite, the more elevated coefficient of thermal expansion of the quality A, the presence of graphite in the porosities of the carbide and the traces of graphite on the friction surface of the mating ring explain the evolution of the frictional coefficient effect for the quality of carbon.

The effect of the load is compensated by the environment. In the case of dry running, the dynamic friction coefficient increases proportionally with the load. However, in a lubricated environment, it diminishes when the load increases. The cavities formed from the pores create greasing reservoirs and zones of locally limited pressure. With increasing contact pressure, the antagonistic carbon face is pressed into the pores, and evicts the coolant.

4. Conclusion

As part of a dynamic sealed application, the coefficients of dynamic friction of different carbon primary ring / porous silicon carbide mating ring couples have been studied.

The factors as the quality of the primary ring, the speed and the load have little influence on the coefficient of dynamic friction. However, a decrease of the coefficient is observed in a lubricated environment with the boron 800 finish for the mating ring.

References

[1] PLACET A., *Etude du couple carbone-céramique de garniture d'étanchéité*, thèse CNAM Picardie, soutenue le 29 avril 1997.

[2] GOLUBIEV A. I., *On the existence of a hydrodynamic film in mechanical seals*, Cambridge, Conf. on Fluid Sealing, England, 1967

Author Index*

* The page numbers refer to the first page of the respecting article

Subject Index*

* The page numbers refer to the first page of the respecting article

P

R

S